A
Statistical
Guide
for the
Ethically
Perplexed

A
STATISTICAL
GUIDE
FOR THE
ETHICALLY
PERPLEXED

LAWRENCE HUBERT

HOWARD WAINER

CRC Press
Taylor & Francis Group
Boca Raton   London   New York

CRC Press is an imprint of the
Taylor & Francis Group, an **informa** business
A CHAPMAN & HALL BOOK

CRC Press
Taylor & Francis Group
6000 Broken Sound Parkway NW, Suite 300
Boca Raton, FL 33487-2742

© 2013 by Taylor & Francis Group, LLC
CRC Press is an imprint of Taylor & Francis Group, an Informa business

No claim to original U.S. Government works

Printed in the United States of America on acid-free paper
Version Date: 20120822

International Standard Book Number: 978-1-4398-7368-7 (Paperback)

Visit the Taylor & Francis Web site at
http://www.taylorandfrancis.com

and the CRC Press Web site at
http://www.crcpress.com

Nullius in verba (take nobody's word for it).
— Motto of the Royal Society

Nullius addictus iurare in verba magistri (I am not bound to believe in the word of any master).
— Horace (First Century B.C.)

There is only one good; knowledge;
And only one evil; ignorance.
— Socrates (470 – 399 B.C.)

Preface

> I have never heard any of your lectures, but from what I can learn I
> should say that for people who like the kind of lectures you deliver,
> they are just the kind of lectures such people like.
> – Artemus Ward (from a newspaper advertisement, 1863)

Our title is taken from the seminal work of the medieval Jewish philosopher Moses Maimonides, *The Guide for the Perplexed* (1904, M. Friedlander, Trans.). This monumental contribution was written as a three-volume letter to a student and was an attempt by Maimonides to reconcile his Aristotelian philosophical views with those of Jewish law. In an analogous way, this book tries to reconcile the areas of statistics and the behavioral (and related social and biomedical) sciences through the standards for ethical practice, defined as being in accord with the accepted rules or standards for right conduct that govern a discipline. The standards for ethical practice are what we try to instill in students through the methodology courses we offer, with particular emphasis on the graduate and undergraduate statistics sequence generally required in all of the sciences. It is our hope that the principal general education payoff for competent statistics instruction is an increase in people's ability to be critical and ethical consumers and producers of the statistical reasoning and analyses they will face over the course of their careers.

Maimonides intended his guide for an educated readership, with the ideas concealed from the masses. He writes in the introduction: "A sensible man should not demand of me, or hope that when we mention a subject, we shall make a complete exposition of it." In a related way, this book is not intended to teach the principles of statistics (that should be done within our course sequences), but to provide a context where the principles learned can be used to reason correctly, and thus, ethically. Many situations, or better, situational vignettes, are encountered throughout the chapters to follow. To give a foretaste of these vignettes, some are mentioned briefly below and in the general order they will be encountered when reading linearly from cover to cover:

— The case of Sally Clark, wrongly convicted in England of killing her two children; this miscarriage of justice was due to an inappropriate assumption of statistical independence and the commission of the "Prosecutor's Fallacy";

— Breast cancer screening though mammograms; understanding Bayes' theorem, test sensitivity and specificity, prior probabilities, and the positive predictive value (for example, what is the probability of having breast cancer if the mammogram is "positive"?);

— How subjective probabilities might be related to the four levels of a "legal burden of proof": "preponderance of the evidence"; "clear and convincing evidence"; "clear, unequivocal, and convincing evidence"; and "proof beyond a reasonable doubt";

— The distinction between "general causation" and "specific causation"; the common legal standard for arguing specific causation as an "attributable proportion of risk" of 50% or more;

— Issues of probability, risk, and gambling; spread betting and point shaving; parimutuel betting; the importance of context and framing in risky choice and decision making;

— "Voodoo Correlations in Social Neuroscience"; the "culling" or search for results in clinical trials and elsewhere, with a subsequent failure to cross-validate what is found; more generally, the problem of "double dipping";

— Henry A. Wallace and the modeling of expert judgment ("What Is In the Corn Judge's Mind?"); the distinction between actuarial and clinical prediction, and the Dawes notion of the "robust beauty of improper linear models";

— The decennial problem posed by the United States census; complete enumeration, as required by the Constitution, versus sampling, plus the political issues involved in the problem of "undercount";

— The unfortunate state of the statistical routines present in the widely used Excel program, and the inability of Microsoft to correct errors pointed out by the statistical community;

— The settlement between the Educational Testing Services

and the Golden Rule Insurance Company, and its devastating consequences for carrying out psychometrically defensible "high-stakes" testing;

— The delay (because of the American Psychological Association) in the publication of the article, "Is Criminal Behavior a Central Component of Psychopathy?"; how construct validation should be done, and the need to define a construct by more than just the Hare Psychopathy Checklist;

— The darker side of psychometrics and statistics: eugenics, forced sterilization, immigration restriction, racial purity laws;

— Weight-of-the-evidence argumentation in presenting and interpreting data, particularly for medical and regulatory issues;

— Inferring causality; Bradford–Hill criteria; historic medical conceptions of disease causality; medical error as the causative factor;

— The ubiquity of Simpson's Paradox in the (mis)interpretation of data; when a relationship that appears to be present at an aggregated level disappears or reverses when disaggregated and viewed within levels;

— Meta-analysis and the controversies it engenders in childhood sexual abuse and other medically relevant research summarizations;

— Statistical sleuthing with formal models: Poisson clumping, Benford's law, survival analysis, Kaplan–Meier curves;

— Difficulties with observational studies; the hormone replacement therapy controversy and related artifactual interpretative issues;

— Ethical considerations in data collection and analysis involving human experimentation; the Nazi Doctors' Trial and the Nuremberg Code; the Tuskegee syphilis study and the Belmont Report; the Declaration of Helsinki and the conduct of foreign clinical trials;

— The *Federal Rules of Evidence* and the court admissibility of expert witnesses and scientific data; the *Daubert* trilogy of Supreme Court decisions;

— The importance of context and framing in the presentation of data; the work of Tversky and Kahneman, and the more recent points made by Gigerenzer and his colleagues ("Helping Doctors and Patients Make Sense of Health Statistics," in the series sponsored by the Association for Psychological Science, *Psychological Science in the Public Interest*).

In addition to the topics just listed, we should note the presence of extensive redactions for seven Supreme Court decisions from the last century. These cases reflect the influence of statistical and psychometric reasoning and interpretation, or misinterpretation in several instances. Court opinions are not commonly excerpted at such length in texts such as this, but these particular opinions represent the intersection of ethics and statistics in a way that other mechanisms cannot. The seven cases are listed chronologically with a short statement of the holdings in each as quoted from the corresponding Wikipedia article, or for the most recent case of *Matrixx Initiatives, Inc. v. Siracusano* (2011), what can be abstracted from the unanimous opinion delivered by Justice Sonia Sotomayor:

Buck v. Bell (1927): The Court upheld a statute instituting compulsory sterilization of the unfit "for the protection and health of the state."

Brown v. Board of Education (1954): Segregation of students in public schools violates the Equal Protection Clause of the Fourteenth Amendment, because separate facilities are inherently unequal.

Loving v. Virginia (1967): The Court declared Virginia's antimiscegenation statute, the *Racial Integrity Act of Virginia* (1924), unconstitutional, thereby ending all race-based legal restriction on marriage in the United States.

Barefoot v. Estelle (1983): There is no merit to petitioner's argument that psychiatrists, individually and as a group, are incompetent to predict with an acceptable degree of reliability that a particular criminal will commit other crimes in the future, and so represent a danger to the community.

McCleskey v. Kemp (1987): Despite statistical evidence of a profound racial disparity in application of the death penalty, such evidence is insufficient to invalidate defendant's death sentence.

Daubert v. Merrell Dow Pharmaceuticals (1993): The *Federal Rules of Evidence* govern the admission of scientific evidence in a trial held in federal court. They require the trial judge to act as a gatekeeper before admitting the evidence, determining that the evidence is scientifically valid and relevant to the case at hand.

Matrixx Initiatives, Inc. v. Siracusano (2011): A plaintiff may state a claim against a pharmaceutical company for securities fraud under the *Securities Exchange Act of 1934* based on the company's failure to disclose reports of adverse events even when the reports do not disclose a "statistically significant" number of such adverse events.

* * *

The intended audience for this book is the typical first-year graduate student in a social, behavioral, or health-related program who might be struggling with the required statistics sequence. It should be viewed as something to be consulted for context and motivation when learning the formal techniques and concepts in a traditional statistics sequence. It is not intended as a popularization for general audiences, similar to the recent and well-received *Proofiness* by Charles Seife; therefore, we have little hope for a book review in venues such as the *New York Times*. It assumes the reader is learning the actual methods of statistics from another source; we don't provide much of an exposure to this basic material. Our emphasis is more on (un)ethical applications and related specious implementations and interpretations.

A major part of our attempt to motivate an interest in the ethical practice of statistics is the inclusion of many historical contexts and vignettes involving probabilistic and statistical reasoning, some of it good and some of it not so good. A variety of excerpted or redacted original material appears in text when copyright permissions could be obtained without financially stressing the authors (for example, the material was in the public domain, appeared in court opinions, was produced by some governmental agency, is available in Wikipedia under the Creative Commons understanding, the copyright has expired, the excerpt was short enough to be considered "fair use," or the permission fees were waived or were very nominal). In most cases, a particular article or source might be obtained by a simple search on a publisher's website, such as for items in the *New York Times*, or though access to archived material at a college or community library, such as for *Science* or the *New York Review of Books*. In many if not most instances, a simple "Google" search on the title enclosed in quotes, plus an author's name, will generate an archived source for the item (for example, anything from *ScienceNews* can be found in this way). Almost all of the Suggested Readings are from the last ten years, reflecting both a currency in the topics mentioned and the likelihood of digital storage somewhere. Also, access to many of the world's newspaper articles, including all of the *New York Times* from 1985 onwards, is available through *Access World News*; many university and community libraries maintain licensed subscriptions. The more academically oriented journals, such as *Science*, can be obtained through JSTOR, which is again widely available even in many smaller libraries.

Because of the extensive length of some redacted court opinions and dissents, several of these sources are not incorporated directly into the main text. Instead, a separate file containing this supplementary material is available at

`cda.psych.uiuc.edu/ethically_perplexed`

In the text proper, these sources are referred to as "Appendix Supplements." The file itself has the name

`ethically_perplexed_supplements.pdf`

and is organized according to the book's table of contents. It also includes all the lists for the Suggested Readings, again grouped by relevant chapter and/or section.

* * *

A major purpose for this book is to make the statistical methods learned in the usual graduate courses more relevant and interesting to students in their day-to-day lives. Ideally, it will also be useful to teachers of statistics and their efforts to impart the importance of such reasoning. Statistical literacy is a crucial skill to have in navigating our supposedly evidence-based society, whether in making personal health or financial decisions or in supporting political and judicial initiatives that promote overall societal well-being. Statistics has developed into the primary means for presenting all forms of risk and benefit information, and in interpreting the vast amounts of evidence given as data generated in all sorts of conceivable social and medical arenas. An informed citizenry needs to understand the questions to ask about some presented numerical information on, say, new drugs to take, new worries to be concerned about, or new dangers to avoid. The push should always be to "show me the data" and then to ask how, by what agency, and for what purpose were the data produced.

* * *

This book has been evolving for some time. It started life as a much shorter contributed chapter with the same name that appeared in the *Handbook of Ethics in Quantitative Methodology*, edited by A. T. Panter and Sonya K. Sterba (2011, pp. 61–124). In a sense this chapter could be seen as a "condensed" version of the book. We thank both editors of the ethics volume for asking us to contribute on a topic relating to what we teach in our behavioral science methodology classes. It has turned into quite an extensive labor of love for both of us, with no signs of abatement in interest on our part even with the appearance of this expanded volume.

* * *

Moses Maimonides, besides being the most important Jewish philosopher of the Middle Ages, was a practicing physician and prolific writer on medical education. The Maimonides Medical Center in Brooklyn, New York, attests to his historical influence on medicine. He is also considered a father of toxicology for his writings in that area. We end this preface with one of his shorter contributions: The Oath (or Invocation) of Maimonides. Although it explicitly articulates the ethical practice of medicine, it also informs the conduct of any profession concerned with human well-being.

The Oath of Maimonides

Thy Eternal Providence has appointed me to watch over the life and health of Thy creatures. May the love for my art actuate me at all times; may neither avarice nor miserliness, nor the thirst for glory, nor for a great reputation engage my mind; for the enemies of Truth and Philanthropy could easily deceive me and make me forgetful of my lofty aim of doing good to Thy children.

May I never see in the patient anything but a fellow creature in pain.

Grant me strength, time, and opportunity always to correct what I have acquired, always to extend its domain; for knowledge is immense and the spirit of man can extend infinitely to enrich itself daily with new requirements. Today he can discover his errors of yesterday, and tomorrow he can obtain a new light on what he thinks himself sure of today.

Oh, God, Thou has appointed me to watch over the life and death of Thy creatures; here am I ready for my vocation.

And now I turn unto my calling.

Maimonides (as quoted by Lucia, 1929, pp. 119–120)

Contents

Contents

List of Figures

List of Tables

Chapter 1

Preamble

> The true foundation of theology is to ascertain the character of God. It is by the art of Statistics that law in the social sphere can be ascertained and codified, and certain aspects of the character of God thereby revealed. The study of statistics is thus a religious service.
> — Florence Nightingale (quoted by F. N. David, *Games, Gods, and Gambling*, 1962, p. 103)

Two major characterizations of the term "ethical" need to be distinguished in this preamble, if only because this book focuses on just one of them. The definition not used is where "ethical" pertains to principles of morality and what is right or wrong in conduct; thus, we speak of an "ethical (or moral) dilemma" as a situation involving an apparent conflict between moral imperatives, where obeying one would result in transgressing another.[1] This is more than we wish to undertake, or even be capable of, in a book devoted to statistical literacy as an assistance to ethical reasoning. The meaning of "ethical" adopted here is one of being in accordance with the accepted rules or standards for right conduct that govern the practice of some profession. The professions we have in mind are statistics and the behavioral sciences, and the standards for ethical practice are what we try to instill in our students through the methodology courses we offer, with particular emphasis on the graduate statistics sequence generally required for all the behavioral sciences. Our hope is that the principal general education payoff for competent statistics instruction is an improvement in people's ability to be critical and ethical consumers and producers of the statistical reasoning and analyses faced in various applied contexts over the course of their careers. Thus, this book is not as much about the good uses of statistics, but more about the specious applications when either statistical ideas are being applied unethically, or some quantitative insight might otherwise help prevent statistical "blunders" by the chronically careless. It

1

may not be unethical to be ignorant of the principles that guide a particular profession, but it is unethical to be ignorant and act as if one is not. As an example, it is unethical to use some statistical program blindly to do something that you don't understand, and know that you don't, but then proceed to interpret the results as if you really did. It is best to keep the adage in mind that if you don't know how to do something, then you don't know how to do it on a computer.[2]

Once individuals pass through a competently taught behaviorally oriented statistics sequence, certain statistical literacy skills should then be available to help guide their ethical application. For someone in a clinical psychology program, for example, and who should now know better, it would be unethical to use projective instruments, such as the Rorschach Test (2010) or Draw-A-Person Test (2010), in clinical practice for diagnosis (because of the prevalence of illusory correlations and unproven validity; see Lilienfeld, Wood, & Garb, 2000); it is unethical to prefer clinical judgment for diagnosis when standard mechanical methods for using intake information exist (for example, by discriminant analysis methods; see Dawes, 1979, 2002; the latter article is entitled "The Ethics of Using or Not Using Statistical Prediction Rules in Psychological Practice and Related Consulting Activities"); it is ethically questionable to attempt equating intact treatment groups using analysis of covariance (due to Lord's Paradox; see Miller and Chapman, 2001); it is unethical to argue for clinical improvement based on regression toward the mean; and so on.[3]

Some ethical reasoning skills are more than just a recognition of a flaw or error, and may require working through a few basic statistical and probabilistic mechanics; for example, a facility with Bayes' theorem is needed when understanding the effects of screening, such as for depression, or how drug companies try to increase specific drug sales with selective and misleading presentation of risks and benefits. Engaging in the ethical practice of statistics in how data are presented and analyzed is ultimately intended to enhance our well-being and that of others by promoting informed decisions and preventing deceitful practice in reporting and interpreting data, especially in how risks and benefits are presented. Obviously, one does not know in advance what might occur in the future, but regardless, a statistically literate citizenry with the ability to reason with the tools learned in a methodology sequence will be needed.[4]

A person might act unethically for many reasons. One could be due to some basic character flaw; another could be due to ignorance of the true state of affairs. We have no suggestions on ameliorating the former, but have undertaken the writing of this book with the aim of illustrating how a deeper understanding of statistics, the science of uncertainty, can help with the latter. Understanding the subtleness of statistical thinking can be of assistance in deciphering the complex world in which we find ourselves, and formulating our behavior in light of the best estimates of what is correct. Such reasoning does not come naturally. Daniel Kahneman won a Nobel Prize for studies demonstrating that humans do not reason logically in many realistic circumstances. His results, joint with the late Amos Tversky, show convincingly why a formal structure for reasoning is necessary. Henceforth, we consider the phrase "ethical reasoning" as a shorthand to mean "careful, logical, correct reasoning from facts to conclusions," with the hope of encouraging ethical behavior in the practice of statistical understanding.[5]

Notes

[1]A well-known literary form of such an ethical or moral dilemma appears in the 1979 William Styron novel, *Sophie's Choice*, and in the 1982 movie of the same name, starring Meryl Streep. A Polish women, Sophie Sawistowski, is arrested by the Nazis and sent to Auschwitz along with her two children. Upon arrival, a sadistic doctor makes Sophie choose which of her children would die immediately by gassing and which would continue to live in the camp. The horrendous guilt that this choice induced eventually led to her suicide at the end of the novel (and of the movie).

[2]An amusing anecdote from an unknown author parallels this adage:

> There was once a group of Biostatisticians and a group of Epidemiologists riding together on a train to joint meetings. All the Epidemiologists had tickets, but the Biostatisticians only had one ticket among them. Inquisitive by nature, the Epidemiologists asked the Biostatisticians how they were going to get away with such a small sample of tickets when the conductor came through. The Biostatisticians said, "Easy. We have methods for dealing with that."

Later, when the conductor came to punch tickets, all the Biostatisticians slipped quietly into the bathroom. When the conductor knocked on the door, the head Biostatistician slipped their one ticket under the door thoroughly fooling the layman conductor.

After the joint meetings were over, the Biostatisticians and the Epidemiologists again found themselves on the same train. Always quick to catch on, the Epidemiologists had purchased one ticket among them. The Biostatisticians (always on the cutting edge) had purchased NO tickets for the trip home. Confused, the Epidemiologists asked the Biostatisticians, "We understand how your methods worked when you had one ticket, but how can you possibly get away with no tickets?" "Easy," replied the Biostatisticians smugly, "we have different methods for dealing with that situation."

Later, when the conductor was in the next car, all the Epidemiologists trotted off to the bathroom with their one ticket and all the Biostatisticians packed into the other bathroom. Shortly, the head Biostatistician crept over to where the Epidemiologists were hiding and knocked authoritatively on the door. As they had been instructed, the Epidemiologists slipped their one ticket under the door. The head Biostatistician took the Epidemiologists' one and only ticket and returned triumphantly to the Biostatistician group. Of course, the Epidemiologists were subsequently discovered and publicly humiliated.

Moral of the story: Do not use statistical methods unless you understand the principles behind them!

[3] The field of clinical psychology seems particularly prone to ethical lapses in reasoning, and in how clinical practice is taught and conducted. A current critical report of the state of clinical psychology is by Baker, McFall, and Shoham, entitled "Current Status and Future Prospects of Clinical Psychology: Toward a Scientifically Principled Approach to Mental and Behavioral Health Care" (*Psychological Science in the Public Interest*, 2008, *9*, 67–103). We list two items from more popular news sources in the Suggested Reading that provide reactions to this report: "Psychology: A Reality Check" (A.2.2: *Nature*, October 15, 2009); "Ignoring the Evidence: Why Do Psychologists Reject Science?" (A.2.3: *Newsweek*, Sharon Begley, October 1, 2009). A much earlier and more popularized account of some of the same issues and criticisms raised in the more academic venue referenced above, is the *New York Times* article by Erica Goode (March 9, 2004), "Defying Psychiatric Wisdom, These Skeptics Say 'Prove It'"; this last item is referenced as A.2.1 in the Suggested Reading. To end the additional Suggested Reading on Clinical Psychology and Psychiatry, A.2.4 lists an article, "The Trauma Trap," by Frederick C. Crews (*New York Review of Books*, March 11, 2004), which reviews Richard McNally's (2003) debunking consolidation, *Remembering Trauma*.

[4] A variety of items listed in the Suggested Reading, Section A.3, on Ethi-

cally Questionable Appeals illustrates well the need to understand how drug companies and others systematically manipulate evidence to support increased sales. One article is retrospective (A.3.2, "Menopause, as Brought to You by Big Pharma," Natasha Singer and Duff Wilson, *New York Times*, December 13, 2009); this article is on the fiasco resulting from the selling of hormone replacement therapy. The article in A.3.7, ("The Menopausal Marketplace," Amanda Spake, *US News and World Report*, November 10, 2002), picks up on the issues raised by hormone replacement therapy, particularly on treating "change of life" as a medical disorder to benefit the lucrative marketing of replacement therapies (and whether or not such marketing is detrimental to a patient). The first article listed in Section A.3, (A.3.1, "Sure, It's Treatable. But Is It a Disorder?"; Natasha Singer, *New York Times*, December 12, 2009), is prospective on the upcoming push to convince males that they now need to worry about P.E. (premature ejaculation), just as they have been concerned with E.D. (erectile dysfunction). A third item (A.3.3) is on the ethically dubious practice of otherwise reputable cancer centers preying on fear and emotion rather than evidence to attract patients ("Cancer Center Ads Use Emotion More Than Fact," Natasha Singer, *New York Times*, December 18, 2009). The fourth article (A.3.4) is online at the W.P. Carey School of Business at Arizona State University (February 13, 2008): "Ask Your Doctor If Direct-to-Consumer Healthcare Advertising is Right for You." It discusses all of the Direct-to-Consumer (DTC) advertising that we are continually bombarded with from Big Pharma (warning: should you have to listen to these ads for more than four hours, call your doctor). The fifth article (A.3.5) by Duff Wilson (April 19, 2010, *New York Times*): "Flavored Tobacco Pellets Are Denounced as a Lure to Young Users," should be self-explanatory. A new product that looks like Tic Tacs is being introduced to hook the young on nicotine. The item in A.3.6 is on the use of salt in our diets, and in how industry giants are pushing back hard on potential regulations and/or recommendations for its (decreased) use, especially for the types of processed foods the country is now dependent on (Michael Moss, "The Hard Sell on Salt," *New York Times*, May 29, 2010). The last five articles form a 2005 series from the *Seattle Times* by Duff Wilson (now writing for the *New York Times*) or Susan Kelleher. The pharmaceutical industry would clearly like to sell us more drugs, whether we really need them, and irrespective of what the side effects might be. Thus, the industry has a major hand in changing the definition of diseases (or making up new ones) that cater to the "worried well" (or what might be labeled as the "suddenly sick" syndrome):

(a) the changing of blood pressure guidelines for the designation of hypertension: "New Blood-Pressure Guidelines Pay Off—For Drug Companies" (June 26, 2005);

(b) the marketing of weight-loss drugs by promoting the Body Mass Index (BMI) to give an indication that any departure from "ideal" is a reason to take medication: "Rush Toward New Weight-Loss Drugs Tramples Patients' Health" (June 27, 2005);

(c) the questionable production of a "new" disease, pre-osteoporosis or osteopenia, that obviously now needs drugs to prevent: "Disease Expands Through Marriage of Marketing and Machines" (June 28, 2005);

(d) the promotion of "conditions," such as deep-vein thrombosis or insomnia, that require drugs: "Many New Drugs Have Strong Dose of Media Hype" (March 16, 2010);

(e) the need for a Viagra counterpart for women with supposed "female sexual dysfunction": "Clash Over 'Little Blue Pill' for Women" (June 30, 2005).

[5]Seven items are presented in the A.1 Suggested Reading section that pertain to several of the general issues raised in this preamble. One item (A.1.1) is tied to the first notion of "ethical" (as used in the phrase, "ethical dilemma") that we noted would not be our primary focus; it is a telling Editorial ("Chance, or Human Judgment?") written for *Science* in 1970 by its then publisher, Dael Wolfle. This was a time when lotteries were being used to establish draft priorities for the Vietnam War. Wolfle addresses the ethical issues of using random mechanisms to decide on who should receive benefits or incur risk, or to instead, rely on human judgment. The second amusing OpEd (A.1.2) from Calvin Trillin (October 13, 2009) relates a story about using quantitative expertise unethically in the service of Wall Street greed alone, which nearly brought down the world economy. The third (A.1.3) is concerned with whether physicians in medical school and elsewhere should be taught about and be concerned with the price of procedures they suggest for their patient's care (Susan Okie, "Teaching Physicians the Price of Care," *New York Times*, May 3, 2010). A.1.4 lists an essay from Arthur Miller in the *New Yorker* (October 21, 1996) on "Why I Wrote 'The Crucible'." It is a quite powerful retrospective about the early 1950s in the United States, including the McCarthy hearings and the House Un-American Activities Committee. Gail Collins in A.1.5 provides an insightful and amusing comment, here on the 50th anniversary of the birth control pill ("What Every Girl Should Know," *New York Times*, May 7, 2010). Two other ethically charged issues are discussed in the last two items in Section A.1: the possible production of synthetic life forms (Nicholas Wade, *New York Times*, May 20, 2010; "Researchers Say They Created a 'Synthetic Cell'"); and the specter of ongoing racial or ethnic profiling: "New York Minorities More Likely to Be Frisked" (Al Baker, *New York Times*, May 12, 2010).

Chapter 2

Introduction

> "Winwood Reade is good upon the subject," said Holmes. "He remarks that, while the individual man is an insoluble puzzle, in the aggregate he becomes a mathematical certainty. You can, for example, never foretell what any one man will do, but you can say with precision what an average number will be up to. Individuals vary, but percentages remain constant. So says the statistician."
> — Sir Arthur Conan Doyle (*The Sign of Four*, 1890)

The vexing practical problems associated with navigating an uncertain world have been with us for millennia, going back to the very beginnings of our written historical record (as an example, see the discussion in Rubin [1971], "Quantitative Commentary in Thucydides," and on the latter's monumental work, *The History of the Peloponnesian War*, dating from the fifth century B.C.).[1] By the 17th century, work began to appear that was statistical (or more accurately, stochastic): the 1650s correspondence between Fermat and Pascal on probability (Devlin, 2008); Huygens' 1656/1657 tract on probability and reasoning in games of chance; and John Graunt's *Natural and political observations mentioned in a following index, and made upon the bills of mortality* (1662), which contained both data and wise suggestions about how to interpret variability, and generally established the value of careful observation of imperfectly collected human data. These isolated incidents were but the start of what developed quickly. In 1710, John Arbuthnot's "Argument for Divine Providence" used Graunt's data to illustrate how to do formal hypothesis testing.

In the same century, Adrian Marie Legendre (1752–1833) used least squares to do astronomical calculations in the face of the uncertainty inherent in such measurements (Legendre, 1805); also, in that golden age of science, Pierre Laplace (1749–1827), Abraham De Moivre (1667–1754), and Jacob Bernoulli (1634–1705), among many others, were developing and using the tools that today help us measure, understand, and control uncertainty. Despite this long preamble, statistics, the science of uncertainty, is primarily a 20th century creation, dating its modern progenitors from Francis Galton (1822–1911), Karl Pearson (1857–1936), Francis Ysidro Edgeworth (1845–1926), George Udny Yule (1871–1951), and then to R. A. Fisher (1890–1962). (The definitive history of statistical thought before 1900 is available in Stigler [1986].)

The emerging field of statistics is marked by the publication of a handful of critically important books. Arguably, the two most important are R. A. Fisher's (1925) *Statistical Methods for Research Workers*, and a half century later, John Tukey's (1977) *Exploratory Data Analysis*. The former laid out the analytic tools for generations of scientists; the latter provided the formal justification for a new way of thinking about statistical methods. Taken as a pair, they provide the basis for modern statistics. The focus of this book is principally on the view of statistics as expressed by Fisher. This is not to denigrate exploratory methods, which we believe are at least as important, and usually a lot more fun, but only because of length restrictions. We leave a parallel discussion of exploratory procedures to other accounts, although we touch on them here and there when we cannot resist the temptation.

Generations of graduate students in the behavioral and social sciences have completed mandatory year-long course sequences in statistics, sometimes with difficulty and possibly with less than positive regard for the content and how it was taught. Prior to the 1960s, such a sequence usually emphasized a cookbook approach where formulas were applied unthinkingly using mechanically operated calculators. The instructional method could be best characterized as "plug and chug," where there was no need to worry about the meaning of what one was doing, only that the numbers could be put in and an answer generated. It was hoped that this process would lead to numbers that could then be looked up in tables; in turn, p-values were sought that were less than the iconic .05, giving some hope of getting an attendant paper published.

The situation began to change for the behavioral sciences in 1963 with the publication of *Statistics for Psychologists* by William Hays. For the first time, graduate students could be provided both the needed recipes and some deeper understanding of and appreciation for the whole enterprise of inference in the face of uncertainty and fallibility. The Hays text is now in its fifth edition, with a shortened title of *Statistics* (1994); the name of Hays itself stands as the eponym for what kind of methodology instruction might be required for graduate students; that is, at the level of Hays, and "cover to cover." Although now augmented by other sources for related computational work (e.g., by SAS, SPSS, or SYSTAT), the Hays text remains a standard of clarity and completeness. Many methodologists have based their teaching on this resource for more than four decades. Hays typifies books that although containing lucid explanations of statistical procedures, are still too often used by students only as a cookbook of statistical recipes.

In teaching graduate statistics, there are multiple goals:

(1) to be capable of designing and analyzing one's own studies, including doing the computational "heavy lifting" oneself, and the ability to verify what others attached to a project may be doing;

(2) to understand and consume other research intelligently, both in one's own area, and more generally as a statistically and numerically literate citizen;

(3) to argue for and justify analyses when questioned by journal and grant reviewers or others, and to understand the basic justification for what was done. For example, an ability to reproduce a formal proof of the central limit theorem is unnecessary, but a general idea of how it is formulated and functions is relevant, and that it might help justify assertions of robustness being made for the methods used. These skills in understanding are not "theoretical" in a pejorative sense, although they do require more thought than just being content to run the SPSS program blindly. They are absolutely crucial in developing both the type of reflective teaching and research careers we would hope to nurture in graduate students, and more generally for the quantitatively literate citizenry we would wish to be made up of our society.

Graduate instruction in statistics requires the presentation of general frameworks and how to reason from these. These frameworks can be conceptual: for example, (a) the Fisherian (*Design of Experiments*, 1935, first edition) view that provided the evidence of success in the Salk Polio vaccine trials where the physical act of randomization lead to credible causal inferences (see Francis et al., 1955; or (b) to the unification given by the notion of maximum likelihood estimation and likelihood ratio tests both for our general statistical modeling as well as for more directed formal modeling in a behavioral science subdomain, such as image processing or cognitive neuroscience. These frameworks can also be based on more quantitatively formal structures: for example, (a) the general linear model and its special cases of multiple regression, analysis of variance (ANOVA), and analysis of covariance (ANCOVA), along with model comparisons through full and reduced models; (b) the general principles behind prediction/selection/correlation in simple two-variable systems, with extensions to multiple-variable contexts; and (c) the various dimensionality reduction techniques of principal components/factor analysis, multidimensional scaling, cluster analysis, and discriminant analysis.

The remainder of this book will attempt to sketch some basic structures typically introduced in the required graduate statistics sequence in the behavioral sciences, along with some necessary cautionary comments on usage and interpretation. The purpose is to provide a small part of the formal scaffolding needed in reasoning ethically about what we see in the course of our careers, both in our own work and that of others, or what might be expected of a statistically literate populace generally. Armed with this deeper understanding, graduates can be expected to deal more effectively with whatever ethically charged situations they might face.

2.1 The (Questionable) Use of Statistical Models

> There is no more common error than to assume that because prolonged and accurate mathematical calculations have been made, the application of the result to some fact of nature is absolutely certain.
> — Alfred North Whitehead (*An Introduction to Mathematics*, 1911)

The form of statistical practice most commonly carried out by those with a mathematical bent (and in contrast to those more concerned with simple manifest forms of data analysis and visualization), is through the adoption of a stochastic model commonly containing latent (unobserved) variables. Here, some data-generating mechanism is postulated, characterized by a collection of parameters and strong distributional assumptions (for example, conditional independence, normality, or homogeneous variability). Based on a given dataset, the parameters are estimated, and usually, the goodness of fit of the model assessed by some statistic. We might even go through a ritual of hoping for nonsignificance in testing a null hypothesis that the model is true, generally through some modified chi-squared statistic heavily dependent on sample size. The cautionary comments of Roberts and Pashler (2000) should be kept in mind that the presence of a good fit does not imply a good or true model. Moreover, models with many parameters are open to the problems engendered by overfitting and of a subsequent failure to cross-validate. We provide the abstract of the Roberts and Pashler (2000) article, "How Persuasive Is a Good Fit? A Comment on Theory Testing," (*Psychological Review, 107*, 358–367):

> Quantitative theories with free parameters often gain credence when they closely fit data. This is a mistake. A good fit reveals nothing about the flexibility of the theory (how much it cannot fit), the variability of the data (how firmly the data rule out what the theory cannot fit), or the likelihood of other outcomes (perhaps the theory could have fit any plausible result), and a reader needs all three pieces of information to decide how much the fit should increase belief in the theory. The use of good fits as evidence is not supported by philosophers of science nor by the history of psychology; there seem to be no examples of a theory supported mainly by good fits that has led to demonstrable progress. A better way to test a theory with free parameters is to determine how the theory constrains possible outcomes (i.e., what it predicts), assess how firmly actual outcomes agree with those constraints, and determine if plausible alternative outcomes would have been inconsistent with the theory, allowing for the variability of the data. (p. 358)

A model-based approach is assiduously avoided throughout this book. It seems ethically questionable to base interpretations about a given dataset and the story that the data may be telling, through a model that is inevitably incorrect. As one highly cherished example in the behavioral sciences, it is now common practice to frame questions of causality through structural equation or path models, and to perform most data analysis tasks through the fitting of various highly parameterized latent variable models.[2] In a devastating critique of this practice, David Freedman in a *Journal of Educational Statistics* article, "As Others See Us: A Case Study in Path Analysis" (1987, *12*, 101–128), ends with this paragraph:

> My opinion is that investigators need to think more about the under-
> lying social processes, and look more closely at the data, without the
> distorting prism of conventional (and largely irrelevant) stochastic
> models. Estimating nonexistent parameters cannot be very fruitful.
> And it must be equally a waste of time to test theories on the ba-
> sis of statistical hypotheses that are rooted neither in prior theory
> nor in fact, even if the algorithms are recited in every statistics text
> without caveat. (p. 128)

Leo Breiman took on the issue directly of relying on stochastic models (or, as he might have said, "hiding behind") in most of contemporary statistics. What Breiman advocates is the adoption of optimization in place of parameter estimation, and of methods that fall under the larger rubric of supervised or unsupervised statistical learning theory. Currently, this approach is best exemplified by the comprehensive text, *The Elements of Statistical Learning: Data Mining, Inference, and Prediction* (2nd ed., 2009) (T. Hastie, R. Tibshirani, & J. Friedman). We give the abstract from Leo Breiman's *Statistical Science* article, "Statistical Modeling: The Two Cultures" (2001, *16*, 199–215):

> There are two cultures in the use of statistical modeling to reach
> conclusions from data. One assumes that the data are generated by
> a given stochastic data model. The other uses algorithmic models and
> treats the data mechanism as unknown. The statistical community
> has been committed to the almost exclusive use of data models. This
> commitment has led to irrelevant theory, questionable conclusions,
> and has kept statisticians from working on a large range of interesting
> current problems. Algorithmic modeling, both in theory and practice,
> has developed rapidly in fields outside statistics. It can be used both
> on large complex datasets and as a more accurate and informative
> alternative to data modeling on smaller datasets. If our goal as a
> field is to use data to solve problems, then we need to move away
> from exclusive dependence on data models and adopt a more diverse
> set of tools. (p. 199)

The view of statistics to be followed in this book is to consider what linear regression models can or cannot do, or the implications of a basic sampling model, but we go no further than least squares treated as an algorithmic optimization process, and a suggestion to adopt various sample reuse methods to gauge stability and assess cross-validation. Remembering the definition of a *deus ex machina*—a plot device in Greek drama whereby a seemingly insoluble problem is suddenly and abruptly solved with the contrived and unexpected intervention of some new character or god—we will not invoke any statistical *deus ex machina* analogues.[3]

Stochastic data models do have a place but not when that is only as far as it goes. When we work solely within the confines of a closed system given by the model, and base all inferences and conclusions under that rubric alone (for example, we claim a causal link because some path coefficient is positive and significant), the ethicality of such a practice is highly questionable. George Box has famously said that "essentially, all models are wrong, but some are useful" (*Empirical Model-Building and Response Surfaces*, 1987, co-authored with Norman R. Draper, p. 424); or Henri Theil's similar quip: "It does require maturity to realize that models are to be used, but not to be believed" (*Principles of Econometrics*, 1971, p. vi). Box was referring to the adoption of a model heuristically to guide a process of fitting data; the point being that we only "tentatively entertain a model," with that model then subjected to diagnostic testing and reformulation, and so on iteratively. The ultimate endpoint of such a process is to see how well the fitted model works, for example, on data collected in the future. Once again, some type of (cross-)validation is essential, which should be the sine qua non of any statistical undertaking.[4]

Notes

[1]Plataean Escape From Circumvallation: Taken from W. Allen Wallis and Harry V. Roberts, *Statistics: A New Approach* (p. 215):

> One of the most common and useful applications of the mode is illustrated by the following example of its employment for military purposes 24 centuries ago:
> The same winter [428 B.C.] the Plataeans, who were still being besieged by the Peloponnesians and Boeotians, distressed by the failure of their provisions, and seeing no hope of relief from Athens,

nor any other means of safety, formed a scheme with the Athenians besieged with them for escaping, if possible, by forcing their way over the enemy's walls [of circumvallation]. ... Ladders were made to match the height to the enemy's wall, which they measured by the layers of bricks, the side turned towards them not being thoroughly whitewashed. These were counted by many persons at once; and though some might miss the right calculation, most would hit upon it, particularly as they counted over and over again, and were no great way from the wall, but could see it easily enough for their purpose. The length required for the ladders was thus obtained, being calculated from the breadth of the brick.

The everyday, contemporary version of this application of the mode is in checking calculations. If a calculation is repeated several times, the value accepted is that which occurs most often, not the median, mean, or any other figure. Even in these cases, a majority, or some more overwhelming preponderance, rather than merely a mode, is usually required for a satisfactory decision.

[2] A current discussion of yet another attempt to use statistical models to infer causality from observational data is in the article by Gina Kolata, "Catching Obesity From Friends May Not Be So Easy" (*New York Times*, August 8, 2011). Kolata reviews the criticisms of causal inferences made using social network models, particularly from the 2009 book by Christakis and Fowler, *Connected: The Surprising Power of Our Social Networks and How They Shape Our Lives—How Your Friends' Friends' Friends Affect Everything You Feel, Think, and Do.* Their basic argument is that homophily, the tendency of individuals to associate and bond with similar others, can somehow be separated from contagion, the spread of some societal ill, such as obesity, through an explicit social network. The quantitative argument rests on the presence of an asymmetry in the relationship of who is a friend of whom, based solely on parameter estimates in a statistical model. Supposedly, this implies causality in who affects whom in the direct sense of a contagious transmission. Thus, contagion through a network is causal for items such as depression, happiness, illegal drug use, smoking, and loneliness. This can all be gleaned directly from observational data through the intermediary of fitting a statistical model; moreover, contagion effects can be cleanly separated from homophily. Freedman's quotation given for path models is just as germane for these network models.

[3] An ancedote told in some of our beginning statistics sequences reflects this practice of postulating a *deus ex machina* to carry out statistical interpretations. Three academics—a philosopher, an engineer, and a statistician—are walking in the woods toward a rather large river that needs to be crossed. The pensive philosopher stops, and opines about whether they really need to cross the river; the engineer pays no attention to the philosopher and proceeds

immediately to chop down all the trees in sight to build a raft; the statistician yells to the other two: "stop, assume a boat."

[4]These comments about models brings to mind an observation from John Tukey (personal communication to HW, October 26, 1996): "In any situation we have three models, in increasing order of complexity: (1) the model we fit to the data; (2) the model we used to think about the data; and (3) the correct model."

Part I

Tools From Probability and Statistics

Chapter 3

Probability Theory: Background and Bayes' Theorem

> It is remarkable that a science which began with the consideration of games of chance should have become the most important object of human knowledge. ... The most important questions of life are indeed, for the most part, really only problems of probability.
> — Pierre-Simon Laplace (*Théorie Analytique des Probabilités*, 1812)

The formalism of thought offered by probability theory is one of the more useful portions of any beginning course in statistics in helping to promote ethical reasoning. As typically presented, we speak of an event represented by a capital letter, say A, and the probability of the event as some number in the range from 0 to 1, written as $P(A)$. The value of 0 is assigned to the "impossible" event that can never occur; 1 is assigned to the "sure" event that will always occur. The driving condition for the complete edifice of all probability theory is one single postulate: for two mutually exclusive events, A and B (where mutually exclusivity implies that both events cannot occur at the same time), $P(A \text{ or } B) = P(A) + P(B)$. As a final beginning definition, we say that two events are independent whenever the probability of the joint event, $P(A \text{ and } B)$, factors as the product of the individual probabilities, $P(A)P(B)$.

The idea of statistical independence and the factoring of the joint event probability immediately provide a formal tool for understanding a number of historical miscarriages of justice. In particular, if two events are not independent, then the joint probability cannot be generated by a simple product of the individual probabilities. A recent example is the case of Sally Clark; she was convicted in England, partially on the basis of an inappropriate assumption of statistical independence, of killing her two children, .

The purveyor of statistical misinformation in this case was Sir Roy Meadow, famous for Meadow's Law: "'One sudden infant death is a tragedy, two is suspicious, and three is murder until proved otherwise' is a crude aphorism but a sensible working rule for anyone encountering these tragedies" (Meadow, 1997, p. 29). We quote part of a news release from the Royal Statistical Society (October 23, 2001):

> The Royal Statistical Society today issued a statement, prompted by issues raised by the Sally Clark case, expressing its concern at the misuse of statistics in the courts.
>
> In the recent highly-publicised case of *R v. Sally Clark*, a medical expert witness drew on published studies to obtain a figure for the frequency of sudden infant death syndrome (SIDS, or 'cot death') in families having some of the characteristics of the defendant's family. He went on to square this figure to obtain a value of 1 in 73 million for the frequency of two cases of SIDS in such a family.
>
> This approach is, in general, statistically invalid. It would only be valid if SIDS cases arose independently within families, an assumption that would need to be justified empirically. Not only was no such empirical justification provided in the case, but there are very strong a priori reasons for supposing that the assumption will be false. There may well be unknown genetic or environmental factors that predispose families to SIDS, so that a second case within the family becomes much more likely.
>
> The well-publicised figure of 1 in 73 million thus has no statistical basis. Its use cannot reasonably be justified as a 'ballpark' figure because the error involved is likely to be very large, and in one particular direction. The true frequency of families with two cases of SIDS may be very much less incriminating than the figure presented to the jury at trial.

Several other examples of a misuse for the idea of statistical independence exist in the legal literature, such as the notorious 1968 jury trial in California, *People v. Collins*. Here, the prosecutor suggested that the jury merely multiply several probabilities together, which he conveniently provided, to ascertain the guilt of the defendant. In overturning the conviction, the Supreme Court of California criticized both the statistical reasoning and the framing of the decision for the jury:

> We deal here with the novel question whether evidence of mathematical probability has been properly introduced and used by the

prosecution in a criminal case. .... Mathematics, a veritable sorcerer in our computerized society, while assisting the trier of fact in the search of truth, must not cast a spell over him. We conclude that on the record before us, defendant should not have had his guilt determined by the odds and that he is entitled to a new trial. We reverse the judgment.

We will return to both the Clark and Collins cases later when Bayes' rule is discussed in the context of conditional probability confusions and the "Prosecutor's Fallacy."[1]

Besides the concept of independence, the definition of conditional probability plays a central role in all our uses of probability theory; in fact, many misapplications of statistical/probabilistic reasoning involve confusions of some sort regarding conditional probabilities. Formally, the conditional probability of some event A given that B has already occurred, denoted $P(A|B)$, is defined as $P(A \text{ and } B)/P(B)$. When A and B are independent, $P(A|B) = P(A)P(B)/P(B) = P(A)$; or in words, knowing that B has occurred does not alter the probability of A occurring. If $P(A|B) > P(A)$, we will say that B is "facilitative" of A; when $P(A|B) < P(A)$, B is said to be "inhibitive" of A. As a small example, suppose A is the event of receiving a basketball scholarship; B, the event of being seven feet tall; and C, the event of being five feet tall. One obviously expects B to be facilitative of A (that is, $P(A|B) > P(A)$); and of C to be inhibitive of A (that is, $P(A|C) < P(A)$). In any case, the size and sign of the difference between $P(A|B)$ and $P(A)$ is an obvious raw descriptive measure of how much the occurrence of B is associated with an increased or decreased probability of A, with a value of zero corresponding to statistical independence.

One convenient device for interpreting probabilities and understanding how events can be "facilitative" or "inhibitive" is through the use of a simple 2×2 contingency table that cross-classifies a set of objects according to the events A and $\bar{A}$, and B and $\bar{B}$ (here, $\bar{A}$ and $\bar{B}$ represent the complements of A and B, which occur when the original events do not). For example, suppose we have a collection of N balls placed in a container; each ball is labeled A or $\bar{A}$, and also B or $\bar{B}$, according to the notationally self-evident table of frequencies below:

	A	$\bar{A}$	Row Sums
B	N_{AB}	$N_{\bar{A}B}$	N_B
$\bar{B}$	$N_{A\bar{B}}$	$N_{\bar{A}\bar{B}}$	$N_{\bar{B}}$
Column Sums	N_A	$N_{\bar{A}}$	N

The process we consider is one of picking a ball blindly from the container, where the balls are assumed to be mixed thoroughly, and noting the occurrence of the events A or $\bar{A}$ and B or $\bar{B}$. Based on this physical idealization of such a selection process, it is intuitively reasonable to assign probabilities according to the proportion of balls in the container satisfying the attendant conditions:

$P(A) = N_A/N; P(\bar{A}) = N_{\bar{A}}/N; P(B) = N_B/N; P(\bar{B}) = N_{\bar{B}}/N;$

$P(A|B) = N_{AB}/N_B; P(B|A) = N_{AB}/N_A;$
$P(\bar{A}|B) = N_{\bar{A}B}/N_B; P(B|\bar{A}) = N_{\bar{A}B}/N_{\bar{A}};$
$P(\bar{B}|A) = N_{A\bar{B}}/N_A; P(A|\bar{B}) = N_{A\bar{B}}/N_{\bar{B}};$
$P(\bar{A}|\bar{B}) = N_{\bar{A}\bar{B}}/N_{\bar{B}}; P(\bar{B}|\bar{A}) = N_{\bar{A}\bar{B}}/N_{\bar{A}}.$

By noting the relationships: $N_B = N_{AB} + N_{\bar{A}B}$; $N_{\bar{B}} = N_{A\bar{B}} + N_{\bar{A}\bar{B}}$; $N_A = N_{AB} + N_{A\bar{B}}$; $N_{\bar{A}} = N_{\bar{A}B} + N_{\bar{A}\bar{B}}$; $N_B + N_{\bar{B}} = N_A + N_{\bar{A}} = N$, a variety of interesting connections can be derived and understood that can assist immensely in our probabilistic reasoning. We present a short numerical example below on how these ideas might be used in a realistic context; several such uses are then expanded upon in the subsections to follow.

As a numerical example of using a 2×2 contingency table to help explicate probabilistic reasoning, suppose we have an assumed population of 10,000, cross-classified according to the presence or absence of Colorectal Cancer (CC) [A: +CC; $\bar{A}$: −CC], and the status of a Fecal Occult Blood Test (FOBT) [B: +FOBT; $\bar{B}$: −FOBT]. Using data from Gerd Gigerenzer, *Calculated Risks* (2002, pp. 104–107), we have the following 2×2 table:

	+CC	−CC	Row Sums
+FOBT	15	299	314
−FOBT	15	9671	9686
Column Sums	30	9970	10,000

The probability $P(+CC \,|\, +FOBT)$ is simply $15/314 = .048$, based on the frequency value of 15 for the cell (+FOBT, +CC), and the +FOBT row sum of 314. The marginal probability, P(+CC), is $30/10,000 = .003$, and thus, a positive FOBT is "facilitative" of a positive CC because .048 is greater than .003. The size of the difference, $P(+CC \mid +FBOT) - P(+CC) = +.045$, may not be large in any absolute sense, but the change does represent a fifteenfold increase over the marginal probability of .003. (But note that if you have a positive FOBT, over 95% ($= \frac{299}{314}$) of the time you don't have cancer; that is, there are 95% false positives.)[2]

Many day-to-day contexts are faced where our decisions might best be made from conditional probabilities, if we knew them, instead of from marginal information. When deciding on a particular medical course of action, for example, it is important to condition on our own circumstances of age, risk factors, family medical history, and our own psychological needs and makeup. A recent and controversial instance of this, where the conditioning information is "age," is reported in the *New York Times* article by Gina Kolata, "Panel Urges Mammograms at 50, Not 40" (November 16, 2009). The failure to consider conditional instead of marginal probabilities is particularly grating for many of us who follow various sporting activities and enjoy second-guessing managers, quarterbacks, sports commentators, and their ilk. As an example, consider the "strike-'em-out-throw-'em-out" double play in baseball, where immediately after the batter has swung and missed at a third strike or taken a called third strike, the catcher throws out a base runner attempting to steal second or third base. Before such a play occurs, announcers routinely state that the runner "will or will not be sent" because the "batter strikes out only some percentage of the time." The issue of running or not shouldn't be based on the marginal probability of the batter striking out but on some conditional probability (for example, how often does the batter strike out when faced with a particular count or type of pitcher). For many other instances, however, we might be content not to base our decisions on conditional information; for example, always wear a seat belt irrespective of the type or length of trip being taken.

A variety of probability results will prove useful throughout our attempt to reason probabilistically and follow the field of statistical inference. We list several of these below, with uses given throughout this book.

(1) For the complementary event, $\bar{A}$, which occurs when A does not, $P(\bar{A}) = 1 - P(A)$.

(2) For events A and B that are not necessarily mutually exclusive,

$$P(A \text{ or } B) = P(A) + P(B) - P(A \text{ and } B) \, .$$

(3) The *rule of total probability*: Given a collection of mutually exclusive and exhaustive events, $B_1, \ldots, B_K$ (that is, all are pairwise mutually exclusive and their union gives the sure event),

$$P(A) = \sum_{k=1}^{K} P(A|B_k)P(B_k) \, .$$

(4) Bayes' theorem (or rule) for two events, A and B:

$$P(A|B) = \frac{P(B|A)P(A)}{P(B|A)P(A) + P(B|\bar{A})P(\bar{A})} \, .$$

(5) Bonferroni inequality: for a collection of events, $A_1, \ldots, A_K$,

$$P(A_1 \text{ or } A_2 \text{ or } \cdots \text{ or } A_K) \leq \sum_{k=1}^{K} P(A_k) \, .$$

(6) $P(A \text{ and } B) \leq P(A \text{ or } B) \leq P(A) + P(B) \, .$
In words, the first inequality results from the event "A and B" being wholly contained within the event "A or B" ("A or B" occurs when A or B or both occur); the second results from the Bonferroni inequality restricted to two events.

(7) $P(A \text{ and } B) \leq \text{minimum}(P(A), P(B)) \leq P(A) \text{ or } \leq P(B) \, .$
In words, the first inequality results from the event "A and B" being wholly contained both within A and within B; the second inequalities are more generally appropriate—the minimum of any two numbers is always less than either of the two numbers.[3]

3.1   The (Mis)assignment of Probabilities

Just think of all the billions of coincidences that don't happen.
— Dick Cavett

Although the assignment of probabilities to events consistent with the disjoint rule may lead to an internally valid system mathematically, there is still no assurance that this assignment is "meaningful," or bears any empirical validity for observable long-run expected frequencies. There seems to be a never-ending string of misunderstandings in the way probabilities can be generated that are either blatantly wrong, or more subtly incorrect, irrespective of the internally consistent system they might lead to. Some of these problems are briefly sketched below, but we can only hope to be representative of a few possibilities, not exhaustive.

One inappropriate way of generating probabilities is to compute the likelihood of some joint occurrence after some of the outcomes are already known. For example, there is the story about the statistician who takes a bomb aboard a plane, reasoning that if the probability of one bomb on board is small, the probability of two is infinitesimal. Or, during World War I, soldiers were actively encouraged to use fresh shell holes as shelter because it was very unlikely for two shells to hit the same spot during the same day. And the Minnesota Twins baseball manager who bats for an individual who earlier in the game hit a home run because it would be very unlikely for him to hit two home runs in the same game. Although these slightly amusing stories may provide obvious misassignments of probabilities, other related situations are more subtle. For example, whenever coincidences are culled or "hot spots" identified from a search of available information, the probabilities that are then regenerated for these situations may not be valid. There are several ways of saying this: when some set of observations is the source of an initial suspicion, those same observations should not be used in a calculation that then tests the validity of the suspicion. In Bayesian terms, you should not obtain the posterior probabilities from the same information that gave you the prior probabilities. Alternatively said, it makes no sense to do formal hypothesis assessment by finding estimated probabilities when the data themselves have suggested the hypothesis in the first place. Some cross-validation strategy is necessary; for example, collecting independent data. Generally, when some process of search or optimization has been used to identify an unusual situation (for example, when a "good" regression equation is found through a step-wise procedure [see Freedman, 1983, for a devastating critique]; when data are "mined" and unusual patterns identified; when DNA databases are searched for "cold-hits" against evidence left at a crime scene; when geographic "hot spots" are identified for, say, some particularly unusual cancer; or when the whole human genome is searched for clues to common diseases), the same methods for assigning probabilities before the particular situation was identified are generally no longer appropriate after the fact.[4]

A second general area of inappropriate probability assessment concerns the model postulated to aggregate probabilities over several events. Campbell (1974, p. 126) cites an article in the *New York Herald Tribune* (May, 1954) stating that if the probability of knocking down an attacking airplane were .15 at each of five defensive positions before reaching the target, then the probability of knocking down the plane before it passed all five barriers would be .75 (5 × .15), this last value being the simple sum of the individual probabilities—and an inappropriate model. If we could correctly assume independence between the Bernoulli trials at each of the five positions, a more justifiable value would be one minus the probability of passing all barriers successfully: $1.0 - (.85)^5 \approx .56$. The use of similar binomial modeling possibilities, however, may be specious—for example, when dichotomous events occur simultaneously in groups (such as in the World Trade Center disaster on 9/11/01); when the success proportions are not valid; when the success proportions change in value over the course of the trials; or when time dependencies are present in the trials (such as in tracking observations above and below a median over time). In general, when wrong models are used to generate probabilities, the resulting values may have little to do with empirical reality. For instance, in throwing dice and counting the sum of spots that result, it is not true that each of the integers from two through twelve is equally likely. The model of what is equally likely may be reasonable at a different level (for example, pairs of integers appearing on the two dice), but not at all aggregated levels. There are some stories, probably apocryphal, of methodologists meeting their demises by making these mistakes for their gambling patrons.

Flawed calculations of probability can have dire consequences within our legal systems, as the case of Sally Clark and related others make clear. One broad and current area of possible misunderstanding of probabilities is in the context of DNA evidence (which is exacerbated in the older and more fallible system of identification through fingerprints).[5] In the use of DNA evidence (and with fingerprints), one must be concerned with the Random Match Probability (RMP): the likelihood that a randomly selected unrelated person from the population would match a given DNA profile. Again, the use of independence in RMP estimation is questionable; also, how does the RMP relate to, and is it relevant for, "cold-hit" searches in DNA databases. In a confirmatory identification case, a suspect is first identified by non-DNA evidence; DNA evidence is then used to corroborate traditional police investigation. In a "cold-hit" framework, the suspect is first identified by a search of DNA databases; the DNA evidence is thus used to identify the suspect as perpetrator, to the exclusion of others, directly from the outset (this is akin to shooting an arrow into a tree and then drawing a target around it). Here, traditional police work is no longer the focus. For a thorough discussion of the probabilistic context surrounding DNA evidence, which extends with even greater force to fingerprints, the article by Jonathan Koehler is recommended ("Error and Exaggeration in the Presentation of DNA Evidence at Trial," *Jurimetrics Journal*, *34*, 1993–1994, 21–39). We excerpt part of the introduction to this article below:

> DNA identification evidence has been and will continue to be powerful evidence against criminal defendants. This is as it should be. In general, when blood, semen or hair that reportedly matches that of a defendant is found on or about a victim of violent crime, one's belief that the defendant committed the crime should increase, based on the following chain of reasoning:
>
> Match Report $\Rightarrow$ True Match $\Rightarrow$ Source $\Rightarrow$ Perpetrator
>
> First a reported match is highly suggestive of a true match, although the two are not the same. Errors in the DNA typing process may occur, leading to a false match report. Second, a true DNA match

usually provides strong evidence that the suspect who matches is indeed the source of the trace, although the match may be coincidental. Finally, a suspect who actually is the source of the trace may not be the perpetrator of the crime. The suspect may have left the trace innocently either before or after the crime was committed.

In general, the concerns that arise at each phase of the chain of inferences are cumulative. Thus, the degree of confidence one has that a suspect is the source of a recovered trace following a match report should be somewhat less than one's confidence that the reported match is a true match. Likewise, one's confidence that a suspect is the perpetrator of a crime should be less than one's confidence that the suspect is the source of the trace.

Unfortunately, many experts and attorneys not only fail to see the cumulative nature of the problems that can occur when moving along the inferential chain, but they frequently confuse the probabilistic estimates that are reached at one stage with estimates of the others. In many cases, the resulting misrepresentations and misinterpretation of these estimates lead to exaggerated expressions about the strength and implications of the DNA evidence. These exaggerations may have a significant impact on verdicts, possibly leading to convictions where acquittals might have been obtained.

This Article identifies some of the subtle, but common, exaggerations that have occurred at trial, and classifies each in relation to the three questions that are suggested by the chain of reasoning sketched above: (1) Is a reported match a true match? (2) Is the suspect the source of the trace? (3) Is the suspect the perpetrator of the crime? Part I addresses the first question and discusses ways of defining and estimating the false positive error rates at DNA laboratories. Parts II and III address the second and third questions, respectively. These sections introduce the "source probability error" and "ultimate issue error" and show how experts often commit these errors at trial with assistance from attorneys on *both* sides. (pp. 21–22)

In 1989, and based on urging from the FBI, the National Research Council (NRC) formed the Committee on DNA Technology in Forensic Science, which issued its report in 1992 (*DNA Technology in Forensic Science*; or more briefly, NRC I). The NRC I recommendation about the cold-hit process was as follows:

The distinction between finding a match between an evidence sample and a suspect sample and finding a match between an evidence sample and one of many entries in a DNA profile databank is important. The chance of finding a match in the second case is considerably higher. ... The initial match should be used as probable cause to obtain a blood sample from the suspect, but only the statistical frequency associated with the additional loci should be presented at trial (to prevent the selection bias that is inherent in searching a databank). (p. 124)

A follow-up report by a second NRC panel was published in 1996 (*The Evaluation of Forensic DNA Evidence*; or more briefly, NRC II), having the following main recommendation about cold-hit probabilities and using the "database match probability" or DMP:

> When the suspect is found by a search of DNA databases, the random-match probability should be multiplied by N, the number of persons in the database. (p. 161)

The term "database match probability" (DMP) is somewhat unfortunate. This is not a real probability but more of an expected number of matches given the RMP. A more legitimate value for the probability that another person matches the defendant's DNA profile would be $1 - (1 - \frac{1}{\text{RMP}})^N$, for a database of size N; that is, one minus the probability of no matches over N trials. For example, for an RMP of 1/1,000,000 and an N of 1,000,000, the above probability of another match is .632; the DMP (not a probability) number is 1.00, being the product of N and RMP. In any case, NRC II made the recommendation of using the DMP to give a measure of the accuracy of a cold-hit match, and did not support the more legitimate "probability of another match" using the formula given above (possibly because it was considered too difficult?):[6]

> A special circumstance arises when the suspect is identified not by an eyewitness or by circumstantial evidence but rather by a search through a large DNA database. If the only reason that the person becomes a suspect is that his DNA profile turned up in a database, the calculations must be modified. There are several approaches, of which we discuss two. The first, advocated by the 1992 NRC report, is to base probability calculations solely on loci not used in the search. That is a sound procedure, but it wastes information, and if too many loci are used for identification of the suspect, not enough might be left for an adequate subsequent analysis. ... A second procedure is to apply a simple correction: Multiply the match probability by the size of the database searched. This is the procedure we recommend. (p. 32)

3.2   The Probabilistic Generalizations of Logical Fallacies Are No Longer Fallacies

> It is the peculiar and perpetual error of the human understanding to
> be more moved and excited by affirmatives than by negatives.
> — Francis Bacon (*Novum Organum*, Aphorism 46, 1620)

In our roles as instructors in beginning statistics, we commonly introduce some simple logical considerations early on that revolve around the usual "if p, then q" statements, where p and q are two propositions. As an example, we might let p be "the animal is a Yellow Labrador Retriever," and q, "the animal is in the order *Carnivora*." Continuing, we note that if the statement "if p, then q" is true (which it is), then logically, so must be the contrapositive of "if not q, then not p"; that is, if "the animal is not in the order *Carnivora*," then "the animal is not a Yellow Labrador Retriever." However, there are two fallacies awaiting the unsuspecting:

denying the antecedent: if not p, then not q (if "the animal is not a Yellow Labrador Retriever," then "the animal is not in the order *Carnivora*");

affirming the consequent: if q, then p (if "the animal is in the order *Carnivora*," then "the animal is a Yellow Labrador Retriever").

Also, when we consider definitions given in the form of "p if and only if q," (for example, "the animal is a domesticated dog" if and only if "the animal is a member of the subspecies *Canis lupus familiaris*"), or equivalently, "p is necessary and sufficient for q," these separate into two parts:

"if p, then q" (that is, p is a sufficient condition for q);

"if q, then p" (that is, p is a necessary condition for q).

So, for definitions, the two fallacies are not present.

In a probabilistic context, we reinterpret the phrase "if p, then q" as B being facilitative of A; that is, $P(A|B) > P(A)$, where p is identified with B and q with A. With such a probabilistic reinterpretation, we no longer have the fallacies of denying the antecedent (that is, $P(\bar{A}|\bar{B}) > P(\bar{A})$), or of affirming the consequent (that is, $P(B|A) > P(B)$). Both of the latter two probability statements can be algebraically shown true using the simple 2×2

cross-classification frequency table and the equivalences among frequency sums given earlier:

(original statement) $P(A|B) > P(A) \Leftrightarrow N_{AB}/N_B > N_A/N \Leftrightarrow$

(denying the antecedent) $P(\bar{A}|\bar{B}) > P(\bar{A}) \Leftrightarrow N_{\bar{A}\bar{B}}/N_{\bar{B}} > N_{\bar{A}}/N \Leftrightarrow$

(affirming the consequent) $P(B|A) > P(B) \Leftrightarrow N_{AB}/N_A > N_B/N \Leftrightarrow$

(contrapositive) $P(\bar{B}|\bar{A}) > P(\bar{B}) \Leftrightarrow N_{\bar{A}\bar{B}}/N_{\bar{A}} > N_{\bar{B}}/N$

To continue the example on Colorectal Cancer and the Fecal Occult Blood Test, we put the appropriate frequencies from the earlier 2×2 table into the expressions just given to show numerically that they hold:

(original statement) $P(A|B) > P(A) \Leftrightarrow .048 = 15/315 >$
30/10, 000 = .003 $\Leftrightarrow$

(denying the antecedent) $P(\bar{A}|\bar{B}) > P(\bar{A}) \Leftrightarrow .998 = 9671/9686 >$
9970/10, 000 = .997 $\Leftrightarrow$

(affirming the consequent) $P(B|A) > P(B) \Leftrightarrow .500 = 15/30 >$
314/10, 000 = .031 $\Leftrightarrow$

(contrapositive) $P(\bar{B}|\bar{A}) > P(\bar{B}) \Leftrightarrow .970 = 9671/9970 >$
9686/10, 000 = .969

Another way of understanding these results is to note that the original statement of $P(A|B) > P(A)$ is equivalent to $N_{AB} > N_A N_B/N$. Or in the usual terminology of a 2×2 contingency table, the frequency in the cell labeled (A, B) is greater than the typical expected value constructed under independence of the attributes based on the row total, N_B, times the column total, N_A, divided by the grand total, N. The other probability results follow from the observation that with fixed marginal frequencies, a 2×2 contingency table has only one degree of freedom. These results derived from the original of B being facilitative for A, $P(A|B) > P(A)$, could have been restated as $\bar{B}$ being inhibitive of A, or as $\bar{A}$ being inhibitive of B.

In reasoning logically about some situation, it would be rare to have a context that would be so cut and dried as to lend itself to the simple logic of "if p, then q," and where we could look for the attendant fallacies to refute some causal claim.[7] More likely, we are given problems characterized by fallible data, and subject to other types of probabilistic processes. For example, even though someone may have some genetic marker that has a greater presence in individuals who have developed some disease (for example, breast cancer and the BRAC1 gene), it is not typically an unadulterated causal necessity. In other words, it is not true that "if you have the marker, then you must get the disease." In fact, many of these situations might be best reasoned through using our simple 2×2 tables; A and $\bar{A}$ denote the presence/absence of the marker; B and $\bar{B}$ denote the presence/absence of the disease. Assuming A is facilitative of B, we could go on to ask about the strength of the facilitation by looking at, say, the difference, $P(B|A) - P(B)$, or possibly, the ratio, $P(B|A)/P(B)$.

The idea of arguing probabilistic causation is, in effect, the notion of one event being facilitative or inhibitive of another. If a collection of "q" conditions is observed that would be the consequence of a single "p," we may be more prone to conjecture the presence of "p." Although this process may seem like merely affirming the consequent, in a probabilistic context this could be referred to as "inference to the best explanation," or as a variant of the Charles Peirce notion of abductive reasoning ("Abductive reasoning," 2010). In any case, with a probabilistic reinterpretation, the assumed fallacies of logic may not be such. Moreover, most uses of information in contexts that are legal (forensic) or medical (through screening), or that might, for example, involve academic or workplace selection, need to be assessed probabilistically.

3.3 Using Bayes' Rule to Assess the Consequences of Screening for Rare Events

> In referring to Bayesians, 'If they would only do as he did and publish posthumously, we should all be saved a lot of trouble.'
> — Maurice Kendall ("On the Future of Statistics—A Second Look," *JRSS(A)*, 1968)

Bayes' theorem or rule was given earlier in a form appropriate for two events, A and B. It allows the computation of one conditional probability, $P(A|B)$, from two other conditional probabilities, $P(B|A)$ and $P(B|\bar{A})$, and the prior probability for the event

A, $P(A)$. A general example might help show the importance of Bayes' rule in assessing the value of screening for the occurrence of rare events:

Suppose we have a test that assesses some relatively rare occurrence (for example, disease, ability, talent, terrorism propensity, drug or steroid usage, antibody presence, being a liar [where the test is a polygraph], or fetal hemoglobin). Let B be the event that the test says the person has "it," whatever that may be; A is the event that the person really does have "it." Two "reliabilities" are needed:

(a) the probability, $P(B|A)$, that the test is positive if the person has "it"; this is referred to as the *sensitivity* of the test;

(b) the probability, $P(\bar{B}|\bar{A})$, that the test is negative if the person doesn't have "it"; this is the *specificity* of the test. The conditional probability used in the denominator of Bayes' rule, $P(B|\bar{A})$, is merely $1 - P(\bar{B}|\bar{A})$, and is the probability of a "false positive."

The quantity of prime interest, the *positive predictive value* (PPV), is the probability that a person has "it" given that the test says so, $P(A|B)$, and is obtainable from Bayes' rule using the specificity, sensitivity, and prior probability, $P(A)$:

$$P(A|B) = \frac{P(B|A)P(A)}{P(B|A)P(A) + (1 - P(\bar{B}|\bar{A}))(1 - P(A))} . \qquad (3.1)$$

To understand how well the test does, the facilitative effect of B on A needs interpretation; that is, a comparison of $P(A|B)$ to $P(A)$, plus an absolute assessment of the size of $P(A|B)$ by itself. Here, the situation is usually dismal whenever $P(A)$ is small (such as when screening for a relatively rare occurrence), and the sensitivity and specificity are not perfect. Although $P(A|B)$ will generally be greater than $P(A)$, and thus B facilitative of A, the absolute size of $P(A|B)$ is commonly so small that the value of the screening may be questionable.

As an example, consider the efficacy of mammograms in detecting breast cancer. In the United States, about 180,000 women are found to have breast cancer each year from among the 33.5 million women who annually have a mammogram. Thus, the probability of a tumor is about $180,000/33,500,000 = .0054$. Mammograms are no more than 90% accurate, implying that

P(positive mammogram | tumor) = .90;

P(negative mammogram | no tumor) = .90.

Because we do not know whether a tumor is present, all we know is whether the test is positive, Bayes' theorem must be used to calculate the probability we really care about, the Positive Predictive Value (PPV). All the pieces are available to use Bayes' theorem to calculate this probability, and we will do so below. But first, as an exercise for the reader, try to estimate the order of magnitude of that probability, keeping in mind that cancer is rare and the test for it is only 90% accurate. Do you guess that if you test positive, you have a 90% chance of cancer? Or perhaps 50%, or 30%? How low must this probability drop before we feel that mammograms may be an unjustifiable drain on resources? Using Bayes' rule, the PPV of the test is .047:

$$P(\text{tumor} \mid \text{positive mammogram}) =$$

$$\frac{.90(.0054)}{.90(.0054) + .10(.9946)} = .047,$$

which is obviously greater than the prior probability of .0054, but still very small in magnitude; that is, more than 95% of the positive tests that arise turn out to be incorrect. Figure 3.1 gives a convenient tree diagram using calculated probabilities to obtain the PPV for mammograms. The values on each branch of the tree should be self-explanatory. The summary figures at the end of each of the four branches are merely the product of the values along its length.

Whether using a test that is wrong 95% of the time is worth doing is, at least partially, an ethical question, because if we decide that it isn't worth doing, what is the fate of the 5% or so of women who are correctly diagnosed? We will not attempt a full analysis, but some factors considered might be economic; 33.5 million mammograms cost about \$3.5 billion, and the 3.5 million women incorrectly diagnosed can be, first, dysfunctionally frightened, and second, they must use up another day for a biopsy, in turn costing at least \$1,000 and adding another \$3.5 billion to the overall diagnostic bill. Is it worth spending \$7 billion to detect

FIGURE 3.1: A tree diagram using calculated probabilities to obtain the positive predictive value for mammograms.

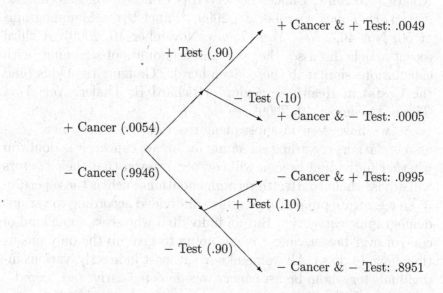

+ Cancer & + Test: .0049

+ Test (.90)

− Test (.10)

+ Cancer & − Test: .0005

+ Cancer (.0054)

− Cancer (.9946)

− Cancer & + Test: .0995

+ Test (.10)

− Test (.90)

− Cancer & − Test: .8951

$$P(\text{Breast Cancer Present} \mid \text{Positive Test}) = \frac{.0049}{.0049 + .0995} = .047$$

180,000 tumors? That is about \$39,000/tumor detected. And, not
to put too fine a point on it, biopsies have their own risks: 1%
yield staph infections, and they too have false positives, implying
that some women end up being treated for nonexistent cancers.
Also, the majority of the cancers detected in the 5% alluded to
above are generally not life-threatening and just lead to the ills
caused by overdiagnosis and invasive overtreatment. The statistics just calculated do not make the decision about whether it is
ethical or not to do mammograms, but such a decision to be ethical should be based on accurate information. Two recent articles
discuss how the Cancer Society may itself be shifting its stance
on screening; the "page one, above the fold" articles are by Gina
Kolata ("In Shift, Cancer Society Has Concerns on Screenings,"
New York Times, October 21, 2009; "Panel Urges Mammograms
at 50, Not 40," *New York Times*, November 16, 2009). A third
recent article discusses the odds and economics of screening (with
calculations similar to those given here): "Gauging the Odds (and
the Costs) in Health Screening" (Richard H. Thaler, *New York
Times*, December 19, 2009).[8]

As we have seen in subsequent reactions to these "new" recommendations regarding screening for breast cancer, it is doubtful
whether individual women will comply, or even that their doctors
will advise them to. Health recommendations such as these pertain
to an aggregate populace, possibly subdivided according to various
demographic categories. But an individual who seeks some kind of
control over breast cancer is not going to give up the only means
they have to do so. All women know, at least indirectly, various individuals for whom breast cancer was detected early, and "cured,"
even though the given cancer may not have been actually harmful. Similarly, all women know about individuals who died after
a cancer had metastasized and before screening located it. A reluctance to give up one possible mechanism for control persists
even in the face of evidence that early detection may not improve
survival chances when modern, focused treatments are used. Several of the concluding paragraphs are given below from an article
("How Should We Screen for Breast Cancer?") that HW wrote for
Significance (2011, *8*(1), 28–30):

More light was shed on this very difficult triage decision in a September 2010 report on the results of a Norwegian study of the efficacy of mammograms. This is the first study that examined the value of the early detection possible through mammograms when coupled with modern treatments such as hormonal therapy and other targeted drugs. In it the researchers compared the breast cancer death rates for women who had early detection with mammograms and those whose cancer was detected later after the tumour had grown enough to be noticed manually. They found that the difference in survival rates were small enough to be chalked-up to chance.

The Norwegian result lends support to the growing concern over the widespread use of mammograms without regard to the often profoundly negative consequences for the many women on the receiving end of a false positive diagnosis. The price of false positives was worthwhile when successful treatment was dependent on early detection. But now modern treatments have seemingly thrown the earlier practice into a cocked hat.

But then, scarcely a week after the Norwegian study was published, the journal *Cancer* published another study, by Swedish researchers, that purported to show that women in their forties whose cancer was detected early by mammogram had a 26% lower death rate than women whose cancers were not detected early. This confusing result just added inertia to the status quo. This is unfortunate, because the Swedish study was marred by a methodological error that invalidates its results.

The Swedish study looked at the probability of death conditioned on the fact that the woman in question had a tumour. For the women who did not have a mammogram the numerator of this probability was the number of women who died of breast cancer and the denominator was the number of women who had a tumour that had grown large enough to have been discovered through self-examination. The women from the mammogram group had a different fraction. Its numerator was the same, the number of women who died from breast cancer. But the denominator was very different, and much more inclusive. It contained all women whose mammograms showed some sort of tumour. Most of those tumours were not destined to grow into anything, and surely some were false positives. Thus the value of the mammograms was much overstated. It is not clear how such a study could be corrected post hoc, but its flaws should have precluded its inclusion in the serious medical literature, where it could mislead those who are unfamiliar with the subtleties of self-selected samples. (p. 30)

The issue that trumps all in the mammogram discussion is what women want, or think they want, which amounts to the same thing. It is doubtful whether a reasoned argument for diminished screening could ever be made politically palatable. To many, a statistical argument for a decrease in screening practice would merely be another mechanism by which insurance companies can deny coverage and make yet more money. To paraphrase a quotation about General Motors, it is not true that "what is good for the Insurance Industry is good for the country," or for that matter, for any single individual living in it. Two very cogent articles on these issues of screening both for individuals and the aggregate are in the *New York Times*: "Screening Debate Reveals Culture Clash in Medicine" by Kevin Sack (November 20, 2009), and "Addicted to Mammograms" by Robert Aronowitz (November 19, 2009).[9]

Gigerenzer and colleagues (see Gigerenzer et al., 2007) have argued for the importance of understanding the PPV of a test, but suggest the use of "natural frequencies" and a simple 2×2 table of the type presented earlier, rather than actual probabilities substituted into Bayes' rule. Based on an assumed population of 10,000, the prior probability of A, plus the sensitivity and specificity values, we have the following 2×2 table:

	tumor	no tumor	Row Sums
+ mammogram	49	995	1044
− mammogram	5	8951	8956
Column Sums	54	9946	10,000

The PPV is then simply $49/1044 = .047$, using the frequency value of 49 for the cell (+ mammogram, tumor) and the + mammogram row sum of 1044.[10]

It might be an obvious statement to make, but in our individual dealings with doctors and the medical establishment generally, it is important for all to understand the PPVs for whatever screening tests we now seem to be constantly subjected to, and thus, the number, $(1 - \text{PPV})$, referring to the false positives; that is, if a patient tests positive, what is the probability that "it" is not actually present. It is a simple task to plot PPV against $P(A)$ from 0 to 1 for any given pair of sensitivity and specificity values. Such a plot can show dramatically the need for highly reliable tests in the presence of low values for $P(A)$ to attain even mediocre PPV values.

Besides a better understanding of how PPVs are determined,

there is a need to recognize that even when a true positive exists, not every disease needs to be treated. In the case of another personal favorite of ours, prostate cancer screening, its low accuracy makes mammograms look good, where the worst danger is one of overdiagnosis and overtreatment, leading to more harm than good (see, for example, Gina Kolata, "Studies Show Prostate Test Saves Few Lives," *New York Times*, March 19, 2009). Armed with this information, we no longer give blood for a PSA screening test. When we so informed our doctors as to our wishes, they agreed completely. The only reason such tests were done routinely was to practice "defensive medicine" on behalf of their clinics, and to prevent possible lawsuits arising from such screening tests not being administered routinely. In other words, clinics get sued for underdiagnosis but not for overdiagnosis and overtreatment.[11]

A good way to conclude this section on issues involving (cancer) screening is to refer the reader to an OpEd article from the *New York Times* ("The Great Prostate Mistake," March 9, 2010) by Richard J. Ablin. Dr. Ablin is a research professor of immunobiology and pathology at the University of Arizona College of Medicine, and President of the Robert Benjamin Ablin Foundation for Cancer Research. Most important for our purposes, he is the individual who in 1970 discovered the PSA test for detecting prostate cancer; his perspective on the issues is therefore unique:[12]

> I never dreamed that my discovery four decades ago would lead to such a profit-driven public health disaster. The medical community must confront reality and stop the inappropriate use of P.S.A. screening. Doing so would save billions of dollars and rescue millions of men from unnecessary, debilitating treatments.

3.3.1 Ethical Issues in Medical Screening

This section has thus far emphasized the screening for breast and prostate cancer as informed by the use of Bayes' rule. Before we leave the topic of medical screening completely, there are several additional issues having possible ethical implications that should at least be raised, if only briefly.

Premarital screening: From the early part of the 20th century, it has been standard practice for states to require a test for syphilis before a marriage license was issued. The rationale for this requirement was so the disease was not passed on to a newborn in the birth canal, with the typical result of blindness, or to an unaffected partner. Besides requiring a test for syphilis, many states in the late 1980s considered mandatory HIV evaluations before marriage licenses were issued. Illinois passed such a law in 1987 that took effect on January 1, 1988, and continued through August of 1989. It was a public health disaster. In the first six months after enactment, the number of marriage licenses issued in Illinois dropped by 22.5%; and of the 70,846 licenses issued during this period, only eight applicants tested positive with a cost of $312,000 per seropositive identified individual. Even for the eight identified as positive, the number of false positives was unknown; the more definitive follow-up Western blot test was not available at that time. This particular episode was the most expensive public health initiative ever for Illinois; the understated conclusion from this experience is that mandatory premarital testing is not a cost-effective method for the control of human immunodeficiency virus infection. For a further discussion of the Illinois experience in mandatory HIV premarital testing, see Turnock and Kelly (1989).

Fagan's nomogram: Several different ways can be used to represent Bayes' rule, in addition to the expressions given in Equations (3.1) and (3.2) (the latter is given in Section 3.4). Based on the more complete characterization of "odds" in Section 3.4, Bayes' rule can be recast as post-test odds,

$$\frac{P(A|B)}{(1 - P(A|B))} \, ,$$

equaling pre-test odds,

$$\frac{P(A)}{(1 - P(A))} \, ,$$

times the likelihood ratio,

$$\frac{P(B|A)}{P(B|\bar{A})} \, .$$

In a letter to the editor of the *New England Journal of Medicine* (1975, *293*, p. 257), T. J. Fagan suggested an easy-to-use nomogram based on this form of Bayes' rule. A simple Google search using "Fagan nomogram" will generate a variety of images for the nomogram, all consisting of three vertical numerical scales (note: the reader may wish to carry out the search at this time as a help in understanding the verbal explanation that follows). From right to left, there are scales for the pre-test probability, the likelihood ratio, and finally, the post-test probability. To use the nomogram, the user needs an estimate of the prior probability of, say, the disease. This is usually related to the prevalence of the disease as modified by the risk factors present for a particular patient. Also, the likelihood ratio must be known for the diagnostic test being used. With these two numerical bits of information, a line is drawn connecting the pre-test probability on the left-most scale through the likelihood value in the center—the extended line intersects the post-test probability at the right-most scale. An interactive version of Fagan's nomogram is available from the Oxford Centre for Evidence-Based Medicine (www.cebm.net). Using the Adobe Shockwave plug-in (loaded automatically when first used), blue arrows are dragged to correspond to a patient's pretest probability and the test's likelihood ratio; the post-test probability is directly read from a red arrow on the far right (this is so convenient, it has been said, that it may be appropriate for "bedside use").

Prenatal screening: The area of prenatal screening inevitably raises ethical issues. Some screening could be labeled quickly as unethical, for example, when selective abortions occur as the result of an ultrasound to determine the sex of a fetus. In other cases, the issues are murkier.[13] For instance, in screening for Down's syndrome because of a mother's age, acting solely on the use of noninvasive biomedical markers with poor selectivity and sensitivity values is questionable; the further screening with more invasive methods, such as amniocentesis, may be justifiable even when considering an accompanying one to two percent chance of the invasive test inducing a miscarriage. At least in the case of screening for Down's syndrome, these trade-offs between invasive screening and the risk of spontaneous miscarriage may no longer exist given a new noninvasive DNA blood test announced in the *British Medical Journal* in January 2011, "Non-invasive Prenatal Assessment of Trisomy 21 by Multiplexed Maternal Plasma DNA Sequencing: Large Scale Validity Study." The article abstract follows:

Objectives: To validate the clinical efficacy and practical feasibility of massively parallel maternal plasma DNA sequencing to screen for fetal trisomy 21 among high risk pregnancies clinically indicated for amniocentesis or chorionic villus sampling.

Design: Diagnostic accuracy validated against full karyotyping, using prospectively collected or archived maternal plasma samples.

Setting: Prenatal diagnostic units in Hong Kong, United Kingdom, and the Netherlands.

Participants: 753 pregnant women at high risk for fetal trisomy 21 who underwent definitive diagnosis by full karyotyping, of whom 86 had a fetus with trisomy 21.

Intervention: Multiplexed massively parallel sequencing of DNA molecules in maternal plasma according to two protocols with different levels of sample throughput: 2-plex and 8-plex sequencing.

Main outcome measures: Proportion of DNA molecules that originated from chromosome 21. A trisomy 21 fetus was diagnosed when the z-score for the proportion of chromosome 21 DNA molecules was greater than 3. Diagnostic sensitivity, specificity, positive predictive value, and negative predictive value were calculated for trisomy 21 detection.

Results: Results were available from 753 pregnancies with the 8-plex sequencing protocol and from 314 pregnancies with the 2-plex protocol. The performance of the 2-plex protocol was superior to that of the 8-plex protocol. With the 2-plex protocol, trisomy 21 fetuses were detected at 100% sensitivity and 97.9% specificity, which resulted in a positive predictive value of 96.6% and negative predictive value of 100%. The 8-plex protocol detected 79.1% of the trisomy 21 fetuses and 98.9% specificity, giving a positive predictive value of 91.9% and negative predictive value of 96.9%.

Conclusion: Multiplexed maternal plasma DNA sequencing analysis could be used to rule out fetal trisomy 21 among high risk pregnancies. If referrals for amniocentesis or chorionic villus sampling were based on the sequencing test results, about 98% of the invasive diagnostic procedures could be avoided.

Costs of screening: All screening procedures have costs attached, if only for the laboratory fees associated with carrying out the diagnostic test. When implemented on a more widespread public health basis, however, screenings may soon become cost-prohibitive for the results obtained. The short-lived premarital HIV screening in Illinois is one example, but new diagnostic screening methods seem to be reported routinely in the medical literature. These then get picked up in the more popular media, possibly with some recommendation for further broad implementation. A societal reluctance to engage in such a process may soon elicit a label of "medical rationing" (possibly, with some further allusion to socialized medicine, or what one can expect under "Obama-care").[14]

Besides initial screening costs and those involved in dealing with follow-up procedures for all the false positives identified, there may also be costs involved in the particular choice among alternatives for a diagnostic procedure. If one strategy has demonstrable advantages but increased costs over another, based on an evidence-based assessment it still may be cost-effective to choose the higher-priced alternative. But if the evidence does not document such an advantage, it would seem fiscally prudent in controlling the increasing societal health-care costs to not choose the more expensive option as the default, irrespective of what professional pressure groups may want and who would profit the most from the specific choices made. A case in point is the use of colonoscopy in preference to sigmoidoscopy. We quote from a short letter to the editor of the *New York Times* by John Abramson (February 22, 2011) entitled "The Price of Colonoscopy":

> Colon cancer screening with colonoscopy—viewing the entire colon—has almost completely replaced more limited sigmoidoscopy, which costs as little as one-tenth as much. Yet studies have repeatedly failed to show that colonoscopy reduces the risk of death from colon cancer more effectively than sigmoidoscopy.

A recent example of a breakthrough in medical screening for lung cancer that may end up being very cost-ineffective was reported in a *News of the Week* article by Eliot Marshall, appearing in *Science* (2010), entitled "The Promise and Pitfalls of a Cancer Breakthrough." It reviews the results of a $250 million study sponsored by the National Cancer Institute (NCI) named the National Lung Screening Trial (NLST). The diagnostic test evaluated was a three-dimensional low-dose helical computed tomography (CT) scan of an individual's lung. Although Harold Varmus commented that he saw "a potential for saving many lives," others saw some of the possible downsides of widespread CT screening, including costs. For example, note the comments from the NCI Deputy Director, Douglas Lowy (we quote from the *Science* news item):[15]

> Lowy, also speaking at the teleconference, ticked off some "disadvantages" of CT screening. One is cost. The price of a scan, estimated

at about \$300 to \$500 per screening, is the least of it. Big expenses ensue, Lowy said, from the high ratio of people who get positive test results but do not have lung cancer. Even if you focus strictly on those with the highest risk—this trial screened smokers and ex-smokers who had used a pack of cigarettes a day for 30 years—"20% to 50%" of the CT scans "will show abnormalities" according to recent studies, said Lowy. According to NCI, about 96% to 98% are false positives. (p. 900)

Besides controlling health-case expenditures by considering the cost-effectiveness of tests, there are other choices involved in who should get screened and at what age. In an article by Gina Kolata in the *New York Times* (April 11, 2011), "Screening Prostates at Any Age," a study is discussed that found men 80 to 85 years old are being screened (using the PSA test) as often as men 30 years younger. Both the American Cancer Society and the American Urological Society discourage screenings for men whose life expectancy is ten years or less; prostate cancer is typically so slow-growing that it would take that long for any benefits of screening to appear. In addition, the United States Preventative Services Task Force recommends that screening should stop at 75. Given the observations we made about prostate screening in the previous section and the OpEd article by Richard Ablin, it appears we have an instance, not of practicing "evidence-based medicine," but a more likely one of "(Medicare) greed-induced medicine."

Informed consent and screening: Before participation in a screening program, patients must give informed consent, with an emphasize on the word "informed." Thus, the various diagnostic properties of the test should be clearly communicated, possibly with the use of Gigerenzer's "natural frequencies"; the risk of "false positives" must be clearly understood, as well as the risks associated with any follow-up invasive procedures. All efforts must be made to avoid the type of cautionary tale reported in Gigerenzer et al. (2007): at a conference on AIDS held in 1987, the former senator from Florida, Lawton Childs, reported that of twenty-two (obviously misinformed about false positives) blood donors in Florida who had been notified they had tested HIV-positive, seven committed suicide.

To inform patients properly about screening risks and benefits, the medical professionals doing the informing must be knowledgeable themselves. Unfortunately, as pointed out in detail by Gigerenzer et al. (2007), there is now ample evidence that many in the medical sciences are profoundly confused. An excellent model for the type of informed dialog that should be possible is given by John Lee in a short "sounding board" article in the *New England Journal of Medicine* (1993, *328*, 438–440), "Screening and Informed Consent." This particular article is concerned with mammograms for detecting breast cancer but the model can be easily extended to other diagnostic situations where informed consent is required. Finally, to show that the type of exemplar dialog that Lee models is not now widespread, we refer the reader to an editorial by Gerd Gigerenzer in *Maturitas* (2010, *67*, 5–6) entitled "Women's Perception of the Benefit of Breast Cancer Screening." The gist of the evidence given in the editorial should be clear from its concluding two sentences: "Misleading women, whether intentionally or unintentionally, about the benefit of mammography screening is a serious issue. All of those in the business of informing women about screening should recall that medical systems are for patients, not the other way around" (p. 6).

The (social) pressure to screen: Irrespective of the evidence for the value for a diagnostic screen, there are usually strong social pressures for us to engage in this behavior. These urgings may comes from medical associations devoted to lobbying some topic, from private groups formed to advocate for some position, or from our own doctors and clinics not wishing to be sued for underdiagnosis. The decision to partake or not in some screening process, should depend on the data-driven evidence of its value, or on the other side, of the potential for harm. For this reason, neither of the authors do any prostate screening. On the other hand, there are many instances where the evidence is present for the value of some ongoing screening procedure. We are both currently engaged in several medications, all to control surrogate endpoints (or test levels), with the promise of keeping one in a reasonable healthy state. Eye drops are used to control eye pressure (and to forestall glaucoma); lisinopril and amlodipine to keep blood pressure under control (and prevent heart attacks); and a statin to keep cholesterol levels down (and again, to avoid heart problems).

The decision to institute or encourage widespread diagnostic screening should be based on evidence that shows effectiveness in relation to all the costs incurred. Part of the national discussion in the United States of evidence-based medical decision making is now taking place for the common screening targets of cervical, prostate, and breast cancer. Until recently it was considered an inappropriate question to ask whether it might be best if we didn't screen and identify a nonlethal cancer, and thus avoid debilitating and unnecessary treatment. A recent survey article by Gina Kolata makes these points well: "Considering When it Might Be Best Not to Know About Cancer" (*New York Times*, October 29, 2011). The United Kingdom is somewhat more advanced than the United States with respect to guidelines when screening programs should be implemented. The British National Health Service has issued useful "appraisal criteria" to guide the adoption of a screening program. The appendix to follow reproduces these criteria.

Appendix: U.K. National Screening Committee Programme Appraisal Criteria

Criteria for appraising the viability, effectiveness and appropriateness of a screening programme —

Ideally all the following criteria should be met before screening for a condition is initiated:

The Condition:

1. The condition should be an important health problem.

2. The epidemiology and natural history of the condition, including development from latent to declared disease, should be adequately understood and there should be a detectable risk factor, disease marker, latent period or early symptomatic stage.

3. All the cost-effective primary prevention interventions should have been implemented as far as practicable.

4. If the carriers of a mutation are identified as a result of screening, the natural history of people with this status should be understood, including the psychological implications.

The Test:

5. There should be a simple, safe, precise and validated screening test.

6. The distribution of test values in the target population should be known and a suitable cut-off level defined and agreed.

7. The test should be acceptable to the population.

8. There should be an agreed policy on the further diagnostic investigation of individuals with a positive test result and on the choices available to those individuals.

9. If the test is for mutations, the criteria used to select the subset of

mutations to be covered by screening, if all possible mutations are not being tested, should be clearly set out.

The Treatment:

10. There should be an effective treatment or intervention for patients identified through early detection, with evidence of early treatment leading to better outcomes than late treatment.

11. There should be agreed evidence-based policies covering which individuals should be offered treatment and the appropriate treatment to be offered.

12. Clinical management of the condition and patient outcomes should be optimised in all health care providers prior to participation in a screening programme.

The Screening Programme:

13. There should be evidence from high quality Randomised Controlled Trials that the screening programme is effective in reducing mortality or morbidity. Where screening is aimed solely at providing information to allow the person being screened to make an "informed choice" (e.g., Down's syndrome, cystic fibrosis carrier screening), there must be evidence from high quality trials that the test accurately measures risk. The information that is provided about the test and its outcome must be of value and readily understood by the individual being screened.

14. There should be evidence that the complete screening programme (test, diagnostic procedures, treatment/intervention) is clinically, socially and ethically acceptable to health professionals and the public.

15. The benefit from the screening programme should outweigh the physical and psychological harm (caused by the test, diagnostic procedures and treatment).

16. The opportunity cost of the screening programme (including testing, diagnosis and treatment, administration, training and quality assurance) should be economically balanced in relation to expenditure on medical care as a whole (i.e., value for money). Assessment against this criteria should have regard to evidence from cost benefit and/or cost effectiveness analyses and have regard to the effective use of available resources.

17. All other options for managing the condition should have been considered (e.g., improving treatment, providing other services), to ensure that no more cost effective intervention could be introduced or current interventions increased within the resources available.

18. There should be a plan for managing and monitoring the screening programme and an agreed set of quality assurance standards.

19. Adequate staffing and facilities for testing, diagnosis, treatment and programme management should be available prior to the commencement of the screening programme.

20. Evidence-based information, explaining the consequences of testing, investigation and treatment, should be made available to potential participants to assist them in making an informed choice.

21. Public pressure for widening the eligibility criteria for reducing the

screening interval, and for increasing the sensitivity of the testing process, should be anticipated. Decisions about these parameters should be scientifically justifiable to the public.

22. If screening is for a mutation, the programme should be acceptable to people identified as carriers and to other family members.

3.4 Bayes' Rule and the Confusion of Conditional Probabilities

> I came to you, Mr. Holmes, because I recognized that I am myself an unpractical man and because I am suddenly confronted with a most serious and extraordinary problem. Recognizing, as I do, that you are the second highest expert in Europe —
>
> Indeed, sir! May I inquire who has the honour to be the first? asked Holmes with some asperity.
>
> To the man of precisely scientific mind the work of Monsieur Bertillon must always appeal strongly.
>
> Then had you not better consult him?
>
> I said, sir, to the precisely scientific mind. But as a practical man of affairs it is acknowledged that you stand alone. I trust, sir, that I have not inadvertently —
>
> Just a little, said Holmes. I think, Dr. Mortimer, you would do wisely if without more ado you would kindly tell me plainly what the exact nature of the problem is in which you demand my assistance.
>
> — Sir Arthur Conan Doyle (*The Hound of the Baskervilles*, 1902)

One way of rewriting Bayes' rule is to use a ratio of probabilities, $P(A)/P(B)$, to relate the two conditional probabilities of interest, $P(B|A)$ (test sensitivity) and $P(A|B)$ (positive predictive value):

$$P(A|B) = P(B|A)\,\frac{P(A)}{P(B)} \,. \tag{3.2}$$

With this rewriting, it is obvious that $P(A|B)$ and $P(B|A)$ will be equal only when the prior probabilities, $P(A)$ and $P(B)$, are the same. Yet, this confusion error is so common in the forensic literature that it is given the special name of the "Prosecutor's Fallacy" (Buchanan, May 16, 2007; "Prosecutor's fallacy," 2010). In the behavioral sciences, the "Prosecutor's Fallacy" is sometimes referred to as the "Fallacy of the Transposed Conditional" or the

"Inversion Fallacy" (see Aitken & Taroni, 2004, pp. 80, 96). In the context of statistical inference, it appears when the probability of seeing a particular data result conditional on the null hypothesis being true, $P(\text{data} \mid H_o)$, is confused with $P(H_o \mid \text{data})$; that is, the probability that the null hypothesis is true given that a particular data result has occurred.

As a case in point, we return to the Sally Clark conviction where the invalidly constructed probability of 1 in 73 million was used to argue successfully for Sally Clark's guilt. Let A be the event of innocence and B the event of two "cot deaths" within the same family. The invalid probability of 1 in 73 million was considered to be for $P(B|A)$; a simple equating with $P(A|B)$, the probability of innocence given the two cot deaths, led directly to Sally Clark's conviction.[16]

We continue with the Royal Statistical Society news release:

> Aside from its invalidity, figures such as the 1 in 73 million are very easily misinterpreted. Some press reports at the time stated that this was the chance that the deaths of Sally Clark's two children were accidental. This (mis-)interpretation is a serious error of logic known as the Prosecutor's Fallacy.
>
> The Court of Appeal has recognised these dangers (*R v. Deen* 1993, *R v. Doheny/Adams* 1996) in connection with probabilities used for DNA profile evidence, and has put in place clear guidelines for the presentation of such evidence. The dangers extend more widely, and there is a real possibility that without proper guidance, and well-informed presentation, frequency estimates presented in court could be misinterpreted by the jury in ways that are very prejudicial to defendants.
>
> Society does not tolerate doctors making serious clinical errors because it is widely understood that such errors could mean the difference between life and death. The case of *R v. Sally Clark* is one example of a medical expert witness making a serious statistical error, one which may have had a profound effect on the outcome of the case.
>
> Although many scientists have some familiarity with statistical methods, statistics remains a specialised area. The Society urges the Courts to ensure that statistical evidence is presented only by appropriately qualified statistical experts, as would be the case for any other form of expert evidence.

The situation with Sally Clark and the Collins case in California (where both involved the Prosecutor's Fallacy) is not isolated.

There was the recent miscarriage of justice in the Netherlands involving a nurse, Lucia de Berk, accused of multiple deaths at the hospitals where she worked. This case aroused the international community of statisticians to redress the apparent injustices visited upon Lucia de Berk. One source for background, although now somewhat dated, is Mark Buchanan at the *New York Times* online opinion pages ("The Prosecutor's Fallacy," May 16, 2007).[17] The Wikipedia article on Lucia de Berk provides the details of the case and the attendant probabilistic arguments, up to her complete exoneration in April 2010 ("Lucia de Berk," 2010).

A much earlier and historically important *fin de siecle* case, is that of Alfred Dreyfus, the much maligned French Jew, and captain in the military, who was falsely imprisoned for espionage. In this case, the nefarious statistician was Alphonse Bertillon (see the Sherlock Holmes epigram that opened this section), who through a very convoluted argument reported a small probability that Dreyfus was "innocent." This meretricious probability had no justifiable mathematical basis and was generated from culling coincidences involving a document, the handwritten *bordereau* (without signature) announcing the transmission of French military information. Dreyfus was accused and convicted of penning this document and passing it to the (German) enemy. The "Prosecutor's Fallacy" (2010) was more or less invoked to ensure a conviction based on the fallacious small probability given by Bertillon. In addition to Émile Zola's well-known article, *J'accuse* ... !, in the newspaper *L'Aurore* on January 13, 1898, it is interesting to note that turn-of-the-century well-known statisticians and probabilists from the French Academy of Sciences (among them Henri Poincaré) demolished Bertillon's probabilistic arguments, and insisted that any use of such evidence needs to proceed in a fully Bayesian manner, much like our present understanding of evidence in current forensic science and the proper place of probabilistic argumentation. A detailed presentation of all the probabilistic and statistical issues and misuses present in the Dreyfus case is given by Champod, Taroni, and Margot (1999). (Also, see the comprehensive text by Aitken and Taroni [2004], *Statistics and the Evaluation of Evidence for Forensic Scientists*, pp. 79–81, 122–126, 153–155, 231–233.)[18]

We observe the same general pattern in all of the miscarriages of justice involving the Prosecutor's Fallacy. A very small reported probability of "innocence" is reported, typically obtained incorrectly either by culling, misapplying the notion of statistical independence, or using an inappropriate statistical model. This probability is calculated by a supposed expert with some credibility in court: a community college mathematics instructor for Collins, Roy Meadow for Clark, Henk Elffers for de Berk, Alphonse Bertillon for Dreyfus. The Prosecutor's Fallacy then takes place, leading to a conviction for the crime. Various outrages ensue from the statistically literate community, with the eventual emergence of some "statistical good guys" hoping to redress the wrongs done: an unnamed court-appointed statistician for the California Supreme Court for Collins, Richard Gill for de Berk, Henri Poincaré (among others) for Dreyfus, the Royal Statistical Society for Clark. After long periods of time, convictions are eventually overturned, typically after extensive prison sentences have already been served. We can only hope to avoid similar miscarriages of justice in cases yet to come by recognizing the tell-tale pattern of occurrences for the Prosecutor's Fallacy.

Any number of conditional probability confusions can arise in important contexts and possibly when least expected. A famous instance of such a confusion was in the O.J. Simpson case, where one conditional probability, say, $P(A|B)$, was equated with another, $P(A|B$ and $D)$. We quote the clear explanation of this obfuscation by Krämer and Gigerenzer (2005):

> Here is a more recent example from the U.S., where likewise $P(A|B)$ is confused with $P(A|B$ and $D)$. This time the confusion is spread by Alan Dershowitz, a renowned Harvard Law professor who advised the O.J. Simpson defense team. The prosecution had argued that Simpson's history of spousal abuse reflected a motive to kill, advancing the premise that "a slap is a prelude to homicide." Dershowitz, however, called this argument "a show of weakness" and said: "We knew that we could prove, if we had to, that an infinitesimal percentage—certainly fewer than 1 of 2,500—of men who slap or beat their domestic partners go on to murder them." Thus, he argued that the probability of the event K that a husband killed his wife if he battered her was small, $P(K|\text{battered}) = 1/2{,}500$. The relevant probability, however, is not this one, as Dershowitz would have us believe. Instead, the relevant probability is that of a man murdering his partner given that he battered her and that she was murdered, $P(K|\text{battered and murdered})$. This probability is about

8/9. It must of course not be confused with the probability that O.J. Simpson is guilty; a jury must take into account much more evidence than battering. But it shows that battering is a fairly good predictor of guilt for murder, contrary to Dershowitz's assertions. (p. 228)

Avoiding the Prosecutor's Fallacy is one obvious characteristic of correct probabilistic reasoning in legal proceedings. A related specious argument on the part of the defense is the "Defendant's Fallacy" (Committee on DNA Technology in Forensic Science, 1992, p. 31). Suppose that for an accused individual who is innocent, there is a one-in-a-million chance of a match (such as for DNA, blood, or fiber). In an area of, say, 10 million people, the number of matches expected is 10 even if everyone tested is innocent. The Defendant's Fallacy would be to say that because 10 matches are expected in a city of 10 million, the probability that the accused is innocent is 9/10. Because this latter probability is so high, the evidence of a match for the accused cannot be used to indicate a finding of guilt, and therefore, the evidence of a match should be excluded. A version of this fallacy appeared (yet again) in the O.J. Simpson murder trial; we give a short excerpt about the Defendant's Fallacy that is embedded in the Wikipedia article on the Prosecutor's Fallacy (2010):

A version of this fallacy arose in the context of the O.J. Simpson murder trial where the prosecution gave evidence that blood from the crime scene matched Simpson with characteristics shared by 1 in 400 people. The defense retorted that a football stadium could be filled full of people from Los Angeles who also fit the grouping characteristics of the blood sample, and therefore the evidence was useless. The first part of the defenses' argument that there are several other people that fit the blood grouping's characteristics is true, but what is important is that few of those people were related to the case, and even fewer had any motivation for committing the crime. Therefore, the defenses' claim that the evidence is useless is untrue.

3.5   Bayes' Rule and the Importance of Baserates

Young man, in mathematics you don't understand things. You just get used to them.
— John von Neumann (as quoted by Gary Zukav, *The Dancing Wu Li Masters*, 1979)

In the formulation of Bayes' rule given in Section 3.3, the two prior probabilities, $P(A)$ and $P(B)$, are also known as "baserates"; that is, in the absence of other information, how often do the events A and B occur. Baserates are obviously important in the conversion of $P(B|A)$ into $P(A|B)$, but as shown by Tversky and Kahneman, and others (for example, Tversky and Kahneman, 1974), baserates are routinely ignored when using various reasoning heuristics. An example is given below on the importance of baserates in eyewitness identification. The example will be made-up for clarity, but the principle it illustrates has far-reaching real-world implications. It will be phrased in the language of "odds," so we first digress slightly to introduce that language.

Given an event, A, the odds in favor of the event occurring (in relation to $\bar{A}$, the event not occurring), is the ratio

$$\frac{P(A)}{P(\bar{A})} = \frac{P(A)}{1 - P(A)} \ .$$

Thus, if $P(A) = 6/7$, the odds in favor of A is $(6/7)/(1/7) = (6/1)$, which is read as 6 to 1, meaning that A occurs 6 times for every single time $\bar{A}$ occurs. Bayes' rule can be restated in terms of odds:

$$O_{odds}(A|B) = O_{odds}(A) \times \Lambda(A|B) \ ,$$

where $\Lambda(A|B)$ is the likelihood ratio:

$$\Lambda(A|B) = \frac{P(B|A)}{P(B|\bar{A})} \ ;$$

$O_{odds}(A|B)$ is the posterior odds of A given B:

$$O_{odds}(A|B) = \frac{P(A|B)}{P(\bar{A}|B)} \ ;$$

and $O_{odds}(A)$, the prior odds of A by itself:

$$O_{odds}(A) = \frac{P(A)}{P(\bar{A})} \ .$$

For our example, we paraphrase an illustration given by Devlin and Lorden (2007, pp. 83–85). A certain town has two taxi companies, Blue Cabs and Black Cabs, having, respectively, 15 and 75 taxis. One night when all the town's 90 taxis were on the streets, a hit-and-run accident occurred involving a taxi. A witness sees the accident and claims a blue taxi was responsible. At the request of the police, the witness underwent a vision test with conditions similar to those on the night in question, indicating the witness could successfully identify the taxi color 4 times out of 5. So, the question: which company is the more likely to have been involved in the accident?

If we let B be the event that the witness says the hit-and-run taxi is blue, and A the event that the true culprit taxi is blue, the following probabilities hold: $P(A) = 15/90$; $P(\bar{A}) = 75/90$; $P(B|A) = 4/5$; and $P(B|\bar{A}) = 1/5$. Thus, the posterior odds are 4 to 5 that the true taxi was blue: $O_{odds}(A|B) = [(15/90)/(75/90)][(4/5)/(1/5)] \approx 4$ to 5. In other words, the probability that the culprit taxi is blue is $4/9 \approx 44\%$. We note that this latter value is much different from the probability (of $4/5 = 80\%$) that the eyewitness could correctly identify a blue taxi when presented with one. This effect is due to the prior odds ratio reflecting the prevalence of black rather than blue taxis on the street.

Some interesting commonalities are present across several forensic and medical domains where a knowledge of Bayes' theorem and the use of prior probabilities (or, baserates) may be crucial to the presentation of science-based recommendations, but which are then subsequently ignored (or discounted) by those very groups to which they are addressed. One area causing a great deal of controversy in the latter part of 2009 was the United States Preventive Services Task Force recommendations on cancer screening in women, particularly regarding when mammograms should start and their frequency. It is clear from the reactions in the media and elsewhere (for example, Congress), that irrespective of what may be reasonable science-based guidelines for a populace of women, on an individual level they will probably have no force whatsoever, despite recent reassuring results that targeted therapy is just as effective at saving lives without early detection.

Another arena in which Bayes' theorem has a role is in assessing and quantifying in a realistic way the probative (that is, legal-proof) value of eyewitness testimony. The faith the legal system has historically placed in eyewitnesses has been shaken by the advent of forensic DNA testing. In the majority of the numerous DNA exonerations occurring over the last twenty years, mistaken eyewitness identifications have been involved. A 2006 article by Wells, Memon, and Penrod ("Eyewitness Evidence: Improving Its Probative Value," in the series *Psychological Science in the Public Interest*), highlights the place that psychology and statistical reasoning have in this endeavor. We quote part of the abstract to give the flavor of the review:[19]

> Decades before the advent of forensic DNA testing, psychologists were questioning the validity of eyewitness reports. Hugo Münsterberg's writings in the early part of the 20th century made a strong case for the involvement of psychological science in helping the legal system understand the vagaries of eyewitness testimony. But it was not until the mid-to-late 1970s that psychologists began to conduct programmatic experiments aimed at understanding the extent of error and the variables that govern error when eyewitnesses give accounts of crimes they have witnessed. Many of the experiments conducted in the late 1970s and throughout the 1980s resulted in articles by psychologists that contained strong warnings to the legal system that eyewitness evidence was being overvalued by the justice system in the sense that its impact on triers of fact (e.g., juries) exceeded its probative (legal-proof) value. Another message of the research was that the validity of eyewitness reports depends a great deal on the procedures that are used to obtain those reports and that the legal system was not using the best procedures. (p. 45)

A third area in which Bayesian notions are crucial to an understanding of what is possible, is in polygraph examinations and the quality of information that they can or cannot provide. Again, what appears to happen is that people want desperately to believe in some rational mechanism for detecting liars and cheats, and thereby increase one's sense of security and control. So, irrespective of the statistical evidence marshalled, and probably because nothing else is really offered to provide even an illusion of control in identifying prevarication, lie detector tests still get done, and a lot of them. An illuminating tale is Fienberg and Stern's "In Search of the Magic Lasso: The Truth About the Polygraph," (2005) and the work of the National Research Council Committee to Review the Scientific Evidence on the Polygraph (2003). We give the abstract of the Fienberg and Stern article below, followed by three telling paragraphs from their concluding section:[20]

> In the wake of controversy over allegations of espionage by Wen Ho Lee, a nuclear scientist at the Department of Energy's Los Alamos National Laboratory, the department ordered that polygraph tests be given to scientists working in similar positions. Soon thereafter, at the request of Congress, the department asked the National Research Council (NRC) to conduct a thorough study of polygraph testing's ability to distinguish accurately between lying and truth-telling across a variety of settings and examinees, even in the face of countermeasures that may be employed to defeat the test. This paper tells some of the story of the work of the Committee to Review the Scientific Evidence on the Polygraph, its report and the reception of that report by the U.S. government and Congress. (p. 249)
>
> At the outset, we explained the seemingly compelling desire for a device that can assist law enforcement and intelligence agencies to identify criminals, spies and saboteurs when direct evidence is lacking. The polygraph has long been touted as such a device. In this article and in the NRC report on which it draws, we explain the limited scientific basis for its use, the deep uncertainty about its level of accuracy and the fragility of the evidence supporting claims of accuracy in any realistic application.
>
> How should society, and the courts in particular, react to such a situation? At a minimum they should be wary about the claimed validity of the polygraph and its alternatives for use in the myriad settings in which they are used or proposed for use. This is especially relevant to current forensic uses of the polygraph. We believe that the courts have been justified in casting a skeptical eye on the relevance and suitability of polygraph test results as legal evidence. Generalizing from the available scientific evidence to the circumstances of a particular polygraph examination is fraught with difficulty. Further, the courts should extend their reluctance to rely upon the polygraph to the many quasiforensic uses that are emerging, such as in sex offender management programs. The courts and the legal system should not act as if there is a scientific basis for many, if any, of these uses. They need to hear the truth about lie detection.
>
> As this paper was going to press in January 2005, the Department of Energy finally announced its proposed revised polygraph rules in the Federal Register. They provide a detailed plan for implementing

the plan outlined in Deputy Secretary McSlarrow's September 2003 testimony. [Note: This was to only do 4,500 lie detector tests rather than the usual 20,000.] But no other federal agency has stepped forward with a plan to curb the use of polygraphs. All of them have heard the truth about polygraphs as we know it, but they have failed to acknowledge it by action. (p. 259)

.We mention one last topic where a knowledge of Bayes' rule might help in arguing within another arena of forensic evidence: the assessment of blood alcohol content (BAC). The United States Supreme Court heard arguments in January 2010 (Briscoe v. Virginia, 2010) about crime analysts being required to make court appearances, and to (presumably) testify about the evidence and its reliability that they present now only in written form. The case was spurred in part by a California woman convicted of vehicular manslaughter with a supposed blood alcohol level two hours after the accident above the legal limit of .08. The woman denied being drunk but did admit to taking two shots of tequila (with Sprite chasers).[21]

There are several statistically related questions pertaining to the use of a dichotomous standard for BAC (usually, .08) as a definitive indication of impairment and, presumably, of criminal liability when someone is injured in an accident. Intuitively, it would seem that the same level of BAC might lead to different levels of impairment conditional on individual characteristics. Also, was this value set based on scientifically credible data? A variety of different BAC tests could be used (for example, urine, blood, saliva, breath, hair); thus, there are all the possible interchangeability and differential reliability issues that this multiplicity implies.

The two most common alternatives to the supposedly most accurate blood test are based on urine and breath. Urine tests indicate the presence of alcohol in a person's system, but it takes up to two hours for the alcohol to show up. A positive urine test does not necessarily mean the person was under the influence of alcohol at the time of the test. Rather, it detects and measures usage within the last several days. Breath alcohol does not directly measure BAC but the amount of supposed "alcohol" in one's breath (as well as all chemically similar compounds and extraneous material such as vomit), and can be influenced by many external factors—cell phones, gasoline, blood, exercise, holding one's breath, and so on. We give in an appendix to this subsection an entry, "Blood Alcohol Testing in Drunk Driving Cases" (2000), posted by a lawyer, Aaron Larson, on the "expertlaw.com" website.

A knowledge of Bayes' theorem and the way in which sensitivity, specificity, the positive predictive value, and the prior probability all operate together may at times be helpful to you or to others in mitigating the effects that a single test may have on one's assessment of culpability. There are many instances where the error rates associated with an instrument are discounted, and it is implicitly assumed that an "observed value" is the "true value." The example of blood alcohol level just discussed seems to be, on the face of it, a particularly egregious example. But there are other tests that could be usefully approached with an understanding of Bayes' rule, such as drug/steroid/human growth hormone use in athletes, blood doping in bicycle racers, polygraph tests for spying/white collar crime, fingerprint or eyewitness (mis)identification, or laser gun usage for speeding tickets. We are not saying that a savvy statistician armed with a knowledge of how Bayes' theorem works can "beat the rap," but it couldn't hurt. Anytime a judgment is based on a single fallible instrument, the value of the positive predictive value assumes a great importance in establishing guilt or innocence.[22]

Appendix: *Blood Alcohol Testing in Drunk Driving Cases*, Aaron Larson; March, 2000

What Are My Chemical Test Rights?

If you are arrested and will be subject to blood alcohol testing, the officer must inform you of your "chemical test" rights. This will typically include a description of what happens to you if you refuse to take a chemical test, and of your right to obtain a blood or urine sample so you can conduct your own test. If the police fail to inform you of the consequences of refusing to take a "chemical test," you may be able to challenge the suspension of your license, resulting from your refusal.

How Is My Blood Alcohol Measured By The Police?

In most states, you have some degree of choice as to how your blood alcohol will be measured. You may be allowed to choose between a breath test and a blood test, and in some states you may be allowed to choose a urine test. Some states require the breath test, but allow you to demand a blood or urine test so you can verify the breath test results.

Breath tests are typically performed using a "breathalyzer" or, in a growing number of jurisdictions, a "datamaster." The "datamaster" is a newer machine, which is much less complicated to use than a breathalyzer. The "breathalyzer" typically requires that the officer read a needle, which indicates your blood alcohol content, and it can be a very subjective determination if you blew a "0.09" or a "0.10." The "datamaster" prints out a ticket with your mechanically determined breath test results, and no interpretation is required.

If you wish to be able to verify your sobriety, the best form of verification is a blood test. Urine tests are the least accurate measurement of your blood alcohol, as your urine often does not reflect your blood alcohol content at a particular point in time. However, if you believe that you are, in fact, over the legal limit, you may wish to gamble on urine testing, if your state grants you that option.

Can Blood Tests Be Inaccurate?

Obviously, lab testing errors can occur, which can render any test result inaccurate. However, the most significant concern for blood testing is whether the test is of the "whole blood" or of the "blood serum." Police crime labs will test the "whole blood," and will produce a "blood alcohol content" figure on that basis. However, many hospitals and clinics test only the "blood serum," resulting in a blood alcohol content figure that can be 25% - 33% higher than a "whole blood" test result. A hospital may also use an alcohol swab before drawing the blood sample, which may contaminate the sample.

What Should I Tell The Police Before I Take A Breath Test?

The police will observe you for a period of time, usually about fifteen minutes, before administering the breath test, to make sure that you have not vomited or put anything in your mouth that might affect the test result.

If you are diabetic, asthmatic, on ulcer medications, or on a number of other medications, you need to tell the police what medications you are taking, and make sure that they write down the list. Certain medical conditions and medications can interfere with the accuracy of your breath test, raising the result.

You need to tell the police if you have any sores, wounds, or infections in your mouth. If you are bleeding into your mouth, your breath test result will be artificially inflated.

You need to tell the police if you have vomited, even if you taste just a little bit of vomit in your mouth. If you still have alcohol in your stomach, even a tiny amount of vomit can substantially increase your "breath test" result. (Even burping or hiccupping can bring alcohol up from the stomach.)

You should indicate if you have any dental bridges or caps, which might be trapping alcohol inside your mouth.

What Happens If I Refuse To Be Tested?

You will face mandatory driver's license sanctions, typically involving adding points to your driving record, and the suspension of your license, under your state's "implied consent" law. The police may also seek a search warrant, allowing them to take a blood sample for testing. The blood sample must be drawn by a professional, and blood sampling usually occurs at a clinic or hospital, not at the police station.

In some states, refusing an alcohol test is a criminal offense, separate from the drunk driving offense.

Are There Times When I Should Refuse To Be Tested?

This is a controversial issue, and depends very much upon your situation, your prior record, and the laws of your state. In some states, some attorneys caution drivers with histories of drunk driving offenses that they may be better off refusing to take an alcohol test than by cooperating. The rationale is that the driver will be better able to defend himself against drunk driving charges if the police don't know his blood alcohol, and that the severe consequences of being convicted as a habitual drunk driver outweigh the mandatory license suspension associated with refusing to take an alcohol test. It should be remembered that the police may apply for a search warrant to obtain a blood sample for testing, if a breath test is refused.

This is not a decision to be made lightly. It is so dependent upon state law, and an individual's driving record, that it is best made after consulting with an attorney. You can do a lot of damage to your driving record, and needlessly lose your license, if you make an ill-advised decision to refuse testing.

What Is "Implied Consent"?

Implied consent is the principle that, by driving your car on the public roadways, you agree to take "chemical tests," such as a blood, breath or urine test, if you are arrested as an impaired driver. If you refuse to take a test after being arrested, the state can add points to your driving record and administratively suspend your driver's license for your refusal—even if you are not charged with drunk driving, or are acquitted.

3.5.1   The (Legal) Status of the Use of Baserates

> The Gileadites seized the fords of the Jordan before the Ephraimites
> arrived. And when any Ephraimite who escaped said, "Let me cross
> over," the men of Gilead would say to him, "Are you an Ephraimite?"
> If he said, "No," then they would say to him, "Then say, 'Shibbo-
> leth'!" And he would say, "Sibboleth," for he could not pronounce
> it right. Then they would take him and kill him at the fords of the
> Jordan. There fell at that time forty-two thousand Ephraimites.
> — Judges 12:5-6

The use of baserates in the context of various legal proceedings, criminal matters, and subject identification has been problematic. The quotation that just opened this section shows the historical range for which baserates have come into consideration in a variety of (quasi-)legal settings. This section reviews several of these areas in more detail.

Shibboleth: This word comes directly from the Old Testament Biblical quotation (Judges 12:5-6) regarding the Gileadites and Ephraimites. It refers to any distinguishing practice, usually one of language, associated with social or regional origin that identifies its speaker as being a member of a group. There are a number of famous shibboleths: German spies during World War II mispronounced the initial "sch" in the Dutch port city's name of Scheveningen (and thereby could be "caught"); during the Battle of the Bulge, American soldiers used knowledge of baseball to tell whether there were German infiltrators in American uniforms; United States soldiers in the Pacific used the word "lollapalooza" to identify the Japanese enemy because a repeat of the word would come back with a pronunciation beginning with "rorra."[23] ("Shibboleth," 2010)

Criminal trials: In the *Federal Rules of Evidence*, Rule 403 implicitly excludes the use of baserates that would be more prejudicial than probative (that is, having value as legal proof). Examples of such exclusions abound but generally involve some judgment as to which types of demographic groups commit which crimes and which ones don't. Rule 403 follows:

> Rule 403. Exclusion of Relevant Evidence on Grounds of Prejudice, Confusion, or Waste of Time: Although relevant, evidence may be excluded if its probative value is substantially outweighed by the danger of unfair prejudice, confusion of the issues, or misleading the jury, or by considerations of undue delay, waste of time, or needless presentation of cumulative evidence.

Racial profiling: Although the Arizona governor vehemently denies that the State's Senate Bill 1070 encourages racial profiling, her argument comes down to officers knowing an illegal alien when they see one, and this will never depend on racial profiling because that, she says, "is illegal." How an assessment of "reasonable suspicion" would be made is left to the discretion of the officers—possibly a shibboleth will be used, such as speaking perfect English without an accent (or as the then governor of the state adjoining Arizona (Arnold Schwarzenegger) said: "I was also going to go and give a speech in Arizona but with my accent, I was afraid they were going to deport me back to Austria."). The reader is referred to the *New York Times* article by Randal C. Archibold ("Arizona Enacts Stringent Law on Immigration," April 23, 2010) that states succinctly the issues involved in Arizona's "Papers, Please" law.[24] *Constitutional protections*: Two constitutional amendments protect the rights of individuals residing in the United States. The first amendment discussed is the Fourteenth, with its three operative clauses:

— The Citizenship Clause provides a broad definition of citizenship, overruling the decision in *Scott v. Sandford* (1857), which held that blacks could not be citizens of the United States. Those who follow current politics might note that this clause makes anyone born in the United States a citizen. Calls for its repeal are heard routinely from the political right, with the usual laments about "tourism babies," or those born to illegal immigrants. Irrespective of the citizenship of the parents, a baby born to someone temporarily in the United States is a United States citizen by default, and therefore, under all the protections of its laws.

— The Due Process Clause prohibits state and local governments from depriving persons of life, liberty, or property without steps being taken to insure fairness.

— The Equal Protection Clause requires the States to provide equal protection under the law to all people within its jurisdiction. This was the basis for the unanimous opinion in *Brown v. Board of Education* (1954) discussed further in Section 9.1.1.

These three clauses are part of only one section of the Fourteenth Amendment, which follows:

> Section 1. All persons born or naturalized in the United States, and subject to the jurisdiction thereof, are citizens of the United States and of the State wherein they reside. No State shall make or enforce any law which shall abridge the privileges or immunities of citizens of the United States; nor shall any State deprive any person of life, liberty, or property, without due process of law; nor deny to any person within its jurisdiction the equal protection of the laws.

Although the "due process" and "equal protection" clauses seem rather definitive, the United States judicial system has found ways to circumvent their application when it was viewed necessary. One example discussed fully in Section 14.3 is the Supreme Court decision in *McCleskey v. Kemp* (1987) on racial disparities in the imposition of the death penalty (in Georgia). But probably the most blatant disregard of "equal protection" was the Japanese-American internment and relocation of about 110,000 individuals living along the United States Pacific coast in the 1940s. These "war relocation camps" were authorized by President Roosevelt on February 19, 1942, with the infamous *Executive Order 9066*. The Supreme Court opinion (6 to 3) in *Korematsu v. United States* (1944) upheld the constitutionality of *Executive Order 9066*. The majority opinion written by Hugo Black argued that the need to protect against espionage outweighed Fred Korematsu's individual rights and the rights of Americans of Japanese descent. In dissent, Justices Robert Jackson and Frank Murphy commented about both the bad precedent this opinion set and the racial issues it presented. We quote part of these two dissenting opinions:

> Murphy: I dissent, therefore, from this legalization of racism. Racial discrimination in any form and in any degree has no justifiable part whatever in our democratic way of life. It is unattractive in any setting, but it is utterly revolting among a free people who have embraced the principles set forth in the Constitution of the United States. All residents of this nation are kin in some way by blood or culture to a foreign land. Yet they are primarily and necessarily a part of the new and distinct civilization of the United States. They must, accordingly, be treated at all times as the heirs of the American experiment, and as entitled to all the rights and freedoms guaranteed by the Constitution.
>
> Jackson: A military order, however unconstitutional, is not apt to last longer than the military emergency. Even during that period, a succeeding commander may revoke it all. But once a judicial opinion rationalizes such an order to show that it conforms to the Constitution, or rather rationalizes the Constitution to show that the Constitution sanctions such an order, the Court for all time has validated the principle of racial discrimination in criminal procedure and of transplanting American citizens. The principle then lies about like a loaded weapon, ready for the hand of any authority that can bring forward a plausible claim of an urgent need. Every repetition imbeds that principle more deeply in our law and thinking and expands it to new purposes.
>
> . . .
>
> Korematsu was born on our soil, of parents born in Japan. The Constitution makes him a citizen of the United States by nativity and a citizen of California by residence. No claim is made that he is not loyal to this country. There is no suggestion that apart from the matter involved here he is not law abiding and well disposed.

Korematsu, however, has been convicted of an act not commonly a crime. It consists merely of being present in the state whereof he is a citizen, near the place where he was born, and where all his life he has lived. ... [H]is crime would result, not from anything he did, said, or thought, different than they, but only in that he was born of different racial stock. Now, if any fundamental assumption underlies our system, it is that guilt is personal and not inheritable. Even if all of one's antecedents had been convicted of treason, the Constitution forbids its penalties to be visited upon him. But here is an attempt to make an otherwise innocent act a crime merely because this prisoner is the son of parents as to whom he had no choice, and belongs to a race from which there is no way to resign. If Congress in peace-time legislation should enact such a criminal law, I should suppose this Court would refuse to enforce it.

Congress passed and President Reagan signed legislation in 1988 apologizing for the internment on behalf of the United States government. The legislation noted that the actions were based on "race prejudice, war hysteria, and a failure of political leadership." Over $1.6 billion was eventually dispersed in reparations to the interned Japanese-Americans and their heirs ("Japanese American internment," 2011).

The other main amendment that has an explicit rights protection as its focus is the Fourth (from the Bill of Rights); its purpose is to guard against unreasonable searches and seizures, and to require a warrant to be judicially sanctioned and supported by "probable cause." The text of the amendment follows:

The right of the people to be secure in their persons, houses, papers, and effects, against unreasonable searches and seizures, shall not be violated, and no Warrants shall issue, but upon probable cause, supported by Oath or affirmation, and particularly describing the place to be searched, and the persons or things to be seized.

Various interpretations of the Fourth Amendment have been made through many Supreme Court opinions. We mention two here that are directly relevant to the issue of law-enforcement application of baserates, and for (racial) profiling: *Terry v. Ohio* (1968) and *Whren v. United States* (1996). The Wikipedia summaries are given in both cases:

Terry v. Ohio ... (1968) was a decision by the United States Supreme

Court which held that the Fourth Amendment prohibition on unreasonable searches and seizures is not violated when a police officer stops a suspect on the street and frisks him without probable cause to arrest, if the police officer has a reasonable suspicion that the person has committed, is committing, or is about to commit a crime and has a reasonable belief that the person "may be armed and presently dangerous." ...

For their own protection, police may perform a quick surface search of the person's outer clothing for weapons if they have reasonable suspicion that the person stopped is armed. This reasonable suspicion must be based on "specific and articulable facts" and not merely upon an officer's hunch. This permitted police action has subsequently been referred to in short as a "stop and frisk," or simply a "Terry stop."

Whren v. United States ... (1996) was a United States Supreme Court decision which "declared that any traffic offense committed by a driver was a legitimate legal basis for a stop," [and] ... "the temporary detention of a motorist upon probable cause to believe that he has violated the traffic laws does not violate the Fourth Amendment's prohibition against unreasonable seizures, even if a reasonable officer would not have stopped the motorist absent some additional law enforcement objective."

In a dissenting opinion in *Terry v. Ohio* (1968), Justice William O. Douglas strongly disagreed with permitting a stop and search without probable cause:

I agree that petitioner was "seized" within the meaning of the Fourth Amendment. I also agree that frisking petitioner and his companions for guns was a "search." But it is a mystery how that "search" and that "seizure" can be constitutional by Fourth Amendment standards, unless there was "probable cause" to believe that (1) a crime had been committed or (2) a crime was in the process of being committed or (3) a crime was about to be committed.

The opinion of the Court disclaims the existence of "probable cause." If loitering were in issue and that was the offense charged, there would be "probable cause" shown. But the crime here is carrying concealed weapons; and there is no basis for concluding that the officer had "probable cause" for believing that that crime was being committed. Had a warrant been sought, a magistrate would, therefore, have been unauthorized to issue one, for he can act only if there is a showing of "probable cause." We hold today that the police have greater authority to make a "seizure" and conduct a "search" than a judge has to authorize such action. We have said precisely the opposite over and over again.

...

There have been powerful hydraulic pressures throughout our history that bear heavily on the Court to water down constitutional guarantees and give the police the upper hand. That hydraulic pressure has probably never been greater than it is today.

Yet if the individual is no longer to be sovereign, if the police can pick him up whenever they do not like the cut of his jib, if they can "seize" and "search" him in their discretion, we enter a new regime. The decision to enter it should be made only after a full debate by the people of this country.

Government institution protections: Although government institutions should protect rights guaranteed by the Constitution, there have been many historical failures. Many of these (unethical) intrusions are statistical at their core, where data are collected on individuals who may be under surveillance only for having unpopular views. To give a particularly salient and egregious example involving the FBI, J. Edgar Hoover, Japanese-American internment, and related topics, we redact the Wikipedia entry on the Custodial Detention Index (under the main heading of "FBI Index," 2011) used by the FBI from the 1930s to the 1970s (with various renamed successor indices, such as Rabble-Rouser, Agitator, Security, Communist, Administrative):

The Custodial Detention Index (CDI), or Custodial Detention List was formed in 1939-1941, in the frame of a program called variously the "Custodial Detention Program" or "Alien Enemy Control."

J. Edgar Hoover described it as having come from his resurrected General Intelligence Division in Washington:

"This division has now compiled extensive indices of individuals, groups, and organizations engaged in subversive activities, in espionage activities, or any activities that are possibly detrimental to the internal security of the United States. The Indexes have been arranged not only alphabetically but also geographically, so that at any rate, should we enter into the conflict abroad, we would be able to go into any of these communities and identify individuals or groups who might be a source of grave danger to the security of this country. These indexes will be extremely important and valuable in a grave emergency."

Congressmen Vito Marcantonio called it "terror by index cards."

...

The Custodial Detention Index was a list of suspects and potential subversives, classified as "A," "B," and "C"; the ones classified as "A" were destined to be immediately arrested and interned at the outbreak of war. Category A were leaders of Axis-related organizations, category B were members deemed "less dangerous" and category C were sympathizers. The actual assignment of the categories was, however, based on the perceived individual commitment to the person's native country, rather than the actual potential to cause harm; leaders of cultural organizations could be classified as "A," members of non-Nazi and pro-Fascist organizations.

The program involved creation of individual dossiers from secretly obtained information, including unsubstantiated data and in some cases, even hearsay and unsolicited phone tips, and information acquired without judicial warrants by mail covers and interception of mail, wiretaps and covert searches. While the program targeted primarily Japanese, Italian, and German "enemy aliens," it also included some American citizens. The program was run without Congress-approved legal authority, no judicial oversight and outside of the official legal boundaries of the FBI. A person against which an accusation was made was investigated and eventually placed on the index; it was not removed until the person died. Getting on the list was easy; getting off of it was virtually impossible.

According to the press releases at the beginning of the war, one of the purposes of the program was to demonstrate the diligence and vigilance of the government by following, arresting and isolating a previously identified group of people with allegedly documented sympathies for Axis powers and potential for espionage or fifth column activities. The list was later used for Japanese-American internment.

Attorney General Francis Biddle, when he found out about the Index, labeled it "dangerous, illegal" and ordered its end. However, J. Edgar Hoover simply renamed it the Security Index, and told his people not to mention it.

USA PATRIOT Act: The attitude present during World War II that resident Japanese-Americans had a proclivity for espionage has now changed after September 11, 2001, to that of Middle Eastern men having a proclivity for committing terrorist acts. The acronym of being arrested because of a DWB ("driving while black") has now been altered to FWM ("flying while Muslim") ("Words concerning anti-black discrimination," December, 2005). Section 412 of the *USA PATRIOT Act* allows the United States Attorney General to detain aliens for up to seven days without bringing charges when the detainees are certified as threats to national security. The grounds for detention are the same "reasonable suspicion" standard of *Terry v. Ohio* (1968). The Attorney General certification must state that there are "reasonable grounds to believe" the detainee will commit espionage or sabotage, commit terrorist acts, try to overthrow the government, or otherwise behave in a way that would endanger national security. After seven days, the detention may continue if the alien is charged with a crime or violation of visa conditions. When circumstances prohibit the repatriation of a person for an immigration offense, the detention may continue indefinitely if recertified by the attorney general every six months. Under the *USA PATRIOT Act*, a person confined for a violation of conditions of United States entry but who cannot be deported to the country of origin, may be indefinitely confined without criminal charges ever being filed.

Profiling, ethnic or otherwise, has been an implicit feature of United States society for some time. The particular targets change, but the idea that it is permissible to act against specific individuals because of group membership does not. In the 1950s there were popular radio and television programs, such as *The FBI in Peace and War* or *I Led 3 Lives* about the double agent Herbert Philbrick. These all focused on the Red menace in our midst, bent on overthrowing our form of government. It is instructive to remember our history whenever a new group is targeted for surveillance, and to note that the *Smith Act of 1940* (also known as the *Alien Registration Act*) is still on the books; the enabling "membership clause" and other conditions in the *Smith Act* follow:

> Whoever knowingly or willfully advocates, abets, advises, or teaches the duty, necessity, desirability, or propriety of overthrowing or destroying the government of the United States or the government of any State, Territory, District or Possession thereof, or the government of any political subdivision therein, by force or violence, or by the assassination of any officer of any such government; or
>
> Whoever, with intent to cause the overthrow or destruction of any such government, prints, publishes, edits, issues, circulates, sells, distributes, or publicly displays any written or printed matter advocating, advising, or teaching the duty, necessity, desirability, or propriety of overthrowing or destroying any government in the United States by force or violence, or attempts to do so; or
>
> Whoever organizes or helps or attempts to organize any society, group, or assembly of persons who teach, advocate, or encourage the overthrow or destruction of any such government by force or

violence; or becomes or is a member of, or affiliates with, any such society, group, or assembly of persons, knowing the purposes thereof

Shall be fined under this title or imprisoned not more than twenty years, or both, and shall be ineligible for employment by the United States or any department or agency thereof, for the five years next following his conviction.

If two or more persons conspire to commit any offense named in this section, each shall be fined under this title or imprisoned not more than twenty years, or both, and shall be ineligible for employment by the United States or any department or agency thereof, for the five years next following his conviction.

As used in this section, the terms "organizes" and "organize," with respect to any society, group, or assembly of persons, include the recruiting of new members, the forming of new units, and the regrouping or expansion of existing clubs, classes, and other units of such society, group, or assembly of persons.

Eyewitness reliability and false confessions: Several troublesome forensic areas exist in which baserates can come into nefarious play. One is in eyewitness testimony and how baserates are crucial to assessing the reliability of a witness's identification. The criminal case reported later in Section 4.2 (Probability and Litigation) of "In Re As.H (2004)" illustrates this point well, particularly as it deals with cross-racial identification, memory lapses, how lineups are done, and so forth. Also, we have the earlier taxicab anecdote of this section. One possibly unexpected use that we turn to next involves baserate considerations in "false confessions." False confessions appear more frequently than we might expect and also in some very high profile cases. The most sensationally reported example may be the Central Park jogger incident of 1989, in which five African- and Hispanic-Americans all falsely confessed. To give a better sense of the problem, a short abstract is given below from an informative review article by Saul Kassin in the *American Psychologist* (2005, *60*, 215–228), entitled "On the Psychology of Confessions: Does Innocence Put Innocents at Risk":

The Central Park jogger case and other recent exonerations highlight the problem of wrongful convictions, 15% to 25% of which have contained confessions in evidence. Recent research suggests that actual innocence does not protect people across a sequence of pivotal decisions: (a) In preinterrogation interviews, investigators commit false-positive errors, presuming innocent suspects guilty; (b) naively believing in the transparency of their innocence, innocent suspects waive their rights; (c) despite or because of their denials, innocent suspects elicit highly confrontational interrogations; (d) certain commonly used techniques lead suspects to confess to crimes they did not commit; and (e) police and others cannot distinguish between uncorroborated true and false confessions. It appears that innocence puts innocents at risk, that consideration should be given to reforming current practices, and that a policy of videotaping interrogations is a necessary means of protection. (p. 215)

To put this issue of false confession into a Bayesian framework, our main interest is in the term, $P(\text{guilty} \mid \text{confess})$. Based on Bayes' rule this probability can be written as

$$\frac{P(\text{confess} \mid \text{guilty})P(\text{guilty})}{P(\text{confess} \mid \text{guilty})P(\text{guilty}) + P(\text{confess} \mid \text{not guilty})P(\text{not guilty})}.$$

The most common interrogation strategy taught to police officers is the 9-step Reid Technique.[25] The proponents of the Reid Technique hold two beliefs: that $P(\text{confess} \mid \text{not guilty})$ is zero, and that they never interrogate innocent people, so the prior probability, $P(\text{guilty})$, is 1.0. Given these assumptions, it follows that if a confession is given, the party must be guilty. There is no room for error in the Reid system; also, training in the Reid system does not increase accuracy of an initial prior assessment of guilt but it does greatly increase confidence in that estimate. We thus have a new wording for an old adage: "never in error and never in doubt."

A number of psychological concerns are present with how interrogations are done in the United States. Innocent people are more likely to waive their *Miranda* rights (so unfortunately, they can then be subjected to interrogation); but somehow this does not seem to change an interrogator's prior probability of guilt.[26] People have a naive faith in the power of their own innocence to set them free. They maintain a belief in a just world where people get what they deserve and deserve what they get. People are generally under an illusion of transparency where they overestimate the extent that others can see their true thoughts. When in doubt, just remember the simple words—"I want a lawyer." (Or, in the idiom of the *Law & Order* series on TV, always remember to "lawyerup.") If an interrogation proceeds (against our recommendation), it is a guilt-presumptive process that unfolds (it is assumed from the outset that P(guilty) is 1.0). False incriminating evidence can be presented to you (in contrast to the U.K, which is surprising because the United Kingdom doesn't have a "Bill of Rights"). Some people who are faced with false evidence may even begin to believe they are guilty. The interrogation process is one of social influence, with all the good cards stacked on one side of the table. It does not even have to be videotaped, so any post-confession argument of psychological coercion is hard to make.

As part of our advice to "lawyer up" if you happen to find yourself in a situation where you could be subjected to interrogation (and regardless of whether you believe yourself to be innocent or not), there is now a stronger need to be verbally clear about invoking one's *Miranda* rights—counterintuitively, you have to be clear and audible in your wish not to talk. The Supreme Court issued the relevant ruling in June 2010. An article reviewing the decision from the *Los Angeles Times* (David G. Savage, "Supreme Court Backs Off Strict Enforcement of Miranda Rights," June 2, 2010) provides a cautionary piece of advice for those of us who might someday fall into the clutches of the criminal system through no fault of our own.

3.5.2 Forensic Evidence Generally

Most of us learn about forensic evidence and how it is used in criminal cases through shows such as *Law & Order*. Rarely, if ever, do we learn about evidence fallibility and whether it can be evaluated through the various concepts introduced to this point, such as baserates, sensitivity, specificity, prosecutor or defendant fallacy, or the positive predictive value. Contrary to what we may come to believe, evidence based on things such as bite marks, fibers, and voice prints are very dubious. As one example, we give the conclusion of a conference presentation by Jean-François Bonastre and colleagues (2003), entitled "Person Authentication by Voice: A Need for Caution":

> Currently, it is not possible to completely determine whether the similarity between two recordings is due to the speaker or to other factors, especially when: (a) the speaker does not cooperate, (b) there is no control over recording equipment, (c) recording conditions are not known, (d) one does not know whether the voice was disguised

and, to a lesser extent, (e) the linguistic content of the message is
not controlled. Caution and judgment must be exercised when applying speaker recognition techniques, whether human or automatic,
to account for these uncontrolled factors. Under more constrained or
calibrated situations, or as an aid for investigative purposes, judicious application of these techniques may be suitable, provided they
are not considered as infallible.

   At the present time, there is no scientific process that enables one
to uniquely characterize a person's voice or to identify with absolute
certainty an individual from his or her voice. (p. 35)

Because of the rather dismal state of forensic science in general,
Congress in 2005 authorized "the National Academy of Sciences
to conduct a study on forensic science, as described in the Senate
report" (H. R. Rep. No. 109-272). The Senate Report (No. 109-
88, 2005) states in part: "While a great deal of analysis exists of
the requirements in the discipline of DNA, there exists little to no
analysis of the remaining needs of the community outside of the
area of DNA. Therefore ... the Committee directs the Attorney
General to provide [funds] to the National Academy of Sciences to
create an independent Forensic Science Committee. This Committee shall include members of the forensics community representing
operational crime laboratories, medical examiners, and coroners;
legal experts; and other scientists as determined appropriate."

   The results of this National Research Council (NRC) study appeared in book form in 2009 from the National Academies Press
(the quotations just given are from this source): *Strengthening
Forensic Science in the United States: A Path Forward.* The Summary of this NRC report provides most of what we need to know
about the state of forensic science in the United States, and what
can or should be done. The material that follows (and continuing
into the endnotes) is an excerpt from the NRC Summary chapter:[27]

   Problems Relating to the Interpretation of Forensic Evidence:
   Often in criminal prosecutions and civil litigation, forensic evidence is offered to support conclusions about "individualization"
   (sometimes referred to as "matching" a specimen to a particular
   individual or other source) or about classification of the source of
   the specimen into one of several categories. With the exception of
   nuclear DNA analysis, however, no forensic method has been rigorously shown to have the capacity to consistently, and with a high
   degree of certainty, demonstrate a connection between evidence and

a specific individual or source. *In terms of scientific basis, the analytically based disciplines generally hold a notable edge over disciplines based on expert interpretation.* [italics added for emphasis] But there are important variations among the disciplines relying on expert interpretation. For example, there are more established protocols and available research for fingerprint analysis than for the analysis of bite marks. There also are significant variations within each discipline. For example, not all fingerprint evidence is equally good, because the true value of the evidence is determined by the quality of the latent fingerprint image. These disparities between and within the forensic science disciplines highlight a major problem in the forensic science community: The simple reality is that the interpretation of forensic evidence is not always based on scientific studies to determine its validity. This is a serious problem. Although research has been done in some disciplines, there is a notable dearth of peer-reviewed, published studies establishing the scientific bases and validity of many forensic methods. (pp. 7–8)

A central idea of this book is that "context counts" and it "counts crucially." It is important both for experts and novices in how a question is asked, how a decision task is framed, and how forensic identification is made. People are primed by context whether as a victim making an eyewitness identification of a perpetrator, or as an expert making a fingerprint match. As an example of the latter, we have the 2006 article by Dror, Charlton, and Péron, "Contextual Information Renders Experts Vulnerable to Making Erroneous Identifications" (*Forensic Science International, 156*, 74–78). We give their abstract below:

We investigated whether experts can objectively focus on feature information in fingerprints without being misled by extraneous information, such as context. We took fingerprints that have previously been examined and assessed by latent print experts to make positive identification of suspects. Then we presented these same fingerprints again, to the same experts, but gave a context that suggested that they were a no-match, and hence the suspects could not be identified. Within this new context, most of the fingerprint experts made different judgments, thus contradicting their own previous identification decisions. Cognitive aspects involved in biometric identification can explain why experts are vulnerable to make erroneous identifications. (p. 74)

Notes

[1]In the volume, *The Evolving Role of Statistical Assessments as Evidence in the Courts* (1988, Stephen Fienberg, Editor), the useful phrase "numerical obscurantism" is defined: "the use of mathematical evidence in a deliberate attempt to overawe, confuse, or mislead the trier of fact" (p. 215).

[2]Besides miscarriages of justice that result from confusions involving conditional probabilities, others have suffered because of failures to understand clearly the fallibility of diagnostic testing. Probably the most famous example of this is the disappearance of Azaria Chamberlain, a nine-week-old Australian baby who disappeared on the night of August 17, 1980, while on a camping trip to Ayers Rock. The parents, Lindy and Michael Chamberlain, contended that Azaria had been taken from their tent by a dingo. After several inquests, some broadcast live on Australian television, Lindy Chamberlain was tried and convicted of murder, and sentenced to life imprisonment. A later chance finding of a piece of Azaria's clothing in an area with many dingo lairs, lead to Lindy Chamberlain's release from prison and eventual exoneration of all charges.

The conviction of Lindy Chamberlain for the alleged cutting of Azaria's throat in the front seat of the family car rested on evidence of fetal hemoglobin stains on the seat. Fetal hemoglobin is present in infants who are six months old or younger—Azaria Chamberlain was only nine weeks old when she disappeared. As it happens, the diagnostic test for fetal hemoglobin is very unreliable, and many other organic compounds can produce similar results, such as nose mucus and chocolate milkshakes, both of which were present in the vehicle (in other words, the specificity of the test [to be defined shortly] was terrible). It was also shown that a "sound deadener" sprayed on the car during its production produced almost identical results for the fetal hemoglobin test.

The Chamberlain case was the most publicized in Australian history (and on a par with the O.J. Simpson trial in the United States). Because most of the evidence against Lindy Chamberlain was later rejected, it is a good illustration of how media hype and bias can distort a trial ("Azaria Chamberlain disappearance," 2011).

[3]Three items are listed in the Suggested Reading, Section B.2 (Probability Issues), that show how probability models can inform work in several areas of scientific study:

B.2.3 ("If Smallpox Strikes Portland ... ," Chris L. Barrett, Stephen G. Eubank, and James P. Smith, *Scientific American*, 2005) is concerned with the modeling of epidemics in conjunction with existing (social) networks, and how best to interfere with and defeat such processes in stemming the spread of a disease;

B.2.4 ("The Treatment," Malcolm Gladwell, *New Yorker*, May 17, 2010)

indirectly surveys the combinatorial search processes involved in drug identification, and the long-shot success issues of culling promising interactions;

B.2.5 ("All Present-Day Life Arose From a Single Ancestor," Tina Hes-
man Saey, *ScienceNews*, June 5, 2010) discusses the (probabilistic) inference
obtained from comparing various evolutionary models that all life on Earth
share a common ancestor.

[4] A particularly problematic case of culling or locating "hot spots" is that
of residential cancer-cluster identification. A readable account is listed in the
Suggested Reading (B.2.2): Atul Gawande, "The Cancer-Cluster Myth," *New
Yorker*, February 8, 1999. For the probability issues that arise in searching
the whole human genome for clues to some condition, see B.2.6: "Nabbing
Suspicious SNPS; Scientists Search the Whole Genome for Clues to Common
Diseases" (Regina Nuzzo, *ScienceNews*, June 21, 2008).

[5] Two informative articles are listed in the Suggested Reading on identifica-
tion error using fingerprints (B.3.1, "Do Fingerprints Lie?", Michael Specter,
New Yorker, May 27, 2002), and DNA (B.3.2, "You Think DNA Evidence is
Foolproof? Try Again," Adam Liptak, *New York Times*, March 16, 2003).

[6] As noted repeatedly by Gigerenzer and colleagues (e.g., Gigerenzer, 2002;
Gigerenzer et al., 2007), it also may be best for purposes of clarity and understanding, to report probabilities using "natural frequencies." For example,
instead of saying that a random match probability is .01, this could be restated
alternatively that for this population, 1 out of every 10,000 men would be expected to show a match. The use of natural frequencies supposedly provides a
concrete reference class for a given probability that then helps interpretation.

[7] But sometimes it can indeed happen that an author commits the fallacy of
affirming the consequent and even when the consequent might be a statement
involving probabilities. As an instance of this, we quote most of the following
rejoinder:

Froman, T., & Hubert, L. J. (1981). A reply to Moshman's critique of
prediction analysis and developmental priority. *Psychological Bulletin, 90,* 188.

Abstract: We point out (a) the implicit logical fallacy of affirming
the consequent in Moshman's argument and (b) the counterintuitive
contention that statistical independence does not rule out developmental priority.

Moshman's (1981) argument about what constitutes evidence for
developmental priority can be rephrased as follows. Suppose A and $\bar{A}$
represent the events of having and not having Concept A; similarly,
B and $\bar{B}$ represent the events of having and not having Concept B.
In terms of probabilities, $P(A) + P(\bar{A}) = 1$ and $P(B) + P(\bar{B}) = 1$.
Moshman's contention is that $P(B) > P(A)$ provides evidence that
Concept B is developmentally prior to Concept A.

Although this argument may appear intuitively reasonable, there
are two major problems with the logic. First of all, the sense of the
formal implication should be reversed: If B is developmentally prior

to A, then $P(B)$ should be larger than $P(A)$. Unfortunately, the sense of Moshman's statement is in the opposite direction and, in this context, reduces to the logical fallacy of affirming the consequent of the above implication. The observations that $P(B)$ is greater than $P(A)$ does not tell us about the truth of the statement that B is developmentally prior to A, although the negation, $P(B) < P(A)$, would provide evidence against such a developmental priority. For example, A and B could represent different informational items that do not depend on one another and are taught at different points in a curriculum. If B is usually taught before A, then we would expect $P(B) > P(A)$. However, this says nothing about a developmental as opposed to an instructional priority between A and B.

Second, Moshman's (1981) contention that "statistical independence does not rule out developmental priority" (p. 186) is counterintuitive. What it implies is that B can be developmentally prior to A, but because A and B are statistically independent, $P(B|A) = P(B|\bar{A})$. Thus, even though B is developmentally prior to A, knowing whether a person has or does not have Concept A tells us nothing about the probability that a person has B. There is no differential prediction possible in assessing performance on A and conversely. We seriously doubt that many developmental psychologists would accept the consequences of this argument.

Moshman, D. (1981). Prediction analysis and developmental priority: A comment on Froman and Hubert. *Psychological Bulletin, 90*, 185 – 187.

[8]The first article is listed in the Suggested Reading (B.4.1); the third is referenced in B.4.7.

[9]These two items are listed in the Suggested Reading, B.4.2 and B.4.3. Also, we reference the more personal OpEd article by Gail Collins, "The Breast Brouhaha," November 18, 2009, in B.4.4; in B.4.6, we list "Mammogram Math" by John Allen Paulos (*New York Times Magazine*, December 10, 2009).

[10]A more popularized discussion of using Bayes' theorem and natural frequencies in medical settings and elsewhere, is the Steven Strogatz column from the *New York Times* ("Chances Are," April 25, 2010); this is listed in the Suggested Reading, B.4.9.

[11]Several additional items in the Suggested Reading, Section B.4, are relevant to screening: B.4.8 references an article by Sandra G. Boodman for the *AARP Bulletin* (Januaary 1, 2010) that summarizes well what its title offers: "Experts Debate the Risks and Benefits of Cancer Screening." A cautionary example of breast cancer screening that tries to use dismal specificity and sensitivity values for detecting the HER2 protein, is listed in B.4.10 (Gina Kolata, "Cancer Fight: Unclear Tests for New Drug," *New York Times*, April 19, 2010). The reasons behind proposing cancer screening guidelines and the con-

temporary emphasis on evidence-based medicine is discussed by Gina Kolata in "Behind Cancer Guidelines, Quest for Data" (*New York Times*, November 22, 2009), B.4.5. Of the five last items in Section B.4 that involve screening, one (B.4.11) discusses how a fallible test for ovarian cancer (based on the CA-125 protein) might be improved using a particular algorithm to monitor CA-125 fluctuations more precisely (Tom Randall, *Bloomberg Businessweek*, May 21, 2010, "Blood Test for Early Ovarian Cancer May Be Recommended for All"). The next three items (B.4.12; B.4.13; B.4.14) are all by Gina Kolata and concern food allergies (or nonallergies, as the case may be) and a promising screening test for Alzheimer's: "Doubt Is Cast on Many Reports of Food Allergies" (*New York Times*, May 11, 2010); and "I Can't Eat That. I'm Allergic" (*New York Times*, May 15, 2010); "Promise Seen for Detection of Alzheimer's" (*New York Times*, June 23, 2010). The final item in B.4.15 discusses a promising alternative to mammogram screening: "Breast Screening Tool Finds Many Missed Cancers" (Janet Raloff, *ScienceNews*, July 1, 2010).

[12]To show the ubiquity of screening appeals, we reproduce a solicitation letter to LH from Life Line Screening suggesting that for only $139, he could get four unnecessary screenings right in Champaign, Illinois, at the Temple Baptist Church:

> Dear Lawrence,
> Temple Baptist Church in Champaign may not be the location that you typically think of for administering lifesaving screenings. However, on Tuesday, September 22, 2009, the nation's leader in community-based preventive health screenings will be coming to your neighborhood.
> Over 5 million people have participated in Life Line Screening's ultrasound screenings that can determine your risk for stroke caused by carotid artery diseases, abdominal aortic aneurysms and other vascular diseases. Cardiovascular disease is the #1 killer in the United States of both men and women—and a leading cause of permanent disability.
> Please read the enclosed information about these painless lifesaving screenings. A package of four painless Stroke, Vascular Disease & Heart Rhythm screenings costs only $139. Socks and shoes are the only clothes that will be removed and your screenings will be completed in a little more than an hour.
> You may think that your physician would order these screenings if they were necessary. However, insurance companies typically will not pay for screenings unless there are symptoms. Unfortunately, 4 out of 5 people that suffer a stroke have no apparent symptoms or warning signs. That is why having a Life Line Screening is so important to keep you and your loved ones healthy and independent.
> "These screenings can help you avoid the terrible consequences of stroke and other vascular diseases. I've seen firsthand what the

devastating effects of stroke, abdominal aortic aneurysms and other vascular diseases can have on people and I feel it is important that everyone be made aware of how easily they can be avoided through simple, painless screenings." — Andrew Monganaro, MD, FACS, FACC (Board Certified Cardiothoracic and Vascular Surgeon)

I encourage you to talk to your physician about Life Line Screening. I am confident that he or she will agree with the hundreds of hospitals that have partnered with us and suggest that you participate in this health event.

We are coming to Champaign for one day only and appointments are limited, so call 1-800-395-1801 now.

Wishing you the best of health,

Karen R. Law, RDMS, RDCS, RVT

Director of Clinical Operations

[13]There is also the fear that increasingly sophisticated prenatal genetic testing will enable people to engineer "designer babies," where parents screen for specific traits and not for birth defects per se. The question about perfection in babies being an entitlement is basically an ethical one; should otherwise healthy fetuses be aborted if they do not conform to parental wishes? To an extent, some of this selection is done indirectly and crudely already when choices are made from a sperm bank according to desired donor characteristics.

[14]One possible mechanism that may be a viable strategy for keeping the cost of screenings under some control is through a clever use of statistics. Depending on what is being assessed (for example, in blood, soil, air), it may be possible to test a "pooled" sample; only when that sample turns out to be "positive" would the individual tests on each of the constituents need to be carried out.

[15]Continued from the main text:

In NLST (National Lung Screening Trial), about 25% of those screened with CT got a positive result requiring followup. Some researchers have seen higher rates. Radiologist Stephen Swensen of the Mayo Clinic in Rochester, Minnesota, says that a nonrandomized study he led in 2005 gave positive results for 69% of the screens. One difference between the Mayo and NLST studies, Swensen says, is that Mayo tracked nodules as small as 1 to 3 millimeters whereas NLST, which began in 2002, cut off positive findings below 4 mm.

One negative consequence of CT screening, Lowy said at the teleconference, is that it triggers follow-up scans, each of which increases radiation exposure. Even low-dose CT scans deliver a "significantly greater" exposure than conventional chest x-rays, said Lowy, noting that, "It remains to be determined how, or if, the radiation doses from screening ... may have increased the risks for cancer during the remaining lifetime" of those screened. Clinical followup may also include biopsy and surgery, Lowy said, "potentially risky procedures that can cause a host of complications."

G. Scott Gazelle, a radiologist and director of the Institute for Technology Assessment at Massachusetts General Hospital in Boston, has been analyzing the likely impacts of lung cancer screening for a decade. He agrees that people are going to demand it—and that "there are going to be a huge number of false positives." He was not surprised at NLST's finding of a lifesaving benefit of 20%. His group's prediction of mortality reduction through CT scans, based on "micromodeling" of actual cancers and data from previous studies, was 18% to 25%, right on target. But Gazelle says this analysis, now under review, still suggests that a national program of CT screening for lung cancer "would not be cost effective." Indeed, the costs seem likely to be three to four times those of breast cancer screening, with similar benefits.

Advocates of screening, in contrast, see the NLST results as vindicating a campaign to put advanced computer technology to work on lung cancer. The detailed images of early tumors in CT scans are "exquisite," says James Mulshine, vice president for research at Rush University Medical Center in Chicago, Illinois, and an adviser to the pro-screening advocacy group, the Lung Cancer Alliance in Washington, D.C. He thinks it should be straightforward to reduce the number of biopsies and surgeries resulting from false positives by monitoring small tumors for a time before intervening. There are 45 million smokers in the United States who might benefit from CT screening, says Mulshine. He asks: Do we provide it, or "Do we tell them, 'Tough luck'?"

[16]Sally Clark's conviction was overturned in 2003, and she was released from prison. Sally Clark died of acute alcohol poisoning in her home four years later in 2007, at the age of 42. Roy Meadow (1933–) is still an active British pediatrician. He rose to fame for his 1977 academic article in the *Lancet* on Munchausen Syndrome by Proxy (MSbP); he is the person who coined the name. He has spent his whole career crusading and testifying against parents, especially mothers, who supposedly wilfully harmed or killed their children. We quote from Lord Howe, the opposition spokesman for health, speaking in the House of Lords on MSbP (February 2003) (Curzon, 2003):

... [O]ne of the most pernicious and ill-founded theories to have gained currency in childcare and social services over the past 10 to 15 years. The theory states that there are parents who induce or fabricate illnesses in their children in order to gain attention for themselves. The name given to it is Münchausen's syndrome by proxy, or factitious or induced illness—FII, as it is now known. It is a theory without science. There is no body of peer-reviewed research to underpin MSBP or FII. It rests instead on the assertions of its inventor and on a handful of case histories. When challenged to produce his research papers to justify his original findings, the inventor of MSBP stated, if you please, that he had destroyed them. (c. 316)

[17]This article is listed in the Suggested Reading, B.1.1.

[18]By all accounts, Bertillon was a dislikable person ("Alphonse Bertillon," 2012). He is best known for the development of the first workable system of identification through body measurements; he named this "anthropometry" (later called "bertillonage" by others). We give a brief quotation about Bertillon from *The Science of Sherlock Holmes* by E. J. Wagner (2006):

And then, in 1882, it all changed, thanks to a twenty-six-year old neurasthenic clerk in the Paris Police named Alphonse Bertillon. It is possible that Bertillon possessed some social graces, but if so, he was amazingly discreet about them. He rarely spoke, and when he did, his voice held no expression. He was bad-tempered and avoided people. He suffered from an intricate variety of digestive complaints, constant headaches, and frequent nosebleeds. He was narrow-minded and obsessive.

Although he was the son of the famous physician and anthropologist Louis Adolphe Bertillon and had been raised in a highly intellectual atmosphere appreciative of science, he had managed to be thrown out of a number of excellent schools for poor grades. He had been unable to keep a job. His employment at the police department was due entirely to his father's influence. But this misanthropic soul managed to accomplish what no one else had: he invented a workable system of identification.

Sherlock Holmes remarks in *The Hound of the Baskervilles*, "The world is full of obvious things which nobody by any chance ever observes." It was Bertillon who first observed the obvious need for a scientific method of identifying criminals. He recalled discussions in his father's house about the theory of the Belgian statistician Lambert Adolphe Jacques Quetelet, who in 1840 had suggested that there were no two people in the world who were exactly the same size in all their measurements. (pp. 97–98)

Bertillonage was widely used for criminal identification in the decades surrounding the turn-of-the-century. It was eventually supplanted by the use of fingerprints, as advocated by Sir Francis Galton in his book, *Finger Prints*, published in 1892. A short extraction from Galton's introduction mentions Bertillon by name:

My attention was first drawn to the ridges in 1888 when preparing a lecture on Personal Identification for the Royal Institution, which had for its principal object an account of the anthropometric method of Bertillon, then newly introduced into the prison administration of France. Wishing to treat the subject generally, and having a vague knowledge of the value sometimes assigned to finger marks, I made inquiries, and was surprised to find, both how much had been done, and how much there remained to do, before establishing their theoretical value and practical utility. (p. 2)

One of the better known photographs of Galton (at age 73) is a Bertillon record from a visit Galton made to Bertillon's laboratory in 1893 (a Google search using the two words "Galton" and "Bertillon" will give the image).

Besides anthropometry, Bertillon contributed several other advances to what would now be referred to as "forensic science." He standardized the criminal "mug shot," and the criminal evidence picture through "metric photography." Metric photography involves taking pictures before a crime scene is disturbed; the photographs had mats printed with metric frames placed on the sides. As in "mug shots," photographs are generally taken of both the front and side views of a scene. Bertillon also created other forensic techniques, for example, forensic document examination (but in the case of Dreyfus, this did not lead to anything good), the use of galvanoplastic compounds to preserve footprints, the study of ballistics, and the dynamometer for determining the degree of force used in breaking and entering.

[19] A very informative *New Yorker* article on eyewitness evidence by Atul Gawande ("Under Suspicion," January 8, 2001) is listed in the Suggested Reading, B.3.3. A more recent news item (B.3.4) from *Nature*, concentrates specifically on how lines-ups are (ill)conducted: "Eyewitness Identification: Line-Ups on Trial" (*Nature*, Laura Spinney, May 21, 2008). We also list a recent news item (B.3.5) from the popular press, on an exoneration of a man imprisoned for thirty-five years based on a faulty lineup identification (M. Stacy, December 17, 2009).

[20] An interesting historical subplot in the development of lie detection involved William Moulton Marston. Marston is usually given credit for promoting the development of an instrument for lie detection based on systolic blood pressure. His doctoral dissertation in experimental psychology at Harvard (1921) was entitled *Systolic Blood Pressure and Reaction-Time Symptoms of Deception and of Constituent Mental States*. It has been suggested (by none other than Marston's son) that it was actually Elizabeth Marston, William's wife, who was the motivation for his work on lie detection and its relation to blood pressure (quoting the son (Lamb, 2001), "when she got mad or excited, her blood pressure seemed to climb"). In any case, Marston lived with two women in a polyamorous relationship—Elizabeth Holloway Marston, his wife, and Olive Byrne. Both these two women served as exemplars and inspirations for Marston's more well-known contribution to American life—the creation of the character and comic strip, *Wonder Woman*, in the early 1940s under the pseudonym of Charles Moulton. Supposedly, it was Elizabeth's idea to create

a female superhero who could triumph not with fists or firepower, but with love. This character would have a Magic Lasso (or a Golden Lasso, or a Lasso of Truth) that would force anyone captured by it to obey and tell the truth. So, besides introducing Wonder Woman and a lie detection instrument to a United States audience, Marston is credited with several additional cultural introductions. For more detail the reader is referred to the Wikipedia article on Marston ("William Moulton Marston," 2010).

We will see more of Marston's influence in Chapter 17 (*The Federal Rules of Evidence*). It was the exclusion of a lie detector exoneration (from Marston) for a convicted criminal, named Frye, that led to the *Frye* standard for the admissibility of expert testimony and evidence ("Frye standard," 2010). The "general acceptance" *Frye* standard held sway for much of the 20th century until superseded, at least at the federal level, by the *Federal Rules of Evidence* and the famous *Daubert* trilogy of rulings from the United States Supreme Court. This history is developed more thoroughly in Chapter 17.

[21]The woman's name is Virginia Hernandez Lopez; see, for example, Adam Liptak, *New York Times* (December 19, 2009), "Justices Revisit Rule Requiring Lab Testimony." In the actual case being orally argued of *Briscoe v. Virginia* (2010), the Court merely sent it back to a lower court in light of a recently decided case (*Melendez-Diaz v. Massachusetts*, 2009), which held that it is unconstitutional for a prosecutor to submit a chemical drug test report without the testimony of the scientist who conducted the test.

A more recent (5-4) Supreme Court ruling in *Bullcoming v. New Mexico* (2011) reaffirmed the *Melendez-Diaz* decision, saying that "surrogate testimony" would not suffice, and substitutes were not acceptable in crime lab testimony. The first paragraph of the syllabus in the *Bullcoming* opinion follows:

> The Sixth Amendment's Confrontation Clause gives the accused "[i]n all criminal prosecutions ... the right ... to be confronted with the witnesses against him." In Crawford v. Washington ... this Court held that the Clause permits admission of "[t]estimonial statements of witnesses absent from trial ... only where the declarant is unavailable, and only where the defendant has had a prior opportunity to cross-examine." Later, in Melendez-Diaz v. Massachusetts ... the Court declined to create a "forensic evidence" exception to Crawford, holding that a forensic laboratory report, created specifically to serve as evidence in a criminal proceeding, ranked as "testimonial" for Confrontation Clause purposes. Absent stipulation, the Court ruled, the prosecution may not introduce such a report without offering a live witness competent to testify to the truth of the report's statements.

[22]Two Suggested Reading items regarding lie detection in the Forensic Issues Section B.3 are relevant to making judgments based on a fallible instru-

ment. One is by Margaret Talbot (B.3.7) on using brain scans to detect lying ("Duped: Can Brain Scans Uncover Lies?," *New Yorker*, July 2, 2007); the other debunks voice-based lie detection (B.3.8): "The Truth Hurts: Scientists Question Voice-Based Lie Detection" (Rachel Ehrenberg, *ScienceNews*, June 22, 2010). A more general review devoted to lie detection by Vrij, Granhag, and Porter appeared in *Psychological Science in the Public Interest* ("Pitfalls and Opportunities in Nonverbal and Verbal Lie Detection," 2010, *11*, 89–121). This article discusses behaviors that are not the best diagnostic indicators of lying. A section in a later chapter on correlation, entitled "illusory correlation," refers to a false but widely held belief in a relationship between two behaviors, for example, the drawing of big eyes in a Draw-A-Person projective test and a person's paranoia. There is an illusory correlation between lying and both gaze aversion and nervousness.

The notion that gaze aversion reflects lying appears in our common idiomatic language in phrases such as "he won't look me in the eye." An editorial accompanying the review article cited above (Elizabeth Loftus, "Catching Liars," 2010, *11*, 87–88), comments directly on the cross-racial problem of using gaze aversion to suggest someone is lying:

> Using gaze aversion to decide that someone is lying can be dangerous for that someone's health and happiness. And—what was news to me—some cultural or ethnic groups are more likely to show gaze aversion. For example, Blacks are particularly likely to show gaze aversion. So imagine now the problem that might arise when a White police officer interviews a Black suspect and interprets the gaze aversion as evidence of lying. This material needs to be put in the hands of interviewers to prevent this kind of cross-racial misinterpretation. (p. 87)

Coupled with a human tendency to engage in confirmation bias when an illusory correlation is believed, and to look for evidence of some type of "tell" such as "gaze aversion," we might once again remind ourselves to "lawyer up" early and often.

The illusory connection between nervousness and lying is so strong it has been given the name of "the Othello error." A passage from the Vrij et al. (2010) review provides a definition:

> A common error in lie detection is to too readily interpret certain behaviors, particularly signs of nervousness, as diagnostic of deception. A common mistake for lie detectors is the failure to consider that truth tellers (e.g., an innocent suspect or defendant) can be as nervous as liars. Truth tellers can be nervous as a result of being accused of wrongdoing or as a result of fear of not being believed, because they too could face negative consequences if they are not believed (C. F. Bond & Fahey, 1987; Ofshe & Leo, 1997). The misinterpretation of signs of nervousness in truth tellers as signs of deceit

is referred to as the *Othello error* by deception researchers (Ekman, 1985/2001), based on Shakespeare's character. Othello falsely accuses his wife Desdemona of infidelity, and he tells her to confess because he is going to kill her for her treachery. When Desdemona asks Othello to summon Cassio (her alleged lover) so that he can testify her innocence, Othello tells her that he has already murdered Cassio. Realizing that she cannot prove her innocence, Desdemona reacts with an emotional outburst, which Othello misinterprets as a sign of her infidelity. The Othello error is particularly problematic in attempting to identify high-stakes lies because of the observer's sense of urgency and a host of powerful cognitive biases that contribute to tunnel-vision decision making (see Porter & ten Brinke, 2009). (p. 98)

[23]Or, asking a person to say "rabbit" to see if he is Elmer Fudd.

[24]As discussed in training videos for Arizona law-enforcement personnel, police can consider a variety of characteristics in deciding whether to ask about an individual's immigration status: does the person speak poor English, look nervous, is he traveling in an overcrowded vehicle, wearing several layers of clothing in a hot climate, hanging out in areas where illegal immigrants look for work, does not have identification, does he try to run away, ... See Amanda Lee Myers, "Seventh Lawsuit Filed Over Ariz. Immigration Law" (Associated Press, July 10, 2010). It is difficult to see how any convincing statistical argument could be formulated that the use of behaviors correlated with ethnicity and race does not provide a *prima facie* case for racial profiling.

[25]A discussion of how police interrogation operates is referenced in the Suggested Reading (B.3.6); it was written (and available online) by Julia Layton (May 18, 2006), "How Police Interrogation Works."

[26]A minimal statement of a Miranda warning is given in the Supreme Court case of *Miranda v. Arizona* (1966): "You have the right to remain silent. Anything you say can and will be used against you in a court of law. You have the right to speak to an attorney, and to have an attorney present during any questioning. If you cannot afford a lawyer, one will be provided for you at government expense" ("The Miranda warning," 2010).

[27]Continued from the main text:

The Need for Research to Establish Limits and Measures of Performance:

In evaluating the accuracy of a forensic analysis, it is crucial to clarify the type of question the analysis is called on to address. Thus, although some techniques may be too imprecise to permit accurate identification of a specific individual, they may still provide useful and accurate information about questions of classification. For example, microscopic hair analysis may provide reliable evidence on some characteristics of the individual from which the specimen was taken, but it may not be able to reliably match the specimen with a specific individual. However, the definition of the appropriate question is only a first step in the evaluation of the performance of a forensic technique. A body of research is required to establish the limits and measures of performance and to address the impact of sources of variability and potential bias. Such research is sorely needed, but it seems to be lacking in most of the forensic disciplines that rely on subjective assessments of matching characteristics. These disciplines need to develop rigorous protocols to guide these subjective interpretations and pursue equally rigorous research and evaluation programs. The development of such research programs can benefit significantly from other areas, notably from the large body of research on the evaluation of observer performance in diagnostic medicine and from the findings of cognitive psychology on the potential for bias and error in human observers.

The Admission of Forensic Science Evidence in Litigation:

Forensic science experts and evidence are used routinely in the service of the criminal justice system. DNA testing may be used to determine whether sperm found on a rape victim came from an accused party; a latent fingerprint found on a gun may be used to determine whether a defendant handled the weapon; drug analysis may be used to determine whether pills found in a person's possession were illicit; and an autopsy may be used to determine the cause and manner of death of a murder victim. ... for qualified forensic science experts to testify competently about forensic evidence, they must first find the evidence in a usable state and properly preserve it. A latent fingerprint that is badly smudged when found cannot be usefully saved, analyzed, or explained. An inadequate drug sample may be insufficient to allow for proper analysis. And, DNA tests performed on a contaminated or otherwise compromised sample cannot be used reliably to identify or eliminate an individual as the perpetrator of a crime. These are important matters involving the proper processing of forensic evidence. The law's greatest dilemma in its heavy reliance on forensic evidence, however, concerns the question of whether—and to what extent—there is *science* in any given forensic science discipline.

Two very important questions should underlie the law's admission of and reliance upon forensic evidence in criminal trials: (1) the extent to which a particular forensic discipline is founded on a reliable scientific methodology that gives it the capacity to accurately analyze evidence and report findings and (2) the extent to which practitioners in a particular forensic discipline rely on human interpretation that could be tainted by error, the threat of bias, or the absence of sound operational procedures and robust performance standards.

These questions are significant. Thus, it matters a great deal whether an expert is qualified to testify about forensic evidence and whether the evidence is sufficiently reliable to merit a fact finder's reliance on the truth that it purports to support. Unfortunately, these important questions do not always produce satisfactory answers in judicial decisions pertaining to the admissibility of forensic science evidence proffered in criminal trials. (pp. 8–9)

Chapter 4

Probability Theory: Application Areas

4.1 Some Probability Considerations in Discrimination and Classification

It is better that ten guilty persons escape than one innocent suffer.
— Sir William Blackstone (*Commentary on the Laws of England*, 1765)

The term *discrimination* (in a nonpejorative statistical sense) can refer to the task of separating groups through linear combinations of variables maximizing a criterion, such as an *F*-ratio. The linear combinations themselves are commonly called Fisher's linear discriminant functions. The related term *classification* refers to the task of allocating observations to existing groups, typically to minimize the cost and/or probability of misclassification. These two topics are intertwined, but here we briefly comment on only the topic of classification. Any applied multivariate analysis course should treat these two topics in much greater depth.

In the simplest situation, we have two populations, π_1 and π_2; π_1 is assumed to be characterized by a normal distribution with mean μ_1 and variance σ_X^2 (the density is denoted by $f_1(x)$); π_2 is characterized by a normal distribution with mean μ_2 and (common) variance σ_X^2 (the density is denoted by $f_2(x)$). Given an observation, say x_0, we wish to decide whether it should be assigned to π_1 or to π_2. Assuming that $\mu_1 \leq \mu_2$, a criterion point c is chosen; the rule then becomes: allocate to π_1 if $x_0 \leq c$, and to π_2 if $> c$. The probabilities of misclassification are given in the following chart:

| | True State | |
	π_1	π_2
Decision π_1	$1 - \alpha$	β
π_2	α	$1 - \beta$

In the terminology of our previous usage in Section 3.3 of Bayes' rule to obtain the positive predictive value of a test, and assuming that π_1 refers to a person having "it," and π_2 to not having "it," the sensitivity of the test is $1 - \alpha$ (true positive); specificity is $1 - \beta$, and thus, β refers to a false positive.

To choose c so that $\alpha + \beta$ is smallest, select the point at which the densities are equal. A more complicated way of stating this decision rule is to allocate to π_1 if $f_1(x_0)/f_2(x_0) \geq 1$; if < 1, then allocate to π_2. Suppose now that the prior probabilities of being drawn from π_1 and π_2 are p_1 and p_2, respectively, where $p_1 + p_2 = 1$. If c is chosen so the Total Probability of Misclassification (TPM) is minimized (that is, $p_1\alpha + p_2\beta$), the rule would be to allocate to π_1 if $f_1(x_0)/f_2(x_0) \geq p_2/p_1$; if $< p_2/p_1$, then allocate to π_2. Finally, to include costs of misclassification, $c(1|2)$ (for assigning to π_1 when actually coming from π_2), and $c(2|1)$ (for assigning to π_2 when actually coming from π_1), choose c to minimize the Expected Cost of Misclassification (ECM), $c(2|1)p_1\alpha + c(1|2)p_1\beta$, by the rule of allocating to π_1 if $f_1(x_0)/f_2(x_0) \geq (c(1|2)/c(2|1))(p_2/p_1)$; if $< (c(1|2)/c(2|1))(p_2/p_1)$, then allocate to π_2.

Using logs, the last rule can be restated:

allocate to π_1 if $\log(f_1(x_0)/f_2(x_0)) \geq \log((c(1|2)/c(2|1))(p_2/p_1))$. The left-hand side is equal to

$$(\mu_1 - \mu_2)(\sigma_X^2)^{-1}x_0 - (1/2)(\mu_1 - \mu_2)(\sigma_X^2)^{-1}(\mu_1 + \mu_2),$$

so the rule can be rephrased further:

allocate to π_1 if

$$x_0 \leq \{(1/2)(\mu_1 - \mu_2)(\sigma_X^2)^{-1}(\mu_1 + \mu_2) -$$

$$\log((c(1|2)/c(2|1))(p_2/p_1))\{\frac{\sigma_X^2}{-(\mu_1 - \mu_2)}\}$$

or

$$x_0 \leq \{(1/2)(\mu_1 + \mu_2) - \log((c(1|2)/c(2|1))(p_2/p_1))\}\{\frac{\sigma_X^2}{(\mu_2 - \mu_1)}\} = c.$$

If the costs of misclassification are equal (that is, $c(1|2) = c(2|1)$), then the allocation rule is based on classification functions: allocate to π_1 if

$$[\frac{\mu_1}{\sigma_X^2}x_0 - (1/2)\frac{\mu_1^2}{\sigma_X^2} + \log(p_1)] - [\frac{\mu_2}{\sigma_X^2}x_0 - (1/2)\frac{\mu_2^2}{\sigma_X^2} + \log(p_2)] \geq 0 .$$

In the terminology of signal detection theory and the general problem of yes/no diagnostic decisions, a plot of sensitivity (true positive probability) on the y-axis against $1-$ specificity on the x-axis as c varies, is an ROC curve (for Receiver Operating Characteristic). This ROC terminology originated in World War II in detecting enemy planes by radar (group π_1) from the noise generated by random interference (group π_2). The ROC curve is bowed from the origin of $(0, 0)$ at the lower-left corner to $(1.0, 1.0)$ at the upper right; it indicates the trade-off between increasing the probability of true positives and the increase of false positives. Generally, the adequacy of a particular diagnostic decision strategy is measured by the area under the ROC curve, with areas closer to 1.0 being better; that is, steeper bowed curves hugging the left wall and the top border of the square box. For a comprehensive introduction to diagnostic processes, see Swets, Dawes, and Monahan (2000b).[1]

4.2   Probability and Litigation

It is now generally recognized, even by the judiciary, that since all evidence is probabilistic—there are no metaphysical certainties— evidence should not be excluded merely because its accuracy can be expressed in explicitly probabilistic terms.
  — Judge Richard A. Posner ("An Economic Approach to the Law of Evidence," *Stanford Law Review*, 1999)

The retirement of Supreme Court Justice John Paul Stevens gave President Obama a second opportunity to nominate a successor who drew the ire of the Republican Party during the confirmation process (similar to the previous such hearing with "the wise Latina woman"). For one who might have enjoyed witnessing a collective apoplexy from the conservative right, we could have suggested that President Obama nominate Jack Weinstein, a sitting federal judge in the Eastern District of New York (Brooklyn), if it weren't for the fact that at 89 he was only one year younger than the retiring Stevens. Weinstein is one of the most respected and influential judges in America ("Jack B. Weinstein," 2010). He has directly organized and presided over some of the most important mass tort cases of the last forty years (for example, Agent Orange, asbestos, tobacco, breast implants, DES, Zyprexa, handgun regulation, and repetitive-stress injuries).[2] For present purposes, our interest is in Weinstein's deep respect for science-based evidence in the judicial process, and in particular, for how he views probability and statistics as an intimate part of that process. He also may be the only federal judge ever to publish an article in a major statistics journal (*Statistical Science*, 1988, *3*, 286–297, "Litigation and Statistics"). This last work developed out of Weinstein's association in the middle 1980s with the National Academy of Science's Panel on Statistical Assessment as Evidence in the Courts. This panel produced the comprehensive Springer-Verlag volume *The Evolving Role of Statistical Assessments as Evidence in the Courts* (1988; Stephen E. Fienberg, Editor).

The importance that Weinstein gives to the role of probability and statistics in the judicial process is best expressed by Weinstein himself (we quote from his *Statistical Science* article):

> The use of probability and statistics in the legal process is not unique to our times. Two thousand years ago, Jewish law, as stated in the Talmud, cautioned about the use of probabilistic inference. The medieval Jewish commentator Maimonides summarized this traditional view in favor of certainty when he noted:
> "The 290th commandment is a prohibition to carry out punishment on a high probability, even close to certainty ... No punishment [should] be carried out except where ... the matter is established in certainty beyond any doubt ... "
> That view, requiring certainty, is not acceptable to the courts. We deal not with the truth, but with probabilities, in criminal as well as civil cases. Probabilities, express and implied, support every factual decision and inference we make in court. (p. 287)

Maimonides' description of the 290th Negative Commandment is given in its entirety in an appendix to this section. According to this commandment, an absolute certainty of guilt is guaranteed by having two witnesses to exactly the same crime. Such a probability of guilt being identically 1 is what is meant by the contemporary phrase "without any shadow of a doubt."

Two points need to be emphasized about this Mitzvah (Jewish commandment). One is the explicit unequalness of costs attached to the false positive and negative errors: "it is preferable that a thousand guilty people be set free than to execute one innocent person." The second is in dealing with what would now be characterized as the (un)reliability of eyewitness testimony. Two eyewitnesses are required, neither is allowed to make just an inference about what happened but must have observed it directly, and exactly the same crime must be observed by both eyewitnesses. Such a high standard of eyewitness integrity might have made the current rash of DNA exonerations unnecessary.

Judge Weinstein's interest in how probabilities could be part of a judicial process goes back some years before the National Research Council Panel. In one relevant opinion from 1978, *United States v. Fatico*, he wrestled with how subjective probabilities might be related to the four levels of a "legal burden of proof"; what level was required in this particular case; and, finally, was it then met. The four (ordered) levels are: preponderance of the evidence; clear and convincing evidence; clear, unequivocal, and convincing evidence; and proof beyond a reasonable doubt. The case in point involved proving that Daniel Fatico was a "made" member of the Gambino organized crime family, and thus could be given a "Special Offender" status. "The consequences of (such) a 'Special Offender' classification are significant. In most cases, the designation delays or precludes social furloughs, release to half-way houses and transfers to other correctional institutions; in some cases, the characterization may bar early parole" (text taken from the opinion). The summary of Weinstein's final opinion in the Fatico case follows:

> In view of prior proceedings, the key question of law now presented is what burden of proof must the government meet in establishing a critical fact not proved at a criminal trial that may substantially enhance the sentence to be imposed upon a defendant. There are no precedents directly on point.

> The critical factual issue is whether the defendant was a "made"
> member of an organized crime family. Clear, unequivocal and con-
> vincing evidence adduced by the government at the sentencing hear-
> ing establishes this proposition of fact.

The text of Weinstein's opinion in the Fatico case explains some
of the connections between subjective probabilities, burdens of
proof, and the need for different levels depending on the particular
case (we note that the numerical values suggested in this opinion
as corresponding to the various levels of proof, appear to be based
only on Judge Weinstein's "best guesses"). We redact part of his
opinion in an appendix to this section.

Other common standards used for police searches or arrests
might also be related to an explicit probability scale. The lowest
standard (perhaps a probability of 20%) would be "reasonable suspicion" to determine whether a brief investigative stop or search
by any governmental agent is warranted (in the 2010 "Papers,
Please" law in Arizona, a "reasonable suspicion" standard is set
for requesting documentation). A higher standard would be "probable cause" to assess whether a search or arrest is warranted, or
whether a grand jury should issue an indictment. A value of, say,
40% might indicate a "probable cause" level that would put it
somewhat below a "preponderance of the evidence" criterion. In
all cases, a mapping of such verbal statements to numerical values
requires "wiggle room" for vagueness, possibly in the form of an
interval estimate rather than a point estimate. As an example of
this variability of assessment in the Fatico case, Judge Weinstein
informally surveyed the judges in his district court regarding the
four different standards of proof. The data are given in Table 4.1
(taken from Fienberg, 1988, p. 204). Reader, we leave it to you to
decide whether you would want Judge 4 or Judge 7 to hear your
case.

4.2.1 Probability of Causation

Judge Weinstein is best known for the mass (toxic) tort cases
he has presided over for the last four decades (for example, asbestos, breast implants, Agent Orange). In all of these kinds of
torts, there is a need to establish, in a legally acceptable fashion,
some notion of causation. There is first a concept of *general causa-*

TABLE 4.1: Probabilities associated with different standards of proof by judges in the Eastern District of New York.

Judge	Prepon-derance	Clear and convincing	Clear, unequivocal, and convincing	Beyond a reasonable doubt	row median
1	50+	60	70	85	65
2	51	65	67	90	66
3	50+	60-70	65-75	80	67
4	50+	67	70	76	69
5	50+	Standard is elusive		90	70
6	50+	70+	70+	85	70
7	50+	60	90	85	72
8	50+	70+	80+	95+	75
9	50.1	75	75	85	75
10	51	Cannot estimate			–
column median	50	66	70	85	

tion concerned with whether an agent can increase the incidence of disease in a group; because of individual variation, a toxic agent will not generally cause disease in every exposed individual. *Specific causation* deals with an individual's disease being attributable to exposure from an agent. (For a further discussion of general and specific causation, see Green, Freedman, and Gordis, 2000.)

The establishment of general causation (and a necessary requirement for establishing specific causation) typically relies on a *cohort study*. We give a definition from the Green et al. (2000) epidemiology chapter of the *Reference Manual on Scientific Evidence*:

> *cohort study*: The method of epidemiologic study in which groups of individuals can be identified who are, have been, or in the future may be differentially exposed to an agent or agents hypothesized to influence the probability of occurrence of a disease or other outcome. The groups are observed to find out if the exposed group is more likely to develop disease. The alternative terms for a cohort study (concurrent study, follow-up study, incidence study, longitudinal study, prospective study) describe an essential feature of the method, which is observation of the population for a sufficient number of person-years to generate reliable incidence or mortality rates

in the population subsets. This generally implies study of a large population, study for a prolonged period (years), or both. (p. 389)

One common way to organize data from a cohort study is through a simple 2×2 contingency table, similar in form to that introduced at the beginning of Chapter 3:

	Disease	No Disease	Row Sums
Exposed	N_{11}	N_{12}	N_{1+}
Not Exposed	N_{21}	N_{22}	N_{2+}

Here, N_{11}, N_{12}, N_{21}, and N_{22} are the cell frequencies; N_{1+} and N_{2+} are the row frequencies. Conceptually, these data are considered generated from two (statistically independent) binomial distributions for the "Exposed" and "Not Exposed" conditions. If we let p_E and p_{NE} denote the two underlying probabilities of getting the disease for particular cases within the conditions, respectively, the ratio $\frac{p_E}{p_{NE}}$ is referred to as the relative risk (RR), and may be estimated with the data as follows:

estimated relative risk $= \widehat{RR} = \frac{\hat{p}_E}{\hat{p}_{NE}} = \frac{N_{11}/N_{1+}}{N_{21}/N_{2+}}$.

A measure commonly referred to in tort litigations is attributable risk (AR), defined as

$AR = \frac{p_E - p_{NE}}{p_E}$, and estimated by

$\widehat{AR} = \frac{\hat{p}_E - \hat{p}_{NE}}{\hat{p}_E} = 1 - \frac{1}{RR}$.

Attributable risk, also known as the "attributable proportion of risk" or the "etiologic fraction," represents the amount of disease among exposed individuals assignable to the exposure. It measures the maximum proportion of the disease attributable to exposure from an agent, and consequently, the maximum proportion of disease that could be potentially prevented by blocking the exposure's effect or eliminating the exposure itself. If the association is causal, AR is the proportion of disease in an exposed population that might be caused by the agent, and therefore, that might be prevented by eliminating exposure to the agent (Green, Freedman, & Gordis, 2000, pp. 351–352).

The common legal standard used to argue for both specific and general causation is an RR of 2.0, or an AR of 50%. At this level, it is "as likely as not" that exposure "caused" the disease (or "as likely to be true as not," or "the balance of the probabilities"). Obviously, one can never be absolutely certain that a particular agent was "the" cause of a disease in any particular individual, but to allow an idea of "probabilistic causation" or "attributable risk" to enter into legal arguments provides a justifiable basis for compensation. It has now become routine to do this in the courts.

We give one illustration where implementing the "as likely as not standard" for attributing possible causation (and compensation) has gone to great technical levels (and which should keep many biostatisticians gainfully employed for years to come). What follows in an appendix to this section is an extensive excerpt from the *Federal Register* concerning the Department of Health and Human Services and its *Guidelines for Determining the Probability of Causation and Methods for Radiation Dose Reconstruction Under the [Energy] Employees Occupational Illness Compensation Program Act of 2000.* This material should give a good sense of how the modeling principles of probability and statistics are leading to ethically defensible compensation models; here, the models used are for all those exposed to ionizing radiation through an involvement with the United States' nuclear weapons industry.

Several points need emphasize about this *Federal Register* excerpt: (1) the calculation of a "probability of causation" is much more sophisticated (and fine-grained) than one based on a simple aggregate 2×2 contingency table where AR is just calculated from the explicit cell frequencies. Statistical models (of the generalized linear model variety) are being used to estimate the AR tailored to an individual's specific circumstances—type of cancer, type of exposure, other individual characteristics; (2) all the models are now implemented (interactively through a graphical user interface) within the Interactive RadioEpidemiological Program (IREP), making obsolete the very cumbersome charts and tables previously used; also, IREP allows a continual updating to the model estimation process when new data become available; (3) it is not just a point estimate for the probability of causation that is used to determine compensation, but rather the upper limit for a 99% confidence interval; this obviously gives a great "benefit of the doubt" to an individual seeking compensation for a presumably radiation-induced disease; (4) as another "benefit of the doubt" calculation, if there are two or more primary cancers, the probability of causation reported will be the probability that at least one of the cancers was caused by the radiation. Generally, this will result in a larger estimate for the probability of causation, and thus to a greater likelihood of compensation; (5) when cancers are identified from secondary sites and the primary site is unknown, the final assignment of the primary cancer site will be the one resulting in the highest estimate for the probability of causation.

4.2.2   Probability Scales and Rulers

The topic of relating a legal understanding of burdens of proof to numerical probability values has been around for a very long time. Fienberg (1988) provides a short discussion of Jeremy Bentham's (1827) suggestion of a "persuasion thermometer," and some contemporary reaction to this idea from Thomas Starkie (1833).[3] We quote:

> Jeremy Bentham appears to have been the first jurist to seriously propose that witnesses and judges numerically estimate their degrees of persuasion. Bentham (1827; Vol. 1, pp. 71–109) envisioned a kind of moral thermometer:
>
> > The scale being understood to be composed of ten degrees—in the language applied by the French philosophers to thermometers, a decigrade scale—a man says, My persuasion is at 10 or 9, etc. affirmative, or at least 10, etc. negative . . .
>
> Bentham's proposal was greeted with something just short of ridicule, in part on the pragmatic grounds of its inherent ambiguity and potential misuse, and in part on the more fundamental ground that legal probabilities are incapable of numerical expression. Thomas Starkie (1833) was merely the most forceful when he wrote:
>
> > The notions of those who have supposed that mere moral probabilities or relations could ever be represented by numbers or space, and thus be subjected to arithmetical analysis, cannot but be regarded as visionary and chimerical. (p. 212)

Several particularly knotty problems and (mis)interpretations when it comes to assigning numbers to the possibility of guilt arise most markedly in eyewitness identification. Because cases involving eyewitness testimony are typically criminal cases, they demand burdens of proof "beyond a reasonable doubt"; thus, the (un)reliability of eyewitness identification becomes problematic when it is the primary (or only) evidence presented to meet this standard. As discussed extensively in the judgment and decision-making literature, there is a distinction between making a subjective estimate of some quantity, and one's confidence in that estimate once made. For example, suppose someone picks a suspect out of a lineup, and is then asked the (Bentham) question, "on a scale of from one to ten, characterize your level of 'certainty'." Does an answer of "seven or eight" translate into a probability of innocence of two or three out of ten? Exactly such confusing situations, however, arise. We give a fairly extensive redaction in an appendix to this section of an opinion from the District of Columbia Court of Appeals in a case named "In re As.H" (2004). It combines extremely well both the issues of eyewitness (un)reliability and the attempt to quantify that which may be better left in words; the dissenting Associate Judge Farrel noted pointedly: "I believe that the entire effort to quantify the standard of proof beyond a reasonable doubt is a search for fool's gold."

4.2.3   The Cases of Vincent Gigante and Agent Orange

Although Judge Weinstein's reputation may rest on his involvement with mass toxic torts, his most entertaining case occurred in the middle 1990s, with the murder-conspiracy and racketeering trial and conviction of Vincent Gigante, the boss of the most powerful Mafia family in the United States. The issue here was assessing the evidence of Gigante's mental fitness to be sentenced to prison, and separating such evidence from the putative malingering of Gigante. Again, Judge Weinstein needed to evaluate the evidence and make a probabilistic assessment ("beyond a reasonable doubt") that Gigante's trial was a "valid" one.

Apart from the great legal theater that the Gigante case provided, Judge Weinstein's most famous trials all involve the Agent Orange defoliant used extensively by the United States in Vietnam in the 1960s. Originally, he oversaw in the middle 1980s the $200 million settlement fund provided by those companies manufacturing the agent. Most recently, Judge Weinstein presided over the dismissal of the civil lawsuit filed on behalf of millions of Vietnamese individuals. The 233-page decision in this case is an incredible "read" about United States polices during this unfortunate period in our country's history (In re "Agent Orange" Product Liability Litigation, 2005). The suit was dismissed not because of poor documentation of the effects of Agent Orange and various ensuing conditions, but because of other legal conditions. The judge concluded that even if the United States had been a Geneva Accord signatory (outlawing use of poisonous gases during war), Agent Orange would not have been banned: "The prohibition extended only to gases deployed for their asphyxiating or toxic effects on man, not to herbicides designed to affect plants that may have unintended harmful side effects on people" (In re "Agent Orange" Product Liability Litigation, 2005, p. 190). The National Academy of Science through its Institute of Medicine, regularly updates what we know about the effects of Agent Orange, and continues to document many associations between it and various disease conditions. The issues in the Vietnamese lawsuit, however, did not hinge on using a probability of causation assessment, but rather on whether, given the circumstances of the war, the United States could be held responsible for what it did in Vietnam in the 1960s.[4]

Appendix: Maimonides' 290th Negative Commandment

"And an innocent and righteous person you shall not slay" — Exodus 23:7. Negative Commandment 290
Issuing a Punitive Sentence Based on Circumstantial Evidence:
The 290th prohibition is that we are forbidden from punishing someone based on our estimation [without actual testimony], even if his guilt is virtually certain. An example of this is a person who was chasing after his enemy to kill him. The pursued escaped into a house and the pursuer entered the house after him. We enter the house after them and find the victim lying murdered, with the pursuer standing over him holding a knife, with both covered with blood. The Sanhedrin may not inflict the death penalty on this pursuer since there were no witnesses who actually saw the murder.
The Torah of Truth (Toras Emess) comes to prohibit his execution with G—d's statement (exalted be He), "Do not kill a person who has not been proven guilty."
Our Sages said in Mechilta: "If they saw him chasing after another to kill him and they warned him, saying, 'He is a Jew, a son of the Covenant! If you kill him you will be executed!' If the two went out of sight and they found one murdered, with the sword in the murderer's hand dripping blood, one might

think that he can be executed. The Torah therefore says, 'Do not kill a person who has not been proven guilty.' "

Do not question this law and think that it is unjust, for there are some possibilities that are extremely probable, others that are extremely unlikely, and others in between. The category of "possible" is very broad, and if the Torah allowed the High Court to punish when the offense was very probable and almost definite (similar to the above example), then they would carry out punishment in cases which were less and less probable, until people would be constantly executed based on flimsy estimation and the judges' imagination. G—d (exalted be He), therefore "closed the door" to this possibility and forbid any punishment unless there are witnesses who are certain beyond a doubt that the event transpired and that there is no other possible explanation.

If we do not inflict punishment even when the offense is most probable, the worst that could happen is that someone who is really guilty will be found innocent. But if punishment was given based on estimation and circumstantial evidence, it is possible that someday an innocent person would be executed. And it is preferable and more proper that even a thousand guilty people be set free than to someday execute even one innocent person.

Similarly, if two witnesses testified that the person committed two capital offenses, but each one saw only one act and not the other, he cannot be executed. For example: One witness testified that he saw a person doing a *melachah* on *Shabbos* and warned him not to. Another witness testified that he saw the person worshipping idols and warned him not to. This person cannot be executed by stoning. Our Sages said, "If one witness testified that he worshipped the sun and the other testified that he worshipped the moon, one might think that they can be joined together. The Torah therefore said, 'Do not kill a person who has not been proven guilty.' "

Appendix: The Redacted Text of Judge Weinstein's Opinion in United States v. Fatico (1978)

Given in the Appendix Supplements.

Appendix: *Guidelines for Determining the Probability of Causation and Methods for Radiation Dose Reconstruction Under the [Energy] Employees Occupational Illness Compensation Program Act of 2000*

Given in the Appendix Supplements.

Appendix: District of Columbia Court of Appeals, "In re As.H" (2004)

Given in the Appendix Supplements.

4.3   Betting, Gaming, and Risk

I know what you're thinkin'. "Did he fire six shots or only five?"
Well, to tell you the truth, in all this excitement I kind of lost track
myself.
— Harry Callahan (*Dirty Harry*)

Antoine Gombaud, better known as the Chevalier de Méré, was
a French writer and amateur mathematician from the early 17th
century. He is important to the development of probability theory
because of one specific thing; he asked a mathematician, Blaise
Pascal, about a gambling problem, named the "problem of points,
"dating from the Middle Ages. The question was one of fairly dividing the stakes among individuals who had agreed to play a
certain number of games, but for whatever reason had to stop before they were finished. Pascal in a series of letters with Pierre
de Fermat, solved this equitable division task, and in the process
laid out the foundations for a modern theory of probability. Pascal and Fermat also provided the Chevalier with a solution to a
vexing problem he was having in his own personal gambling. Apparently, the Chevalier had been very successful in making even
money bets that a six would be rolled at least once in four throws
of a single die. But when he tried a similar bet based on tossing
two dice 24 times and looking for a double-six to occur, he was
singularly unsuccessful in making any money. The reason for this
difference between the Chevalier's two wagers was clarified by the
formalization developed by Pascal and Fermat for such games of
chance. This formalization is briefly reviewed below, and then used
to discuss the Chevalier's two gambles as well as those occurring
in various other casino-type games.[5]

We begin by defining several useful concepts: a simple experiment, sample space, sample point, event, elementary event:

A *simple experiment* is some process that we engage in that
leads to one single outcome from a set of possible outcomes
that could occur. For example, a simple experiment could consist of rolling a die once, where the set of possible outcomes is
$\{1, 2, 3, 4, 5, 6\}$ (note that curly braces will be used consistently to
denote a set). Or, two dice could be tossed and the number of

spots occurring on each die noted; here, the possible outcomes are integer number pairs: $\{(a, b) \mid 1 \leq a \leq 6; 1 \leq b \leq 6\}$. Flipping a single coin would give the set of outcomes, $\{H, T\}$, with "H" for "heads" and "T" for "tails"; picking a card from a normal deck could give a set of outcomes containing 52 objects, or if we were interested only in the particular suit for a card chosen, the possible outcomes could be $\{H, D, C, S\}$, corresponding to heart, diamond, club, and spade, respectively.

The set of possible outcomes for a simple experiment is the *sample space* (which we denote by the script letter $\mathcal{S}$). An object in a sample space is a *sample point*. An *event* is defined as a subset of the sample space, and an event containing just a single sample point is an *elementary event*. A particular event is said to occur when the outcome of the simple experiment is a sample point belonging to the defining subset for that event.

As a simple example, consider the toss of a single die, where $\mathcal{S}$ = $\{1, 2, 3, 4, 5, 6\}$. The event of obtaining an even number is the subset $\{2, 4, 6\}$; the event of obtaining an odd number is $\{1, 3, 5\}$; the (elementary) event of tossing a 5 is a subset with a single sample point, $\{5\}$, and so on.

For a sample space containing K sample points, there are 2^K possible events (that is, there are 2^K possible subsets of the sample space). This includes the "impossible event" (usually denoted by $\emptyset$), characterized as that subset of $\mathcal{S}$ containing no sample points and which therefore can never occur; and the "sure event," defined as that subset of $\mathcal{S}$ containing all sample points (that is, $\mathcal{S}$ itself), which therefore must always occur. In our single die example, there are $2^6 = 64$ possible events, including $\emptyset$ and $\mathcal{S}$.

The motivation for introducing the idea of a simple experiment and sundry concepts is to use this structure as an intuitively reasonable mechanism for assigning probabilities to the occurrence of events. These probabilities are usually assigned through an assumption that sample points are equally likely to occur, assuming we have characterized appropriately what is to be in $\mathcal{S}$. Generally, only the probabilities are needed for the K elementary events containing single sample points. The probability for any other event is merely the sum of the probabilities for all those elementary events defined by the sample points making up that particular event. This last fact is due to the disjoint set property of probability introduced at the beginning of the last chapter. In the specific instance in which the sample points are equally likely to occur, the probability assigned to any event is merely the number of sample points defining the event divided by K. As special cases, we obtain a probability of 0 for the impossible event, and 1 for the sure event.

The use of the word *appropriately* in characterizing a sample space is important to keep in mind whenever we wish to use the idea of being equally likely to generate the probabilities for all the various events. For example, in throwing two dice and letting the sample space be $\mathcal{S} = \{(a,b) \mid 1 \leq a \leq 6; 1 \leq b \leq 6\}$, it makes sense, assuming that the dice are not "loaded," to consider the 36 integer number pairs to be equally likely. When the conception of what is being observed changes, however, the equally-likely notion may no longer be "appropriate." For example, suppose our interest is only in the sum of spots on the two dice being tossed, and let our sample space be $\mathcal{S} = \{2, 3, \ldots, 12\}$. The eleven integer sample points in this sample space are not equally likely; in fact, it is a common exercise in an elementary statistics course to derive the probability distribution for the objects in this latter sample space based on the idea that the underlying 36 integer number pairs are equally likely. To illustrate, suppose our interest is in the probability that a "sum of seven" appears on the dice. At the level of the sample space containing the 36 integer number pairs, a "sum of seven" corresponds to the event $\{(1,6),(6,1),(2,5),(5,2),(3,4),(4,3)\}$. Thus, the probability of a "sum of seven" is 6/36; there are six equally-likely sample points making up the event and there are 36 equally-likely integer pairs in the sample space. Although probably apocryphal, it has been said that many would-be probabilists hired by gambling patrons in the 17th century, came to grief when they believed that every stated sample space had objects that could be considered equally likely, and communicated this fact to their employers as an aid in betting.

One particularly helpful use of the sample space/event concepts is when a simple experiment is carried out multiple times (for, say, N replications), and the outcomes defining the sample space are the ordered N-tuples formed from the results obtained for the individual simple experiments. The Chevalier who rolls a single die four times, generates the sample space

$$\{(D_1, D_2, D_3, D_4) \mid 1 \le D_i \le 6, 1 \le i \le 4\},$$

that is, all 4-tuples containing the integers from 1 to 6. Generally, in a replicated simple experiment with K possible outcomes on each trial, the number of different N-tuples is K^N (using a well-known arithmetic multiplication rule). Thus, for the Chevalier example, there are $6^4 = 1296$ possible 4-tuples, and each such 4-tuple should be equally likely to occur (given the "fairness" of the die being used; so, no "loaded" dice are allowed). To define the event of "no sixes rolled in four replications," we would use the subset (event)

$$\{(D_1, D_2, D_3, D_4) \mid 1 \le D_i \le 5, 1 \le i \le 4\},$$

containing $5^4 = 625$ sample points. Thus, the probability of "no sixes rolled in four replications" is $625/1296 = .4822$. As we will see formally below, the fact that this latter probability is strictly less than $1/2$ gives the Chevalier a distinct advantage in playing an even money game defined by his being able to roll at least one six in four tosses of a die.

The other game that was not as successful for the Chevalier, was tossing two dice 24 times and betting on obtaining a double-six somewhere in the sequence. The sample space here is $\{(P_1, P_2, \ldots, P_{24})\}$, where $P_i = \{(a_i, b_i) \mid 1 \le a_i \le 6; 1 \le b_i \le 6\}$, and has 36^{24} possible sample points. The event of "not obtaining a double-six somewhere in the sequence" would look like the sample space just defined except that the $(6, 6)$ pair would be excluded from each P_i. Thus, there are 35^{24} members in this event. The probability of "not obtaining a double-six somewhere in the sequence" is

$$\frac{35^{24}}{36^{24}} = \left(\frac{35}{36}\right)^{24} = .5086 .$$

Because this latter value is greater than $1/2$ (in contrast to the previous gamble), the Chevalier would now be at a disadvantage making an even money bet.

The best way to evaluate the perils or benefits present in a wager is through the device of a discrete random variable. Suppose X denotes the outcome of some bet; and let $a_1, \ldots, a_T$ represent the T possible payoffs from one wager, where positive values reflect gain and negative values reflect loss. In addition, we know the probability distribution for X; that is, $P(X = a_t)$ for $1 \leq t \leq T$. What one expects to realize from one observation on X (or from one play of the game) is its expected value,

$$E(X) = \sum_{t=1}^{T} a_t P(X = a_t).$$

If $E(X)$ is negative, we would expect to lose this much on each bet; if positive, this is the expected gain on each bet. When $E(X)$ is 0, the term "fair game" is applied to the gamble, implying that one neither expects to win or lose anything on each trial; one expects to "break even." When $E(X) \neq 0$, the game is "unfair" but it could be unfair in your favor ($E(X) > 0$), or unfair against you ($E(X) < 0$).

To evaluate the Chevalier's two games, suppose X takes on the values of $+1$ and -1 (the winning or losing of one dollar, say). For the single die rolled four times, $E(X) = (+1)(.5178) + (-1)(.4822) = .0356 \approx .04$. Thus, the game is unfair in the Chevalier's favor because he expects to win a little less than four cents on each wager. For the 24 tosses of two dice, $E(X) = (+1)(.4914) + (-1)(.5086) = -.0172 \approx -.02$. Here, the Chevalier is at a disadvantage. The game is unfair against him, and he expects to lose about two cents on each play of the game.

Besides using the expectation of X as an indication of whether a game is fair or not, and in whose favor, the variance of X is an important additional characteristic of any gamble. The larger the variance, the more one would expect a "boom or bust" scenario to take over, with the possibility of wild swings in the sizes of the gains or losses. But if one cannot play a game having a large variance multiple times, then it doesn't make much difference if one has a slight positive favorable expectation. There is another story, probably again apocryphal, of a man with a suitcase of money who for whatever reason needed twice this amount or it really didn't matter if he lost it all. He goes into a casino and bets it all at once at a roulette table—on red. He either gets twice his money on this one play or loses it all; in the latter case as we noted, maybe it doesn't matter; for example, because he previously borrowed money, the mob will place a "hit" on him if he can't come up with twice the amount that he had to begin with. Or recently, consider the hugely successful negative bets that Goldman Sachs and related traders (such as John Paulson) made on the toxic derivatives they had themselves created (in the jargon, they held a "short position" where one expects the price to fall and to thereby make money in the process).

A quotation from the author of the 1995 novel *Casino*, Nicholas Pileggi, states the issue well for casinos and the usual games of chance where skill is irrelevant (for example, roulette, slots, craps, keno, lotto, or blackjack [without card counting]); all are unfair and in the house's favor:

> A casino is a mathematics palace set up to separate players from their money. Every bet in a casino has been calibrated within a fraction of its life to maximize profit while still giving the players the illusion they have a chance. (p. 5)

The negative expectations may not be big in any absolute sense, but given the enormous number of plays made, and the convergent effects of the law of large numbers, casinos don't lose money, period. The next time an acquaintance brags about what a killing he or she made in the casino on a game involving no skill, you can just comment that the game must not have been played long enough.[6]

4.3.1 Spread Betting

The type of wagering that occurs in roulette or craps is often referred to as fixed-odds betting; you know your chances of winning when you place your bet. A different type of wager is spread betting, invented by a mathematics teacher from Connecticut, Charles McNeil, who became a Chicago bookmaker in the 1940s. Here, a payoff is based on the wager's accuracy; it is no longer a simple "win or lose" situation. Generally, a spread is a range of outcomes, and the bet itself is on whether the outcome will be above or below the spread. In common sports betting (for example, NCAA college basketball), a "point spread" for some contest is typically advertised by a bookmaker. If the gambler chooses to bet on the "underdog," he is said to "take the points" and will win if the underdog's score plus the point spread is greater than that of the favored team; conversely, if the gambler bets on the favorite, he "gives the points" and wins only if the favorite's score minus the point spread is greater than the underdog's score. In general, the announcement of a point spread is an attempt to even out the market for the bookmaker, and to generate an equal amount of money bet on each side. The commission that a bookmaker charges will ensure a livelihood, and thus, the bookmaker can be unconcerned about the actual outcome.

Several of the more notorious sports scandals in United States history have involved a practice of "point shaving," where the perpetrators of such a scheme try to prevent a favored team from "covering" a published point spread. This usually involves a sports gambler and one or more players on the favored team. They are compensated when their team fails to "cover the spread"; and those individuals who have bet on the underdog, win. Two famous examples of this practice in college basketball are the Boston College point shaving scandal of 1978/9, engineered by the gangsters Henry Hill and Jimmy Burke, and the CCNY scandal of 1950/1 involving organized crime and 33 players from some seven schools (CCNY, Manhattan College, NYU, Long Island University, Bradley University (Peoria), University of Kentucky, and the University of Toledo). More recently, there is the related 2007 NBA betting scandal surrounding a referee, Tim Donaghy.[7]

In an attempt to identify widespread corruption in college basketball, Justin Wolfers investigated the apparent tendency for favored NCAA teams nationally not to "cover the spread." His article in the *American Economic Review* (2006, *96*, 279–283) is provocatively entitled "Point Shaving: Corruption in NCAA Basketball." We quote the discussion section of this article to give a sense of what Wolfers claims he found in the data:

> These data suggest that point shaving *may* be quite widespread, with an indicative, albeit rough, estimate suggesting that around 6 percent of strong favorites have been willing to manipulate their performance. Given that around one-fifth of all games involve a team favored to win by at least 12 points, this suggests that around 1 percent of all games (or nearly 500 games through my 16-year sample) involve gambling related corruption. This estimate derives from analyzing the extent to which observed patterns in the data are con-

sistent with the incentives for corruption derived from spread betting; other forms of manipulation may not leave this particular set
of footprints in the data, and so this is a lower bound estimate of the
extent of corruption. Equally, the economic model suggests a range of
other testable implications, which are the focus of ongoing research
(Wolfers, 2006).

My estimate of rates of corruption receives some rough corrobo-
ration in anonymous self-reports. Eight of 388 Men's Division I basketball players surveyed by the NCAA reported either having taken
money for playing poorly or having knowledge of teammates who
had done so.

A shortcoming of the economic approach to identifying corruption is that it relies on recognizing systematic patterns emerging
over large samples, making it difficult to pinpoint specific culprits.
Indeed, while the discussion so far has proceeded as if point shaving
reflected a conspiracy between players and gamblers, these results
might equally reflect selective manipulation by coaches of playing
time for star players. Further, there need not be any shadowy gamblers offering bribes, as the players can presumably place bets themselves, rendering a coconspirator an unnecessary added expense.

The advantage of the economic approach is that it yields a clear
understanding of the incentives driving corrupt behavior, allowing
policy conclusions that extend beyond the usual platitudes that "increased education, prevention, and awareness programs" are required
(NCAA, 2004, p. 5). The key incentive driving point shaving is that
bet pay-offs are discontinuous at a point—the spread—that is (or
should be) essentially irrelevant to the players. Were gamblers restricted to bets for which the pay-off was a linear function of the
winning margin, their incentive to offer bribes would be sharply reduced. Similarly, restricting wagers to betting on which team wins
the game sharply reduces the incentive of basketball players to accept any such bribes. This conclusion largely repeats a finding that is
now quite well understood in the labor literature and extends across
a range of contexts—that highly nonlinear pay-off structures can
yield rather perverse incentives and, hence, undesirable behaviors.
(p. 283)

Another more recent article on this same topic is by Dan Bernhardt and Steven Heston (*Economic Inquiry*, 2010, *48*, 14–25) entitled "Point Shaving in College Basketball: A Cautionary Tale for
Forensic Economics." As this title might suggest, an alarmist position about the rampant corruption present in NCAA basketball
is not justified. An alternative explanation for the manifest "point
shaving" is the use of strategic end-game efforts by a basketball

team trying to maximize its probability of winning (for example, when a favored team is ahead late in the game, the play may move from a pure scoring emphasis to one that looks to "wind down the clock"). The first paragraph of the conclusion section of the Bernhardt and Heston article follows:

> Economists must often resort to indirect methods and inference to uncover the level of illegal activity in the economy. Methodologically, our article highlights the care with which one must design indirect methods in order to distinguish legal from illegal behavior. We first show how a widely reported interpretation of the patterns in winning margins in college basketball can lead a researcher to conclude erroneously that there is an epidemic of gambling-related corruption. We uncover decisive evidence that this conclusion is misplaced and that the patterns in winning margins are driven by factors intrinsic to the game of basketball itself. (p. 24)

The use of spreads in betting has moved somewhat dramatically to the world financial markets, particularly in the United Kingdom. We suggest the reader view an article from the *Times (London)* (April 10, 2009) by David Budworth entitled "Spread-Betting Fails Investors in Trouble." Even though it emphasizes what is occurring in the United Kingdom, it still provides a cautionary tale for the United States as well. The moral might be that just because someone can create something to bet on (think CDOs [Collateralized Debt Obligations] and Goldman Sachs) doesn't mean that it is necessarily a good idea to do so.

4.3.2   Parimutuel Betting

The term *parimutuel betting* (based on the French for "mutual betting") characterizes the type of wagering system used in horse racing, dog tracks, jai alai, and similar contests where the participants end up in a rank order. It was devised in 1867 by Joseph Oller, a Catalan impresario (he was also a bookmaker and founder of the Paris Moulin Rouge in 1889; see "Joseph Oller," 2010; "Parimutel betting," 2010). Very simply, all bets of a particular type are first pooled together; the house then takes its commission and the taxes it has to pay from this aggregate; finally, the payoff odds are calculated by sharing the residual pool among the winning bets. To explain using some notation, suppose there are T

contestants and bets are made of $W_1, W_2, \ldots, W_T$ on an outright "win." The total pool is $T_{pool} = \sum_{t=1}^{T} W_t$. If the commission and tax rate is a proportion, R, the residual pool, R_{pool}, to be allocated among the winning bettors is $R_{pool} = T_{pool}(1 - R)$. If the winner is denoted by $t*$, and the money bet on the winner is W_{t*}, the payoff per dollar for a successful bet is R_{pool}/W_{t*}. We refer to the odds on outcome $t*$ as

$$(\frac{R_{pool}}{W_{t*}} - 1) \text{ to } 1 .$$

For example, if $\frac{R_{pool}}{W_{t*}}$ had a value of 9.0, the odds would be 8 to 1: you get 8 dollars back for every dollar bet plus the original dollar.

Because of the extensive calculations involved in a parimutuel system, a specialized mechanical calculating machine, named a totalizator, was invented by the mechanical engineer George Julius, and first installed at Ellerslie Race Track in New Zealand in 1913 (see "George Julius," 2010; "Tote board," 2010). In the 1930s, totalizators were installed at many of the race tracks in the United States (for example, Hialeah Park in Florida and Arlington Race Track and Sportsman's Park in Illinois). All totalizators came with "tote" boards giving the running payoffs for each horse based on the money bet up to a given time. After the pools for the various categories of bets were closed, the final payoffs (and odds) were then determined for all winning bets.

In comparison with casino gambling, parimutuel betting pits one gambler against other gamblers, and not against the house. Also, the odds are not fixed but calculated only after the betting pools have closed (thus, odds cannot be turned into real probabilities legitimately; they are empirically generated based on the amounts of money bet). A skilled horse player (or "handicapper") can make a steady income, particularly in the newer Internet "rebate" shops that return to the bettor some percentage of every bet made. Because of lower overhead, these latter Internet gaming concerns can reduce their "take" considerably (from, say, 15% to 2%), making a good handicapper an even better living than before.

4.3.3 Psychological Considerations in Gambling

As shown in the work of Tversky and Kahneman (for example, Tversky & Kahneman, 1981), the psychology of choice is dictated to a great extent by the framing of a decision problem; that is, the context into which a particular decision problem is placed. The power of framing in how decision situations are assessed, can be illustrated well though an example and the associated discussion provided by Tversky and Kahneman (1981):

Problem 1 [$N = 152$]: Imagine that the United States is preparing for the outbreak of an unusual Asian disease, which is expected to kill 600 people. Two alternative programs to combat the disease have been proposed. Assume that the exact scientific estimate of the consequences of the programs are as follows:
If Program A is adopted, 200 people will be saved. [72 percent]
If Program B is adopted, there is 1/3 probability that 600 people will be saved, and 2/3 probability that no people will be saved. [28 percent]
Which of the two programs would you favor?
The majority choice in this problem is risk averse: the prospect of certainly saving 200 lives is more attractive than a risky prospect of equal expected value, that is, a one-in-three chance of saving 600 lives.

A second group of respondents was given the cover story of problem 1 with a different formulation of the alternative programs, as follows:
Problem 2 [$N = 155$]:
If Program C is adopted, 400 people will die. [22 percent]
If Program D is adopted, there is 1/3 probability that nobody will die, and 2/3 probability that 600 people will die. [78 percent]
Which of the two programs would you favor?
The majority choice in problem 2 is risk taking: the certain death of 400 people is less acceptable than the two-in-three chance that 600 will die. The preferences in problems 1 and 2 illustrate a common pattern: choices involving gains are often risk averse and choices involving losses are often risk taking. However, it is easy to see that the two problems are effectively identical. The only difference between them is that the outcomes are described in problem 1 by the number of lives saved and in problem 2 by the number of lives lost. The change is accompanied by a pronounced shift from risk aversion to risk taking. (p. 453)

The effects of framing can be very subtle when certain conscious or unconscious (coded) words are used to provide a salient context that influences decision processes. A recent demonstration of this in the framework of our ongoing climate-change debate is given by Hardisty, Johnson, and Weber (2010) in *Psychological Science*. The article has the interesting title, "A Dirty Word or a Dirty World? Attribute Framing, Political Affiliation, and Query Theory." The abstract follows:

We explored the effect of attribute framing on choice, labeling charges for environmental costs as either an earmarked tax or an offset. Eight hundred ninety-eight Americans chose between otherwise identical products or services, where one option included a surcharge for emitted carbon dioxide. The cost framing changed preferences for self-identified Republicans and Independents, but did not affect Democrats' preferences. We explain this interaction by means of query theory and show that attribute framing can change the order in which internal queries supporting one or another option are posed. The effect of attribute labeling on query order is shown to depend on the representations of either taxes or offsets held by people with different political affiliations. (p. 86)

Besides emphasizing the importance of framing in making decisions, Tversky and Kahneman developed a theory of decision making, called prospect theory, to model peoples' real-life choices, which are not necessarily the optimal ones (Kahneman & Tversky, 1979). Prospect theory describes decisions between risky alternatives with uncertain outcomes when the probabilities are generally known. One particular phenomenon discussed at length in prospect theory is loss aversion, or the tendency to strongly avoid loss as opposed to acquiring gains. In turn, loss aversion leads to risk aversion, or the reluctance of people to choose gambles with an uncertain payoff rather than others with a more certain but possibly lower expected payoff. For example, an investor who is risk averse might choose to put money into a fixed-interest bank account or a certificate of deposit rather than into some stock with the potential of high returns but also with a chance of becoming worthless.

The notion of risk aversion has been around since antiquity. Consider the legend of Scylla and Charybdis, two sea monsters of Greek mythology situated on opposite sides of the Strait of Messina in Italy, between Calabria and Sicily. They were placed close enough to each other that they posed an inescapable threat to passing ships, so avoiding Scylla meant passing too close to Charybdis and conversely. In Homer's *Odyssey*, Odysseus is advised by Circe to follow the risk-adverse strategy of sailing closer to Scylla and losing a few men rather than sailing closer to the whirlpools created by Charybdis that could sink his ship. Odysseus sailed successfully past Scylla and Charybdis, losing six sailors to Scylla —

> they writhed
> gasping as Scylla swung them up her cliff and there
> at her cavern's mouth she bolted them down raw —
> screaming out, flinging their arms toward me,
> lost in that mortal struggle. (Homer, Trans. 1996, p. 279)

The phrase of being "between a rock and a hard place" is a more modern version of being "between Scylla and Charybdis."

The most relevant aspect of any decision-making proposition involving risky alternatives is the information one has, both on the probabilities that might be associated with the gambles and what the payoffs might be. In the 1987 movie, *Wall Street*, the character playing Gordon Gekko states: "The most valuable commodity I know of is information." The value that information has is reflected in a great many ways: by laws against "insider trading" (think Martha Stewart); the mandatory injury reports and the not-likely-to-play announcements by the sports leagues before games are played; the importance of counting cards in blackjack to obtain some idea of the number of high cards remaining in the deck (and to make blackjack an unfair game in your favor); massive speed trading on Wall Street designed to obtain a slight edge in terms of what the market is doing currently (and to thereby "beat out" one's competitors with this questionably obtained edge); the importance of correct assessments by the credit rating agencies (think of all the triple-A assessments for the Goldman Sachs toxic collateralized debt obligations and what that meant to the buyers of these synthetic financial instruments); and finally, in the case against Goldman Sachs, the bank supposedly knew about the toxicity of what it sold to their clients and then made a huge profit betting against what they sold (the proverbial "short position"). The movie quotation given from *Dirty Harry* that began this section illustrates the crucial importance of who has information and who doesn't. At the end of Callahan's statement to the bank robber as to whether he felt lucky, the bank robber says: "I gots to know!" Harry puts the .44 Magnum to the robber's head and pulls the trigger; Harry knew that he had fired six shots and not five.

The availability of good information is critical in all the decisions we make under uncertainty and risk, both financially and in terms of our health. When buying insurance, for example, we knowingly engage in loss-adverse behavior. The information we have on the possible downside of not having insurance usually outweighs any consideration that insurance companies have an unfair game going in their favor. When deciding to take new drugs or undergo various medical procedures, information is again crucial in weighing risks and possible benefits—ask your doctor if he or she has some information that is right for you—and coming to a decision that is "best" for us (consider, for example, the previous discussion in Section 3.3 about undergoing screenings for various kinds of cancer).

At the same time that we value good information, it is important to recognize when available "information" really isn't of much value and might actually be counterproductive, for example, when we act because of what is most likely just randomness or "noise" in a system. An article by Jeff Sommer in the *New York Times* (March 13, 2010) has the intriguing title, "How Men's Overconfidence Hurts Them as Investors." Apparently, men are generally more prone to act (trade) on short-term financial news that is often only meaningless "noise." Men are also more confident in their abilities to make good decisions, and are more likely to make many more high-risk gambles.[8]

For many decades, the financial markets have relied on rating agencies, such as Moody's, Standard & Poor's, and Fitch, to provide impeccable information to guide wise investing, and for assessing realistically the risk being incurred. We are now learning that we can no longer be secure in the data the rating agencies produce. Because rating agencies have made public the computer programs and algorithms they use, banks have learned how to "reverse-engineer" the process to see how the top ratings might be obtained (or better, scammed). In the Goldman Sachs case, for example, the firm profited from the misery it helped create through the inappropriate high ratings given to its toxic CDOs. As Carl Levin noted as Chair of the Senate Permanent Subcommittee on Investigations: "A conveyor belt of high-risk securities, backed by toxic mortgages, got AAA ratings that turned out not to be worth the paper they were printed on" (Andringa, April 22, 2010). The rating agencies have been in the position of the "fox guarding the hen house." The reader is referred to an informative editorial that appeared in the *New York Times* ("What About the Raters?", May 1, 2010) dealing with rating agencies and the information they provide.

By itself, the notion of "insurance" is psychologically interesting; the person buying insurance is willingly giving away a specific amount of money to avoid a more catastrophic event that might happen even though the probability of it occurring might be very small. Thus, we have a bookie "laying off" bets made with him or her to some third party; a blackjack player buying insurance on the dealer having a "blackjack" when the dealer has an ace showing (it is generally a bad idea for a player to buy insurance); or individuals purchasing catastrophic health insurance but paying the smaller day-to-day medical costs themselves. Competing forces are always at work between the insurer and the insured. The insurer wishes his "pool" to be as large as possible (so the central limit theorem can operate), and relatively "safe"; thus, the push to exclude high-risk individuals is the norm, and insuring someone with pre-existing conditions is always problematic. The insured, on the other hand, wants to give away the least money to buy the wanted protection. As one final item to keep in mind, we should remember that insurance needs to be purchased before and not after the catastrophic event occurs. In late 2010, there was a national cable news story about the person whose house burned down as the county firetrucks stood by. The person felt very put upon and did not understand why they just let his house burn down; he had offered to pay the $75 fire protection fee (but only after the house stated to burn). The cable news agencies declared a "duty to rescue," and the failure of the fire trucks to act was "manifestly immoral." Well, we doubt it because no life was lost, only the property, and all because of a failure to pay the small insurance premium "up front." For a discussion of this incident, see the article by Robert Mackey, "Tennessee Firefighters Watch Home Burn" (*New York Times*, October 6, 2010)

A second aspect of insurance purchase with psychological interest is how to estimate the probability of some catastrophic event. Insurers commonly have a database giving an estimated value over those individuals they may consider insuring. This is where the actuaries and statisticians make their worth known; how much should the insurance companies charge for a policy so the company would continue to make money. The person to be insured has no easy access to any comparable database and merely guesses a value or more usually, acts on some vague "gut feeling" as to what one should be willing to pay to avoid the catastrophic downside. The person being insured has no personal relative frequency estimate on which to rely.

Assessing risks when no database is available to an insuring body is more problematic. If every one were honest about these situations, it might be labeled as subjectively obtained, or more straightforwardly, a "guess." This may be "gussied up" slightly with the phrase "engineering judgment," but at its basis it is still a guess. Richard Feynman, in his role on the Rogers Commission investigating the Challenger accident of 1986, commented that "engineering judgment" was making up numbers according to the hallowed tradition of the "dry lab." Here, one makes up data as opposed to observation and experimentation. You work backwards to the beginning from the result you want to obtain at the end. For shuttle risk, the management started with a level of risk that was acceptable and worked backwards until they got the probability estimate that gave this final "acceptable" risk level.[9]

Notes

[1]A popular summary of this more comprehensive article appears in *Scientific American*, *283*(4), 82–87: "Better Decisions through Science" (B.1.2).

[2]A tort is a civil wrong; tort law concerns situations where a person's behavior has harmed someone else.

[3]This is the same Bentham known for utilitarianism, and more amusingly, for the "auto-icon." A short section from the Wikipedia article on "Jeremy Bentham" (2010) describes the auto-icon:

> As requested in his will, Bentham's body was dissected as part of a public anatomy lecture. Afterward, the skeleton and head were preserved and stored in a wooden cabinet call the "Auto-icon," with the skeleton stuffed out with hay and dressed in Bentham's clothes. Originally kept by his disciple Thomas Southwood Smith, it was

acquired by University College London in 1850. It is normally kept on public display at the end of the South Cloisters in the main building of the college, but for the 100th and 150th anniversaries of the college, it was brought to the meeting of the College Council, where it was listed as "present but not voting."

The Auto-icon has a wax head, as Bentham's head was badly damaged in the preservation process. The real head was displayed in the same case for many years, but became the target of repeated student pranks, including being stolen on more than one occasion. It is now locked away securely.

[4]In the Suggested Reading of Section B.5 (Agent Orange and Judge Weinstein), we list seven items. Five deal with the path that the Agent Orange litigation took in Judge Weinstein's court since the early 1980s:

B.5 – Agent Orange and Judge Weinstein

B.5.1 – "Vietnam Agent Orange Suit By Veterans Is Going to Trial" (Ralph Blumenthal, *New York Times*, May 6, 1984)

B.5.2 – "Lack of Military Data Halts Agent Orange Study" (Philip M. Boffey, *New York Times*, September 1, 1987)

B.5.3 – "Agent Orange, the Next Generation; In Vietnam and America, Some See a Wrong Still Not Righted" (William Glaberson, *New York Times*, August 8, 2004)

B.5.4 – "Agent Orange Lawsuits Dismissed" (William Glaberson, *New York Times*, November 19, 2004)

B.5.5 – "Civil Lawsuit on Defoliant in Vietnam Is Dismissed" (William Glaberson, *New York Times*, March 11, 2005)

B.5.6 – "High Caliber Justice" (Robert Kolker, *New York Magazine*, April 5, 1999)

B.5.7 – "Defiant Judge Takes on Child Pornography Law" (A. G. Sulzberger, *New York Times*, May 21, 2010)

The penultimate article listed as B.5.6 is a profile of Judge Weinstein. The final item (B.5.7) regards a recent kerfuffle that Judge Weinstein raised with respect to mandatory sentences for the possession of child pornography. The issue concerns what a judge can do when disagreeing with an imposed mapping (created by legislative action) between "the crime" and "the amount of time to serve." In the past, Judge Weinstein has refused to preside over many drug trials when he felt his hands would be tied inappropriately regarding final sentencing.

[5]For some further background on the Chevalier and the history of probability, see "Antoine Gombaud," 2010; "Problem of points," 2010; Maxwell, 1999.

[6]We give two short anecdotes that may be helpful in motivating the material in this section:

Charles Marie de La Condamine (1701–1774) is best known for answering the question as to whether the earth was flat or round. He based his answer (which was "round") on extensive measurements taken at the equator in Ecuador and in Lapland. For our purposes, however, he will be best known for giving the French philosopher Voltaire a gambling tip that allowed him to win 500,000 francs in a lottery. Condamine noted to Voltaire that through a miscalculation, the sum of all the ticket prices for the lottery was far less than the prize. Voltaire bought all the tickets and won (Wallechinsky & Wallace, 1981, pp. 393–394).

Joseph Jagger (1830–1892) is known as "the man who broke the bank at Monte Carlo." In reality, he was a British engineer working in the Yorkshire cotton manufacturing industry, and very knowledgeable about spindles that were "untrue." Jagger speculated that a roulette wheel did not necessarily "turn true," and the outcomes not purely random but biased toward particular outcomes. We quote a brief part of the Wikipedia entry on Joseph Jagger that tells the story:

> Jagger was born in September 1829 in the village of Shelf near Halifax, Yorkshire. Jagger gained his practical experience of mechanics working in Yorkshire's cotton manufacturing industry. He extended his experience to the behaviour of a roulette wheel, speculating that its outcomes were not purely random sequences but that mechanical imbalances might result in biases toward particular outcomes.
>
> In 1873, Jagger hired six clerks to clandestinely record the outcomes of the six roulette wheels at the Beaux-Arts Casino at Monte Carlo, Monaco. He discovered that one of the six wheels showed a clear bias, in that nine of the numbers (7, 8, 9, 17, 18, 19, 22, 28 and 29) occurred more frequently than the others. He therefore placed his first bets on 7 July 1875 and quickly won a considerable amount of money, £14,000 (equivalent to around 50 times that amount in 2005, or £700,000, adjusted for inflation). Over the next three days, Jagger amassed £60,000 in earnings with other gamblers in tow emulating his bets. In response, the casino rearranged the wheels, which threw Jagger into confusion. After a losing streak, Jagger finally recalled that a scratch he noted on the biased wheel wasn't present. Looking for this telltale mark, Jagger was able to locate his preferred wheel and resumed winning. Counterattacking again, the casino moved the frets, metal dividers between numbers, around daily. Over the next two days Jagger lost and gave up, but he took his remaining earnings, two million francs, then about £65,000 (around £3,250,000 in 2005), and left Monte Carlo never to return.

[7]For further background, see "Boston College basketball point shaving scandal of 1978-79," 2010; "CCNY point shaving scandal," 2010; "Point shaving," 2010; "Spread betting," 2010.

[8]This article is listed in the Suggested Reading, B.7.3.

[9]The exact Feynman quote is "As far as I can tell, 'engineering judgment' means they're just going to make up numbers" (Feynman, 1988, p. 183, footnote).

Chapter 5

Correlation

> Some people hate the very name statistics, but I find them full of beauty and interest. Whenever they are not brutalized, but delicately handled by the higher methods, and are warily interpreted, their power of dealing with complicated phenomena is extraordinary.
> – Sir Francis Galton (*Natural Inheritance*, 1889)

The association between two variables measured on the same set of objects is commonly referred to as their correlation and often measured by the Pearson product moment correlation coefficient. Specifically, suppose $Z_{X_1}, \ldots, Z_{X_N}$ and $Z_{Y_1}, \ldots, Z_{Y_N}$ refer to z-scores (that is, having mean zero and variance one) calculated for our original observational pairs, (X_i, Y_i), $i = 1, \ldots, N$; then the correlation between the original variables, r_{XY}, is defined as

$$r_{XY} = (\frac{1}{N}) \sum_{i=1}^{N} Z_{X_i} Z_{Y_i} \, ,$$

or the average product of the z-scores. As usually pointed out early in any statistics course, r_{XY} measures the linearity of any relation that might be present; thus, if some other (nonlinear) form of association exists, different means of assessing the latter are needed.

In any reasoning based on the presence or absence of a correlation between two variables, it is imperative that graphical mechanisms be used in the form of scatterplots. One might go so far to say that if only the value of r_{XY} is provided and nothing else, we have a *primae facie* case for statistical malpractice. Scatterplots are of major assistance in a number of ways: (1) to ascertain the degree to which linearity might be the type of association present between the variables; this assessment could take the form of directly imposing various scatterplot smoothers and using these to help characterize the association present, if any; (2) to identify outliers or data points that for whatever reason are not reflective of the general pattern exhibited in the scatterplot, and to help figure out why; (3) to provide a graphical context for assessing the influence of a data point on a correlation, possibly by the size and/or color of a plotting symbol, or contour lines indicating the change in value for the correlation that would result if it were to be removed.

One of the more shopworn adages we hear in any methodology course is that "correlation does not imply causation." It is usually noted that other "lurking" or third variables might affect both X and Y, producing a spurious association; also, because r_{XY} is a symmetric measure of association, there is no clue in its value as to the directionality of any causal relationship. For example, we have had some recent revisions in our popular views on the positive effects of moderate drinking; it may be that individuals who otherwise lead healthy lifestyles also drink moderately. In a football sports context, "running the ball" does not cause winning; it is more likely that winning causes "running the ball." Teams that get an early lead try to run the ball frequently because it keeps the clock running and decreases the time for an opponent to catch up. Or in politics, it is commonly assumed that the more money one raises, the greater the chances are for election; however, it may be the other way around where greater electability leads to more donations.

The adage of "correlation does not imply causation" has several Latin variants: we have the fallacy of *post hoc, ergo propter hoc* (after this, therefore because of this); or alternatively, *cum hoc, ergo propter hoc* (with this, therefore because of this). Just because something may occur after an event does not imply that it is therefore causative (or, in other [logical] words introduced earlier in Section 3.2, temporality is a necessary but not sufficient condition to invoke causality). Although the fallacy may be obvious when stated, a number of us seem to fall prey to it rather often: when cancer survivors attribute their good outcomes to previous positive mental attitudes; mothers who blame their child's autism on earlier receiving the mumps-measles-rubella vaccine; people in a clinical trial who are actually receiving the "placebo," getting miraculously better; or observing that the more money spent on medical care, the sicker one becomes, so obviously, doctors should be avoided to improve one's health.

In any multiple variable context, it is possible to derive the algebraic restrictions present among any subset of the variables based on the correlations among all the variables. The simplest case involves three variables, say X, Y, and W. From the basic formula for the partial correlation between X and Y "holding W constant," an *algebraic* restriction is present on r_{XY} given the values of r_{XW} and r_{YW}:

$$r_{XW}r_{YW} - \sqrt{(1 - r_{XW}^2)(1 - r_{YW}^2)} \leq$$

$$r_{XY} \leq r_{XW}r_{YW} + \sqrt{(1 - r_{XW}^2)(1 - r_{YW}^2)}.$$

Note that this is not a probabilistic statement (that is, it is not a confidence interval); it says that no dataset exists where the correlation r_{XY} lies outside of the upper and lower bounds provided by $r_{XW}r_{YW} \pm \sqrt{(1 - r_{XW}^2)(1 - r_{YW}^2)}$. As a numerical example, suppose X and Y refer to height and weight, respectively, and W is a measure of age. If, say, the correlations, r_{XW} and r_{YW} are both .8, then $.28 \leq r_{XY} \leq 1.00$. In fact, if a high correlation value of .64 were observed for r_{XY}, should we be impressed by the magnitude of the association between X and Y? Probably not; if the partial correlation between X and Y "holding W constant" were computed with $r_{XY} = .64$, a value of zero would be obtained. All of the observed high association between X and Y can be attributed to their association with the developmentally related variable. Conversely, if X and Y are both uncorrelated with W (so $r_{XW} = r_{YW} = 0$), then no restrictions are placed on r_{XY}; the algebraic inequalities reduce to a triviality: $-1 \leq r_{XY} \leq +1$. These very general restrictions on correlations have been known for a long time and appear, for example, in Yule's first edition (1911) of *An Introduction to the Theory of Statistics* under the title, "Conditions of Consistence Among Correlation Coefficients." Also, see the chapter, "Fallacies in the Interpretation of Correlation Coefficients," in this same volume.

A related type of algebraic restriction for a correlation is present when the distribution of the values taken on by the variables include ties. In the extreme, consider a 2×2 contingency table, and the fourfold point correlation; this is constructed by using a 0/1

coding of the category information on the two attributes and calculating the usual Pearson correlation. Because of the nonuniform
marginal frequencies present in the 2×2 table, the fourfold correlation cannot extend over the complete ± 1 range. The achievable
bounds possible can be computed (Carroll, 1961); and it therefore
may be of some interest descriptively to see how far an observed
fourfold correlation is away from its achievable bounds, and possibly, even to normalize the observed value by such a bound.[1]

The bounds of ± 1 on a Pearson correlation can be achieved only
by datasets demonstrating a perfect linear relationship between
the two variables. Another measure that achieves the bounds of
± 1 whenever the datasets have merely consistent rank orderings
is Guttman's (weak) monotonicity coefficient, μ_2:

$$\mu_2 = \frac{\sum_{i=1}^{n} \sum_{h=1}^{n} (x_h - x_i)(y_h - y_i)}{\sum_{i=1}^{n} \sum_{h=1}^{n} |x_h - x_i||y_h - y_i|} \, ,$$

where (x_h, y_h) denote the pairs of values being "correlated" by μ_2.
The coefficient, μ_2, expresses the extent to which values on one
variable increase in a particular direction as the values on another
variable increases, without assuming that the increase is exactly
according to a straight line. It varies between -1 and $+1$, with
$+1$ $[-1]$ reflecting a perfect monotonic trend in a positive [nega-
tive] direction. The adjective "weak" refers to the untying of one
variable without penalty. In contrast to the Pearson correlation,
μ_2 can equal $+1$ or -1, even though the marginal distributions
of the two variables differ from one another. When the Pearson
correlation is $+1.00$ or -1.00, μ_2 will have the same value; in all
other cases, the absolute value of μ_2 will be higher than that of the
Pearson correlation including the case of a fourfold point correlation. Here, μ_2 reduces to Yule's Q (which is a special case of the
Goodman–Kruskal gamma statistic for a 2×2 contingency table
[a measure of rank-order consistency] that we come back to in the
subsection on measures of nonlinear association).[2]

Several other correlational pitfalls seem to occur in various
forms whenever we try to reason through datasets involving multiple variables. We briefly mention six of these areas in the sections
to follow.

5.1 Illusory Correlation

> There is a strong correlation between belief in evolution and liberal views on government control, pornography, prayer in schools, abortion, gun control, economic freedom, and even animal rights.
> — Phyllis Schlafly

An illusory correlation is present whenever a relation is seen in data where none exists. Common examples would be between membership in some minority group and rare and typically negative behavior; the endurance of stereotypes and an overestimation of the link between group membership and certain traits; or the connection between a couple adopting a child and the subsequent birth of their own. Illusory correlations seem to depend on the novelty or uniqueness of the variables considered. Four decades ago, Chapman and Chapman (1967, 1969) studied such false associations in relation to psychodiagnostic signs seen in projective tests. For example, in the "Draw-A-Person" test, a client draws a person on a blank piece of paper ("Draw-A-Person," 2010). Although some psychologists believe that drawing a person with big eyes is a sign of paranoia, such a correlation is illusory but very persistent. When data that are deliberately uncorrelated are presented to college students, the same diagnostic signs are found that some psychologists still believe in. This notion of illusory correlation has been around since the early 1900s; see, for example, Yule's first edition (1911) of *An Introduction to the Theory of Statistics*, and the chapter entitled "Illusory Associations."[3]

Several faulty reasoning relatives exist for the notion of an illusory correlation. One is *confirmation bias*, where there are tendencies to search for, interpret, and remember information only in a way that confirms one's preconceptions or working hypotheses ("Confirmation bias," 2010). Confirmation bias is related to the "I'm not stupid" fallacy that rests on the belief that if one is mistaken, one must therefore be stupid, and we generally believe that we are not stupid; witness the prosecutor who refuses to drop charges against an obviously innocent suspect because otherwise, he or she would need to admit error and wasted effort (Tavris & Aronson, 2007, pp. 229–235). At an extreme, there is the trap of *apophènia*, or seeing patterns or connections in random or meaningless data ("Apophenia," 2010). A subnotion is *pareidolia*, where vague and random stimuli, often images or sounds, are perceived as significant, for example, the Virgin Mary is seen on a grilled cheese sandwich ("Pareidolia," 2010).[4] One particular problematic realization of apophenia is in epidemiology when residential cancer clusters are identified that rarely if ever result in identifiable causes. What seems to be occurring is sometimes labeled the Texas sharpshooter fallacy, where a Texas sharpshooter fires at the side of a barn and then draws a bullseye around the largest cluster of bullet holes ("Texas sharpshooter fallacy," 2010). In identifying residential cancer clusters, we tend to notice multiple cancer patients on the same street and then define the population base around these. A particularly well-presented popular article on these illusory associations entitled "The Cancer-Cluster Myth," is by Atul Gawande in the February 8th 1999, *New Yorker*.[5]

Illusory relations occur commonly in our day-to-day interactions with others. We have the *clinician's fallacy* due to the self-selected and biased sample of individuals whom a clinician actually sees in practice ("Natural history of disease," 2010); thus, we have the (incorrect) inference of a uniformity of serious adult trauma for any instance of childhood sexual abuse (see McNally, 2003, *Remembering Trauma*, for a comprehensive discussion). As a slight variant, the *clinician's illusion* is the overestimation of how chronic certain psychological problems may be, just because of selective exposure to a chronic and biased sample (Cohen & Cohen, 1984). Whenever we come in contact with various biasedly sampled groups of individuals in our professions as doctors, lawyers, educators, and so on, it is judicious not to immediately attribute possibly incorrect causal associations to a group's general character and/or demographics (for example, podiatrists who claim that most women wear shoes at least one size too small, or clinical psychologists who assert the prevalence of childhood sexual abuse). Also, there is a need to guard against *detection bias*, where one is sensitized to see or look for what might not actually be present ("Bias (statistics)," 2010). For instance, could an increase in the number of diagnoses of autism in the 1990s be attributed to the priming effect provided by the Dustin Hoffman/Tom Cruise movie, *Rain Man*?[6]

A 2009 book by Steven Feldman has the intriguing title of *Compartments: How the Brightest, Best Trained, and Most Caring People Can Make Judgements That Are Completely and Utterly Wrong*. Feldman's cautions are based on his experiences as a dermatologist, but they still suggest several general principles behind how we all make inferences about what we see. The general argument is that people can look at one objective reality, but depending on the context they bring to it, perceive totally different things. Feldman gives one example involving the loss of efficacy of tropical cortisone treatment that was so common it was given the name of *tachyphylaxis*—"the more you use the medicine, the less it works" (p. 16). The reality of it, however, is that patients just don't take their medication. *Tachyphylaxis* is a myth based on a false assumption.

Besides making incorrect inferences based on things we don't observe (for example, the lack of compliance in taking medication), there are things we do observe that should not be trusted. For example, we must be careful about inferences that depend on what news appears on the air, or on what responses we receive from volunteers in a survey. The representativeness of what is encountered should always be kept in mind. Forms of selection bias appear constantly because what is observed or heard results from its being different than what usually happens. Making inferences based on out-of-the-ordinary events is generally not a good idea.

5.2 Ecological Correlation

An ecological correlation is one calculated between variables that are group averages of some sort; this is in contrast to obtaining a correlation between variables measured at the level of individuals. Several issues are faced immediately with the use of ecological correlations: they tend to be a lot higher than individual-level correlations, and assuming that what is seen at the group level also holds automatically at the level of the individual is so pernicious that it has been labeled the "ecological fallacy" by Selvin (1958). The specific instance developed by Selvin concerns the 19th cen-

tury French sociologist Émile Durkheim and his contention that
suicide was promoted by the social conditions inherent in Protestantism. This individual-level inference is not justified from the
data Durkheim actually had at the aggregated level of country,
which did show a relationship between the levels of Protestantism
and suicide. As is true in interpreting any observational study,
confounding variables may exist; here, it is that that Protestant
countries differ from Catholic countries in many ways other than
religion. Durkheim's data do not link individual-level suicide with
the practice of any particular religious faith; and to do so is to fall
prey to the ecological fallacy.

The term "ecological correlation" was popularized by William
Robinson (1950), although the idea had been around for some
time (for example, see the 1939 article by E. L. Thorndike, "On
the Fallacy of Imputing the Correlations Found for Groups to the
Individuals or Smaller Groups Composing Them"). Robinson computed a correlation of .53 between literacy rate and the proportion
of the population born outside the United States for the 48 states
of the 1930 census. At the individual level, however, the correlation
was −.11, so immigrants were on average less literate than their
native counterparts. The high ecological correlation of .53 was due
to immigrants settling in states with a more literate citizenry.

A recent discussion of ecological correlation in our present political climate is the entertaining article in the *Quarterly Journal
of Political Science* by Gelman et al. (2007), "Rich State, Poor
State, Red State, Blue State: What's the Matter with Connecticut?" An expansion of this article to book form is Gelman et al.
(2010), *Red State, Blue State, Rich State, Poor State: Why Americans Vote the Way They Do.* The data discussed in these sources
seem to suggest that voters act against their own best interests
when viewed at an aggregate state level but not when considered
at an individual level. For the last three presidential elections of
2000, 2004, and 2008, wealthier states with high per capita incomes counterintuitively tended to vote Democratic; poorer states
tended to vote Republican. For example, consider the situation
in 2004: the Republican candidate Bush won the fifteen poorest
states, but the Democrat Kerry won nine of the eleven wealth-
iest states; moreover, 62% of voters having annual incomes over

$200,000 voted for Bush, but only 36% of voters having incomes of $15,000 or less voted for Bush.

Using a relation seen at an aggregate level to infer individual-level behavior is referred to as "ecological inference." It is a risky endeavor that requires the use of models with generally unverifiable assumptions, such as the "constancy" hypothesis below. This will be illustrated by an example paraphrased from David Freedman's 2001 article, "Ecological Inference," appearing in the *International Encyclopedia of the Social & Behavioral Sciences*. Our interest is in the estimation of support for a specific candidate among Hispanic and non-Hispanic voters. For each electoral precinct, the fraction, x, of voters who are Hispanic is known, as is the fraction, y, of voters for the candidate. The problem is to estimate the fraction of Hispanic voters for the candidate, which is unknown because of ballot secrecy. A regression equation is fitted to the data having the form

$$y_i = a + bx_i + \epsilon_i \ ,$$

where x_i is the fraction of Hispanic voters in precinct i, y_i is the vote fraction for the candidate, and ϵ_i is the error term. Least squares estimates of a and b are denoted by $\hat{a}$ and $\hat{b}$. Here, $\hat{a}$ is the height of the regression line at $x = 0$, corresponding to precincts having no Hispanic voters; $\hat{a} + \hat{b}$ is the height of the regression line at $x = 1$, and interpretable as the fraction of Hispanic voters supporting the candidate. (Note that no constraints are imposed on the estimates, $\hat{a}$ and $\hat{b}$, so they could lie outside of the zero-to-one interval that must constrain any voter fraction.)

The strategy just illustrated for estimating the fraction of Hispanic voters supporting the candidate goes back to Leo Goodman and his 1953 article, "Ecological Regressions and Behavior of Individuals" (*American Sociological Review*, *18*, 663–664). In our example, data are available for groups defined by area of residence, but the inference applies to groups defined by ethnicity. Justifying the statistical procedure requires invoking the "constancy assumption"—voting preferences within ethnic groups do not systematically depend on the ethnic composition of the area of residence. Strong conditions such as the constancy assumption are generally unverifiable but must be assumed true. This is not a very satisfying solution to the problem of ecological inference. As Goodman himself notes:

> In summary, we would like to emphasize that *in general* ecological regression methods cannot be used to make inferences about individual behavior. However, *under very special circumstances* the analysis of the regression between ecological variables may be used to make such inferences. (p. 664; italics in the original)

Another major competitor for ecological inference from Duncan and Davis should be mentioned; it was published adjacent to Goodman's original 1953 article, and entitled "An Alternative to Ecological Correlation" (*American Sociological Review, 18,* 665–666). It relies on algebraic bounding strategies to bracket the desired fraction to be estimated. Unfortunately, the bounds are generally not very tight. As an example of how the bounding strategy might work for our Hispanic voter illustration, let p be the unknown fraction of Hispanic voters for the candidate, and q the unknown fraction of non-Hispanics for the candidate. The known fraction of the population that is Hispanic is denoted by R; the known fraction of the population voting for the candidate is S. Thus,

$$Rp + (1 - R)q = S .$$

Knowing that both p and q must lie between the bounds of zero and one leads to the following bounds:

$$\frac{S - (1 - R)}{R} \leq p \leq \frac{S}{R} ;$$

$$\frac{S - R}{1 - R} \leq q \leq \frac{S}{1 - R} .$$

Depending on the values taken on by R and S, constraints may or may not be placed on the possible values for p and q. For example, when $R < S$ and $S + R < 1$, no constraints are placed on p; if $S < R$ and $S + R > 1$, no constraints are placed on q. The one main advantage of this "method of bounds" when they do lead to constraints for p and q, is that no constancy assumption is needed. The main disadvantage, as noted earlier, is that even when constraints are imposed, they may not be very confining.

One of the more recent and controversial approaches to ecological inference is Gary King's *A Solution to the Ecological Inference Problem* (1997; Princeton University Press). This strategy, given the acronym EI for Ecological Inference, is a mixture of the regression approach and the Duncan and Davis boundary strategy. But again, the driving force behind EI must be a version of a constancy assumption. As noted in Chapter 2 (Introduction), models requiring strong assumptions to function correctly are to be eschewed—and we will do so here. Our cautionary notes on ecological inference will be left at that, with no attempt to delve further into a criticism or advocacy of EI.

In addition to the general difficulties of ecological inference (and the fallacy), aggregation causes problems across several disciplines. Aggregation bias in econometrics refers to how aggregation changes the micro-level structural relationships among the economic variables of interest; explicitly, aggregation bias is a deviation of the macro-level parameters from the average of the corresponding micro-level parameters. Or in psychology, the models we fit and evaluate at the group level may be very different from what is operative at the level of the individual subject. A good cautionary article on the topic of aggregation in psychology is Ashby, Maddox, and Lee, "On the Dangers of Averaging Across Subjects When Using Multidimensional Scaling or the Similarity-Choice Model" (*Psychological Science*, 1994, 5, 144–151). A later chapter on meta-analysis characterizes the method as an "empirical bulldozer," where all individual differences are obliterated. What is analyzed are averages at a group level, but these cannot do complete justice to what occurs internally. Zero overall effects could well mask who does great, who does badly, and who just doesn't change at all. In short, the fundamental difficulty with ecological inference is that many different possible relationships at an individual level are capable of generating the same results at an aggregate level. No deterministic solution exists for the ecological inference problem—individual-level information is irretrievably lost by the process of aggregation.

A problem related to ecological correlation is the modifiable areal unit problem, where differences in spatial units used in the aggregation can cause wide variation in the resulting correlations, ranging anywhere from plus to minus one. Generally, the manifest association between variables depends on the size of areal units used, with increases as areal unit size gets larger. A related "zone" effect concerns the variation in correlation caused by reaggregating data into different configurations at the same scale. Obviously, the existence of modifiable areal units has serious implications for our abilities to reason with data: when strong relationships exist between variables at an individual level, these can be obscured through aggregation; conversely, aggregation can lead to an apparently strong association when none is actually present. A thorough discussion of the modifiable areal unit problem appears in Yule and Kendall (1950, pp. 310–313).

5.3   Restriction of Range for Correlations

When a psychological test is used to select personnel based on the achievement of a certain cut-score, an unusual circumstance may occur. The prediction of job performance after selection is typically much poorer than what one might have expected beforehand. In one of the more well-known papers in all of Industrial and Organizational Psychology, Taylor and Russell (1939) offered an explanation of this phenomenon by noting the existence of a restriction of range problem: in a group selected on the basis of some test, the correlation between test and performance must be lower than it would be in an unselected group. Based on the assumption of bivariate normality between job performance and the selection test, Taylor and Russell provided tables and charts for estimating what the correlation would be in an unselected population from the value seen between test and performance in the selected population.

An issue related to the restriction of range in its effect on correlations is the need to deal continually with fallible measurement. Generally, the more unreliable our measures, the lower (or more attenuated) the correlations. The field of psychometrics has for many decades provided a mechanism for assessing the effects of fallible measurement through its "correction for attenuation": the correlation between "true scores" for our measures is the observed

FIGURE 5.1: A graph depicting restriction of range affecting the correlation between undergraduate grade-point average (UGPA) and the scores on the Law School Admission Test (LSAT).

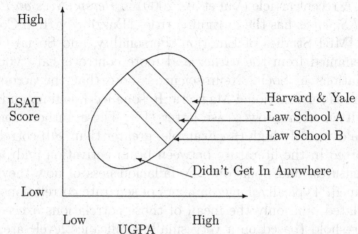

correlation divided by the square roots of their reliabilities. Various ways are available for estimating reliability, so implementing attenuation corrections is an eminently feasible enterprise. Another way of stating this correction is to note that any observed correlation must be bounded above by the square root of the product of the reliabilities. Obviously, if reliabilities are not very good, observed correlations can never be very high.[7]

Another type of range restriction problem appears in Figure 5.1. We observe the empirical fact of a negative correlation between Law School Admission Test (LSAT) scores and undergraduate grade-point average (UGPA) within almost all law schools. Does this mean that the worse you perform in college courses, the better you will do on the LSAT? Well, no; it is because if you performed very well on both, you went to Harvard, and if you performed poorly on both, you didn't get into law school. So at all other law schools, there were admittees who did relatively better on one than on the other. The Figure 5.1 graph of the LSAT scores versus UGPA shows bands running from upper left to lower right representing each law school, with the better schools higher up on both; the overall picture, however, is a very positive data swirl with the lower segment not admitted.

5.4   Odd Correlations

A recent article (Vul et al., 2009) in *Perspectives on Psychological Science*, has the intriguing title, "Puzzlingly High Correlations in fMRI Studies of Emotion, Personality, and Social Cognition" (renamed from the earlier and more controversial "Voodoo Correlations in Social Neuroscience"; note that the acronym fMRI stands for functional Magnetic Resonance Imaging, and is always written with a lower-case letter "f"). These authors comment on the extremely high (for example, greater than .80) correlations reported in the literature between brain activation and personality measures, and point out the fallaciousness of how they were obtained. Typically, huge numbers of separate correlations were calculated, and only the mean of those correlations exceeding some threshold (based on a very small significance level) are reported. It is tautological that these correlations selected for size must then be large in their average value. With no cross-validation attempted to see the shrinkage expected in these measures on new samples, we have sophistry at best. Any of the usual understanding of yardsticks provided by the correlation or its square, the proportion of shared variance, are inappropriate. In fact, as noted by Vul et al. (2009), these inflated mean correlations typically exceed the upper bounds provided by the correction for attenuation based on what the reliabilities should be for the measures being correlated.[8]

When a correlation reported in the literature seems odd, it is incumbent on a literate consumer of such information to understand why.[9] Sometimes it is as simple as noting the bias created by the selection process as in the fMRI correlations, and that such selection is not being mitigated by any cross-validation. Or, possibly, inflated or deflated association measures may occur because of the use of ecological correlations or modifiable areal units, restriction of range, the fallibility of the behavioral measures, the presence of a nonlinear relationship, and so on. The reason behind apparent correlational artifacts can be subtle and require a careful explication of the processes leading to the measures being correlated and on what objects. For instance, if correlations are being monitored over time, and the group on which the correlations are based changes composition, the effects could be dramatic. Such composition changes might be one of different sex ratios, immigrant influxes, economic effects on the available workforce, or age. One particularly unusual example is discussed by Dawes (1975) on the relation between graduate admission variables and future success. Because admission criteria tend to be compensatory (where good values on certain variables can make up for not-so-good values on others), the covariance structure among admissions variables in the selected group is unusual in that it involves negative correlations. As argued nicely by Dawes, it must be the case that the variables used to admit graduate students have low correlation with future measures of success.[10]

5.5 Measures of Nonlinear Association

An overall concern with the use of the simple correlation coefficient is that it measures linearity only, and then through a rather indirect measure of shared variance defined by the coefficient of determination, the squared correlation. Specifically, there is no obvious operational measure of strength of relation defined in terms of the given sample at hand, which in turn could be given a transparent probabilistic meaning with respect to the latter. One century-old suggestion is to use the Spearman correlation, which is equivalent to the Pearson correlation computed on ranks. Although it is true that a perfect monotonic relation between two variables turns into one that is perfectly linear when ranks are used, the strength of such an association measure is now a somewhat unsatisfying shared variance between ranks.

An alternative notion of rank correlation is based on the number of inversions in rank ordering for the two variables, X and Y, taken over all object pairs. Suppose (x_i, y_i) and (x_j, y_j) are the observed measures for two objects, i and j. If $x_i > x_j$ but $y_i < y_j$, we have an "inversion"; when $x_i > x_j$ and $y_i > y_j$, a "non-inversion" exists. A simple measure of rank-order association is the Goodman–Kruskal (G-K) γ (gamma) coefficient obtained over the $N(N-1)/2$ object pairs. The G-K γ coefficient is bounded between plus and minus 1.0, and can be given a convenient and transparent probabilistic meaning with respect to the given sample: if we choose two objects at random, and consider the ordering provided by untied values on X and Y, γ is the probability of a noninversion minus the probability of an inversion. A number of variations on γ can be given depending on how ties are treated and counted, for example, the indices suggested by Kendall and Somers, typically computed by all the commercial statistical software packages, such as SYSTAT, SPSS, or SAS. These variations will commonly retain some type of transparent operational meaning with respect to probability interpretations on the given sample at hand.

The G-K γ measure is appropriate only for a contingency table in which the two cross-classification attributes consist of ordered categories. A more general measure that relates two arbitrary attributes considered to be nominal where both have assumed unordered categories, was also proposed by Goodman and Kruskal and labeled by the Greek letter lambda, λ (the Goodman–Kruskal (G-K) Index of Predictive Association). We define it in terms of an $R \times C$ contingency table having the following form:

	A_1	A_2	$\cdots$	A_C	Row Sums
B_1	N_{11}	N_{12}	$\cdots$	N_{1C}	$N_{1\cdot}$
B_2	N_{21}	N_{22}	$\cdots$	N_{2C}	$N_{2\cdot}$
$\vdots$	$\vdots$	$\vdots$		$\vdots$	$\vdots$
B_R	N_{R1}	N_{R2}	$\cdots$	N_{RC}	$N_{R\cdot}$
Column Sums	$N_{\cdot 1}$	$N_{\cdot 2}$	$\cdots$	$N_{\cdot C}$	$N_{\cdot\cdot} \equiv N$

Suppose a process is initiated where an object is picked from the table and the row event that occurs is noted (that is, $B_1, \ldots, B_R$). Then based on this knowledge of the row event, say B_r, we guess the column event by choosing that column with the highest frequency within row B_r. An error of prediction is made with probability

$$\frac{N_{r\cdot} - \max_{1 \leq c \leq C} N_{rc}}{N_{r\cdot}},$$

and using the rule of total probability, the overall error of prediction is

$$\sum_{r=1}^{R} \left(\frac{N_{r\cdot} - \max_{1 \le c \le C} N_{rc}}{N_{r\cdot}} \right) \left(\frac{N_{r\cdot}}{N_{\cdot\cdot}} \right) = 1 - \frac{\sum_{r=1}^{R} \max_{1 \le c \le C} N_{rc}}{N_{\cdot\cdot}} ,$$

and denoted by $P_{error|row}$ = probability of an error in predicting the column category given knowledge of the row category. If an object is picked from the table and we are asked to make a best prediction of column category without any further information, the column category with the largest frequency would be used. An error in prediction is made with probability

$$\frac{N_{\cdot\cdot} - \max_{1 \le c \le C} N_{\cdot c}}{N_{\cdot\cdot}} = 1 - \frac{\max_{1 \le c \le C} N_{\cdot c}}{N_{\cdot\cdot}} ,$$

denoted by P_{error} = probability of an error in predicting the column category. These two probabilities can be used to form a proportional reduction in error measure, $\lambda_{A|B}$ (predicting a column category $(A_1, \ldots, A_C)$ from a row category $(B_1, \ldots, B_R)$):

$$\lambda_{A|B} = \frac{P_{error} - P_{error|row}}{P_{error}} =$$

$$\frac{\left(\sum_{r=1}^{R} \max_{1 \le c \le C} N_{rc} \right) - \max_{1 \le c \le C} N_{\cdot c}}{N_{\cdot\cdot} - \max_{1 \le c \le C} N_{\cdot c}} .$$

If $\lambda_{A|B}$ is zero, then the maximum of the column marginal frequencies, $\max_{1 \le c \le C} N_{\cdot c}$, is the same as the sum of the maximum column frequencies within rows. In other words, no differential predictions of a column event are made based on knowledge of the row category for an observation.

The G-K λ measure is asymmetric, and $\lambda_{A|B}$ is not necessarily the same as $\lambda_{B|A}$; for example, one measure could be zero and the other positive. In general, there is no necessary relation between $\lambda_{A|B}$ and $\lambda_{B|A}$ and the usual chi-square association statistic. The latter is a nontransparent measure of relationship in a contingency table with unordered attributes that increases proportionately with increasing sample size; a λ measure has the transparent interpretation in terms of the differential predicability of one attribute from another. Given that only nominal attributes are required, it is universally appropriate for just about any task of relating two variables, irrespective of the levels of measurement they might have.

5.6   Intraclass Correlation

A different type of correlational measure, an intraclass correlation coefficient (ICC), can be used when quantitative measurements are made on units organized into groups, typically of the same size. It measures how strongly units from the same group resemble each other. Here, we will emphasize only the case where group sizes are all 2, possibly representing data on a set of N twins, or two raters assessing the same N objects. The basic idea generalizes, however, to an arbitrary number of units within each group.

Early work on the ICC from R. A. Fisher (1925, pp. 176–210) and his contemporaries (e.g., Pearson, Lee, & Bramley-Moore, 1899; Pearson, et al., 1901) conceptualized the problem as follows: let (x_i, x_i'), $1 \leq i \leq N$, denote the N pairs of observations (thus, we have N groups with two measurements in each). The usual correlation coefficient cannot be computed, however, because the order of the measurements within a pair is unknown (and arbitrary). As an alternative, we first double the number of pairs to $2N$ by including both (x_i, x_i') and (x_i', x_i). The Pearson correlation is then computed using the $2N$ pairs to obtain an ICC.

As a more convenient and generalizable version of the ICC computations, we adopt the Model II (random effects) analysis-of-variance model:

$$Y_{ij} = \mu + \alpha_i + \epsilon_{ij} \, ,$$

where Y_{ij} is the jth observation in the ith group, μ is the overall mean, α_i is a random variable indicating an effect shared by all values in group i, and ϵ_{ij} is a random variable representing error. The two random variables, α_i and ϵ_{ij}, are assumed uncorrelated within and between themselves with expected values of zero, and variances of σ_α^2 and σ_ϵ^2, respectively. The population ICC parameter is given by

$$\frac{\sigma_\alpha^2}{\sigma_\alpha^2 + \sigma_\epsilon^2} \, ,$$

and estimated by a ratio: (Mean Square Between − Mean Square Within) divided by (Mean Square Between + Mean Square Within).

In studying heritability, we need two central terms:[11]
Phenotype: the manifest characteristics of an organism that result
from both the environment and heredity; these characteristics can
be anatomical or psychological, and are generally the result of an
interaction between the environment and heredity.
Genotype: the fundamental hereditary (genetic) makeup of an organism; as distinguished from (phenotypic) physical appearance.

Based on the random effects model (Model II) for describing
a particular phenotype, symbolically we have: Phenotype(P) =
Genotype(G) + Environment(E), or in terms of variances, Var(P)
= Var(G) + Var(E), assuming that the covariance between G and
E is zero. The ICC in this case is the heritability coefficient,

$$H^2 = \frac{\text{Var}(G)}{\text{Var}(P)}.$$

Heritability estimates are often misinterpreted, even by those
who should know better. In particular, heritability refers to the
proportion of variation between individuals in a population influenced by genetic factors. Thus, because heritability describes the
population and not the specific individuals within it, it can lead to
an aggregation fallacy when one tries to make an individual-level
inference from a heritability estimate. For example, it is incorrect to say that because the heritability of a personality trait is,
say, .6, that therefore 60% of a specific person's personality is inherited from parents and 40% comes from the environment. The
term "variation" in the phrase "phenotypic variation" is important to note. If a trait has a heritability of .6, it means that of
the observed phenotypic variation, 60% is due to genetic variation. It does not imply that the trait is 60% caused by genetics in
a given individual. Nor does a heritability coefficient imply that
any observed differences between groups (for example, a supposed
15 point I.Q. test score difference between blacks and whites) is
genetically determined. As noted explicitly in Stephen Jay Gould's
The Mismeasure of Man (1996), it is a fallacy to assume that a
(high) heritability coefficient allows the inference that differences
observed between groups must be genetically caused. As Gould
succinctly states: "[V]ariation among individuals within a group
and differences in mean values between groups are entirely separate phenomena. One item provides no license for speculation

about the other" (p. 187). For an in-depth and cogent discussion of the distinction between heritability and genetic determination, the reader is referred to Ned Block, "How Heritability Misleads About Race" (*Cognition*, 1995, *56*, 99–128).

Notes

[1] The ±1 bounds on a Pearson correlation being achievable only when a perfect linear relation exists between the two variables results from using the Cauchy–Schwartz inequality to normalize the raw cross-product statistic to obtain the Pearson correlation. Another type of inequality (usually referred to using the mathematical trio of Hardy–Littlewood–Polya) says that a cross-product statistic is at a maximum when the entries being multiplied have the same ordering from smallest to largest. This is the basis for Guttman's monotonicity coefficient.

[2] The type of algebraic restrictions given earlier between the correlations in a three-variable system can be generalized considerably to an arbitrary p-variable context. A matrix $\mathbf{A}_{p \times p}$ that represents a covariance matrix among a collection of p random variables is positive semidefinite (p.s.d); and conversely, any positive semidefinite matrix represents the covariance matrix for a collection of random variables. We partition $\mathbf{A}$ to isolate its last row and column as

$$\mathbf{A} = \begin{pmatrix} \mathbf{B}_{(p-1) \times (p-1)} & \mathbf{g}_{(p-1) \times 1} \\ \mathbf{g}' & a_{pp} \end{pmatrix} .$$

$\mathbf{B}$ is the $(p-1) \times (p-1)$ covariance matrix among the first $p-1$ variables; $\mathbf{g}$ is $(p-1) \times 1$ and contains the cross-covariance between the the first $p-1$ variables and the pth; a_{pp} is the variance for the pth variable.

Based on the observation that determinants of p.s.d. matrices are nonnegative, and a result on expressing determinants for partitioned matrices (that we do not give here), it must be true that

$$\mathbf{g}' \mathbf{B}^{-1} \mathbf{g} \leq a_{pp} ,$$

or if we think correlations rather than merely covariances (so the main diagonal of $\mathbf{A}$ consists of all ones):

$$\mathbf{g}' \mathbf{B}^{-1} \mathbf{g} \leq 1 .$$

Given the correlation matrix $\mathbf{B}$, the possible values the correlations in $\mathbf{g}$ could have are in or on the ellipsoid defined in $p-1$ dimensions by $\mathbf{g}' \mathbf{B}^{-1} \mathbf{g} \leq 1$. The important point is that we do not have a "box" in $p-1$ dimensions containing the correlations with sides extending the whole range of ± 1; instead, some restrictions are placed on the observable correlations defined by the size of the correlations in **B**. For example, when $p = 3$, a correlation between variables X_1 and X_2 of $r_{12} = 0$ gives the "degenerate" ellipse of a circle for constraining the correlation values between X_1 and X_2 and the third variable X_3 (in a two-dimensional r_{13} versus r_{23} coordinate system); for $r_{12} = 1$, the ellipse flattens to a line in this same two-dimensional space.

[3]Excerpts from this chapter follow as they appeared in Yule and Kendall (1950):

> This peculiar result indicates that, although a set of attributes independent of A and B will not affect the association between them, the existence of an attribute C with which they are both associated may give an association in the population at large which is illusory in the sense that it does not correspond to any real relationship between them. If the association between A and C, B and C are of the same sign, the resulting association between A and B will be positive; if of opposite signs, negative.
>
> The cases which we discussed at the beginning of this chapter are instances in point. In the first illustration we saw that it was possible to argue that the positive associations between *vaccination* and *hygienic conditions*, *exemption from attack* and *hygienic conditions*, led to an illusory association between *vaccination* and *exemption from attack*. Similarly, the question was raised whether the positive association between *grandfather* and *grandchild* may not be due to the positive associations between *grandfather* and *father*, and *father* and *child*. (p. 37)
>
> Illusory associations may also arise in a different way through the personality of the observer or observers. If the observer's attention fluctuates, he may be more likely to notice the presence of A when he notices the presence of B, and *vice versa*; in such a case A and B (so far as the record goes) will both be associated with the observer's attention C, and consequently an illusory association will be created. Again, if the attributes are not well defined, one observer may be more generous than another in deciding when to record the presence of A and also the presence of B, and even one observer may fluctuate in the generosity of his marking. In this case the recording of A and the recording of B will both be associated with the generosity of the observer in recording their presence, C, and an illusory association between A and B will consequently arise, as before. (p. 38)

[4]A related idea is *pyromancy*, divination by the reading of shapes appearing in fire or flames ("Pyromancy," 2010; "Divination," 2010).

[5]This article is listed in the Suggested Reading, Section B.2.2.

[6]Any discussion of clinical (mis)inference should mention the justifiably

well-known if now somewhat dated *Science* article by D. L. Rosenhan, "On Being Sane in Insane Places" (1973, *179*, 250–258). This article reports a study of several normal people who were admitted to clinical inpatient facilities and their subsequent experiences in obtaining treatment and release. As one general observation, it may be true that psychiatric diagnoses are more in the minds of the observers than they are a valid summary of characteristics displayed by the observed. Or stated differently, psychiatric diagnosis may say less about the patient and more about the environment and circumstances in which the patient is placed. Once again, context counts and it counts crucially. To phrase this statistically, a failure to detect sanity in a pseudopatient may be due to physician bias and an inclination to call a healthy person sick (a false positive) rather than a sick person healthy (a false negative). Possibly, this is because it is more dangerous to misdiagnosis illness than health, and therefore, it is best to suspect illness even among the healthy.

[7]The attenuation issues for correlations have been discussed in applied statistics texts for some time. For example, there is a section entitled "The Attenuation Effect" in Yule and Kendall (1950, pp. 313–315), *An Introduction to the Theory of Statistics.*

[8]An amusing critique of fMRI studies that fail to correct for multiple comparisons and control false positives is listed in the Suggested Reading, B.2.1. It involves the scan of a dead salmon's brain and its response to human emotions ("Trawling the Brain," Laura Sanders, December 19, 2009, *ScienceNews*).

[9]For other instances of odd or nonsense correlations, see Yule and Kendall (1950), *An Introduction to the Theory of Statistics.*

[10]We might also note at this point that John Flannagan is sometimes credited with doing the first legitimate validity study (H. Gulliksen, personal communication to HW, 1967). In the 1940s, Flannagan gave aptitude tests to a large sample of Army recruits. The army ignored the test results, but Flannagan kept track of the subsequent recruit success in various training programs. The correlation with test scores was high. This convinced Army Command to use the test scores subsequently for training assignments. The correlation with success then went down. (Note the similarity with the LSAT example given earlier.)

[11]Any comprehensive dictionary should give definitions for the terms "phenotype" and "genotype"; for example, see *American Heritage Dictionary of the English Language* (2000/2006) and *American Heritage Science Dictionary* (2005).

Chapter 6

Prediction

> The race is not always to the swift, nor the battle to the strong, but
> that's the way to bet.
> — Damon Runyon (*Runyon on Broadway*, 1950)

The attempt to predict the values on a criterion (dependent)
variable by a function of predictor (independent) variables is typically approached by simple or multiple regression, for one or more
than one predictor, respectively. The most common combination
rule is a linear function of the independent variables obtained by
least squares; that is, the linear combination that minimizes the
sum of the squared residuals between the actual values on the dependent variable and those predicted from the linear combination.
In the case of simple regression, scatterplots again play a major
role in assessing linearity of the relation, the possible effects of
outliers on the slope of the least-squares line, and the influence
of individual observations in its calculation. Regression slopes, in
contrast to the correlation, are neither scale invariant nor symmetric in the dependent and independent variables. One usually
interprets the least-squares line as one of expecting, for each unit
change in the independent variable, a regression slope change in
the dependent variable.[1]

Several topics in prediction arise continually when we attempt
to reason correctly and therefore ethically with fallible multivariable data. We discuss five such areas in the subsections to follow:
regression toward the mean, methods involved in using regression
for prediction that incorporate corrections for unreliability, differential prediction effects in selection based on tests, interpreting
and making inferences from regression weights, and the distinction
between actuarial (statistical) and clinical prediction.

6.1   Regression Toward the Mean

> I suspect that the regression fallacy is the most common fallacy in
> the statistical analysis of economic data.
> — Milton Friedman (1992, "Do Old Fallacies Ever Die?")

Regression toward the mean is a phenomenon that will occur
whenever dealing with fallible measures with a less-than-perfect
correlation. The word "regression" was used by Galton in his well-
known 1886 article, "Regression Towards Mediocrity in Hereditary
Stature." Galton showed that heights of children from very tall
or short parents regress toward mediocrity (that is, toward the
mean) and exceptional scores on one variable (parental height)
are not matched with such exceptionality on the second (child
height). This observation is purely due to the fallibility for the
various measures and the concomitant lack of a perfect correlation
between the heights of parents and their children.[2]

Regression toward the mean is a ubiquitous phenomenon, and
given the name "regressive fallacy" whenever cause is ascribed
where none exists. Generally, interventions are undertaken if processes are at an extreme (for example, a crackdown on speeding
or drunk driving as fatalities spike, treatment groups formed from
individuals who are seriously depressed, or individuals selected because of extreme good or bad behaviors). In all such instances,
whatever remediation is carried out will be followed by some lessened value on a response variable. Whether the remediation was
itself causative is problematic to assess given the universality of
regression toward the mean.

There are many common instances where regression may lead
to invalid reasoning: I went to my doctor and my pain has now
lessened; I instituted corporal punishment and behavior has improved; he was jinxed by a *Sports Illustrated* cover because subsequent performance was poorer (also known as the "sophomore
jinx"); although he hadn't had a hit in some time, he was "due,"
and the coach played him; and so on. More generally, any time one
optimizes with respect to a given sample of data by constructing
prediction functions of some kind, there is an implicit use and reliance on data extremities. In other words, the various measures of

goodness of fit or prediction calculated need to be cross-validated either on new data or by a clever sample reuse strategy such as the well-known jackknife or bootstrap procedures. The degree of "shrinkage" we see in our measures based on this cross-validation is an indication of the fallibility of our measures and the (in)adequacy of the given sample sizes.

The misleading interpretive effects engendered by regression toward the mean are legion, particularly when we wish to interpret observational studies for some indication of causality. There is a continual violation of the traditional adage that "the rich get richer and the poor get poorer," in favor of "when you are at the top, the only way is down." Extreme scores are never quite as extreme as they first appear. Many of these regression artifacts are discussed in the cautionary source, *A Primer on Regression Artifacts* (Campbell & Kenny, 1999), including the various difficulties encountered in trying to equate intact groups by matching or analysis of covariance. Statistical equating creates the illusion but not the reality of equivalence. As summarized by Campbell and Kenny, "the failure to understand the likely direction of bias when statistical equating is used is one of the most serious difficulties in contemporary data analysis" (p. 85).

The historical prevalence of the regression fallacy is considered by Stephen Stigler in his 1997 article entitled "Regression Towards the Mean, Historically Considered" (*Statistical Methods in Medical Research, 6*, 103–114). Stigler labels it "a trap waiting for the unwary, who were legion" (p. 112). He relates a story that we excerpt below about a Northwestern University statistician falling into the trap in 1933:

> The most spectacular instance of a statistician falling into the trap was in 1933, when a Northwestern University professor named *Horace* Secrist unwittingly wrote a whole book on the subject, *The Triumph of Mediocrity in Business*. In over 200 charts and tables, Secrist "demonstrated" what he took to be an important economic phenomenon, one that likely lay at the root of the great depression: a tendency for firms to grow more mediocre over time. Secrist was aware of Galton's work; he cited it and used Galton's terminology. The preface even acknowledged "helpful criticism" from such statistical luminaries as HC Carver (the editor of the *Annals of Mathematical Statistics*), Raymond Pearl, EB Wilson, AL Bowley, John Wishart and Udny Yule. How thoroughly these statisticians were in-

formed of Secrist's work is unclear, but there is no evidence that they were successful in alerting him to the magnitude of his folly (or even if they noticed it). Most of the reviews of the book applauded it. But there was one dramatic exception: in late 1933 Harold Hotelling wrote a devastating review, noting among other things that "The seeming convergence is a statistical fallacy, resulting from the method of grouping. These diagrams really prove nothing more than that the ratios in question have a tendency to wander about." (p. 112)

Stigler goes on the comment about the impact of the Secrist-Hotelling episode for the recognition of the importance of regression toward the mean:

> One would think that so public a flogging as Secrist received for his blunder would wake up a generation of social scientists to the dangers implicit in this phenomenon, but that did not happen. Textbooks did not change their treatment of the topic, and if there was any increased awareness of it, the signs are hard to find. In the more than two decades between the Secrist-Hotelling exchange in 1933 and the publication in 1956 of a perceptively clear exposition in a textbook by W Allen Wallis and Harry Roberts, I have only encountered the briefest acknowledgements. (p. 113)

A variety of phrases seem to get attached whenever regression toward the mean is probably operative. We have the "winner's curse," where someone is chosen from a large pool (such as of job candidates), who then doesn't live up to expectations; or when we attribute some observed change to the operation of "spontaneous remission." As Campbell and Kenny noted, "many a quack has made a good living from regression toward the mean" (p. 48). Or, when a change of diagnostic classification results upon repeat testing for an individual given subsequent one-on-one tutoring (after being placed, for example, in a remedial context). More personally, there is "editorial burn-out" when someone is chosen to manage a prestigious journal at the apex of a career, and things go quickly downhill from that point.

6.2   Actuarial Versus Clinical Prediction

Where there is fog in the pulpit, there will be dense fog in the Congregation.
— Glenn Firebaugh (*Seven Rules for Social Research*, 2008)

Paul Meehl in his iconic 1954 monograph, *Clinical Versus Statistical Prediction: A Theoretical Analysis and a Review of the Evidence*, created quite a stir with his convincing demonstration that mechanical methods of data combination, such as multiple regression, outperform (expert) clinical prediction. The enormous amount of literature produced since the appearance of this seminal contribution has uniformly supported this general observation; similarly, so have the extensions suggested for combining data in ways other than by multiple regression, for example, by much simpler unit weighting schemes (Wainer, 1976), or those using other prior weights. It appears that individuals who are conversant in a field are better at selecting and coding information than they are at actually integrating it. Combining such selected information in a more mechanical manner will generally do better than the person choosing such information in the first place.[3] This conclusion can be pushed further: if we formally model the predictions of experts using the same chosen information, we can generally do better than the experts themselves. Such formal representations of what a judge does are referred to as "paramorphic."[4]

In an influential review article, Dawes (1979) discussed proper and improper linear models and argued for the "robust beauty of improper linear models." A proper linear model is one obtained by an optimization process, usually least squares. Improper linear models are not "optimal" in this latter sense and typically have their weighting structures chosen by a simple mechanism, for example, by random or unit weighting. Again, improper linear models generally outperform clinical prediction, but even more surprisingly, improper models typically outperform proper models in cross-validation. What seems to be the reason is the notorious instability of regression weights with correlated predictor variables, even if sample sizes are very large. Generally, we know that simple averages are more reliable than individual observations, so it may not be so surprising that simple unit weights are likely to do better on cross-validation than those found by squeezing "optimality" out of a sample. Given that the sine qua non of any prediction system is its ability to cross-validate, the lesson may be obvious: statistical optimality with respect to a given sample may not be the best answer when we wish to predict well.[5]

The idea that statistical optimality may not lead to the best predictions seems counterintuitive, but as argued by Roberts and Pashler (2000), just the achievement of a good fit to observations does not necessarily mean we have found a good model. In fact, because of the overfitting of observations, choosing the model with the absolute best fit is apt to result in poorer predictions. The more flexible the model, the more likely it is to capture not only the underlying pattern but unsystematic patterns such as noise. A single general-purpose tool with many adjustable parameters is prone to instability and greater prediction error as a result of high error variability. An observation by John von Neumann is particularly germane: "With four parameters, I can fit an elephant, and with five, I can make him wiggle his trunk."[6] More generally, this notion that "less is more" is difficult to fathom, but as Gigerenzer and others have argued (for example, Gigerenzer & Brighton, 2009), it is clear that simple heuristics can at times be more accurate than complex procedures. All of the work emanating from the idea of the "robust beauty of improper linear models" may force a reassessment of the normative ideals of rationality might be. Most reduce to simple cautions about overfitting one's observations and then hoping for better predictions because an emphasis has been placed on immediate optimality instead of the longer-run goal of cross-validation.

Appendix: Henry A. Wallace and the Modeling of Expert Judgments

There are several historical connections between Henry A. Wallace, one of Franklin D. Roosevelt's vice presidents (1940–1944), and the formal (paramorphic) modeling of the prediction of experts, and applied statistics more generally. Wallace wrote an article (1923) that appeared in the *Journal of the American Society of Agronomy* (*15*, 300–304) entitled "What Is in the Corn Judge's

Mind?" The data used in this study were ratings of possible yield for some 500 ears of corn from a number of experienced corn judges. In addition to the ratings, measurements were taken on each ear of corn over six variables: length of ear, circumference of ear, weight of kernel, filling of the kernel at the tip (of the kernel), blistering of kernel, and starchiness. Also, because all the ears were planted in 1916, one ear to a row, the actual yields for the ears were available as well.

The method of analysis for modeling both the expert judgments of yield and actual yield was through the new method of path coefficients just developed by Sewall Wright in 1921 ("Correlation and Causation," *Journal of Agricultural Research*, *20*, 557–585). The results were final "scorecards" for how the judges and the actual yield values could be assessed by the six factors (each was normalized to a total of 100 "points"):

JUDGES' SCORE CARD:
Length – 42.0
Circumference – 13.6
Weight of kernel – 18.3
Filling of kernel at tip – 13.3
Blistering of kernel – 6.4
Absence of starchiness – 6.4
Total – 100.00
ACTUAL YIELD SCORE CARD:
Length – 7.7
Circumference – 10.0
Weight of kernel – 50.0
Filling of kernel at tip – 18.0
Blistering of kernel – 9.0
Absence of starchiness – 5.3
Total – 100.00

In rather understated conclusions, Wallace comments:

> It is interesting to note that while the simple correlation coefficients indicate that the judges took into account blistering of kernel as a damaging factor, the path coefficients indicate that they looked on blistering as beneficial. The long ears with heavy kernels for which the judges had such a fondness tended to be freer from blistering than the short ears with light kernels and for that reason it appears on the surface that the judges did not like blistering. But when other

factors are held constant, it is found that there is a slight tendency for the judges to favor blistering. Doubtless this was carelessness on the part of these particular judges. (p. 302)

The contrast between the yield score card and the judges' score card is interesting.

It will be noted that the tendency of the judges is to emphasize more than anything else, length of ear, whereas Mother Nature, judging merely from these two years' work with one variety of corn, lays her outstanding emphasis on weight of kernel. Over a period of years it may be that the judges are well warranted in making it a prime requisite that a seed ear in the central part of the Corn Belt should at least be eight inches long. But in case of an emergency, in a season when seed corn is scarce, it is probable that so far as that particular year is concerned, length of ear can be disregarded altogether. The important thing would seem to be to discard those ears carrying light kernels, especially if they have pointed tips, are blistered, and are starchy.

That the corn judges did not know so very much about the factors which make for yield is indicated by the fact that their scores were correlated with yield to the extent of only .2. The difficulty seems to be that they laid too much emphasis on length of ear and possibly also on some fancy points, which caused them to neglect placing as much emphasis on sound, healthy kernel characteristics as they should.

By using Wright's methods of path coefficients, it should be possible in the future to work out in very definite fashion, what really is in the minds of experienced corn judges. It is suggested that the things which really are in their minds are considerably different from the professed score card. It is realized of course that when the judges are working on samples of corn all of which is of show quality, that length of ear will not be so large a factor as it was in the case of this study when the ears were field run, varying from less than five inches to more than ten inches in length. It should be interesting to make another study to determine just what is in the minds of the corn judges when they are judging single ear samples at a corn show.

That corn judging is to some extent a profession with recognized standards is indicated by the fact that the correlation coefficient between the scores of different judges working on the same 500 ears of field, run corn averaged around .7. Inasmuch as corn judging still has a vogue in some of our Corn Belt states, it would seem to be worth while to determine just what is in different corn judges' minds. It would be especially interesting to have corn judges from central Iowa, central Illinois, and central Indiana work on the same 500 ears and then make up by means of path coefficients their true score cards. (pp. 303–304)

In addition to a political career, Wallace had a life-long and avid interest in statistical computing. We give an essay by David Alan Grier, *The Origins of Statistical Computing*, posted on the American Statistical Association website:

The lab in the Department of Agriculture inspired two Iowans, George Snedecor and Henry A. Wallace, to experiment with punched-card statistical computations. Henry Wallace eventually rose to prominence as the Vice President of the United States, but during the 1920s, he was the publisher of his family's farm journal, *Wallaces' Farmer*. He was also a self-taught statistician and was interested in the interplay of biology and economics in farm management. During the 1910s, he learned the methods of correlation studies and least squares regression by reading Yule's book, *An Introduction to the Theory of Statistics* (London: Griffin, 1911). Finding in that book no easy method for solving the normal equations for regression, Wallace devised his own, using an idea that Gauss had applied to an astronomical problem.

In 1923, Henry A. Wallace learned of the new statistics lab at the Department of Agriculture while he was visiting his father, Harry Wallace, who was then the Secretary of Agriculture. Intrigued with the machines, he borrowed a tabulator at a Des Moines insurance firm and taught himself how to use the device to calculate correlations.. He would punch data cards and would then take them to the offices of the insurance company for tabulating. During the first years of the 1920s, he published ever more sophisticated statistical studies in the pages of *Wallaces' Farmer*, studies that must have baffled many of his loyal readers, who tended to be modestly educated farmers. The last, published in January 1923, was a detailed study of land values in the state.

The study of Iowa land values marked the maturity of Wallace's statistical ability. By the time he published it, Wallace had become a friend of George Snedecor, who taught the statistics courses at Wallace's alma mater, then named Iowa State College. Impressed with Wallace's knowledge of least squares, Snedecor invited him to teach an advanced course on those methods to college faculty. This class, which met for 10 consecutive Saturdays over the fall and winter of 1924, ended with a demonstration of punched-card calculation. After the class, Snedecor helped Wallace prepare a manuscript on his algorithm for solving normal equations. They jointly published the manuscript in 1925 with the title "Correlation and Machine Calculation."

The title of Wallace's and Snedecor's pamphlet tends to mislead modern readers. For the most part, the machines to which it refers are desk calculators, not tabulating machinery. Part of Wallace's methods were easily adapted to tabulating machines. By computing sums of squares and sums of cross-products, a mechanical tabulator could quickly produce a set of normal equations. The same tabulator, however, could not be easily used to solve these equations. It was extremely awkward, if not impossible, to use a 1920s vintage tabulator to solve matrix arithmetic problems. Such problems were solved by human computers who used desk calculators.

Inspired by Wallace, Snedecor devoted much effort to acquiring tabulating machines for his university. He was able to secure them in the Fall of 1927 and established a statistical computing lab within the Department of Mathematics. This first lab seems to have been a cooperative effort by several college departments and may have been partly supported by local IBM officials, who were interested in placing their equipment at universities. IBM helped many schools establish computing labs at that time. The first was at Cornell, which leased tabulating machines to form a lab in 1926. Next came Iowa State College, Columbia University, and the University of Michigan, who acquired these machines in 1927. Shortly thereafter came the University of Texas, Harvard University, Stanford University, and the University of Tennessee

6.3   Incorporating Reliability Corrections in Prediction

If your mother says she loves you, check it out.
— Adage from the Chicago City News Bureau

In prediction, two aspects of variable unreliability have consequences for ethical reasoning. One is in estimating a person's true score on a variable; the second is in how regression might be handled when there is measurement error in the independent and/or dependent variables. In both of these instances, there is an implicit underlying model for how any observed score, X, might be constructed additively from a true score, T_X, and an error score, E_X, where E_X is typically assumed uncorrelated with T_X: $X = T_X + E_X$. When we consider the distribution of an observed variable over, say, a population of individuals, there are two sources of variability present in the true and the error scores. If we are interested primarily in structural models among true scores, then some correction must be made because the common regression models implicitly assume that variables are measured without error.

The estimation, $\hat{T}_X$, of a true score from an observed score, X, was derived using the regression model by Kelley in the 1920s (Kelley, 1947), with a reliance on the algebraic equivalence that the squared correlation between observed and true score is the reliability. If we let $\hat{\rho}$ be the estimated reliability, Kelley's equation can be written as

$$\hat{T}_X = \hat{\rho}X + (1 - \hat{\rho})\bar{X} \,,$$

where $\bar{X}$ is the mean of the group to which the individual belongs. In other words, depending on the size of $\hat{\rho}$, a person's estimate is partly due to where the person is in relation to the group— upward if below the mean, downward if above. The application of this statistical tautology in the examination of group differences provides such a surprising result to the statistically naive that this equation has been labeled "Kelley's Paradox" (Wainer, 2005, pp. 67–70).

In addition to obtaining a true score estimate from an obtained score, Kelly's regression model also provides a standard error of estimation (which in this case is now referred to as the standard error of measurement). An approximate 95% confidence interval on an examinee's true score is given by

$$\hat{T}_X \pm 2\hat{\sigma}_X((\sqrt{1-\hat{\rho}})\sqrt{\hat{\rho}}) \,,$$

where $\hat{\sigma}_X$ is the (estimated) standard deviation of the observed scores. By itself, the term $\hat{\sigma}_X((\sqrt{1-\hat{\rho}})\sqrt{\hat{\rho}})$ is the standard error of measurement, and is generated from the usual regression formula for the standard error of estimation but applied to Kelly's model predicting true scores. The standard error of measurement most commonly used in the literature is not Kelly's but rather $\hat{\sigma}_X\sqrt{1-\hat{\rho}}$, and a 95% confidence interval taken as the observed score plus or minus twice this standard error. An argument can be made that this latter procedure leads to "reasonable limits" (after Gulliksen, 1950) whenever $\hat{\rho}$ is reasonably high, and the obtained score is not extremely deviant from the reference group mean. Why we should assume these latter preconditions and not use the more appropriate procedure to begin with, reminds us of a

Bertrand Russell quotation (1919, p. 71): "The method of postulating what we want has many advantages; they are the same as
the advantages of theft over honest toil."

There are several remarkable connections between Kelley's work
in the first third of the twentieth century and the modern theory
of statistical estimation developed in the last half of the century.
In considering the model for an observed score, X, to be a sum of a
true score, T, and an error score, E, plot the observed test scores on
the x-axis and their true scores on the y-axis. As noted by Galton
in the 1880s (Galton, 1886), any such scatterplot suggests two
regression lines. One is of true score regressed on observed score
(generating Kelley's true score estimation equation given in the
text); the second is the regression of observed score being regressed
on true score (generating the use of an observed score to directly
estimate the observed score). Kelley clearly knew the importance
for measurement theory of this distinction between two possible
regression lines in a true-score versus observed-score scatterplot.
The quotation given below is from his 1927 text, *Interpretation of
Educational Measurements*. The reference to the "last section" is
where the true score was estimated directly by the observed score;
the "present section" refers to his true score regression estimator:

> This tendency of the estimated true score to lie closer to the mean
> than the obtained score is the principle of regression. It was first
> discovered by Francis Galton and is a universal phenomenon in cor-
> related data. We may now characterize the procedure of the last and
> present sections by saying that in the last section regression was not
> allowed for and in the present it is. If the reliability is very high,
> then there is little difference between [the two methods], so that
> this second technique, which is slightly the more laborious, is not
> demanded, but if the reliability is low, there is much difference in
> individual outcome, and the refined procedure is always to be used
> in making individual diagnoses. (p. 177)

Kelley's preference for the refined procedure when reliability is
low (that is, for the regression estimate of true score) is due to
the standard error of measurement being smaller (unless reliability is perfect); this is observable directly from the formulas given
earlier. There is a trade-off in moving to the regression estimator of the true score in that a smaller error in estimation is paid
for by using an estimator that is now biased. Such trade-offs are

common in modern statistics in the use of "shrinkage" estimators (for example, ridge regression, empirical Bayes methods, James–Stein estimators). Other psychometricians, however, apparently just don't buy the trade-off; for example, see Gulliksen (*Theory of Mental Tests*; 1950); Gulliksen wrote that "no practical advantage is gained from using the regression equation to estimate true scores" (p. 45). We disagree—who really cares about bias when a generally more accurate prediction strategy can be defined?

What may be most remarkable about Kelley's regression estimate of true score is that it predates the work in the 1950s on "Stein's Paradox" that shook the foundations of mathematical statistics. A readable general introduction to this whole statistical kerfuffle is the 1977 *Scientific American* article by Bradley Efron and Carl Morris, "Stein's Paradox in Statistics" (*236*(5), 119-127). When reading this popular source, keep in mind that the class referred to as James–Stein estimators (where bias is traded off for lower estimation error) includes Kelley's regression estimate of the true score. We give an excerpt below from Stephen Stigler's 1988 Neyman Memorial Lecture, "A Galtonian Perspective on Shrinkage Estimators" (*Statistical Science*, 1990, 5, 147-155), that makes this historical connection explicit:

> The use of least squares estimators for the adjustment of data of course goes back well into the previous century, as does Galton's more subtle idea that there are two regression lines. ... Earlier in this century, regression was employed in educational psychology in a setting quite like that considered here. Truman Kelley developed models for ability which hypothesized that individuals had true scores ... measured by fallible testing instruments to give observed scores ... ; the observed scores could be improved as estimates of the true scores by allowing for the regression effect and shrinking toward the average, by a procedure quite similar to the Efron–Morris estimator. (p. 152)

Before we leave the topic of true score estimation by regression, we might also note what it does not imply. When considering an action for an individual where the goal is to help make, for example, the right level of placement in a course or the best medical treatment and diagnosis, then using group membership information to obtain more accurate estimates is the appropriate course to follow. But if we are facing a contest, such as awarding scholarships, or offering admission or a job, then it is inappropriate (and ethically questionable) to search for identifiable subgroups that a particular person might belong to and then adjust that person's score accordingly. Shrinkage estimators are "group blind." Their use is justified for whatever population is being observed; it is generally best for accuracy of estimation to discount extremes and "pull them in" toward the (estimated) mean of the population.

In the topic of errors-in-variables regression, we try to compensate for the tacit assumption in regression that all variables are measured without error. Measurement error in a response variable does not bias the regression coefficients per se, but it does increase standard errors and thereby reduces power. This is generally a common effect: unreliability attenuates correlations and reduces power even in standard ANOVA paradigms. Measurement error in the predictor variables biases the regression coefficients. For example, for a single predictor, the observed regression coefficient is the "true" value multiplied by the reliability coefficient. Thus, without taking account of measurement error in the predictors, regression coefficients will generally be underestimated, producing a biasing of the structural relationship among the true variables. Such biasing may be particularly troubling when discussing econometric models where unit changes in observed variables are supposedly related to predicted changes in the dependent measure; possibly the unit changes are more desired at the level of the true scores.

Milton Friedman's 1992 article entitled "Do Old Fallacies Ever Die?" (*Journal of Economic Literature, 30*, 2129-2132), gives a downbeat conclusion regarding errors-in-variables modeling (this is also the source of the epigram in the opening of the section on regression toward the mean):

> Similarly, in academic studies, the common practice is to regress a variable Y on a vector of variables X and then accept the regression coefficients as supposedly unbiased estimates of structural parameters, without recognizing that all variables are only proxies for the variables of real interest, if only because of measurement error, though generally also because of transitory factors that are peripheral to the subject under consideration. I suspect that the regression fallacy is the most common fallacy in the statistical analysis of economic data, alleviated only occasionally by consideration of the bias introduced when "all variables are subject to error." (p. 2131)

6.4   Differential Prediction Effects in Selection

One area in which prediction is socially relevant is in selection based on test scores, whether for accreditation, certification, job placement, licensure, educational admission, or other high-stakes endeavors. Most discussions about fairness of selection are best phrased as regression models relating a performance measure to a selection test and whether the regressions are the same over all identified groups of relevance (e.g., ethnic, gender, or age). Specifically, are slopes and intercepts the same? If so or if not, how does this affect the selection mechanism being implemented, and can it be considered fair? It is safe to say that depending on the pattern of data within groups, all sorts of things can happen. Generally, an understanding of how a regression/selection model works with this kind of variation is necessary for a numerically literate discussion of its intended or unintended consequences. To give a greater sense of the complications that can arise, a redaction is given below from Allen and Yen (1979/2002):

> When regression equations are used in selection procedures and regression lines differ across groups, questions of fairness can arise. For example, suppose that, in predicting a criterion such as college grades for two groups of examinees, the regression lines are found to be parallel but generally higher for one group than the other. ...
>
> The regression equation that was produced in the combined group, when compared with within-group regression equations, consistently overpredicts criterion scores for group 2 and underpredicts criterion scores for group 1. In effect, the combined regression equation favors the low-scoring group rather than the high-scoring group. This effect suggests that, if there are group differences in the level of criterion scores in the regression problem, using a combined-group or the higher group's regression equation can help the 'disadvantaged' group.
>
> If there are group differences in the level of predictor scores, a combined-group regression equation can underpredict the lower group's criterion scores. ... The combined group regression line, when compared with within-group predictions, overpredicts criterion scores for most members of group 1 and underpredicts criterion scores for most members of group 2. Using the combined-group regression line in this situation would hurt the disadvantaged group (that is, the group with lower predictor scores).

> When regression equations differ across groups, we cannot state (without causing an argument) which procedure is more fair—the use of different regression lines for the two groups or the use of the regression line based on the combined group. If different equations are used, examinees in one group can complain that, to be attributed with the same criterion scores, they need a higher predictor score than those in the other group. In other words, two examinees with the same high school grades could have different predicted college grades solely because they belong to different groups. If the regression equation based on the combined groups is used, some examinees can complain that group membership is a valid part of the prediction and their criterion scores are being underpredicted. (pp. 205–207)

The practical consequences of these differential prediction effects were made evident in an employment discrimination suit in which the plaintiffs claimed that women were underpaid. The evidence supporting this claim was a regression analysis in which salary was regressed on a composite index of job qualifications. The regression line for men was higher than that for women, indicating that for equally qualified candidates, men were paid more. The defendants countered by reversing the regression, conditioning on salary, and showed that for the same salary, the employer could get a more qualified man (Conway & Roberts, 1983). Would the judge have been able to reach an ethically and scientifically correct decision without a deep understanding of regression? Well, yes. He ruled for the plaintiffs because the law protects against unequal pay by sex. But it does not protect employers' "rights" to get the most for their salary dollars.

6.5   Interpreting and Making Inferences From Regression Weights

Mathematics has given economics rigor, but alas, also mortis.
— Robert Heilbroner

An all-too-common error in multivariable systems is to overinterpret the meaning of the obtained regression weights. For example, in a model that predicts freshmen college grades from Scholastic Aptitude Test (SAT) scores and high school grade-point average (HSGPA), two highly correlated predictor variables, it has often been argued that because the regression weight is higher for HS-GPA than for SAT, the latter can be eliminated. In fact, both variables are correlated about 0.7 with freshmen grades, and because of their high intercorrelation, their regression weights are wildly unstable. Another instructive and evocative example of the dangers of such interpretations grew out of a large regression analysis done during World War II (John Tukey, personal communication to HW, January 20, 2000). To understand how the various aspects of their flights affected the accuracy of Allied bombers on missions over Germany, a regression analysis was used to predict accuracy of bombing as a function of many variables. After gathering large amounts of data, a model was built. Variables showing no relation to bombing accuracy were eliminated and a final model derived and tested successfully on a neutral data sample. The final model showed a positive regression weight on the variable "number of interceptors encountered." In other words, when the Germans sent up fighters to intercept the bombers, their accuracy improved! This result, however, should not be interpreted causally. A closer examination showed that when the weather was overcast, visibility was so impaired that the bombers couldn't hit anything. In such situations the Germans didn't bother sending up interceptors, relying on ground fire entirely. It was only when the weather was clear that interceptors were launched. When a new variable, "visibility," was added to the model, the regression weight associated with the variable "number of interceptors encountered" changed sign and became negative. The lesson to be learned is that we can never be sure all relevant variables are included, and when we add one, the size and even the direction of the regression weights can change.

Although multiple regression can be an invaluable tool in many arenas, the interpretive difficulties that result from the interrelated nature of the independent variables must always be kept in mind. As in the World War II example just described, depending on the variables included, the structural relations among the variables can change dramatically. At times, this malleability can be put to either good or bad use. For example, in applying regression models to argue for employment discrimination (such as in pay, promotion, or hiring), the multivariable system present could be problematic in arriving at a "correct" analysis. Depending on the variables included, some variables may "act" for others (as "proxies") or be used to hide (or at least, mitigate) various effects. If a case for discrimination rests on the size of a coefficient for some polychotomous variable that indicates group membership (according to race, sex, age, and so on), it may be possible to change its size depending on what variables are included or excluded from the model, and their relation to the polychotomous variable. In short, based on how the regressions are performed and one's own (un)ethical predilections, different conclusions could be produced from what is essentially the same dataset.[7]

In considering regression in econometric contexts, our interest is typically not in obtaining any deep understanding of the interrelations among the independent variables, or in the story that might be told. The goal is usually more pragmatic and phrased in terms of predicting a variable reflecting value and characterized in some numerical way (for example, as in money or performance statistics). The specific predictor variables used are of secondary importance; what is central is that they "do the job." One recent example of success for quantitative modeling is documented by Michael Lewis in *Moneyball* (2003), with its focus on data-driven decision making in baseball. Instead of relying on finding major league ball players using the hordes of fallible scouts visiting interminable high-school and college games, one adopts quantitative measures of performance, some developed by the quantitative guru of baseball, Bill James. *Moneyball* relates the story of the Oakland Athletics and their general manager, Billy Beane, and how a successful team, even with a limited budget, could be built on the basis of statistical analysis and insight, and not on intuitive judgments from other baseball personnel (such as from coaches, scouts, or baseball writers). (We note that a movie version of *Moneyball* opened in September 2011 with Brad Pitt in the role of Billy Beane.)

A contentious aspect of using regression and other types of models to drive decision making arises when "experts" are overridden (or their assessments second-guessed and discounted, or their livelihoods threatened) by replacing their judgments with those provided by an equation. One particularly entertaining example is in the prediction of wine quality in the Bordeaux or elsewhere. Here, we have wine experts such as Robert Parker (of the *Wine Advocate*), pitted against econometricians such as Orley Ashenfelter (of Princeton). One good place to start is with the *Chance* article by Ashenfelter, Ashmore, and LaLonde, "Bordeaux Wine Vintage Quality and the Weather" (*8*(4), 1995, 7–14). As the article teaser states: "Statistical prediction of wine prices based on vintage growing-season characteristics produces consternation among wine 'experts'." We also note an earlier article from the *New York Times* by Peter Passell (March 4, 1990), with the cute double-entendre title "Wine Equation Puts Some Noses Out of Joint."

6.6 The (Un)reliability of Clinical Prediction

> Prosecutors in Dallas have said for years—any prosecutor can convict
> a guilty man. It takes a great prosecutor to convict an innocent man.
> – Melvyn Bruder (*The Thin Blue Line*, 1988)

This last section on prediction concerns the (un)reliability of clinical (behavioral) prediction, particularly for violence, and includes two extensive redactions in the Appendix Supplements: one is the majority opinion in the Supreme Court case of *Barefoot v. Estelle* (1983) and an eloquent Justice Blackmun dissent; the second is an *amicus curiae* brief in this same case from the American Psychiatric Association on the accuracy of clinical prediction of future violence. Both of these documents are detailed, self-explanatory, and highly informative about our current lack of ability to make clinical assessments that lead to accurate and reliable predictions of future behavior. To set the background for the *Barefoot v. Estelle* case, the beginning part of the *amicus curiae* brief follows; a redaction of the remainder of the brief, as already noted, is given in an appendix at the end of the section.

Brief for American Psychiatric Association as *Amicus Curiae* Supporting Petitioner, Barefoot v. Estelle

Petitioner Thomas A. Barefoot stands convicted by a Texas state court of the August 7, 1978 murder of a police officer—one of five categories of homicides for which Texas law authorizes the imposition of the death penalty. Under capital sentencing procedures established after this Court's decision in Furman v. Georgia, the "guilt" phase of petitioner's trial was followed by a separate sentencing proceeding in which the jury was directed to answer three statutorily prescribed questions. One of these questions—and the only question of relevance here—directed the jury to determine: whether there is a probability that the defendant would commit criminal acts of violence that would constitute a continuing threat to society. The jury's affirmative response to this question resulted in petitioner being sentenced to death.

The principle evidence presented to the jury on the question of petitioner's "future dangerousness" was the expert testimony of two psychiatrists, Dr. John T. Holbrook and Dr. James Grigson, both of whom testified for the prosecution. Petitioner elected not to testify in his own defense. Nor did he present any evidence or testimony, psychiatric or otherwise, in an attempt to rebut the state's claim that he would commit future criminal acts of violence.

Over defense counsel's objection, the prosecution psychiatrists were permitted to offer clinical opinions regarding petitioner, including their opinions on the ultimate issue of future dangerousness, even though they had not performed a psychiatric examination or evaluation of him. Instead, the critical psychiatric testimony was elicited through an extended hypothetical question propounded by the prosecutor. On the basis of the assumed facts stated in the hypothetical, both Dr. Holbrook and Dr. Grigson gave essentially the same testimony.

First, petitioner was diagnosed as a severe criminal sociopath, a label variously defined as describing persons who "lack a conscience," and who "do things which serve their own purposes without regard for any consequences or outcomes to other people." Second, both psychiatrists testified that petitioner would commit criminal acts of violence in the future. Dr. Holbrook stated that he could predict petitioner's future behavior in this regard "within reasonable psychiatric certainty." Dr. Grigson was more confident, claiming predictive accuracy of "one hundred percent and absolute."

The prosecutor's hypothetical question consisted mainly of a cataloguing of petitioner's past antisocial behavior, including a description of his criminal record. In addition, the hypothetical question contained a highly detailed summary of the prosecution's evidence introduced during the guilt phase of the trial, as well as a brief statement concerning petitioner's behavior and demeanor during the period from his commission of the murder to his later apprehension by police.

In relevant part, the prosecutor's hypothetical asked the psychiatrists to assume as true the following facts: First, that petitioner had been convicted of five criminal offenses—all of them nonviolent, as far as the record reveals—and that he had also been arrested and charged on several counts of sexual offenses involving children. Second, that petitioner had led a peripatetic existence and "had a bad reputation for peaceful and law abiding citizenship" in each of eight communities that he had resided in during the previous ten years. Third, that in the two-month period preceding the murder, petitioner was unemployed, spending much of his time using drugs, boasting of his plans to commit numerous crimes, and in various ways deceiving certain acquaintances with whom he was living temporarily. Fourth, that petitioner had murdered the police officer as charged, and that he had done so with "no provocation whatsoever" by shooting the officer in the head "from a distance of no more than six inches." And fifth, that subsequent to the murder, petitioner was observed by one witness, "a homosexual," who stated that petitioner "was not in any way acting unusual or that anything was bothering him or upsetting him ..."

Testimony of Dr. Holbrook:

Dr. Holbrook was the first to testify on the basis of the hypothetical question. He stated that the person described in the question exhibited "probably six or seven major criterias (sic) for the sociopath in the criminal area within reasonable medical certainty." Symptomatic of petitioner's sociopathic personality, according to Dr. Holbrook, was his consistent "antisocial behavior" from "early life into adulthood," his willingness to take any action which "serves [his] own purposes" without any regard for the "consequences to other people," and his demonstrated failure to establish any "loyalties to the normal institutions such as family, friends, politics, law or religion."

Dr. Holbrook explained that his diagnosis of sociopathy was also supported by petitioner's past clinical violence and "serious threats of violence," as well as an apparent history of "escaping or running away from authority" rather than "accepting a confrontation in the legal way in a court of law." And finally, Dr. Holbrook testified that petitioner had shown a propensity to "use other people through lying and manipulation ... " According to Dr. Holbrook, by use of such manipulation the sociopath succeeds in "enhancing [his] own ego image ... It makes [him] feel good."

After stating his diagnosis of sociopathy, Dr. Holbrook was asked whether he had an "opinion within reasonable psychiatric certainty as to whether or not there is a probability that the Thomas A. Barefoot in that hypothetical will commit criminal acts of violence in the future that would constitute a continuing threat to society?" Without attempting to explain the implied clinical link between his diagnosis of petitioner and his prediction of future dangerousness, Dr. Holbrook answered simply: "In my opinion he will."

Testimony of Dr. Grigson:

On the basis of the prosecutor's hypothetical question, Dr. Grigson diagnosed petitioner as "a fairly classical, typical, sociopathic personality disorder" of the "most severe category." The most "outstanding characteristic" of persons fitting this diagnosis, according to Dr. Grigson, is the complete "lack of a conscience." Dr. Grigson stated that such persons "repeatedly break the rules, they con, manipulate and use people, [and] are only interested in their own self pleasure [and] gratification."

Although Dr. Grigson testified that some sociopathic individuals do not pose a continuing threat to society, he characterized petitioner as "your most severe sociopath." Dr. Grigson stated that persons falling into this special category are "the ones that . . . have complete disregard for another human being's life." Dr. Grigson further testified that "there is not anything in medicine or psychiatry or any other field that will in any way at all modify or change the severe sociopath."

The prosecutor then asked Dr. Grigson to state his opinion on the ultimate issue—"whether or not there is a probability that the defendant . . . will commit criminal acts of violence that would constitute a continuing threat to society?" Again, without explaining the basis for his prediction or its relationship to the diagnosis of sociopathy, Dr. Grigson testified that he was "one hundred percent" sure that petitioner "most certainly would" commit future criminal acts of violence. Dr. Grigson also stated that his diagnosis and prediction would be the same whether petitioner "was in the penitentiary or whether he was free."

The psychiatrist featured so prominently in the opinions for *Barefoot v. Estelle* and the corresponding American Psychiatric Association *amicus* brief, James Grigson, played the same role repeatedly in the Texas legal system. For over three decades before his retirement in 2003, he testified when requested at death sentence hearings to a high certainty as to "whether there is a probability that the defendant would commit criminal acts of violence that would constitute a continuing threat to society." An affirmative answer by the sentencing jury imposed the death penalty automatically, as it was on Thomas Barefoot; he was executed on October 30, 1984. When asked if he had a last statement to make, he replied:

> Yes, I do. I hope that one day we can look back on the evil that we're doing right now like the witches we burned at the stake. I want everybody to know that I hold nothing against them. I forgive them all. I hope everybody I've done anything to will forgive me. I've been praying all day for Carl Levin's wife to drive the bitterness from her heart because that bitterness that's in her heart will send her to Hell just as surely as any other sin. I'm sorry for everything I've ever done to anybody. I hope they'll forgive me.

James Grigson was expelled in 1995 from the American Psychiatric Association and the Texas Association of Psychiatric Physicians for two chronic ethics violations: making statements in testimony on defendants he had not actually examined, and for predicting violence with 100% certainty. The press gave him the nickname of "Dr. Death."[8] The role that he played was similar to that of Roy Meadow, the villain in the Sally Clark case described earlier who crusaded against mothers supposedly abusing their children (remember the Munchausen Syndrome by Proxy). In Grigson's case, it was sociopaths he wanted put to death, as opposed to receiving just life imprisonment (without parole).

Barefoot v. Estelle has another connection with our earlier discussions about the distinctions between actuarial and clinical prediction, and where the former is commonly better than the latter. There is evidence mentioned in the APA brief that actuarial predictions of violence carried out by statistically informed laymen might be better than those of a clinician. This may be due to a bias that psychiatrists might (unsuspectingly) have in overpredicting violence because of the clients they see or for other reasons related to their practice. There is a pertinent passage from the court opinion (not given in our redactions):

> That psychiatrists actually may be less accurate predictors of future violence than laymen, may be due to personal biases in favor of predicting violence arising from the fear of being responsible for the erroneous release of a violent individual. ... It also may be due to a tendency to generalize from experiences with past offenders on bases that have no empirical relationship to future violence, a tendency that may be present in Grigson's and Holbrook's testimony. Statistical prediction is clearly more reliable than clinical prediction ... and prediction based on statistics alone may be done by anyone.

The two psychiatrists mentioned in *Barefoot v. Estelle*, James Grigson and John Holbrook, appeared together repeatedly in various capital sentencing hearings in Texas during the later part of the 20th century. Although Grigson was generally the more outrageous of the two with predictions of absolute certitude based on a sociopath diagnosis, Holbrook was similarly at fault ethically. This pair of psychiatrists of Texas death penalty fame might well be nicknamed "Dr. Death" and "Dr. Doom." They were both culpable in the famous exoneration documented in the award winning film by Errol Morris, *The Thin Blue Line*. To tell this story, we give the summary of the Randall Dale Adams exoneration from the

Northwestern University School of Law and its Center on Wrongful Convictions (written by Robert Warden with Michael L. Radelet):

Sentenced to death in 1977 for the murder of a police officer in Dallas, Texas, Randall Dale Adams was exonerated as a result of information uncovered by film-maker Errol Morris and presented in an acclaimed 1988 documentary, *The Thin Blue Line.*

Patrolman Robert Wood was shot to death during a traffic stop on November 28, 1976, by sixteen-year-old David Ray Harris, who framed Adams to avoid prosecution himself. Another factor in the wrongful conviction was the surprise—and partly perjured— testimony of three eyewitnesses whose existence had been concealed from the defense until the witnesses appeared in the courtroom. A third factor was a statement Adams signed during interrogation that the prosecution construed as an admission that he had been at the scene of the crime.

The day before the murder, Adams was walking along a Dallas street after his car had run out of gasoline. Harris happened by, driving a stolen car. He offered Adams a ride and the two wound up spending the afternoon and evening together, drinking beer, smoking marijuana, pawning various items Harris had stolen, and going to a drive-in movie theater to watch porn movies. Adams then returned to a motel where he was staying.

Shortly after midnight, Wood and his partner, Teresa Turko, spotted Harris driving a blue car with no headlights. The officers stopped the car and, as Wood approached the driver's side, Harris shot him five times. Wood died on the spot. As the car sped off, Turko fired several shots, but missed. She did not get a license number. She seemed certain that there was only one person in the car—the driver.

Harris drove directly to his home in Vidor, 300 miles southeast of Dallas. Over the next several days, he bragged to friends that he had "offed a pig" in Dallas. When police in Vidor learned of the statements, they took Harris in for questioning. He denied having had anything to do with the murder, claiming he had said otherwise only to impress his friends. But when police told him that a ballistics test established that a pistol he had stolen from his father was the murder weapon, Harris changed his story. He now claimed that he had been present at the shooting, but that it had been committed by a hitchhiker he had picked up—Adams.

Adams, an Ohio native working in Dallas, was taken in for questioning. He denied any knowledge of the crime, but he did give a detailed statement describing his activities the day before the murder. Police told him he had failed a polygraph test and that Harris had passed one, but Adams remained resolute in asserting his innocence.

Although polygraph results are not admissible in Texas courts, the results provided some rationale for questioning Harris's story. However, when a police officer is murdered, authorities usually demand the most severe possible punishment, which in Texas, and most other United States jurisdictions, is death. Harris was only sixteen and ineligible for the death penalty; Adams was twenty-seven and thus could be executed.

At trial before Dallas County District Court Judge Don Metcalfe and a jury, Turko testified that she had not seen the killer clearly, but that his hair was the color of Adams's. She also said that the killer wore a coat with a fur collar. Harris had such a coat, but Adams did not.

Adams took the stand and emphatically denied having any knowledge of the crime. But then the prosecution sprang two surprises. The first was the introduction of Adams's purported signed statement, which police and prosecutors claimed was a confession, although it said only—falsely, according to Adams—that when he was in the car with Harris, they had at one point been near the crime scene. The second was the testimony of three purported eyewitnesses whose existence had until then been unknown to the defense. One of these witnesses, Michael Randell, testified that he had driven by the scene shortly before the murder and, in the car that had been stopped by the officers, had seen two persons, one of whom he claimed was Adams. The other two witnesses, Robert and Emily Miller, had happened by at about the same time, but claimed to have seen only one person in the car—Adams.

Because the eyewitnesses were called only to rebut Adams's testimony, prosecutors claimed that Texas law did not require them to inform the defense of their existence before they testified. The weekend after their surprise testimony, however, the defense learned that Emily Miller had initially told police that the man she had seen appeared to be Mexican or a light-skinned African American. When the defense asked to recall the Millers to testify, the prosecution claimed that the couple had left town. In fact, the Millers had only moved from one part of Dallas to another. When the defense asked to introduce Emily Miller's statement, Judge Metcalfe would not allow it. He said it would be unfair to impeach her credibility when she was not available for further examination.

The jury quickly returned a verdict of guilty and turned to sentencing. Under Texas law, in order for Adams to be sentenced to death, the jury was required to determine, among other things, whether there was "beyond a reasonable doubt [a] probability" that he or she would commit future acts of violence. To establish that Adams met that oxymoronic criterion, the prosecution called Dr. James Grigson, a Dallas psychiatrist known as "Dr. Death," and Dr.

John Holbrook, former chief of psychiatry for the Texas Department of Corrections.

Although the American Psychiatric Association has said on several occasions that future dangerousness was impossible to predict, Grigson and Holbrook testified that Adams would be dangerous unless executed. Grigson testified similarly in more than 100 other Texas cases that ended in death sentences. After hearing the psychiatrists, Adams's jury voted to sentence him to death. Twenty one months later, at the end of January 1979, the Texas Court of Criminal Appeals affirmed the conviction and death sentence. Judge Metcalfe scheduled the execution for May 8, 1979.

Adams was three days away from execution when United States Supreme Court Justice Lewis F. Powell Jr. ordered a stay. Powell was troubled that prospective jurors with moral qualms about the death penalty had been excluded from service, even though they had clearly stated that they would follow the Texas law.

To most observers—including, initially, Dallas District Attorney Henry Wade (of Roe v. Wade fame) the Supreme Court's language meant that Adams was entitled to a new trial. But a few days later Wade announced that a new trial would be a waste of money. Thus, he said, he was asking Governor Bill Clements to commute Adams's sentence to life in prison. When the governor promptly complied, Wade proclaimed that there now would be no need for a new trial. Adams, of course, thought otherwise, but the Texas Court of Criminal Appeals agreed with Wade. As a result of the governor's action, said the court, "There is now no error in the case."

In March 1985, Errol Morris arrived in Dallas to work on a documentary about Grigson—"Dr. Death." Morris's intent had not been to question the guilt of defendants in whose cases Grigson had testified but only to question his psychiatric conclusions. When Morris met Adams, the focus of the project changed.

Morris learned from Randy Schaffer, a volunteer Houston lawyer who had been working on the case since 1982, that Harris had not led an exemplary life after helping convict Adams. Harris had joined the Army and been stationed in Germany, where he had been convicted in a military court of a series [of] burglaries and sent to prison in Leavenworth, Kansas. A few months after his release, Harris had been convicted in California of kidnapping, armed robbery, and related crimes.

After his release from prison in California, and five months after Morris arrived in Dallas, Harris tried to kidnap a young woman named Roxanne Lockard in Beaumont, Texas. In an effort to prevent the abduction, Lockard's boyfriend, Mark Mays, exchanged gunfire with Harris. Mays was shot to death and Harris was wounded. For the Mays murder—a crime that would not have occurred if Dallas authorities convicted the actual killer of Officer Wood eight years earlier—Harris was sentenced to death.

Meanwhile, Morris and Schaffer discovered that Officer Turko had been hypnotized during the investigation and initially had acknowledged that she had not seen the killer—facts that the prosecution had illegally withheld from the defense. Morris and Schaffer also found that robbery charges against the daughter of eyewitness Emily Miller had been dropped after Miller agreed to identify Adams as Wood's killer. The new information, coupled with the fact that Miller initially had described the killer as Mexican or African American, became the basis for a new trial motion.

In 1988, during a three-day hearing on the motion before Dallas District Court Judge Larry Baraka, Harris recanted. "Twelve years ago, I was a kid, you know, and I'm not a kid anymore, and I realize I've been responsible for a great injustice," Harris told Baraka. "And I felt like it's my responsibility to step forward, to be a man, to admit my part in it. And that's why I'm trying to correct an injustice."

On December 2, 1988, Judge Baraka recommended to the Texas Court of Criminal Appeals that Adams be granted a new trial, and two months later he wrote a letter to the Texas Board of Pardons and Paroles recommending that Adams be paroled immediately. The board refused, but on March 1 the Texas Court of Criminal Appeals unanimously concurred with Baraka that Adams was entitled to a new trial. Three weeks later, Adams was released on his own recognizance, and two days after that, Dallas District Attorney John Vance, who had succeeded Wade, dropped all charges.

Harris was never tried for the murder of Officer Woods. On June 30, 2004, he was executed for the Mays murder.

A later chapter of this book discusses in detail the *Federal Rules of Evidence* and the admissibility of expert witnesses and scientific data. The central case discussed in that chapter is *Daubert v. Merrell Dow Pharmaceuticals* (1993) that promulgates the *Daubert* standard for admitting expert testimony in federal courts. The majority opinion in *Daubert* was written by Justice Blackmun, the same justice who wrote the dissent in *Barefoot v. Estelle*. The court stated that Rule 702 of the *Federal Rules of Evidence* was the governing standard for admitting scientific evidence in trials held in federal court (and now in most state courts as well). Rule 702, Testimony by Experts, states:

> If scientific, technical, or other specialized knowledge will assist the trier of fact to understand the evidence or to determine a fact in issue, a witness qualified as an expert by knowledge, skill, experience,

training, or education, may testify thereto in the form of an opinion or otherwise, if (1) the testimony is based upon sufficient facts or data, (2) the testimony is the product of reliable principles and methods, and (3) the witness has applied the principles and methods reliably to the facts of the case.

We give a redaction of that part of the Wikipedia article on *Daubert v. Merrell Dow Pharmaceuticals* devoted to the discussion of the *Daubert* standard governing expert testimony. We doubt that clinical predictions of violence based on a sociopath diagnosis would be admissible under the *Daubert* standard.

> The Standard Governing Expert Testimony: Three key provisions of the Rules governed admission of expert testimony in court. The first was *scientific knowledge*. This means that the testimony must be scientific in nature, and that the testimony must be grounded in "knowledge." Of course, science does not claim to know anything with absolute certainty; science "represents a *process* for proposing and refining theoretical explanations about the world that are subject to further testing and refinement." The "scientific knowledge" contemplated by Rule 702 had to be arrived at by the scientific method.
>
> Second, the scientific knowledge must *assist the trier of fact* in understanding the evidence or determining a fact in issue in the case. The trier of fact is often either a jury or a judge; but other fact-finders may exist within the contemplation of the federal rules of evidence. To be helpful to the trier of fact, there must be a "valid scientific connection to the pertinent inquiry as a prerequisite to admissibility." Although it is within the purview of scientific knowledge, knowing whether the moon was full on a given night does not typically assist the trier of fact in knowing whether a person was sane when he or she committed a given act.
>
> Third, the Rules expressly provided that the judge would make the threshold determination regarding whether certain scientific knowledge would indeed assist the trier of fact in the manner contemplated by Rule 702. "This entails a preliminary assessment of whether the reasoning or methodology underlying the testimony is scientifically valid and of whether that reasoning or methodology properly can be applied to the facts in issue." This preliminary assessment can turn on whether something has been tested, whether an idea has been subjected to scientific peer review or published in scientific journals, the rate of error involved in the technique, and even general acceptance, among other things. It focuses on methodology and principles, not the ultimate conclusions generated.
>
> The Court stressed that the new standard under Rule 702 was rooted in the judicial process and intended to be distinct and separate

from the search for scientific truth. "Scientific conclusions are subject to perpetual revision. Law, on the other hand, must resolve disputes finally and quickly. The scientific project is advanced by broad and wide-ranging consideration of a multitude of hypotheses, for those that are incorrect will eventually be shown to be so, and that in itself is an advance." Rule 702 was intended to resolve legal disputes, and thus had to be interpreted in conjunction with other rules of evidence and with other legal means of ending those disputes. Cross examination within the adversary process is adequate to help legal decision makers arrive at efficient ends to disputes. "We recognize that, in practice, a gate-keeping role for the judge, no matter how flexible, inevitably on occasion will prevent the jury from learning of authentic insights and innovations. That, nevertheless, is the balance that is struck by Rules of Evidence designed not for the exhaustive search for cosmic understanding but for the particularized resolution of legal disputes."

As noted in the various opinions and *amicus* brief given in *Barefoot v. Estelle*, the jury in considering whether the death penalty should be imposed, has to answer affirmatively one question: whether there was a probability that the defendant would commit criminal acts of violence that would constitute a continuing threat to society. The use of the word "probability" without specifying any further size seems odd to say the least, but Texas courts have steadfastly refused to delimit it any further. So, presumably a very small probability of future violence would be sufficient for execution if this small probability could be proved "beyond a reasonable doubt."

The point of much of this section has been to emphasize that actuarial evidence about future violence involving variables such as age, race, or sex, is all there really is in making such predictions. More pointedly, the assignment of a clinical label, such as "sociopath," adds nothing to an ability to predict, and to suggest that it does is to use the worst "junk science," even though it may be routinely assumed true in the larger society. All we have to rely on is the usual psychological adage that the best predictor of future behavior is past behavior. Thus, the best predictor of criminal recidivism is a history of such behavior, and past violence suggests future violence. The greater the amount of past criminal behavior or violence, the more likely that such future behavior or violence will occur (a behavioral form of a "dose-response" relationship).

At its basis, this is statistical evidence of such a likely occurrence and no medical or psychological diagnosis is needed or useful.

Besides the specious application of a sociopath diagnosis to predict future violence, after the Supreme Court decision in *Estelle v. Smith* (1981) such a diagnosis had to be made on the basis of a hypothetical question and not on an actual psychological examination of the defendant. In addition to a 100% incontrovertible assurance of future violence, offering testimony without actually examining a defendant proved to be Grigson's eventual downfall and one reason for the expulsion from his professional psychiatric societies. This prevention of an actual examination of a defendant by the Supreme Court case, *Estelle v. Smith* (1981), also involved James Grigson. Ernest Smith, indicted for murder, had been examined by Grigson in jail and who determined he was competent to stand trial. In the psychiatric report on Smith, Grigson termed him "a severe sociopath" but gave no other statements as to future dangerousness. Smith was sentenced to death based on the sociopath label given by Grigson. In *Estelle v. Smith* the Supreme Court held that because of the well-known case of *Miranda v. Arizona* (1966), the state could not force a defendant to submit to a psychiatric examination for the purposes of sentencing because it violated a defendant's Fifth Amendment rights against self-incrimination and the Sixth Amendment right to counsel. Thus, the examination of Ernest Smith was inadmissible at sentencing. From that point on, predictions of violence were made solely on hypothetical questions and Grigson's belief that a labeling as a sociopath was sufficient to guarantee future violence on the part of a defendant, and therefore, the defendant should be put to death.

The offering of a professional psychiatric opinion about an individual without direct examination is an ethical violation of the Goldwater Rule, named for the Arizona Senator who ran for President in 1964 as a Republican. Promulgated by the American Psychiatric Association in 1971, it delineated a set of requirements for communication with the media about the state of mind of individuals. The Goldwater Rule was the result of a special September/October 1964 issue of *Fact:* magazine, published by the highly provocative Ralph Ginzburg. The issue title was "The Unconscious of a Conservative: Special Issue on the Mind of Barry Goldwater," and reported on a mail survey of 12,356 psychiatrists, of whom

2,417 responded: 24% said they did not know enough about Goldwater to answer the question; 27% said he was mentally fit; 49% said he was not. Much was made of Goldwater's "two nervous breakdowns," because such a person should obviously never be President because of a risk of recurrence under stress that might then lead to pressing the nuclear button.

Goldwater brought a $2 million libel suit against *Fact:* and its publisher, Ginzburg. In 1970 the United States Supreme Court decided in Goldwater's favor giving him $1 in compensatory damages and $75,000 in punitive damages. More important, it set a legal precedent that changed medical ethics forever. For an updated discussion of the Goldwater Rule, this time because of the many psychiatrists commenting on the psychological makeup of the former chief of the International Monetary Fund, Dominique Strauss-Kahn, after his arrest on sexual assault charges in New York, see Richard A. Friedman's article, "How a Telescopic Lens Muddles Psychiatric Insights" (*New York Times*, May 23, 2011).

Appendix: Continuation of the American Psychiatric Association *Amicus Curiae* Brief, Barefoot v. Estelle

Given in the Appendix Supplements.

Appendix: Opinion and Dissent in the United States Supreme Court, Barefoot v. Estelle (Decided, July 6, 1983)

Given in the Appendix Supplements.

Notes

[1]Several topics involving prediction do not (necessarily) concern linear regression, and because of this, no extended discussion of these is given in this chapter. One area important for the legal system is Sex Offender Risk Assessment, and the prediction of recidivism for committing another offense. Several such instruments are available, with most relying on some simple point counting system based on descriptive information about the subject and previous crimes (and readily available in a subject's "jacket"). One of the easiest to implement is the Rapid Risk Assessment for Sexual Offender Recidivism (or

the more common acronym, RRASOR, and pronounced "razor"). It is based on four items:

Prior Sex Offense Convictions: 0, 1, 2, or 3 points for 0, 1, 2, or 3+ prior convictions, respectively; Victim Gender: only female victims (0 points); only male victims (1 point); Relationship to Victim: only related victims (0 points); any unrelated victim (1 point); Age at Release: 25 or more (0 points); 18 up to 25 (1 point).

As the RRASOR author Hanson (1997) notes in validation work, those with a score of zero had a recidivism rate of 6.5% after 10 years; for those who scored 5, the rate was 73% after 10 years: R. K. Hanson (1997), *The Development of a Brief Actuarial Risk Scale for Sexual Offense Recidivism*, Ottawa: Solicitor General of Canada.

Another approach to prediction that we do not develop is in the theory behind chaotic systems, such as the weather. A hallmark of such dynamic prediction problems is an extreme sensitivity to initial conditions, and a general inaccuracy in prediction even over a relatively short time frame. The person best known for chaos theory is Edward Lorenz and his "butterfly effect," where a very small differences in the initial conditions for a dynamical system (for example, a butterfly flapping its wings somewhere in Latin America), may produce large variations in the long-term behavior of the system ("Butterfly effect," 2010).

[2]As an example that might better suit the interests of our students than inherited physical stature, you could point out that highly intelligent women tend to marry men who are less intelligent than they are. But just as the complaints start to arise, you could turn the observation around—highly intelligent men also tend to marry women who are less so.

[3]A 2005 article by Robyn Dawes in the *Journal of Clinical Psychology* (*61*, 1245–1255) has the intriguing title "The Ethical Implications of Paul Meehl's Work on Comparing Clinical Versus Actuarial Prediction Methods." Dawes' main point is that given the overwhelming evidence we now have, it is unethical to use clinical judgment in preference to the use of statistical prediction rules. We quote from the abstract:

> Whenever statistical prediction rules ... are available for making a relevant prediction, they should be used in preference to intuition. ....
> Providing service that assumes that clinicians "can do better" simply based on self-confidence or plausibility in the absence of evidence that they can actually do so is simply unethical. (p. 1245)

[4]For a popular and fairly recent discussion of the clinical/expert versus statistical/computer comparison, see the Suggested Reading, B.1.3: "Maybe We Should Leave That Up to the Computer" (Douglas Heingartner, *New York Times*, July 18, 2006).

[5]We also note that improved cross-validation may at times be obtained

without going to the extremes of ignoring data in estimation, by using shrunken (empirical Bayes) estimates.

[6] Attributed to John von Neumann by Enrico Fermi, as quoted by Freeman Dyson in "A Meeting With Enrico Fermi" (*Nature*, *427*, p. 297).

[7] To give an anecdote about multiple regression not always being the silver bullet some think it is, one of our colleagues many years ago was working on a few creative schemes for augmenting the income of his poverty-stricken quantitative psychology graduate students, and came up with the following idea: he would look for opportunities for statistical consulting anywhere he could, with the offer of him doing the consulting for free as long as the client would then pay the graduate students to carry out the suggested analyses. The first client advertised in the local newspaper for someone with quantitative expertise to assist in a project being carried out in conjunction with a law enforcement agency. The project involved trying to predict blood alcohol level from various dichotomous behavioral indicators for a driver during the operation of a car. Our colleague got $500 for the graduate student to analyze the dataset, which contained an objective blood alcohol level plus a number of 0/1 variables (because of the proprietary nature of the project, the actual meaning of the variables was unknown; they were numbered arbitrarily).

The graduate student ran a version of stepwise regression on a mainframe version of BMDP. Three variables seemed to have some ability to predict blood alcohol level; all the others were more-or-less washouts. Our colleague communicated the variable numbers that seemed to have some explanatory power to the project head. Thereupon, a variety of expletives (none deleted) were used. How dare he just come up with these three when there was obviously such a treasure trove of other subtle behavioral information available in the dataset just waiting to be found! Phrases such as "statistical charlatan" were freely used, but the graduate student still got to keep the $500.

The three dichotomous variables with some explanatory power were: whether (or not) the driver hit a building; whether (or not) the driver waved at a police officer when passing by; whether (or not) the driver was asleep at the side of the road with the car running. Although not subtle, these seem strong indicators to us.

[8] See the article by Lisa Belkin, "The Law; Expert Witness Is Unfazed by 'Dr. Death' Label" (*New York Times*, June 10, 1988).

Chapter 7

The Basic Sampling Model and Associated Topics

I know of scarcely anything so apt to impress the imagination as the wonderful form of cosmic order expressed by the 'Law of Frequency of Error.' The law would have been personified by the Greeks and deified, if they had known of it. It reigns with serenity and in complete self-effacement, amidst the wildest confusion. The huger the mob, and the greater the apparent anarchy, the more perfect is its sway. It is the supreme law of Unreason. Whenever a large sample of chaotic elements are taken in hand and marshaled in the order of their magnitude, an unsuspected and most beautiful form of regularity proves to have been latent all along.
— Sir Francis Galton (*Natural Inheritance*, 1889)

We begin by refreshing our memories about the distinctions between *population* and *sample*, *parameters* and *statistics*, and *population distributions* and *sampling distributions*. Someone who has successfully completed a graduate sequence in statistics should know these distinctions very well. Here, only a simple univariate framework is considered explicitly, but an obvious and straightforward generalization exists for the multivariate context.

A *population* of interest is posited, and operationalized by some random variable, say X. In this *Theory World* framework, X is characterized by *parameters*, such as the expectation of X, $\mu = E(X)$, or its variance, $\sigma^2 = V(X)$. The random variable X has a *(population) distribution*, which is often assumed normal. A *sample* is generated by taking observations on X, say, $X_1, \ldots, X_n$, considered independent and identically distributed as X; that is, they are exact copies of X. In this *Data World* context, statistics are functions of the sample and therefore characterize the sample: the sample mean, $\hat{\mu} = \frac{1}{n} \sum_{i=1}^{n} X_i$; the sample variance, $\hat{\sigma}^2 = \frac{1}{n} \sum_{i=1}^{n} (X_i - \hat{\mu})^2$, with some possible variation in dividing by $n - 1$ to generate an unbiased estimator for σ^2. The statistics, $\hat{\mu}$ and $\hat{\sigma}^2$, are *point estimators* of μ and σ^2. They are random variables by themselves, so they have distributions referred to as *sampling distributions.* The general problem of statistical inference is to ask what sample statistics, such as $\hat{\mu}$ and $\hat{\sigma}^2$, tell us about their population counterparts, μ and σ^2. In other words, can we obtain a measure of accuracy for estimation from the sampling distributions through, for example, confidence intervals?

Assuming that the population distribution is normally distributed, the sampling distribution of $\hat{\mu}$ is itself normal with expectation μ and variance σ^2/n. Based on this result, an approximate 95% confidence interval for the unknown parameter μ can be given by

$$\hat{\mu} \ \pm \ 2.0\frac{\hat{\sigma}}{\sqrt{n}} \ .$$

Note that it is the square root of the sample size that determines the length of the interval (and not the sample size per se). This is both good news and bad. Bad, because if you want to double precision, you need a fourfold increase in sample size; good, because sample size can be cut by four with only a halving of precision.

Even when the population distribution is not originally normally distributed, the central limit theorem (CLT) (that is, the "Law of Frequency of Error," as noted by the opening epigram for this chapter) says that $\hat{\mu}$ is approximately normal in form and becomes exactly so as n goes to infinity. Thus, the approximate confidence interval statement remains valid even when the underlying distribution is not normal. Such a result is the basis for many claims of robustness; that is, when a procedure remains valid even if the assumptions under which it was derived may not be true, as long as some particular condition is satisfied; here, the condition is that the sample size be reasonably large. Although how large is big enough for a normal approximation to be adequate depends generally on the form of the underlying population distribution, a glance at a "t-table" will show that when the degrees of freedom are greater than 30, the values given are indistinguishable from those for the normal. Thus, we surmise that sample sizes above 30 should generally be large enough for the CLT to be used.

Besides the robustness of the confidence interval calculations for μ, the CLT also encompasses the law of large numbers (LLN). ,

As the sample size increases, the estimator, $\hat{\mu}$, gets closer to μ, and converges to μ at the limit as n goes to infinity. This is seen most directly in the variance of the sampling distribution for $\hat{\mu}$, which becomes smaller as the sample size gets larger.

The basic results obtainable from the CLT and LLN that averages are both less variable and more normal in distribution than individual observations, and that averages based on larger sample sizes will show less variability than those based on smaller sample sizes, have far-ranging and sometimes subtle influences on our reasoning skills. For example, suppose we would like to study organizations, such as schools, health care units, or governmental agencies, and have a measure of performance for the individuals in the units, and the average for each unit. To identify those units exhibiting best performance (or, in the current jargon, "best practice"), the top 10%, say, of units in terms of performance are identified; a determination is then made of what common factors might characterize these top-performing units. We are pleased when we are able to isolate one very salient feature that most units in this top tier are small. We proceed on this observation and advise the breaking up of larger units. Is such a policy really justified based on these data? Probably not, if one also observes that the bottom 10% are also small units. That smaller entities tend to be more variable than the larger entities seems to vitiate a recommendation of breaking up the larger units for performance improvement. Evidence that the now-defunct "small schools movement," funded heavily by the Gates Foundation, was a victim of the "square root of n law" was presented by Wainer (2009, pp. 11–14).

Sports is an area in which there is a great misunderstanding and lack of appreciation for the effects of randomness. A reasonable model for sports performance is one of "observed performance" being the sum of "intrinsic ability" (or true performance) and "error," leading to a natural variability in outcome either at the individual or the team level. Somehow it appears necessary for sports writers, announcers, and other pundits to give reasons for what is most likely just random variability. We hear of team "chemistry," good or bad, being present or not; individuals having a "hot hand" (or a "cold hand," for that matter); someone needing to "pull out of a slump"; why there might be many .400 hitters early in the season but not later; a player being "due" for a hit; free-throw failure because of "pressure"; and so on. Making decisions based on natural variation being somehow "predictive" or "descriptive" of the truth, is not very smart, to say the least. But it is done all the time—sports managers are fired and CEOs replaced for what may be just the traces of natural variability.

People asked to generate random sequences of numbers tend to underestimate the amount of variation that should be present; for example, there are not enough longer runs and a tendency to produce too many short alternations. In a similar way, we do not see the naturalness in regression toward the mean, where extremes are followed by less extreme observations just because of fallibility in observed performance. Again, causes are sought. We hear about multi-round golf tournaments where a good performance on the first day is followed by a less adequate score the second (due probably to "pressure"); or a bad performance on the first day followed by an improved performance the next (the golfer must have been able to "play loose"). Or in baseball, at the start of a season an underperforming Derek Jeter might be under "pressure" or too much "media scrutiny," or subject to the difficulties of performing in a "New York market." When individuals start off well but then appear to fade, it must be because people are trying to stop them ("gunning" for someone is a common expression). One should always remember that in estimating intrinsic ability, individuals are unlikely to be as good (or as bad) as the pace they are on. It is always a better bet to vote against someone eventually breaking a record, even when they are "on a pace" to so do early in the season. This may be one origin for the phrase "sucker bet"—a gambling wager where your expected return is significantly lower than your bet.

Another area where one expects to see a lot of anomalous results is when the dataset is split into ever-finer categorizations that end up having few observations in them, and thus subject to much greater variability. For example, should we be overly surprised if Albert Pujols doesn't seem to bat well in domed stadiums at night when batting second against left-handed pitching? The pundits look for "causes" for these kinds of extremes when they should just be marveling at the beauty of natural variation and the effects of sample size. A similar and probably more important misleading effect occurs when our data are on the effectiveness of some medical treatment, and we try to attribute positive or negative results to ever-finer-grained classifications of the clinical subjects.

Random processes are a fundamental part of nature and ubiquitous in our day-to-day lives. Most people do not understand them, or worse, fall under an "illusion of control" and believe they have influence over how events progress. Thus, there is an almost mystical belief in the ability of a new coach, CEO, or president to "turn things around." Part of these strong beliefs may result from the operation of regression toward the mean or the natural unfolding of any random process. We continue to get our erroneous beliefs reconfirmed when cause is attributed when none may actually be present. As humans we all wish to believe we can affect our future, but when events have dominating stochastic components, we are obviously not in complete control. There appears to be a fundamental clash between our ability to recognize the operation of randomness and the need for control in our lives.

An appreciation for how random processes might operate can be helpful in navigating the uncertain world we live in. When investments with Bernie Madoff give perfect 12% returns, year after year, with no exceptions and no variability, alarms should go off. If we see a supposed scatterplot of two fallible variables with a least-squares line imposed but where the actual data points have been withdrawn, remember that the relationship is not perfect. Or when we monitor error in quality assurance and control for various manufacturing or diagnostic processes (for example, application of radiation in medicine), and the tolerances become consistently beyond the region where we should generally expect the process to vary, a need to stop and recalibrate may be necessary. It is generally important to recognize that data interpretation may be a long-term process, with a need to appreciate variation appearing around a trend line. Thus, the immediacy of some major storms does not vitiate a longer-term perspective on global climate change. Remember the old meteorological adage: climate is what you expect; weather is what you get. Relatedly, it is important to monitor processes we have some personal responsibility for (such as our own lipid panels when we go for physicals), and to assess when unacceptable variation appears outside of our normative values.

Besides having an appreciation for randomness in our day-to-day lives, there is also a flip side: if you don't see randomness when you probably should, something is amiss. The Bernie Madoff example noted above is a salient example, but there are many such deterministic traps awaiting the gullible. When something seems just too good to be true, most likely it isn't true. A recent ongoing case in point involves the Dutch social psychologist, Diederik Stapel, and the massive fraud he committed in the very best psychology journals in the field. A news item by G. Vogel in *Science* (2011, *334*, p. 579) has the title, "Psychologist Accused of Fraud on 'Astonishing Scale'." Basically, in dozens of published articles and doctoral dissertations he supervised, Stapel never failed to obtain data showing the clean results he expected to see at the outset. As any practicing researcher in the behavioral sciences knows, this is just too good to be true. We give a short quotation from the *Science* news item commenting on the Tilberg University report on the Stapel affair (authored by a committee headed by the well-known Dutch psycholinguist, Willem Levelt):

> Stapel was "absolute lord of the data" in his collaborations ... many of Stapel's datasets have improbable effect sizes and other statistical irregularities, the report says. Among Stapel's colleagues, the description of data as too good to be true "was a heartfelt compliment to his skill and creativity." (p. 579)

The report discusses the presence of consistently large effects being found; few missing data and outliers; hypotheses rarely refuted. Journals publishing Stapel's articles did not question the omission of details about the source of the data. As understated by Levelt, "We see that the scientific checks and balances process has failed at several levels" (Callaway, 2011, p. 15). In a related article in the *New York Times* by Benedict Carey (November 2, 2011), "Fraud Case Seen as a Red Flag for Psychology Research," the whole field of psychology is now taken to task, appropriately we might add, in how research has generally been done and evaluated in the field. Part of the Levelt Committee report that deals explicitly with data and statistical analysis is redacted below:

> *The data were too good to be true; the hypotheses were almost always confirmed; the effects were improbably large; missing data, or impossible, out-of-range data, are rare or absent.*

This is possibly the most precarious point of the entire data fraud. Scientific criticism and approach failed on all fronts in this respect. The falsification of hypotheses is a fundamental principle of science, but was hardly a part of the research culture surrounding Mr. Stapel. The only thing that counted was verification. However, anyone with any research experience, certainly in this sector, will be aware that most hypotheses that people entertain do not survive. And if they do, the effect often vanishes with replication. The fact that Mr. Stapel's hypotheses were always confirmed should have caused concern, certainly when in most cases the very large "effect sizes" found were clearly out of line with the literature. Rather than concluding that this was all improbable, instead Mr. Stapel's experimental skills were taken to be phenomenal. "Too good to be true" was meant as a genuine compliment to his skill and creativity. Whereas all these excessively neat findings should have provoked thought, they were embraced. If other researchers had failed, that was assumed to be because of a lack of preparation, insight, or experimental skill. Mr. Stapel became the model: the standard. Evidently only Mr. Stapel was in a position to achieve the precise manipulations needed to make the subtle effects visible. People accepted, if they even attempted to replicate the results for themselves, that they had failed because they lacked Mr. Stapel's skill. However, there was usually no attempt to replicate, and certainly not independently. The few occasions when this did happen, and failed, were never revealed, because the findings were not publishable.

In other words, scientific criticism has not performed satisfactorily on this point. Replication and the falsification of hypotheses are cornerstones of science. Mr. Stapel's verification factory should have aroused great mistrust among colleagues, peers and journals.

As a supervisor and dissertation advisor, Mr. Stapel should have been expected to promote this critical attitude among his students. Instead, the opposite happened. A student who performed his own replications with no result was abandoned to his fate rather than praised and helped.

Strange, improbable, or impossible data patterns; strange correlations; identical averages and standard deviations; strange univariate distributions of variables.

The actual data displayed several strange patterns that should have been picked up. The patterns are related to the poor statistical foundation of Mr. Stapel's data fabrication approach (he also tended to make denigrating remarks about statistical methods). It has emerged that some of the fabrication involved simply "blindly" entering numbers based on the desired bivariate relationships, and by cutting and pasting data columns. This approach sometimes gave rise to strange data patterns. Reordering the data matrix by size of a given variable sometimes produces a matrix in which one column is identical to another, which is therefore the simple result of cutting and pasting certain scores. It was also possible for a variable that would normally score only a couple of per cent "antisocial," for no reason and unexpectedly suddenly to show "antisocial" most of the time. Independent replication yielded exactly the same averages and standard deviations. Two independent variables that always correlated positively, conceptually and in other research, now each had the right expected effects on the dependent variable, but correlated negatively with each other. There was no consistent checking of data by means of simple correlation matrices and univariate distributions. It is to the credit of the whistle blowers that they did discover the improbabilities mentioned above.

Finally, a lamentable element of the culture in social psychology and psychology research is for everyone to keep their own data and not make them available to a public archive. This is a problem on a much larger scale, as has recently become apparent. Even where a journal demands data accessibility, authors usually do not comply ... Archiving and public access to research data not only makes this kind of data fabrication more visible, it is also a condition for worthwhile replication and meta-analysis. (pp. 13-15)

7.0.1   Complete Enumeration versus Sampling in the Census

The basic sampling model implies that when the size of the population is effectively infinite, this does not affect the accuracy of our estimate, which is driven solely by sample size. Thus, if we want a more precise estimate, we need only draw a larger sample.[1] For some reason, this confusion resurfaces and is reiterated every ten years when the United States Census is planned, where the issue of complete enumeration, as demanded by the Constitution, and the problems of undercount are revisited. We begin with a short excerpt from a *New York Times* article by David Stout (April 2, 2009), "Obama's Census Choice Unsettles Republicans." The quotation it contains from John Boehner in relation to the 2010 Census is a good instance of the "resurfacing confusion"; also, the ethical implications of Boehner's statistical reasoning skills should be fairly clear.

Mr. Boehner, recalling that controversy [from the early 1990s when Mr. Groves pushed for statistically adjusting the 1990 Census to make up for an undercount], said Thursday that "we will have to watch closely to ensure the 2010 Census is conducted without attempting similar statistical sleight of hand."

There has been a continuing and decades-long debate about the efficacy of using surveys to correct the Census for an undercount. The arguments against surveys are based on a combination of partisan goals and ignorance. Why? First, the Census is a big, costly, and complicated procedure. And like all such procedures, it will have errors. For example, there will be errors where some people are counted more than once, such as an affluent couple with two homes being visited by census workers in May in one and by different workers in July at the other, or they are missed entirely. Some people are easier to count than others. Someone who has lived at the same address with the same job for decades, and who faithfully and promptly returns census forms, is easy to count. Someone else who moves often, is a migrant laborer or homeless and unemployed, is much harder to count. There is likely to be an undercount of people in the latter category. Republicans believe those who are undercounted are more likely to vote Democratic, and so if counted, the districts they live in will get increased representation that is more likely to be Democratic. The fact of an undercount can be arrived at through just logical considerations, but its size must be estimated through surveys. Why is it we can get a better estimate from a smallish survey than from an exhaustive census? The answer is that surveys are, in fact, small. Thus, their budgets allow them to be done carefully and everyone in the sampling frame can be tracked down and included (or almost everyone).[2] A complete enumeration is a big deal, and even though census workers try hard, they have a limited (although large) budget that does not allow the same level of precision. Because of the enormous size of the census task, increasing the budget to any plausible level will still not be enough to get everyone. A number of well-designed surveys will do a better job at a fraction of the cost.

The Supreme Court ruling in *Department of Commerce v. United States House of Representatives* (1999) seems to have resolved the issue of sampling versus complete enumeration in a Solomon-like manner. For purposes of House of Representatives apportionment, complete enumeration is required with all its problems of "undercount." For other uses of the Census, however, "undercount" corrections that make the demographic information more accurate are permissable And these corrected estimates could be used in differential resource allocation to the states. Two items are given in an appendix below: a short excerpt from the American Statistical Association *amicus* brief for this case, and the syllabus from the Supreme Court ruling.

Appendix: Brief for American Statistical Association as Amicus Curiae, Department of Commerce v. United States House of Representatives (1999)

Friend of the Court brief from the American Statistical Association —

ASA takes no position on the appropriate disposition of this case or on the legality or constitutionality of any aspect of the 2000 Census. ASA also takes no position in this brief on the details of any proposed use of statistical sampling in the 2000 Census.

ASA is, however, concerned to defend statistically designed sampling as a valid, important, and generally accepted scientific method for gaining accurate knowledge about widely dispersed human populations. Indeed, for reasons explained in this brief, properly designed sampling is often a better and more accurate method of gaining such knowledge than an inevitably incomplete attempt to survey all members of such a population. Therefore, in principal, statistical sampling applied to the Census "has the potential to increase the quality and accuracy of the count and to reduce costs." ... There are no sound scientific grounds for rejecting all use of statistical sampling in the 2000 Census.

As its argument in this brief, ASA submits the statement of its Blue Ribbon Panel that addresses the relevant statistical issues. ASA respectfully submits this brief in hopes that its explanation of these points will be helpful to the Court.

Appendix: Department of Commerce v. United States House of Representatives (1999)

Syllabus from the Supreme Court ruling: The Constitution's Census Clause authorizes Congress to direct an "actual Enumeration" of the American public every 10 years to provide a basis for apportioning congressional representation among the States. Pursuant to this authority, Congress has enacted the Census Act (Act), ... delegating the authority to conduct the decennial census to the Secretary of Commerce (Secretary). The Census Bureau (Bureau), which is part of the Department of Commerce, announced a plan to use two forms of statistical sampling in the 2000 Decennial Census to address a chronic and apparently growing problem of "undercounting" of some identifiable groups, including certain minorities, children, and renters. In early 1998, two sets of plaintiffs filed separate suits challenging the legality and constitutionality of the plan. The suit in No. 98-564 was filed in the District Court for the Eastern District of Virginia by four counties and residents of 13 States. The suit in No. 98-404 was filed by the United States House of Representatives in the District Court for the District of Columbia. Each of the courts held that the plaintiffs satisfied the requirements for Article III standing, ruled that the Bureau's plan for the 2000 Census violated the Census Act, granted the plaintiffs' motion for summary judgment, and permanently enjoined the planned use of statistical sampling to determine the population for congressional apportionment purposes. On direct appeal, this Court consolidated the cases for oral argument.

Held:

1. Appellees in No. 98-564 satisfy the requirements of Article III standing. In order to establish such standing, a plaintiff must allege personal injury fairly traceable to the defendant's allegedly unlawful conduct and likely to be redressed by the requested relief. ... A plaintiff must establish that there exists no genuine issue of material fact as to justiciability or the merits in order to prevail on a summary judgment motion. ... The present controversy is justiciable because several of the appellees have met their burden of proof regarding their standing to bring this suit. In support of their summary judgment motion, appellees submitted an affidavit that demonstrates that it is a virtual certainty that Indiana, where appellee Hofmeister resides, will lose a House seat under the proposed Census 2000 plan. That loss undoubtedly satisfies the injury-in-fact requirement for standing, since Indiana residents' votes will be diluted by the loss of a Representative. ... Hofmeister also meets the second and third standing requirements: There is undoubtedly a "traceable" connection between the use of sampling in the decennial census and Indiana's expected loss of a Representative, and there is a substantial likelihood that the requested relief—a permanent injunction against the proposed uses of sampling in the census—will redress the alleged injury. Appellees have also established standing on the basis of the expected effects of the use of sampling in the 2000 Census on intrastate redistricting. Appellees have demonstrated that voters in nine counties, including several of the appellees, are substantially likely to suffer intrastate vote dilution as a result of the Bureau's plan. Several of the States in which the counties are located require use of federal decennial census population numbers for their state legislative redistricting, and States use the population numbers generated by the federal decennial cen-

sus for federal congressional redistricting. Appellees living in the nine counties therefore have a strong claim that they will be injured because their votes will be diluted vis-à-vis residents of counties with larger undercount rates. The expected intrastate vote dilution satisfies the injury-in-fact, causation, and redressibility requirements.

2. The Census Act prohibits the proposed uses of statistical sampling to determine the population for congressional apportionment purposes. In 1976, the provisions here at issue took their present form. Congress revised 13 U. S. C. §141(a), which authorizes the Secretary to "take a decennial census ... in such form and content as he may determine, including the use of sampling procedures." This broad grant of authority is informed, however, by the narrower and more specific §195. As amended in 1976, §195 provides: "Except for the determination of population for purposes of [congressional] apportionment ... the Secretary shall, if he considers it feasible, authorize the use of ... statistical ... 'sampling' in carrying out the provisions of this title." Section 195 requires the Secretary to use sampling in assembling the myriad demographic data that are collected in connection with the decennial census, but it maintains the longstanding prohibition on the use of such sampling in calculating the population for congressional apportionment. Absent any historical context, the "except/shall" sentence structure in the amended §195 might reasonably be read as either permissive or prohibitive. However, the section's interpretation depends primarily on the broader context in which that structure appears. Here, that context is provided by over 200 years during which federal census statutes have uniformly prohibited using statistical sampling for congressional apportionment. The Executive Branch accepted, and even advocated, this interpretation of the Act until 1994.

3. Because the Court concludes that the Census Act prohibits the proposed uses of statistical sampling in calculating the population for purposes of apportionment, the Court need not reach the constitutional question presented.

7.1 Multivariable Systems

In multivariate analysis, it is important to remember that there is systematic covariation possible among the variables, and this has a number of implications for how we proceed. Automated analysis methods that search through collections of independent variables to locate the "best" regression equations (for example, by forward selection, backward elimination, or the hybrid of stepwise regression) are among the most misused statistical methods available in software packages. They offer a false promise of blind theory-building without user intervention, but the incongruities present in their use are just too great for this to be a reasonable strategy of data analysis: (a) one does not necessarily end up with the "best" prediction equations for a given number of variables; (b) different implementations of the process don't necessarily end up with the same equations; (c) given that a system of interrelated variables is present, the variables not selected cannot be said to be unimportant; (d) the order in which variables enter or leave in the process of building the equation does not necessarily reflect their importance; (e) all of the attendant significance testing and confidence interval construction methods become completely inappropriate (Freedman, 1983).

Several methods, such as the use of Mallow's C_p statistic for "all possible subsets (of the independent variables) regression," have some possible mitigating effects on the heuristic nature of the blind methods of stepwise regression. They offer a process of screening all possible equations to find the better ones, with compensation for the differing numbers of parameters that need to be fit. Although these search strategies offer a justifiable mechanism for finding the "best" according to ability to predict a dependent measure, they are somewhat at cross-purposes for how multiple regression is typically used in the behavioral sciences. What is important is the structure among the variables as reflected by the regression, and not so much squeezing the very last bit of variance accounted for from our data. More pointedly, if we find a "best" equation with fewer than the maximum number of available independent variables present, and we cannot say that those not chosen are less important than those that are, then what is the point?

A more pertinent analysis was demonstrated by Efron and Gong (1983) in which they bootstrapped the entire model-building process. They showed that by viewing the frequency with which each independent variable finds its way into the model, we can assess the stability of the choice of variables. Examining the structure of the independent variables through, say, a principal component analysis, will alert us to irreducible uncertainty due to high covariance among predictors. This is always a wise step, done in conjunction with bootstrapping, but not instead of it.

The implicit conclusion of the last argument extends more generally to the newer methods of statistical analysis that seem continually to demand our attention, for example, in hierarchical linear modeling, nonlinear methods of classification, or procedures that involve optimal scaling. When the emphasis is solely on getting better "fit" or increased prediction capability, and thereby, modeling "better," the methods may not be of much use in "telling the story" any more convincingly, and that should be the ultimate purpose of any analysis procedure we choose. Also, as Roberts and Pashler (2000) conclude rather counterintuitively, "goodness of fit" does not necessarily imply "goodness-of-model."

Even without the difficulties presented by a multivariate system when searching through the set of independent variables, there are several admonitions to keep in mind when dealing with a single equation. The most important may be to remember that regression coefficients cannot be interpreted in isolation for their importance using their size, even when based on standardized variables (such as those calculated from z-scores). That one coefficient is larger than another does not imply it is therefore more important. For example, consider the task of comparing the relative usefulness of the Scholastic Aptitude Test (SAT) scores and high school grade-point averages (HSGPAs) in predicting freshmen college grades. Both independent variables are highly correlated; so when grades are predicted with SAT scores, a correlation of about 0.7 is found. Correlating the residuals from this prediction with HSGPAs, gives a small value. It would be a mistake, however, to conclude from this that SAT is a better predictor of college success than HSGPA. If the order of analysis is reversed, we would find that HSGPA correlates about 0.7 with freshmen grades, and the residuals from this analysis have only a small correlation with SAT score.

The notion of importance can be explored by comparing models with and without certain variables present, and comparing the changes in variance-accounted-for that ensue. Similarly, the various significance tests for the regression coefficients are not really interpretable independently; for example, a small number of common factors may underlie all the independent variables, and thus generate significance for all the regression coefficients. In its starkest form, we have the one, two, and three asterisks scattered around in a correlation matrix, suggesting an ability to evaluate each correlation by itself without consideration of the multivariable system that the correlation matrix reflects in its totality. Finally, for a single equation, the size of the squared multiple correlation (R^2) gets inflated by the process of optimization, and needs to be adjusted, particularly when sample sizes are small. One beginning option is to use the commonly generated Wherry "adjusted R^2," which makes the expected value of R^2 zero when the true squared multiple correlation is itself zero. Note that the name of "Wherry's shrinkage formula" is a misnomer because it is not a measure based on any process of cross-validation. A cross-validation strategy is now routine in software packages, such as SYSTAT, using the "hold out one-at-a-time" type of mechanism. Given the current ease of implementation, such cross-validation processes should be routinely performed.

7.1.1 Multivariable Systems and Unidimensional Rankings

Understanding the problems that may arise when multivariable systems are used to construct unidimensional scales is a basic statistical literacy skill. We are continually bombarded with ratings on everything imaginable, such as the *US News & World Report* college rankings ("Best colleges"), ratings of the best hospitals, or monthly *Consumer Reports* on a bewildering array of topics from cars to computers and everything in between. In many of these cases, there are substantial statistical flaws and/or misinterpretations behind what is presented that could, at best, be characterized as "misleading." A recent account of several issues having to do with college rankings, and which generalizes to other such scales, is the Malcom Gladwell article in the *New Yorker* (February 14 and 21, 2011) entitled "The Order of Things." As a complement to this Gladwell article, this section reviews several statistical issues to consider when making sense out of what a ranking process provides.

We start with a story about a college instructor. A common demand in academics is an adherence to a "truth in teaching" perspective that requires grading practices for a course to be spelled out in advance and in a syllabus. The following wording is typical: "your grade for this course will depend on two midterm exams and a final; the final will account for 50% of your grade, with 25% for each of the two midterms." The algorithmic process for assigning final grades might be to first standardize the three exam scores, weight the two midterms by .25 and the final by .5, and then sum. Chosen cut-scores would dictate the final letter grades assigned. After the grades had been given, a lawyer-to-be asked why he received a lowly D minus; he had done extremely well on both midterms and decided not to study very hard for the final because of other course commitments. He felt that he had earned at least a B no matter what happened on the final. Being somewhat confrontational, he demanded "proof" that 50% of his grade was determined by his outstanding midterm scores.

The statistician consulted by the instructor had some simple advice: give the future barrister a B plus; revise your syllabus so next time you just list the algorithmic process for determining final grades; and make no statement about certain scores determining certain percentages of the final grades. The percentage statements just can't be done so simply whenever there is a system of correlated (exam) scores. He went on to opine that if a weighting strategy for the scores were developed so that the variance in a set of midterm scores reflected 25% of the total score, it would most likely have the effect of overweighting pure random variability (the specific error) in the midterm scores.

The manner in which the *US News* college rankings are constructed is akin to the trouble our instructor got into using a weighting scheme. Seven variables are now used for the college rankings with the *US News* percentage weights given in parentheses:

1. Undergraduate academic reputation (22.5%); 2. Graduation and freshman retention rates (20%); 3. Faculty resources (20%); 4. Student selectivity (15%); 5. Financial resources (10%); 6. Graduation rate performance (7.5%); 7. Alumni giving (5%).

In effect, the *US News* rankings seem driven by the one underlying factor of wealth (much as our exam scores may have been driven by an underlying factor of "subject matter knowledge"). The (former) President of Penn State, Graham Spanier, is quoted in the Gladwell article to this effect: "What I find more than anything else is a measure of wealth: institutional wealth, how big is your endowment, what percentage of alumni are donating each year, what are your faculty salaries, how much are you spending per student."

Generally, it would be better to have an assessment of what a college adds to those individuals who attend, rather than just what the individuals themselves bring with them. In economic terms, colleges should be evaluated by their production functions—what is the output for all combinations of input.

Several difficulties may be encountered in using any simple weighted average to obtain a unidimensional ranking. A few of these are discussed below:

(a) The variables aggregated to obtain a ranking need to be commensurable (that is, numerically comparable); otherwise, those variables with larger variances dominate the construction of any final unidimensional scale. The most common mechanism for ensuring commensurability is through a z-score transformation so each variable has a mean of zero and a standard deviation (and variance) of one. The *US News* approach to commensurability is not through z-scores but by a transform of each variable to a point scale from 1 to 100 (with Harvard always getting 100 points). Although this moves in the right direction for commensurability, if the point scales have different variances, some variables will still have more of a determining effect on the weighted aggregate that generates the final ratings.

(b) Any multivariate analysis course places great emphasis on linear combinations of variables, and the formulas introduced at the outset are useful tools for evaluating ranking systems. These typically are expressions for means, variances, and covariances (correlations) of arbitrary linear combinations of sets of variables, both for an assumed collection of random variables characterizing a population and for an observed data matrix obtained on these random variables. Thus, there are general mechanisms available for studying the types of weighting systems represented by the *US News* aggregation system. For example, the effects of changing weighting systems can be studied through how the constructed scales would correlate; also, several other statistical tools can be used: scatterplots, measures of nonlinear and rank correlation, consistency of generated rankings, changes in the variances of the generated scores, and sensitivity of rankings to the changes in aggregation weights. As done now, the *US News* ranking is a "take it or leave it" proposition. Not only is the particular choice of variables a *fait accompli*, the numerical aggregation system is a fixed entity as well (based, we suspect, on the hunches of Robert Morse, the director of all *US News* rankings). It is statistically (and ethically) questionable for such a closed system to have the putatively large influence it does on the United States system of higher education.

(c) The one glaring omission in the types of ranking systems used by *US News* is the lack of a defensible criterion variable. All we have are input measures; there are no viable output (or criterion) measures. If a criterion was available, the use of an arbitrary aggregation of input variables could be partially mitigated by the adoption of a defensible weighting mechanism through an optimization process. For example, multiple regression provides that linear combination of the predictor variables correlating the highest with the criterion measure. Also, regression models are available that allow transformations (such as monotonic or nonlinear) of the dependent and independent variables. So, not only is a justifiable aggregation possible for the original variables, a mechanism is also available that permits optimizing variable transformations to be included as well.

(d) The absence of a viable criterion measure does not preclude considering other optimization mechanisms for aggregating sets of variables. The principal components of a correlation matrix, for example, provide a collection of possible weighting schemes having several nice properties. If there are p original variables available, then there are p principal components, where each component provides a set of weights; that is, each principal component defines a linear combination of the original variables; the components as a group "repackage" the information present in the original variables. The p components can be ordered according to the variances of the resulting linear combinations. Thus, the first principal component maximizes the variance of a linear combination over all possible aggregation schemes where the sum of the squared weights is normalized to 1.0. The second principal component is another linear combination, uncorrelated with the first component, but which has maximum variance over all aggregation schemes using a sum of squared weights of 1.0; and so on for the remaining components; all are uncorrelated with the earlier components and maximize variance. The sum of the variances for all the components is p, which is also the sum of the unit variances for each of the original variables (each is assumed to be standardized). Stated in other words, if the ratio of the variance of the first component to p is large, then we might argue that most of what is "going on" in the original variables can be seen in the one linear combination provided by the first component. This first component gives a mechanism for generating a ranking that does not depend on an arbitrary weighting scheme made up only in one's head.

(e) There are other advantages in considering the principal components for the given set of variables. First, the scores on the first component immediately define a ranking based on numerical values obtained by a process of maximizing the variability of the aggregate; this might provide a defensible weighting strategy satisfying the statistical *literati*. How the variables were chosen in the first place may still be questionable, but once this is decided, the method of weighting is constructed through the transparent optimizing of a variance criterion. A principal component analysis itself can offer another justification for a unidimensional ranking, or conversely, be used to argue that more dimensions are needed. If the proportion of variance explained by the first component is high (for example, 80% or more), the remaining components may not have much more to offer. One could argue that most of what we know about our variables can be explained by a single unidimensional scale generated by the first principal component. On the other hand, if several components are needed to attain a sufficient total "variance-accounted-for," the wisdom of relying on a single scale is called into question. When correlations between the given variables are all positive (which might be expected, for example, if a strong single factor of "wealth" underlies them all), the first component weights will all be positive. The weighted average thus obtained has only nonnegative weights, a consequence of the Perron–Frobenius Theorem ("Perron–Frobenius theorem," 2010). If more than one component seems necessary, however, various plus and minus weighting structures might be needed, along with a foray into factor rotation methods to "clean up" a solution. A good applied multivariate analysis course will go into all these methods in greater depth.

In an ideal statistical world, the set of variables obtained on a collection of entities could be used to rank-order the entities lexicographically, where one entity ranked higher than another would have all of the values on its variables be better (or, at least, no worse) than those for the second. Unfortunately, the world is never this simple. Some justification should lie behind the rankings someone provides. In many instances, however, that is just not the case, even though numbers are produced to convince us that something important and real is present. Different aspects of this same problem will appear in other chapters but in different guises, for example, when actuarial versus clinical prediction is discussed, or when the issue is raised about the unitary trait of "intelligence," referred to as "g," which when looked at closely is just the first principal component.[3]

7.2   Graphical Presentation

The graphical method has considerable superiority for the exposition of statistical facts over the tabular. A heavy bank of figures is grievously wearisome to the eye, and the popular mind is as incapable of drawing any useful lessons from it as of extracting sunbeams from cucumbers.
— Arthur B. & Henry Farquhar (*Economic and Industrial Delusions*, 1891)

The importance of scatterplots in evaluating the association between variables was reiterated in our earlier discussions of correlation and prediction. Generally, graphical and other visual methods of data analysis are central to an ability to tell what the data may be reflecting and what conclusions are warranted. In a time when graphical presentation may have been more expensive than it is now, it was common to only use summary statistics, even when various reporting rules were followed. For example, you should never present just a measure of central tendency without a corresponding measure of dispersion; or, in providing the results of a poll, always give the margin of error (usually, the 95% confidence interval) to reflect the accuracy of the estimate based on the sample size being used. If data are not nicely unimodal, however, more is needed than just means and variances. Both "stem-and-leaf" and "box-and-whisker" plots are helpful in this regard and should be routinely used for data presentation.

Several egregious uses of graphs for misleading presentations were documented many years ago in the very popular book by Darrell Huff, *How to Lie with Statistics* (1954), and updated in Wainer's 1984 commentary from the *American Statistician*, "How to Display Data Badly" (also, see the first chapter in Wainer, 1997/2000). Both of these deal with visual representation and how graphs can be used to distort; for example, by truncating the bottoms of line or bar charts so differences are artificially magnified, or using two- and three dimensional objects to compare values on a unidimensional variable where images do not scale the same way as do univariate quantities. Tufte (1983) has lamented the poor use of graphics that relies on "chart junk" for questionable visual effect, or gratuitous color or three-dimensions in bar graphs that do not represent anything real at all. In extending some of these methods of misrepresentation to the use of maps, it is particularly easy to deceive given the effects of scale level usage, ecological correlation, and the modifiable areal unit problem. What should be represented generally in our graphs and maps must be as faithful as possible to the data represented, without the distracting application of unnecessary frills that do not communicate any information of value.

One particularly insidious use of a graphical format almost always misleads: the double y-axis plot. In this format, there are two vertical axes, one on the left and one on the right depicting two completely different variables—say, death rates over time for smokers shown on the left axis (time is on the horizontal axis), and death rates for nonsmokers shown on the right axis. Because the scale on the two vertical axes are independent, they can be chosen to show anything the graph maker wants. Compare the first version in Figure 7.1 (after the Surgeon General's report on the dangers of smoking) with the second in Figure 7.2 prepared by someone attentive to the needs of Big Tobacco that uses the double y-axis format. Few other graphic formats lend themselves so easily to the misrepresentation of quantitative phenomena.

In providing data in the form of matrices, such as subject by variable, we might consider the use of "heat maps," where numerical values, assumed commensurable over variables, are mapped into color spectra reflecting magnitude. The further imposing of orderings on rows and columns to group similar patches of color

FIGURE 7.1: A graph showing that smokers die sooner than nonsmokers: smoking appears to subtract about seven years from life. Figure adapted from Wainer (2000, p. 6).

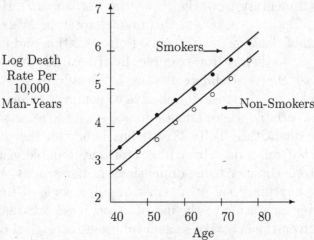

together can lead to useful data displays. A survey of the history of heat maps, particularly as developed in psychology, has been given by Wilkinson and Friendly (2009). This latter article could well be mandatory reading in any part of a statistics course concerned with accurate and informative graphical data presentation. Also, see Bertin (1973/1983); Tufte (1983, 1990, 1997); Tukey (1977); Wainer (1997/2000, 2005, 2009).

7.3   Problems With Multiple Testing

> As is your sort of mind, so is your sort of search; you'll find what you desire.
> — Robert Browning (1812–1889)

A difficulty encountered with the use of automated software analyses is that of multiple testing, where the many significance values provided are all given as if each were obtained individually without regard for how many tests were performed. This situation gets exacerbated when the "significant" results are then culled, and only these are used in further analysis. A good case in point was

FIGURE 7.2: A graph showing that the Surgeon General reports aging is the primary cause of death. Figure adapted from Wainer (2000, p. 8).

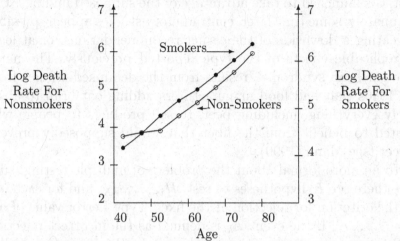

reported earlier in the section on odd correlations where highly inflated correlations get reported in fMRI studies because an average is taken only over those correlations selected to have reached significance according to a stringent threshold. Such a context is a clear violation of a dictum given in many beginning statistics classes: you cannot legitimately test a hypothesis on the same data that first suggested it.[4]

Exactly the same issue manifests itself, although in a more subtle, implicit form, in the modern procedure known as data mining. Data mining consists of using powerful graphical methods to view and search through high-dimensional datasets of moderate-to-large size, looking for interesting features. When such a feature is uncovered, it is isolated and saved—a finding! Implicit in the search, however, are many comparisons that the viewer makes and decides are not interesting. Because the searching and comparing is done in real time, it is difficult to keep track of how many "insignificant" comparisons were discarded before alighting on a significant one. Without knowing how many, we cannot judge the significance of the interesting features found without an independent confirmatory sample. Such independent confirmation is all too rarely done.

Uncontrolled data mining and multiple testing on some large (longitudinal) datasets can also lead to results that might best be labeled with the phrase "the oat bran syndrome." Here, a promising association is identified; the relevant scientists appear in the media and on various cable news shows; and an entrepreneurial industry is launched to take advantage of the supposed findings. Unfortunately, some time later, contradictory studies appear, possibly indicating a downside of the earlier recommendations, or at least no replicable effects of the type reported previously. The name "the oat bran syndrome" results from the debunked studies from the 1980s that had food manufacturers adding oat bran to absolutely everything, including beer, to sell products to people who wanted to benefit from the fiber that would supposedly prevent cancer (see Mann, 1990).

To be more formal about the problem of multiple testing, suppose there are K hypotheses to test, $H_1, \ldots, H_K$, and for each, we set the criterion for rejection at the fixed Type I error value of α_k, $k = 1, \ldots, K$. If the event A_k is defined as the incorrect rejection of H_k (that is, rejection when it is true), the Bonferroni inequality gives

$$P(A_1 \text{ or } \cdots \text{ or } A_K) \leq \sum_{k=1}^{K} P(A_k) = \sum_{k=1}^{K} \alpha_k .$$

Noting that the event $(A_1 \text{ or } \cdots \text{ or } A_K)$ can be verbally restated as one of "rejecting incorrectly *one or more* of the hypotheses," the experiment-wise (or overall) error rate is bounded by the sum of the K α values set for each hypothesis. Typically, we let $\alpha_1 = \cdots = \alpha_K = \alpha$, and the bound is then $K\alpha$. Thus, the usual rule for controlling the overall error rate through the Bonferroni correction sets the individual αs at some small value such as $.05/K$; the overall error rate is then guaranteed to be no larger than $.05$.

The problem of multiple testing and the failure to practice "safe statistics" appears in both blatant and more subtle forms. For example, companies may suppress unfavorable studies until those to their liking occur. A possibly apocryphal story exists about toothpaste companies promoting fluoride in their products in the 1950s and who repeated studies until large effects could be reported for their "look Ma, no cavities" television campaigns. This may be somewhat innocent advertising hype for toothpaste, but when drug or tobacco companies engage in the practice, it is not so innocent and can have a serious impact on our collective health. It is important to know how many things were tested to assess the importance of those reported. For example, when given only those items from some inventory or survey that produced significant differences between groups, be very wary!

People sometimes engage in a number of odd behaviors when doing multiple testing. We list a few of these below in summary form:

(a) It is not legitimate to do a Bonferroni correction post hoc; that is, find a set of tests that lead to significance, and then evaluate just this subset with the correction;

(b) Scheffé's method (and relatives) are the only true post-hoc procedures to control the overall error rate. An unlimited number of comparisons can be made (no matter whether identified from the given data or not), and the overall error rate remains constant;

(c) You cannot legitimately look at your data and then decide which planned comparisons to do;

(d) Tukey's method is not post hoc because you actually plan to do all possible pairwise comparisons;

(e) Even though the comparisons you might wish to test are independent (such as those defined by orthogonal comparisons), the problem of inflating the overall error rate remains; similarly, in performing a multifactor analysis of variance (ANOVA) or testing multiple regression coefficients, all of the tests carried out should have some type of control imposed on the overall error rate;

(f) It makes little sense to perform a multivariate analysis of variance before you go on to evaluate each of the component variables. Typically, a multivariate analysis of variance (MANOVA) is completely noninformative as to what is really occurring, but people proceed in any case to evaluate the individual univariate ANOVAs irrespective of what occurs at the MANOVA level; we may accept the null hypothesis at the overall MANOVA level but then illogically ask where the differences are at the level of the individual variables. Plan to do the individual comparisons beforehand, and avoid the uninterpretable overall MANOVA test completely.

We cannot leave the important topic of multiple comparisons without at least a mention of what is now considered the most powerful method currently available: the False Discovery Rate (Benjamini & Hochberg, 1995). But even this method is not up to the most vexing of problems of multiplicity. We have already mentioned data mining as one of these; a second problem arises in the search for genetic markers. A typical paradigm in this crucial area is to isolate a homogeneous group of individuals, some of whom have a genetic disorder and others do not, and then to see if one can determine which genes are likely to be responsible. One such study is currently being carried out with a group of 200 Mennonites in Pennsylvania (see Wojciechowski, R., Bailey-Wilson, J. E., & Stambolian, D. (n.d.), Macular degeneration in Mennonites, Unpublished raw data; also see Wojciechowski, Bailey-Wilson, & Stambolian, 2010). Macular degeneration is common among the Mennonites, and this sample was chosen so that 100 of them had macular degeneration and a matched sample of 100 did not. The genetic structure of the two groups was very similar, and so the search was on to see which genes were found much more often in the group that had macular degeneration than in the control group. This could be determined with a t-test. Unfortunately, the usefulness of the t-test was diminished considerably when it had to be repeated for more than 100,000 separate genes. The Bonferroni inequality was no help, and the False Discovery Rate, while better, was still not up to the task. The search still goes on to find a better solution to the vexing problem of multiplicity.[5]

7.4 Issues in Repeated-Measures Analyses

The analysis of repeated measures generally needs special treatment in that the usual models are not very trustworthy and can lead to erroneous conclusions. The starting place is commonly a Mixed Model III ANOVA with a fixed treatment factor and a subject factor considered random. To model repeated observations justifying the usual F-ratio test statistic of Mean-Square Treatments to Mean-Square Interaction, an assumption is made that the observations within a subject are correlated. The use of the usual F-ratio, however, requires that all these correlations be the same irrespective of which pair of treatments is considered (an assumption of "compound symmetry"). The compound symmetry assumption may be reasonable when the treatment times are randomly assigned, but if the responses are obtained sequentially, then possibly not. Treatments further apart in time are typically less correlated because of fatigue, boredom, familiarity, and so on. Unfortunately, there is strong evidence of nonrobustness in the use of the equicorrelation assumption when it is not true, with too many false rejections of the null hypothesis of no treatment differences.

A way around this nonrobustness is implemented in many software packages. If we knew the structure of all the variances and covariances among the treatments, we could obtain a parameter, say, θ, that would give an appropriate correction for the degrees of freedom of the F-distribution against which to compare the calculated F-ratio; that is, we would use $F_{\theta(B-1),\theta(A-1)(B-1)}$, where there are B treatments and A subjects. Although θ is unknown, there are two possible strategies to follow: estimate θ with Huynh–Feldt procedures (as is done, for example, in SYSTAT); or use the greatest reduction possible with the discounting bound of $1/(B-1)$ (that is, the Geisser–Greenhouse result of $1/(B-1) \leq \theta$, also as done when an analysis is carried out using SYSTAT). So, if a rejection occurs with the Geisser–Greenhouse method, it would also occur for the Huynh–Feldt estimation, or if you knew and used the actual value of θ.

Another approach to repeated measures, called "profile analysis," uses Hotelling's T^2 statistic and/or MANOVA on difference scores. In fact, the only good use of a usually noninformative MANOVA may be in a repeated-measures analysis. Three types of questions are commonly asked in a profile analysis: Are the profiles parallel to each other? Are the profiles coincident? And, are the profiles horizontal? When done well, a profile analysis can give an informative interpretation of repeated-measures information with an associated graphical presentation.

Two possible issues with repeated measures should be noted. First, it is assumed that the responses from our subjects are commensurable over the variables measured. If not, an artificial transformation could be considered such as to z-scores, but by so doing, the test for horizontal profiles is not meaningful because the associated test statistic is identically zero. Second, the number of subjects versus the number of measurement times may prevent carrying out a Hotelling T^2 comparison in a profile analysis (but not, say, a correction based on a Huynh–Feldt estimated θ). Generally, if there are more time points than subjects, one of the degrees of freedom in the F-distribution used for the T^2 comparison is negative, and thus, the test is meaningless.

Repeated measurements obtained on the same subjects generally require a different approach than do singly occurring measurements. As discussed above, there are various ways of not taking advantage of the usual compound symmetry assumption; for example, carrying out the analyses of variance with Huynh–Feldt or Geisser–Greenhouse corrections, or using alternative Hotelling T^2 or MANOVA approaches. The issues, however, go deeper than just these types of split-plot experimental designs, and the special circumstances that repeated measures offer are commonly just ignored.

Anyone analyzing repeated measures needs to remember that the variance of the difference between two means, say $\bar{X}$ and $\bar{Y}$, is not the same when $\bar{X}$ and $\bar{Y}$ are based on independent samples. In particular, suppose $\bar{X}$ is obtained for the observations $X_1, \ldots, X_N$, and $\bar{Y}$ for $Y_1, \ldots, Y_N$. When the samples are independent, the variance of the difference $\bar{X} - \bar{Y}$, $S^2_{\bar{X}-\bar{Y}}$, can be estimated as $S^2_{\bar{X}} + S^2_{\bar{Y}}$, where $S^2_{\bar{X}} \equiv S^2_X/N$, $S^2_{\bar{Y}} \equiv S^2_Y/N$, and S^2_X and S^2_Y are the sample variances for $X_1, \ldots, X_N$ and $Y_1, \ldots, Y_N$, respectively. In the repeated-measures context, the variance of $\bar{X} - \bar{Y}$ can be estimated as $S^2_{\bar{X}} + S^2_{\bar{Y}} - 2(S_{XY}/N)$, where S_{XY} is the sample covariance between the observations $X_1, \ldots, X_N$ and $Y_1, \ldots, Y_N$. Thus, we have a difference in the term, $-2(S_{XY}/N)$, which in most instances will be a negative correction when the X and Y observations are positively related. In other words, the variance of the difference, $\bar{X} - \bar{Y}$, will generally be less in the context of repeated measures compared to independent samples.

In some areas of neuroimaging, the repeated-measures nature of the data is just ignored; we have pixels (or voxels) that are spatially arranged (and subject to various types of spatial autocorrelation) that move through time (and subject again to various types of temporal autocorrelation). In these frameworks where the repeated measures are both spatial and temporal, it is not sufficient to just use the various multivariate general linear model extensions that assume all error terms are independent and identically distributed (as suggested by some best-selling fMRI handbooks; for example, see Huettel, Song, & McCarthy, 2004, pp. 336–348). Geographers have struggled with this type of spatial and temporal modeling for decades, and have documented the issues extensively; a good current source is Haining (2003), but there are many such texts that would be helpful to the neuroimager in search of a justifiable modeling approach.

A related repeated-measures topic is in the time-series domain, where some variable is observed temporally. Substantial modeling efforts have involved the Box–Jenkins approach of using ARIMA (autoregressive-integrated-moving-average) models. A more subtle question in this context is to assess the effects of an intervention on the progress of such a time series. In the case of single-subject designs, where a subject serves as his or her own control, the issue of evaluating interventions is central (see Kazdin, 1982). A particularly elegant approach to this problem has been developed by Edgington (see Edgington & Onghena, 2007, Chapter 11: "N-of-1 Designs"), where intervention times are chosen randomly. The same logic of analysis is possible as in a Fisherian (1971) approach to analyzing an experiment where the various units have been assigned at random to the conditions.

7.5 Matching and Blocking

One of the main decision points in constructing an experimental design is whether to block or match subjects, and then within blocks randomly assign subjects to treatments. Alternatively, subjects could be randomly assigned to conditions without blocking. As discussed earlier, it is best to control for initial differences beforehand. Intact groups can't be equated legitimately after the fact through methods such as analysis of covariance or post-hoc matching. But the question here is whether blocking makes sense over the use of a completely randomized design. This choice can be phrased more formally by comparing the test statistics appropriate for a two-independent or a two-dependent samples t-test.

The principle derived from this specific comparison generalizes to more complicated designs.

Suppose we have two equal-sized samples of size N, $X_1, \ldots, X_N$ and $Y_1, \ldots, Y_N$. When the two samples are independent, the two-independent samples t-statistic has the form

$$\frac{\bar{X} - \bar{Y}}{\sqrt{(S_X^2 + S_Y^2)/(N - 1)}} \, ,$$

where S_X^2 and S_Y^2 are the sample variances; this statistic is compared to a t-distribution with $2(N - 1)$ degrees of freedom. When the samples are dependent and X_i and Y_i are repeat observations on the ith subject, the paired t-statistic has the form

$$\frac{\bar{X} - \bar{Y}}{\sqrt{S_D^2/(N - 1)}} \, ,$$

where S_D^2 is the sample variance of the difference scores. Here, the paired t-statistic is compared to a t-distribution with $N - 1$ degrees of freedom. We note the relation $S_D^2 = S_X^2 + S_Y^2 - 2S_{XY}$, where S_{XY} is the sample covariance between X and Y.

In the initial design of an experiment, there may be a choice: match subjects and assign members within a pair to the treatments, or just assign all subjects randomly to the two treatments without matching. Generally, if the matching variable is not very important in that the sample covariance is not that large (and positive), to compensate for the halving of the degrees of freedom in going from $2(N - 1)$ to $(N - 1)$, it only hurts to match. To compensate for the loss of degrees of freedom and make the paired t-statistic sufficiently larger than the independent sample t-statistic, the variance of the differences, S_D^2, in the denominator of the paired t-statistic must be sufficiently smaller compared to $S_X^2 + S_Y^2$ in the denominator of the independent samples t-statistic. Unfortunately, unless one has some estimate of the covariance of X and Y, the choice of design must be based on a guess. A dictum, however, may still be gleaned: don't block or match on variables that have no possible (positive and relatively strong) relation to the type of responses being measured.

In an observational study, covariate information is routinely collected to make comparable what is not. Randomization, in the limit, does this automatically. But samples in the real world are always finite; also, because every randomized experiment can easily turn into an observational study by subject dropout, it is always prudent to collect covariate information even with randomized designs. This allows some hope of post-hoc adjustment if, for example, data are missing from the experimental and control groups because of the treatments administered. It also allows us to check on the efficiency of the randomization. Fisher, when asked whether he would adjust using the covariates if he discovered that the randomization left the groups unbalanced, replied, "No,"; he would toss out that treatment assignment and re-randomize (Paul Rosenbaum, personal communication to HW, January 21, 2002).

7.6 Randomization and Permutation Tests

An important benefit from designing an experiment with random assignment of subjects to conditions, possibly with blocking in various ways, is that the method of analysis through randomization tests is automatically provided. As might be expected, the original philosophy behind this approach is due to R. A. Fisher (1971), but it also has been developed and generalized extensively by others (see Edgington & Onghena, 2007). In Fisher's time, and although randomization methods may have been the preferred strategy, approximations were developed based on the usual normal theory assumptions to serve as computationally feasible alternatives. But with this view, our standard methods are just approximations to what the preferred analyses should be. A short quotation from Fisher's *The Design of Experiments* (1971) makes this point well:

> In these discussions it seems to have escaped recognition that the physical act of randomisation, which, as has been shown, is necessary for the validity of any test of significance, affords the means, in respect of any particular body of data, of examining the wider hypothesis in which no normality of distribution is implied. The arithmetical procedure of such an examination is tedious, and we shall only give the results of its application ... to show the possibility of an independent check on the more expeditious methods in common use. (p. 45)

A randomization (or permutation) test uses the given data to generate an exact null distribution for a chosen test statistic. The observed test statistic for the way the data actually arose is compared to this null distribution to obtain a p-value, defined as the probability (if the null distribution were true) of an observed test statistic being as or more extreme than what it actually was. Three situations lead to the most common randomization tests: K-dependent samples, K-independent samples, and correlation. When ranks are used instead of the original data, all of the common nonparametric tests arise. In practice, null randomization distributions are obtained either by complete enumeration, sampling (a Monte Carlo strategy), or through various kinds of large sample approximations (for example, normal or chi-squared distributions).

Permutation tests can be generalized beyond the usual correlational framework or that of K-dependent or K-independent samples. Much of this work falls under a rubric of combinatorial data analysis (CDA), where the concerns are generally with comparing various kinds of complete matrices (such as proximity or data matrices) using a variety of test statistics. The most comprehensive source for this material is Hubert (1987), but the basic matrix comparison strategies are available in a number of places, for example, see discussions of the "Mantel Test" in many packages in R (as one example, see the "Mantel–Hubert general spatial cross-product statistic" in the package, `spdep`). Even more generally, one can at times tailor a test statistic in nonstandard situations and then implement a permutation strategy for its evaluation through the principles developed in CDA.

The idea of repeatedly using the sample itself to evaluate a hypothesis or to generate an estimate of the precision of a statistic, can be placed within the broader category of resampling statistics or sample reuse. Such methods include the bootstrap, jackknife, randomization and permutation tests, and exact tests (for example, Fisher's exact test for 2×2 contingency tables). Given the incorporation of these techniques into conveniently available software, such as R, there are now many options for gauging the stability of the results of one's data analysis.

7.7 Pitfalls of Software Implementations

Fast cars, fast women, fast algorithms ... what more could a man want?
— Joe Mattis

Most statistical analyses are performed using packages such as SYSTAT, SPSS, or SAS. Because these systems are blind to the data being analyzed and the questions asked, it is up to the user to know some of the pitfalls to avoid. For example, the fact that an analysis of covariance is easy to do does not mean that is should be done or that it is possible to legitimately equate intact groups statistically. Even though output may be provided, this doesn't automatically mean it should be used. Cases in point are the inappropriate reporting of indeterminate factor scores, the gratuitous number of decimal places typically given, Durbin–Watson tests when the data are not over time, uninformative overall MANOVAs, and nonrobust tests for variances. We mention two more general traps we've seen repeatedly:

(a) In the construction of items or variables, the numbers assigned may at times be open to arbitrary coding. For instance, instead of using a 1 to 10 scale, where "1" means "best" and "10" "worst," the keying could be reversed so "1" means "worst" and "10" best. When an intercorrelation matrix is obtained among a collection of variables subject to this kind of scoring arbitrariness, it is possible to obtain some impressive (two-group) structures in methods of multidimensional scaling and cluster analysis that are merely artifacts of the keying and not of any inherent meaning in the items themselves. In these situations, it is common to "reverse score or reverse code" a subset of the items in the hope of obtaining an approximate "positive manifold" for the correlation matrix, characterized by few if any negative correlations that can't be attributed to sampling error.

(b) Certain methods of analysis (for example, most forms of multidimensional scaling, K-means and mixture model cluster analysis, and some strategies involving optimal scaling) are prone to local optima where results are presented but not the best possible ones according to the goodness-of-fit measure being optimized.

The strategies used in the optimization cannot guarantee global optimality because of the structure of the functions being optimized (for example, those that are highly nonconvex). One standard method of local optimality exploration is to start (repeatedly and randomly) a specific analysis method, observe how severe the local optima problem is for a given dataset, and then choose the best analysis found for reporting a final result. Unfortunately, none of the current packages offer these random-start options for all the methods that may be prone to local optima (for a good example involving K-means clustering, see Steinley, 2003). These local optimality difficulties are one reason for allowing more than the closed and proprietary analysis systems in graduate statistics instruction that can't be modified, and hence the move toward using environments that can, such as R and MATLAB.

(c) Methods of analysis involving optimization often use iterative algorithms that converge to a solution even though it may be only a local optimum. Generally, various convergence criteria are set by default in the software; for example, maximum number of iterations allowed, minimal change in the stepsize used by the algorithm, or minimal change in the value of the loss criterion being optimized. In any event, it is up to the user to know when a premature termination of an algorithm has occurred, and to be able to change the default convergence criteria values to insure that a "real" solution has been achieved. Again, attempts to explain prematurely terminated results show that you really don't know what you are doing. In an era when computer time was expensive, one had to be very stingy about setting the defaults to ensure that only a limited amount of computational effort could be expended. Now, running our machines for a few (more) hours has no real cost attached. So, if you see a message such as "maximum number of iterations exceeded (or reached)," don't ignore it. Instead, change the default limits until the message disappears for the solution achieved.

Closed statistical systems allow analyses to be done with little or no understanding of what the "point-and-clicks" are really giving. At times, this may be more of an impediment to clear reasoning than assistance. The user does not need to know much before being swamped with copious amounts of output, with little help on how to wade through the results or to engage in further exploration (for example, in investigating local minima or carrying out alternative analyses). One of the main reasons for now employing one of the newer statistical environments (such as R or MATLAB) is that they do not rely on pull-down menus. Instead, they are built up from functions that take various inputs and provide outputs, but you need to know what to ask for and the syntax of the function being used. Also, the source code for the routines is available and can be modified if some variant of an analysis is desired. This again assumes more than a superficial understanding of how the methods work; nevertheless, these are valuable skills to have when attempting to reason from data. The R environment has become the *lingua franca* for framing cutting-edge statistical development and analysis, and is becoming a major computational tool we need to develop in the graduate-level statistics sequence. It is also open-source and free, so there are no additional instructional costs incurred with the adoption of R.

7.7.1 The Unfortunate Case of Excel

The documented inadequacies of Excel are legion, as is Microsoft's inability to correct flaws when they are pointed out. A good place to learn about Excel's statistical failings is a special section on Microsoft Excel 2007 in the journal *Computational Statistics & Data Analysis* (2008c, *52*, 4568-4606), beginning with a scathing editorial by McCullough (2008b, pp. 4568–4569). Several of the points raised in this collection of articles will be noted below, usually with direct quotations from the articles themselves. One overarching conclusion would be that relying solely on Excel's statistical routines needlessly puts people at risk.

(1) As part of this special section on Excel, McCullough (2008a) has an article entitled "Microsoft Excel's 'Not the Wichmann–Hill' Random Number Generators." As the title suggests, it concerns the Wichmann–Hill (WH) Random Number Generator (RNG). A few excerpts from the article's concluding section follow:

> Twice Microsoft has attempted to implement the dozen lines of code that define the Wichmann and Hill (1982) RNG, and twice Microsoft has failed, apparently not using standard methods for verifying that an RNG has been correctly implemented. Consequently, users of Excel's "rand" function have been using random numbers from an

unknown and undocumented RNG of unknown period that is not known to pass any standard tests of randomness. Given Microsoft's first failure to implement the WH RNG correctly, Microsoft should have taken special care to ensure that it was done correctly the second time. Microsoft did not do this. The second failure demonstrates clearly that the first failure (and the second, as well) was not some sort of "unfortunate accident" but, rather, a systemic failure on Microsoft's part. (p. 4591)

It will be interesting to see what Microsoft chooses to do. Will Excel users have to wait until Excel 2010 to get good random numbers, or will Microsoft offer a patch for Excel 2007? Will Microsoft once again attempt to implement the Wichmann–Hill RNG and, if so, will the third time be a charm? Or will Microsoft offer an RNG more suitable to the 21st century and, if so, will Microsoft program it correctly? Whatever Microsoft chooses to do, it should ensure that users have the ability to verify the RNG, since Microsoft has demonstrated that users cannot rely on Microsoft's claims to have done it correctly.

In the meantime, Excel users will have to content themselves with an unknown RNG of unknown period that is not known to pass any standard battery of tests for randomness. At least now, users will not mistakenly believe that they are using WH RNs. Finally, Microsoft should do Messrs. Wichmann and Hill the courtesy of taking their names off the RNG in Excel. (p. 4592)

(2) The most general indicting article is by McCullough and Heiser (2008), "On the Accuracy of Statistical Procedures in Microsoft Excel 2007." We quote from the abstract and concluding subsection:

No statistical procedure in Excel should be used until Microsoft documents that the procedure is correct; it is not safe to assume that Microsoft Excel's statistical procedures give the correct answer. Persons who wish to conduct statistical analyses should use some other package. (p. 4570)

The improvements in Excel 2003 gave hope that Microsoft might finally be committed to providing reliable statistical functionality in Excel. That Microsoft once again decided to allow errors to remain unfixed, or fixed incorrectly, in the Excel 2007 release dashes those hopes. Regardless of the reasons for Microsoft's inability to program Excel correctly, Microsoft has had years to fix these errors, and has proven itself unable to do so. It is clear that when Microsoft claims to have "fixed" a procedure that is known to be in error, Microsoft's claim cannot be safely believed. It is also clear that there are procedures in Excel that are in error, and Microsoft either does not know of the error or does not choose to warn users that the procedure is in error. Since it is generally impossible for the average user to distinguish between the accurate and inaccurate Excel functions, the only safe course for Excel users is to rely on no statistical procedure in Excel unless Microsoft provides tangible evidence of its accuracy consisting of:

1. test data with known input and output;
2. description of the algorithm sufficient to permit a third party to use the test data to reproduce Excel's output; and
3. a bona fide reference for the algorithm.

If Microsoft does not perform these actions for each statistical procedure in Excel, then there are only two safe alternatives for the user who is concerned about the accuracy of his statistical results: the user can perform all these actions himself, or simply not use Excel. (pp. 4577–4578)

(3) The article by Su (2008), "It's Easy to Produce Chartjunk Using Microsoft® Excel 2007 but Hard to Make Good Graphs," is concerned with how Excel violates the principles of good statistical graphics with its usual default settings, which all but the most well-informed probably just use (and the well informed should be using some other statistical package in the first place).[6] We give part of the concluding section:

A properly-designed chart can help people get the most information out of data and an excellent graphic tool helps people to achieve this task without much labour. The purpose of default settings in a graphic tool is to make it easy to produce good graphics that accord with the principles of statistical graphics. If the defaults do not embody these principles, then the only way to produce good graphics is to be sufficiently familiar with the principles of statistical graphics (Tufte, 1990, 1997; Cleveland, 1993, 1994; Wainer, 1997; Spence, 2001; Few, 2004).

This paper has shown that the default chart types in Excel 2007 do not embody these appropriate principles. Instead, these charts create chartjunk that hinder peoples' ability to comprehend the data. Some users have developed add-ins and instructions to redress the malfunctions of the Excel default chart settings and thereby enable users to produce better graphics (O'Day, 2007; Peltier, 2007; Vidman, 2007). Nevertheless, those who want to use Excel are advised to get to know the principles of good graphing well enough so that they know how to choose the appropriate options to override the defaults. Microsoft should overhaul its graphics engine so that it embodies the principles of statistical graphics and makes it easy for nonexperts to produce good graphs. (p. 4580)

(4) A final area of concern is the generation of statistical distributions, as discussed in the article by Yalta (2008), "The Accuracy of Statistical Distributions in Microsoft® Excel 2007." We again provide part of the concluding section of the article:

> It is our understanding that the algorithms for the computation of various statistical distributions in Excel 2007 can be inaccurate and/or unstable, and therefore can be unsafe to use. In particular, for the binomial, Poisson, inverse standard normal, inverse beta, inverse student's t, and inverse F distributions, it is possible to obtain results with zero accurate digits. Our results also show that the alternative Gnumeric and OpenOffice.org Calc programs, which employ dissimilar subroutines for the computation of statistical distributions, provide better accuracy in general in comparison to Excel 2007. In particular, Gnumeric can uniformly return exact values with at least six digits of accuracy for probabilities as small as 10^{-300} in all of our tests except one, and this is already fixed within a few weeks after we contacted the developers about the problem. Calc has important numerical difficulties for the computation of the quantiles of various distributions including the inverse chi-square, inverse beta, inverse t, and inverse F distributions. Once notified about the problems, Calc developers expressed their intention to correct these flaws with the upcoming OpenOffice.org version 2.4. (p. 4585)
>
> It is a known fact that Excel is commonly used in a wide range of decision making processes from options trading to research in physical laboratories. Offering statistical functionality in a computer program is a serious matter and it brings important responsibilities to the software vendor. Microsoft has repeatedly shown its lack of interest to this concernment by releasing new versions of Excel without first correcting the problems documented by different authors on various different occasions. Because of Microsoft's lack of commitment to accuracy, it is now possible to find on the Internet various users' custom scripts and macros for proper computation of statistical distributions in Excel. It is unclear when, if at all, Microsoft will properly fix Excel's inaccurate procedures for all of which there are free, well-known, and reliable alternatives. Meanwhile, researchers should continue to avoid using the statistical functions in Excel 2007 for any scientific purpose. (p. 4585)

The Excel saga points out an interesting lesson about open-source software, such as Gnumeric and R; the community seems to be self-correcting when errors are pointed out. For proprietary and closed systems, however, such as Excel, no such mechanism exists, and there is no real incentive to change, as long as the software still sells (which it does). If one wishes to use a spreadsheet program to do statistics, we suggest the free, open-source Gnumeric. As noted on their website, Gnumeric is "a friend of R"—something like the Good Housekeeping Seal of Approval (established in 1909, and still being awarded to this day with its promise of "Replacement or Refund if Defective").

7.8 Sample Size Selection

A topic that seems particularly difficult for beginning students to grasp is that of choosing the sample size to use in an experiment. At times the issue is moot, such as when some number of subject hours are allocated from the subject pool in a beginning psychology course and no more. In other instances where there is the possibility of recruiting paid subjects, it may be incumbent upon the experimenter to have a defensible rationale for how many subjects are necessary for the study to have a reasonable likelihood of success. A formal process for determining sample size becomes particularly crucial when obtaining data from even one subject is very expensive (for example, in neuroimaging), or when subjects are paid from a grant and allocation of funds for that purpose must be justified, or when competing clinical trials require differential allocation from a common pool of patients.[7]

The choice of sample size almost always involves a trade-off between the size of the effect the experimenter wishes to detect with high probability, and the maximum number of subjects available. It may be that the limited number of possible subjects makes even carrying out the experiment unnecessary because the size of effect likely to be present will not be detectable with the limited number of subjects available. It is also important to remember that measure fallibility reduces power and the ability to detect effects of interest. Other things being equal, measures having low reliability require more subjects to detect an effect of a given size.

Because subjects are generally expensive, experimenters sometimes try to "piggyback" data collection efforts on another study, or within the same study require subjects to complete multiple inventories or tasks. Such economy of effort may be reasonable to follow, but an experimenter must be aware of the possible unintended consequences of multiple measurement. Even when told that various parts of a data collection session are separate, subjects try to be consistent in presentation of self (possibly without much awareness). In other words, context effects can again be important. Whether the responses obtained on one instrument would be the same without the "prime" of another may be unanswerable without an explicit experimental design to get at the question. Here, various types of counterbalancing for the order of presentation could be studied explicitly for the type of carry-over effects that might be present in the data collection. In these latter kinds of studies, the whole apparatus of statistical experimental design comes into play, such as balanced-incomplete-block designs, fractional factorials, or Latin squares.[8]

Sample size decisions are difficult because the precision of measurement increases as the square root of the sample size. So if you need to double the precision, you need to multiply the size of the sample, and hence the cost of data collection, by four. When the phenomenon is rare, large samples and high costs are necessary. If resolution of the question is important enough, society has sometimes decided to pay the cost. The most often cited example of this is the 1954 Salk experimental trial in which the efficacy of the Salk vaccine to prevent polio was confirmed (Francis et al., 1955). Polio, though a very feared disease, was nonetheless relatively rare, with about 70 children per 100,000 in the population contracting it each year. Preliminary results suggested that the Salk vaccine could reduce that figure by roughly two-thirds. There were, however, variations year to year, and without an adequate control it would be difficult to know whether the observed reduction was caused by the vaccine or because it was just a lucky year. To detect differences of the size expected, an experiment was run in which the treatment and the control group each had 200,000 members (there was also a third group of children not vaccinated from families that would not agree to participate, but that is not germane to our main point). It was found that 142 children contracted polio in the control group, but only 56 in the treatment group. This was a statistically significant difference and the experiment ended. (For a further discussion of the Salk trial and relevant references, see Section 15.5.2: An Introductory Oddity: The 1954 Salk Polio Vaccine Trials.)

The Salk experiment illustrates another important ethical issue. In experiments of this kind, a good estimate of the size of the treatment effect can never be obtained, for as soon as we have gathered enough evidence to confirm that the treatment is superior, it is unethical to continue the control condition. We don't know exactly how much better the Salk vaccine was than the control treatment, and never will. Also, and although it is rarely discussed or even understood, vaccination never confers complete immunity against the targeted disease. The 56 individuals who contacted polio in the treatment group mentioned above demonstrate this clearly. Again, when asked to engage in the health-related practice of obtaining a vaccination (for flu, smallpox, or whatever), all the attendant error rate information should be made available to understand better why someone might prefer *not* to be vaccinated. The 1976 Swine Flu inoculation disaster under President Ford is a good cautionary tale to keep in mind; for a contemporary discussion of this episode, see the article by Robert Mackey, "Lessons From the Non-Pandemic of 1976," *New York Times*, April 30, 2009).

7.8.1 Are Large Clinical Trials in Rapidly Lethal Diseases Usually Unethical?

The choice of sample size appears to be fairly mechanical once an effect size is specified that one wishes to detect at a given probability level. However, when delving further into the ethics behind sample size selection in medical trials, the issues become murkier. As a poignant illustration, we give a "personal paper" written by David Horrobin, having the title given to this subsection, that appeared in the *Lancet* the same year he died (2003). As a medical scientist, Horrobin was a controversial figure for his advocacy of evening primrose oil for a variety of medical purposes. As one indication of this controversy, a negative Horrobin obituary that ran in the *British Medical Journal* in 2003 generated the most responses (both pro and con) for any obituary in the history of the journal

(Davies, 2003; Richmond, 2003). For our purposes, however, there is no need to pursue this part of Horrobin's career to appreciate the emotionally moving perspective he brings to sample size selection in clinical trials for lethal diseases, his included.

Medical statisticians have lately been fulminating about the issue of underpowered clinical trials (1, 2), arguing that such trials are not only unscientific but also unethical. The main argument is that patients entering clinical trials are exposed to risk and frequently volunteer for reasons that are in part altruistic. If they enter a study which, because of a lack of power, cannot be expected to yield a reliable result, then the trial organisers are guilty of unethically exploiting that altruism.

Several technical statistical arguments have been proposed to counter this view (3–6). One trial represents only part of the evidence to assess the value of an intervention: even an underpowered trial that gives inconclusive results can contribute to development of a more complete picture. From a Bayesian perspective, therefore, an underpowered trial is not wasted. Moreover, the making of power calculations is a dark art, which makes assumptions that are often unjustified. Only rarely do the actual effect sizes, the placebo effects, the variance of change, and the nature of the population recruited, match the assumptions made before the trial. Choice of the power wanted, whether it is 80%, 90%, or 87.65% is arbitrary and a matter of convention (7).

How can ethical judgments be made on the basis of a process for which there is little evidence base and experience shows that there are large discrepancies between the expectations of theory and the realities of practice? These uncertainties are such that power calculations are just as likely to lead to an overpowered as to an underpowered trial. If we follow the ethical line of reasoning, an overpowered trial is as reprehensible as an underpowered one because it will lead to unnecessary exposure of patients to a less effective treatment and to adverse effects of treatment. Unfortunately, getting things just right is in practice much more difficult than the writers of statistical textbooks and neat theoretical papers, devoid of prospective real world experimental studies, would care to admit.

Quite apart from the statistical arguments, however, there is one situation in which I believe that large trials are unethical. It is this: in diseases that can be rapidly lethal, say those which kill a high proportion of patients affected in a period ranging from days to perhaps 2-3 years, I submit that large clinical trials are unethical. The pros and cons about such trials do not seem to have been put forward. I believe that, whether I am right or wrong, the issues deserve widespread discussion, one that is at present not taking place.

I have been involved in biomedical research, much of it clinical research, for around 40 years. I am thoroughly acquainted with the many important ethical and statistical issues that impinge on clinical trials. I am neither a trained ethicist nor a trained statistician, but over the years I have acquired reasonable familiarity with and understanding of most of the key issues. Or at least I thought I had. But 2 years ago everything changed.

After losing to ill health no more than a total of 10 days in 40 years, I became ill with a febrile illness and a mysterious rash, which cleared up after 10 days or so. I began to lose weight. A couple of months later the rash returned, inguinal, cervical, and axillary glands became painlessly enlarged, and I could feel at least two large abdominal masses. Advanced mantle cell lymphoma was diagnosed. I was told that median survival from first symptoms was around 2-3 years. However, in my case, the multiple large tumours, marrow involvement, systemic symptoms, and rapid loss of around 20% of my bodyweight, meant that I could not realistically expect to live much more than 6 months.

And so I entered a universe parallel to the one in which I had lived for 40 years. I became a patient and suddenly saw everything from the other side. I lived in a realm in which much of my time was spent talking to or otherwise communicating with other cancer patients. I scanned every relevant biomedical database I could find seeking information about mantle cell lymphoma. Like any lay person, I surfed the net and found an astonishing array of both sense and nonsense. I developed a whole new attitude to clinical trials and experimental treatments. I was threatened with death but I wanted to live. I looked at effect sizes and power calculations in a wholly new way.

Anyone who organises or undertakes clinical trials understands the relation between effect size, trial size, and statistical power. The larger the effect, the smaller the trial needed to show statistical significance. A very few drug interventions, such as penicillin for pneumococcal pneumonia, are so effective in almost all patients that no placebo-controlled trials are needed to show efficacy. Many drugs show statistically significant benefits with trial sizes of 20 or 30 patients. With such drugs the prescribing physicians can know that most treated patients will show a response that can be reasonably attributed to the drug by both patient and doctor. But as effect sizes become smaller and trial sizes climb over a hundred, it becomes more and more difficult for anyone to know whether what happens to the individual was caused by the drug.

To my dismay I soon learned that in oncology, with few exceptions, effect sizes were very small. To show these effects, trials had to be very large. I also learned from my fellow patients that the real consequences of this situation were rarely spelled out to those volunteering for such trials in terms they could understand. I thought long and hard about this situation and came to the conclusion that, as presently organised, many oncology trials are unethical. Similar considerations apply to any other rapidly lethal disease. My reasons are set out in the rest of this paper.

The first reaction of most people on receiving a diagnosis of a lethal disease is that they want to live. Above all they want to find the best available treatment. Only at a much later stage, when they have in some part come to terms with the fact that they might have to die in the near future—and that there might not be a so-called best treatment, or indeed any even half decent treatment—do they begin to think at all about altruism. Any desire to contribute to the welfare of others usually comes only after hope has come close to being extinguished. The idea that altruism is an important consideration for most patients with cancer is a figment of the ethicist's and statistician's imagination. Of course, many people, when confronted with the certainty of their own death, will say that even if they cannot live they would like to contribute to understanding so that others will have a chance of survival. But the idea that this is a really important issue for most patients is nonsense. What they want is to survive, and to do that they want the best treatment. Altruism, in the sense of volunteering for a clinical trial whose outcome is uncertain, comes a very poor second. I believe that patients who are asked to volunteer for large trials in cancer or other rapidly lethal diseases are being misled. Most such trials cannot be justified on ethical grounds.

First, patients entering large clinical trials have little chance of benefit. If a trial has to be large, say more than 100 patients, it is large only because the expected effect size is very small. That means that most patients entering the trial have little or no chance of receiving benefit. With the toxic nature of many oncology treatment regimens, there may well be a substantial chance of harm. As far as I can learn from my many contacts with fellow patients over the past 2 years, although the risk of harm is usually well described in patient information leaflets, almost nothing adequate is ever said about the assumed effect size and the real chance of benefit. Almost all patients volunteering for most trials in oncology are doomed: at best they can expect little benefit. They are not usually being properly told about this low expectation.

Second, large trials greatly escalate time delays and costs. Anyone undertaking clinical trials in the present environment knows how difficult it is to get a small trial going in a single centre. Both the administrative and ethical issues can be formidable—let alone the clinical and financial ones. But the involvement of more than one centre escalates all the difficulties in a way that has the characteristics of an exponential curve. As a result, large trials in multiple centres almost always take much longer to get going, take longer to complete, and are enormously more expensive than small single centre trials. They have also become major sources of revenue for many institutions. Because large trials have become the norm, all professionals taking part are now reconciled to the idea that such trials will take forever and will cost the earth. As a result, most patients entering most oncology trials will be dead before the results are known. But the institutions in which they are being treated probably benefit greatly financially. Most patient information leaflets do not tell them either fact. This omission is unethical.

Third, the high cost of large trials means that they can be done only on patent-protected new chemical entities. Because large trials cost such very large amounts of money, they are rarely funded by charitable or public money. When they are so funded, there are so many vested and competing interests to be reconciled that delays become extraordinary, which is certainly no help to patients. What the high cost usually means is that only commercial interests can afford to pay for the trial. And since such companies have to seek a return for investment, trials will be conducted for only a tiny part of the wide range of potential cancer therapies. That part consists of new chemical entities that have a remaining patent life of at the least 10, and preferably 15, years. Cancer patients are, of course, not told that such a small part of potential therapies is open to them. Nor are they told that researchers in most institutions, when considering which trials to take part in, are heavily influenced by the size of the financial contribution from the commercial sponsor. There is distressingly little altruism there.

Fourth, large trials greatly restrict the numbers of treatments that can be tested. For any disease, that the numbers of patients available for clinical trials are restricted is self evident. For all sorts of reasons the realistic universe of patients is much smaller than the actual universe of patients with any particular disease. Thus, a large trial that recruits many patients will considerably reduce the numbers of patients available for other trials. Thus, the number of therapies that can be tested is reduced.

For the past 20 years I have been working in the pharmaceutical industry. Although everyone in the industry will deny it, and I doubt whether there is documentary evidence of this statement anywhere, I know that several of the larger firms use overpowered trials as a way of keeping competitors out of that particular subject. Especially with less common cancers, if a company, by manipulating the power calculations, can recruit for a trial several times more patients than is necessary, then they will gain a clear competitive advantage by mak-

ing it more difficult for rivals to recruit. This practice is unethical, but it happens, and the statisticians who are enthusiasts for power in clinical trials are often unwitting accomplices of these unethical commercial practices. [Italics added to this paragraph by the current authors for emphasis.]

Any scientifically or medically qualified person who develops a lethal cancer rapidly learns many things. Two of them were especially surprising to me. First, as in my own case, the usual effects of standard treatments are all-too-often both toxic and of minimum therapeutic value. Occasionally patients do very well, but the outlook for most is gloomy. Moreover, the evidence base is near useless as a guide to what is likely to happen in one's own case, partly because the exclusion and inclusion criteria for trials are often so narrowly drawn that most individuals are unlikely to fit them. Another contributing factor is that effect sizes are so small that the numbers needed-to-treat to get one durable response may well be over 30 and often even higher. So, for the individual, treatment is indeed a lottery. In view of the frequently severe adverse events, usually much more predictable and reliable in their occurrence than is a therapeutic response, a decision on the patient's part not to be treated is not irrational. I learned that few patients are made aware of this fact: that is unethical.

The second surprising thing that I learned is that, for most cancers, there are many potential treatments, many of which are not toxic. Contrary to general orthodox medical opinion, most such potential treatments are neither fringe nor irrational. They are based on solid biochemical in-vitro work, on reliable work in animals, and occasionally on a few well documented case histories. But they have not been adequately tested in well designed trials, and most of them never will be. The reason for that has nothing to do with their scientific rationale or the strength of the evidence: it is simply that they are unpatentable or difficult to patent. Without patent protection, in the present climate, such potential remedies will never be tested.

Take the example of my own illness, mantle cell lymphoma. This disease is quite well understood. Most, and perhaps all malignant mantle cells have a striking rise of cyclin D1, a factor that drives cell division (8–10). This increase is usually due to a translocation of the IgG promotor to the cyclin D1 gene. Cyclin D1 values can be suppressed by several well known agents, including the antifungal drug clotrimazole, the polyunsaturated fatty acid eicosapentaenoate, and the antidiabetic thiazolidinediones (11–13). The publications in which the effects are described are in leading cancer journals and come from institutions such as Harvard, which are hardly on the fringe. No trials have been done on any of these agents. There are also rational but more complex cases to be made for the possibility of treating mantle cell lymphoma with cyclo-oxygenase inhibitors, or thalidomide, or certain groups of nutrients; but that any of these will be properly tested in this disease is unlikely (14). I was fortunate to be able to discuss these unproven approaches with sympathetic haematologists and to devise a regimen, which so far seems to have been helpful. But most patients are not in that position.

This is a depressing example of the law of unintended consequences. 50 years ago, good scientific evidence of a potential therapeutic effect would quickly have generated a small clinical trial in one or two centres with perhaps 30 or 40 patients. Such a trial would have cost almost nothing. It would certainly have missed small or marginal effects, but it would not have missed the sort of large effect that most patients want.

Unfortunately, now, such an approach has become impossible. Ethics committees, clinical trial regulations, and research costing by administrators would put the cost of even a small pilot study up to £100,000 (about US $160,000) or more. Without patent protection, that anyone would fund such a pilot study is very unlikely. And so, as is easily demonstrable by reviewing publications, scores of compounds that might have a therapeutic effect will never be tested. The escalation of costs has therefore drastically reduced the range of compounds from which new treatments can be drawn. If a compound is not protected by patent it will not be developed, which would not matter if current research in oncology were producing large benefits in common cancers.

But despite huge expenditure, success has largely eluded us. The few outstanding successes in rare cancers cannot hide the overall failure. This situation has to mean that there is a real possibility that standard approaches are wrong, and that we have no firm rational basis for predicting the directions from which success might come. Our best hope of changing our practice is to test as many different approaches and compounds as possible, looking for substantial effects. But the demand for large trials that are adequately powered to detect small effects and that are undertaken in many centres, and a substantial bureaucracy, has effectively killed this possibility.

I submit that there is a strong basis for the claim that large trials are indeed unethical in patients with diseases that have a high probability of killing the patient within 2-3 years. Patients with lethal diseases want to get better, not to have their lives extended by a few weeks or months at great cost in toxicity and time in treatment. The only reasonable approach, if this goal is to be achieved, is to scan a wide range of diverse therapies. Certainly, the mainstream ideas that we have been having for the past 50 years have not done a great deal for us. The only way in which our approach might change is that we largely abandon large-scale trials looking for small effects and instead do large numbers of small trials, often in single centres, looking for large effects. I suspect that patient participation in the trial process would become much more enthusiastic. Most people are more interested in the remote chance of a cure, than in the certainty of toxicity and the near certainty of no useful response.

References:

1 Halpern SD, Karlawish JH, Berlin JA. The continuing unethical conduct of underpowered clinical trials. *JAMA* 2002; **288**: 358–62.

2 Altman DG. Statistics and ethics in medical research, III: how large a sample? *BMJ* 1980; **281**: 1336–38.

3 Edwards SJ, Lilford RJ, Bralmholtz D, Jackson J. Why "underpowered" trials are not necessarily unethical. *Lancet* 1997; **350**: 804–07.

4 Janosky JE. The ethics of underpowered clinical trials. *JAMA* 2002; **288**: 2118.

5 Hughes JR. The ethics of underpowered clinical trials. *JAMA* 2002; **288**: 2118.

6 Lilford RJ. The ethics of underpowered clinical trials. *JAMA* 2002; **288**: 2118–19.

7 Horrobin DF. Statistics: peer review and ethical review of power calculations. *BMJ* 2002; **325**: 491a.

8 Oka K, Ohno T, Yamaguchi M, et al. PRADl/Cyclin D1 gene overexpression in mantle cell lymphoma. *Leuk Lymphoma* 1996; **21**: 37–42.

9 Dreyling MH, Bullinger L, Ott G, et al. Alterations of the cyclin Dl/pl6-pRB pathway in mantle cell lymphoma. *Cancer Res* 1997; **57**: 4608–14.

10 Rimokh R, Berger F, Bastard C, et al. Rearrangement of CCND1 (BCLl/PRAD1) 3A untranslated region in mantle-cell lymphomas and t(llql3)-associated leukemias. *Blood* 1994; **83**: 3689–96.

11 Palakurthi SS, Fluckiger R, Aktas H, et al. Inhibition of translation initiation mediates the anticancer effect of die ti-3 polyunsaturated fatty acid eicosapentaenoic acid. *Cancer Res* 2000; **60**: 2919–25.

12 Aktas H, Fluckiger R, Acosta JA, Savage JM, Palakurthi SS, Halperin JA. Depletion of intracellular Ca2+ stores, phosphorylation of eIF2alpha, and sustained inhibition of translation initiation mediate the anticancer effects of clotrimazole. *Proc Natl Acad Sci USA* 1998; **95**: 8280–85.

13 Palakurthi SS, Aktas H, Grubissich LM, Mortensen RM, Halperin JA. Anticancer effects of thiazolidinediones are independent of peroxisome proliferator-activated receptor gamma and mediated by inhibition of translation initiation. *Cancer Res* 2001; **61**: 6213–18.

14 Horrobin DF. A low toxicity maintenance regime for mantle cell lymphoma and other malignancies with cyclin dI overexpression: role of eicosapentaenoic acid. *Med Hypoth* (in press).

Notes

[1]Courts have been distrustful of sampling versus complete enumeration, and have been so for a long time. A case in 1955, for example, involved Sears, Roebuck, and Company and the City of Inglewood (California). The Court ruled that a sample of receipts was inadequate to estimate the amount of taxes that Sears had overpaid. Instead, a costly complete audit or enumeration was required. For a further discussion of this case, see R. Clay Sprowls, "The Admissibility of Sample Data into a Court of Law: A Case History," *UCLA Law Review, 4*, 222–232, 1956–1957.

[2]A sampling frame is the list of all those in the population that can be sampled.

[3]A clever solution to this problem of rankings was proposed by a group of statisticians at Bell Labs in 1987: "Analysis of Data From the *Places Rated Almanac*" (Richard A. Becker, Lorraine Denby, Robert McGill, & Allan R. Wilks; *American Statistician, 41*, 169-186). Becker and colleagues were searching for the best places to live in the United States. They constructed a dozen or so indices (for example, affordability of housing, availability of good medical care, jobs, public safety, culture, and quality of public schools). Each index might be composed of multiple variables. Obviously, different people have different needs, and so would weight each of the various indices differently. A young person might be interested most in job availability and the quality of public schools. An older person might rate the availability of high-quality medical care and cultural assets highest. No one set of weightings would suffice for everyone. So what they proposed was to search for a set of weights that would rank a particular place first. For many places it was possible to find such a set. For some places such as Yorba Linda, California, no set of weights would work; it was always dominated by other places. But the set of weights that would yield a first place finish for a particular place were rarely unique. You could usually move a tiny distance away from the optimal set and still maintain dominance. So they proposed considering all n indices as axes in n-space, and the set of weights as a point in that space. The volume defined by the set of weights that made a place dominate all others was used to characterize the size of the population that would find that place best. The places were then rated by the size of these volumes; the place associated with the largest volume was rated as the best place to live. This approach could be improved through a survey of all prospective users of this ranking to obtain their actual weights, but without doing this, the methodology provides one

way of ranking multidimensional objects that may be of value to prospective users.

[4] An article in the Suggested Reading, B.2.1, from *ScienceNews* by Laura Sanders ("Trawling the Brain", December 19, 2009) provides a cautionary lesson for anyone involved with the interpretation of fMRI research. A dead salmon's brain can display much of the same beautiful red-hot areas of activity in response to emotional scenes flashed to the (dead) salmon that would be expected for (alive) human subjects.

[5] The Suggested Reading in B.2.6 discusses the probability issues involved with searching through the whole genome: "Nabbing Suspicious SNPS: Scientists Search the Whole Genome for Clues to Common Diseases" (Regina Nuzzo, *ScienceNews*, June 21, 2008).

[6] Scatterplots in Excel generally look odd because of the different types of spacing for the x and y axes. The x-axis uses numerical labels only nominally so the numerical size of the labels is not reflected by the spacing; conversely, the y-axis is spaced equally so the numerical size of the labels does reflect the actual spacing shown.

[7] For an amusing story about sample size, we repeat the preamble discussion by Anu Garg for the March 1, 2010, *A Word A Day* (AWAD):

Recently I visited London to attend a wedding. The bride had graduated from Oxford, and among the invitees were some of her fellow graduates and a professor. During the long ceremony, we intermittently chatted about London weather, Gordon Brown, Queen Lizzie, and language.

Among other things, we talked about the differences between British and American English. I recalled reading about the inroads American English is making even in the UK, so I decided to carry out an experiment to find to what extent American English had "corrupted" English English.

I told them that sometimes the British write certain numerals (e.g., 1 and 7) differently from how they're written in the US, and asked them to write a short sentence so I could see if there were other differences in the script.

I quickly thought of a sentence for them to write:

"Her favorite flavors were in the gray catalog, she realized."

I said it aloud and the five Oxonians and the Oxford don kindly wrote it down on their napkins (serviettes). I collected the napkins and then told them about the experiment—it had nothing to do with handwriting. In reality, that sentence had five words that could be written with American or British spellings (favorite/favourite, flavor/flavour, gray/grey, catalog/catalogue, realize/realise).

Of the six people who participated in the experiment, three spelled (spelt) everything the British way. The other three had one or more words spelled in American English.

What does this experiment prove? Not much, according to my 12-year-old daughter, "Your sample size is too small."

[8] The Kansas City preventive patrol experiment from the early 1970s shows how unintended priming and context effects can compromise the results ob-

tained from an otherwise well-designed study. In 1972 and 1973 an experiment was undertaken in Kansas City to study the effect that police patrols had on crime rates. Patrols were varied within 15 police beats. Routine preventive patrol was eliminated in five beats, labeled "reactive" beats (meaning officers entered these areas only in response to calls from residents). Normal, routine patrol was maintained in five "control" beats. In five "proactive" beats, patrols were intensified by two to three times the norm. The experiment asked the following questions:

— Would citizens notice changes in the level of police patrol?

— Would different levels of visible police patrol affect recorded crime or the outcome of victim surveys?

— Would citizen fear of crime and attendant behavior change as a result of differing patrol levels?

— Would their degree of satisfaction with the police change?

Information was gathered from victimization surveys, reported crime rates, arrest data, a survey of local businesses, attitudinal surveys, and trained observers who monitored police–citizen interaction. The results were clear— patrols didn't seem to matter. Crime rates didn't change, and no one seemed to notice that anything had changed. But these results were cast into doubt by observations made by Al Reiss and his team of Yale sociologists who monitored the experiment. They found that the police felt strongly that their presence mattered and so although they abided by the letter of the rules of the experimental design, they made some informal modifications on the fly. Specifically, while they did not patrol the reactive areas, they did patrol their borders continuously. When they had to go someplace else in the city, they would drive through a reactive area whenever possible. And when they went off-duty for lunch or other break, they would drive to a restaurant in a reactive area. So, the real difference in patrol activity was much smaller than might be supposed.

A redesign was proposed in which the "border effects" (for example, the effect of a reactive zone being next to a proactive zone) were estimated. This could be done by surrounding a reactive zone with proactive zones, another that was surrounded by normal zones, and a third surrounded by reactive zones. This would be repeated for each of the treatment groups. In this way, the boundary effects could be estimated and adjusted for, yielding a more accurate estimate of the patrol effect. This design required a municipality much larger than Kansas City and was never carried out. (For more discussion, see "Kansas City preventive patrol experiment," 2011; Kelling et al., 1974; Police Foundation, 2011.)

Chapter 8

Psychometrics

> The *native intelligence* hypothesis is dead.
> — C. C. Brigham (1934, developer of the SAT)

Psychometrics is a branch of psychology that deals with the design, administration, scoring, and interpretation of objective tests developed for the measurement of psychologically relevant variables, such as aptitude, personality, job performance, or academic achievement. Because of the socially relevant implications of high-stakes testing in areas such as job selection and promotion, licensure, college admission, accreditation, and graduation and other competency certification, the issues of fairness and discrimination surrounding testing are often discussed in the media, and increasingly, in the courts.[1]

The summer of 2009 saw the country engaged in confirmation hearings for the proposed new United States Supreme Court Justice, Sonia Sotomayor, where part of the controversy was in how she ruled on the use of testing for promotion to lieutenant in the City of New Haven Fire Department, and whether the City of New Haven's invalidation of the exam results based on the apparent disparate impact was discriminatory. We note that for the New Haven exam, the pass rate for African-American applicants was half of that for whites. Given the quasi-legal judgment of "disparate impact" whenever passing rate for a minority group is less than 80% of that for the majority, there is a *primae facie* case for disparate impact. As in other legal contexts, this should have triggered a study of whether acceptable alternatives exist, and if not, the original test usage could be upheld. Why this remedy was not applied in this instance is unclear.

Because of the importance of testing in many of our societal endeavors, it is a relevant statistical literacy skill to have some basic understanding of the concerns of both traditional true score theory

(TST), and its emphasis on assessing reliability and validity, and the newer item response theory (IRT), where there is an attempt to scale jointly subjects and assessment items through variants of logistic regression models. Once items are calibrated from some norming sample by estimating parameters characterizing the logistic regressions, the item pool can serve a number of purposes: (1) subsets of items can be used to construct fixed-length tests with subjects' performance estimated by how they respond to the calibrated items; (2) psychometrically equivalent tests can be constructed from the item pool, or distinct tests constructed from the pool can be equated, supposedly making it immaterial to a subject as to which test is actually taken; (3) a computer adaptive testing (CAT) process can be initiated where items are administered according to how a respondent has answered earlier items. In theory, CAT holds the promise of greater efficiency in testing by allowing instruments to be shorter and/or estimating the trait of interest better.

8.1   Traditional True Score Theory Concepts of Reliability and Validity

Several psychometric concepts are important to understand in any informed discussion about the use of tests and how good or bad they may be. We begin with the two key traditional concepts in TST of reliability and validity.

There are several ways to think about these two fundamental concepts. The modern approach is in terms of evidence in which a test score is considered as evidence in support of some claim made about the person who obtained that score. Validity is then the relevance of that evidence for that claim, and reliability is simply the amount of such evidence. This approach makes explicit two important ideas. First, that a test score by itself has no validity; it is only the inferences or the claims whose validity are of interest. And second, without reliability (so reliability is zero), there is no validity. This latter idea, as obvious as it seems, is not always grasped by educational theorists with a meager understanding of psychometric theory (Moss, 1994).

A reliable measure is one that measures something consistently. In theory, as well as sometimes in practice, *test-retest* reliability can be assessed by computing the correlation between repeated administrations of the same instrument over the same subjects. Or, if two equivalent tests are available, possibly constructed from an IRT-calibrated item pool and given to the same subjects, the correlation between such scores is an *equivalent-forms* reliability. The homogeneity of a single test form is referred to as *internal consistency* and can be assessed by correlating performance on two halves of the test to give a *split-half* reliability; this latter value can then be readjusted with the (nonlinear) Spearman–Brown prophecy formula to give the reliability of a full-length test. More commonly, and as computed by all the commercial statistical software packages (for example, SYSTAT, SPSS, SAS), there is Cronbach's alpha, the mean of all possible split-half coefficients. If we formally define reliability within a true score plus error model as the ratio of true score variance to observed score variance, alpha can be considered a lower bound to reliability.

Validity refers to the extent to which the score on an instrument represents what it is supposed to, thus justifying the inferences that are made from that score. It is usually evaluated by a process. Given other criterion measures that may be available, *concurrent* validity refers to the correlation with measures collected at the same time; *predictive* validity to measures collected later; and *construct* validity to relating the current measures to others that should be part of its theoretical context. *Content* validity is more direct and refers to items demonstrably coming from the domain we wish to assess, or that are *bona fide* skills necessary to perform some task.

As discussed by Cronbach and Meehl in their justifiably well-known 1955 article, "Construct Validity in Psychological Tests" (*Psychological Bulletin, 52*, 281–302), the most difficult form of validity to establish is construct validity. In lay terms, a validation process has to be put into place to argue effectively that we are really measuring the construct we think we are measuring. A recent example of the difficulties inherent in construct validation has appeared in the popular media, and involves the notion of psychopathy—a personality disorder indicated by a pattern of lying, exploitativeness, heedlessness, arrogance, sexual promiscuity, low self-control, and lack of empathy and remorse, all of this combined with an ability to appear normal.

The *Diagnostic and Statistical Manual of Mental Disorders (DSM-IV)* (1994) does not include psychopathy as part of its classification scheme for personality disorders. In fact, the notion of psychopathy has been defined *de facto* by one specific 20-item instrument developed by Robert Hare (2003), the Psychopathy Checklist–Revised (PCL-R). The issue now being raised about the PCL-R is the degree to which criminal behavior is a component crucial to psychopathy. Or, as put by the Australian psychiatrist John Ellard in 1989: "Why has this man done these terrible things? Because he is a psychopath. And how do you know that he is a psychopath? Because he has done these terrible things" (p. 128). We refer the reader to an article from the *New York Times* by Benedict Carey, "Academic Battle Delays Publication by 3 Years" (June 11, 2010) that traces some of the sordidness of the current saga; the abstract of the delayed article is appended below (Skeem, J. L., & Cooke, D. J. [2010]. "Is Criminal Behavior a Central Component of Psychopathy? Conceptual Directions for Resolving the Debate," *Psychological Assessment, 22*, 433–445):

The development of the Psychopathy Checklist-Revised (PCL-R) has fueled intense clinical interest in the construct of psychopathy. Unfortunately, a side effect of this interest has been conceptual confusion and, in particular, the conflating of measures with constructs. Indeed, the field is in danger of equating the PCL-R with the theoretical construct of psychopathy. A key point in the debate is whether criminal behavior is a central component, or mere downstream correlate, of psychopathy. In this article, the authors present conceptual directions for resolving this debate. First, factor analysis of PCL-R items in a theoretical vacuum cannot reveal the essence of psychopathy. Second, a myth about the PCL-R and its relation to violence must be examined to avoid the view that psychopathy is merely a violent variant of antisocial personality disorder. Third, a formal, iterative process between theory development and empirical validation must be adopted. Fundamentally, constructs and measures must be recognized as separate entities, and neither reified. Applying such principles to the current state of the field, the authors believe the evidence favors viewing criminal behavior as a correlate, not a component, of psychopathy. (p. 433)

8.2   Test Fairness

Some confusions about the fairness of a test continually reappear in the popular culture that could be clarified with an educated understanding of how tests are generally constructed and intended to function. Because one particular definable subgroup performs less well on a test than another does not automatically imply that the test is "biased" or "unfair" to the subgroup. We quote part of a passage from Linn and Drasgow (1987) that makes this point particularly well:

> [P]sychological tests *do not measure innate abilities or aptitudes.* Instead, they assess a test taker's current repertoire of knowledge and skills. An individual's repertoire of knowledge and skills is certainly affected by his or her "environment"; consequently, we would expect differences in mean test performance for groups whose environments differ. ... The key point is that *unequal environments imply unequal educational achievements and a well-constructed test should reflect this fact.* (p. 13; latter italics added)

As another way of phrasing this argument, if definable subgroups differ in background environments that should be related to what a test is measuring, similarity of performance for subgroups does not necessarily indicate a fairer or less biased instrument; to the contrary, it is more reflective of a test that may not be very good.

An instructive example from the 1980s illustrates the pernicious effects of succumbing to legal pragmatism instead of following sound psychometric principles of test construction. It is the 1984 agreement (some would say, capitulation) between the Illinois Department of Insurance and its partner, the Educational Testing Services, with the Golden Rule Insurance Company (hence, the name "Golden Rule" settlement), and how the Illinois insurance licensing examinations would henceforth be constructed. Items were to be classified into two types:

> Type I—those items for which the correct-answer rates for black Examinees, white Examinees, and all Examinees are not lower than forty percent (40%) at the .05 level of statistical significance, and (b) the correct-answer rates of black Examinees and white Examinees differ by no more than fifteen (15) percentage points at the .05 level of

statistical significance; or Type II—all other items (Linn & Drasgow, 1987, p. 13).

The agreement states that ETS should assemble the licensing exams to be "in accordance with the subject matter coverage and weighting of the applicable content outline," but whenever possible, Type I items are to be used, and within this group, preference given to those with the smallest differences for between-group correct-answer rates. When not enough Type I items are available to satisfy content outline constraints, Type II items would be included with preference to those with correct-answer rates differing the least.

Linn and Drasgow (1987) employ cogent psychometric reasoning in discussing the possible unintended consequences of particular testing policies. We state some of their conclusions about the psychometrically degrading effects of implementing a Golden Rule settlement in the construction of tests generally: (1) because of the 40% provision, the elimination of difficult items with small differences in correct-answer rates could actually increase rather than decrease the average difference on the total test score; (2) the 15 percentage point rule for item difficulty does not have the effect of eliminating items biased according to the accepted psychometric definition: an item is biased if the item response curves are not the same in the subpopulations; or alternatively worded, an item is biased if test takers from different groups have unequal probabilities of a correct response even though they have equal standing on the attribute being measured. On the contrary, the use of item difficulty differences to indicate biased items actually misclassifies unbiased items as biased; also, it is insensitive to bias when it might exist; (3) usage of the Golden Rule procedure can have devastating effects on the two most important test properties of reliability and validity. Reliability is reduced because of favoring items with poor discriminating power; validity is degraded because it eliminates those items providing the best measurement of the attribute of interest.

One of the consequences of the Golden Rule settlement is the precedent it provides for all uses of testing, whether for licensing or other purposes. Its widespread adoption, as noted by Linn and Drasgow (1987), would degrade the testing enterprise and not improve it. As Gregory Anrig (1987), then President of ETS admitted, the Golden Rule Settlement was a major error in judgment. To the present, either the Golden Rule or its disguised surrogates (for example, banning test construction based on high point-biserial [item to total] correlations that reflect correct-answer differences in a different way), degrade test validity and reliability, and need to be argued against in court whenever they arise.

A somewhat different take on the issue of fairness in the use of tests is the practice of "race-norming," the adjustment of scores on a standardized test through separate transformations for different racial groups (see Kilborn, 1991). This type of within-group norming was promulgated in the 1980s as a way of reaching federal equal employment opportunity and affirmative action goals; it was particularly encouraged by the Department of Labor and the United States Employment Services in its promotion of the General Aptitude Test Battery (GATB) to state employment services. Irrespective of whatever admirable social goals were envisioned with the use of race norming, the Civil Rights Act of 1991 prohibits discriminatory use of test scores. The new Title VII provision, section 106, reads in full:

> It shall be an unlawful employment practice for a respondent, in connection with the selection or referral of applicants or candidates for employment or promotion, to adjust the scores of, use different cutoff scores for, or otherwise alter the results of employment related tests on the basis of race, color, religion, sex or national origin.

Although the issue of race norming may now be moot, there are still interesting questions it raises that aren't fully addressed in the psychometric literature. Specifically, is more always better? For example, suppose only a certain level of test performance is really necessary to carry out a job well. It would seem that other criteria might then be used to select from the pool making the cut, such as workplace diversity. Also, if one scores too high, it might signal a potential dissatisfaction with the numbing nature of the job being proffered. One of us had an acquaintance in college who was denied employment at the local drug store lunch counter because she scored too high on an administered (verbally loaded) instrument; she obviously would get bored very quickly and move on from that job, leaving the employer with a need to train yet another new lunch counter assistant. Or consider the test-savvy colleague of LH who argued that anything above a combined score of 1200 on the GRE Math and Verbal tests was complete excess; moreover, if the combined score were too high, the graduate student was bound to be "squirrelly" (with exemplars to make the case). This observation also helps in coming to grips with one's own (possibly) depressed GRE scores.

8.3 Quotidian Psychometric Insights

> The relationship between the intelligence of subjects and the volume of their head ... is very real and has been confirmed by all methodical investigators, without exception. ... As these works include observations on several hundred subjects, we conclude that the preceding proposition [of a correlation between head size and intelligence] must be considered as incontestable.
> — Alfred Binet (*La fatigue intellectuelle*, 1898 [as quoted by Stephen J. Gould, *The Mismeasure of Man*])

A number of observations can be made about some everyday occurrences and how they might be informed by basic psychometric reasoning:

First, difference or gain scores tend to be unreliable and more so the higher the correlation between the constituent scores. Given the attenuating characteristics of unreliability and the general reduction in power, it may be generally problematic to use gain scores in empirical investigations. This is a major hurdle for teacher evaluation programs that propose basing merit increases on student gain scores, using "value-added models" (Braun & Wainer, 2007).

Second, the standard error of a correlation is about $\frac{1}{\sqrt{n}}$; thus, correlations based on small sample sizes need to be fairly large to argue that they reflect more than random variability for a population where the true correlation might, plausibly, be zero.

Third, constructing a test by item selection to optimize an index such as Cronbach's alpha will tend to produce a homogeneous test satisfying a single common factor model. This is because items having correlations of similar size with other items will be culled, and alpha increases with the average correlation between items. However, the implication in the other direction is not correct, that a high value for alpha reflects unidimensionality; unidimensionality depends on the pattern of inter-item covariances, whereas the value of alpha depends on their size.

Fourth, as discussed at some length by Thurstone (1933) in *The Theory of Multiple Factors*, the computations suggested by Spearman for the single common factor model need separation from Spearman's theory that intelligence is such a common factor, referred to as "g." In fact, one might go on to observe that "g" (viz., Guttman's First Law of Intelligence) may just be an artifact of correlations generally being positive among intellective-type tests; here, as stated by the Perron–Frobenius Theorem ("Perron–Frobenius theorem," 2010), weights defining the first principal component must all be positive, providing a simple weighting of the tests. To reify this weighted composite as "g," without additional supporting evidence, seems a bit of a stretch. (For a comprehensive and detailed review of the measurement of intelligence along with its history, including Spearman's own revisions to include more than a single notion of what intelligence might be, see Carroll, 1982.)

Stephen Jay Gould's popular book, *The Mismeasure of Man* (1996), takes on "the argument that intelligence can be meaningfully abstracted as a single number capable of ranking all people on a linear scale of intrinsic and unalterable mental worth" (p. 20). Along the way, several observations about factor analysis are made that deserve reporting:

> IQ, a linear scale first established as a rough, empirical measure, is easy to understand. Factor analysis, rooted in abstract statistical theory and based on the attempt to discover "underlying" structure in large matrices of data, is, to put it bluntly, a bitch. (p. 268)

> Factor analysis, despite its status as pure deductive mathematics, was invented in a social context, and for definite reasons. And, though its mathematical basis is unassailable, its persistent use as a device for learning about the physical structure of intellect has been mired in deep conceptual errors from the start. The principal error, in fact, has involved a major theme of this book: reification—in this case, the notion that such a nebulous, socially defined concept as intelli-

gence might be identified as a "thing" with a locus in the brain and a definite degree of heritability—and that it might be measured as a single number, thus permitting a unilinear ranking of people according to the amount of it they possess. By identifying a mathematical factor axis with a concept of "general intelligence," Spearman and Burt provided a theoretical justification for the unilinear scale that Binet had proposed as a rough empirical guide. (pp. 268–269)

From the midst of an economic depression that reduced many of its intellectual elite to poverty, an America with egalitarian ideas (however rarely practiced) challenged Britain's traditional equation of social class with innate worth. Spearman's *g* had been rotated away, and general mental worth evaporated with it. (p. 334)

Fifth, the unreliability of a measure has a number of statistical consequences: it attenuates the correlation with other tests so any study of a measure's validity needs to take this into consideration; it reduces the power of common statistical analyses, such as analysis of variance; it biases regression coefficients in regression models. In all cases, the more unreliable a measure, the more difficult it will be for it to have value in any scientific or practical investigation.

Sixth, the standard TST idea of reliability can be put into the framework of random effects analysis of variance. These components-of-variance models are commonly discussed in the year-long graduate sequence in statistics, and extensions are now being touted as the "next big thing" in hierarchical linear modeling (HLM). In turn, components-of-variance models lead to the estimation of heritability coefficients in behavioral genetics. Here, the "observed score" variance being additively decomposed as "true score" variance plus uncorrelated "error" variance, is reinterpreted as "phenotypic variance," being the sum of "genotypic variance" and uncorrelated "environmental variance." Thus, some basic proficiency in how reliability is defined and approached generally should also help in reasoning with the sometimes controversial topic of heritability (because the various heritability coefficients are just ratios of true-score to observed-score variances), and thinking through how HLM studies are reported, with the inclusion of random effects to account for the type of hierarchical sampling that has occurred. A more extensive discussion of heritability appeared earlier in Section 5.6, emphasizing intraclass correlations.

Finally, although personality tests may be subject to the same TST ideas of reliability and validity, the status of IRT modeling in personality is somewhat different than for intellective measurement. For the latter, it may be reasonable for a positive response probability to increase consistently with an increase in the latent quantity being estimated. These "dominance" models have IRT functions that increase monotonically with respect to the trait being assessed. Personality items, on the other hand, may be better represented by unfolding models characterized by single-peaked and nonmonotonic IRT functions centered at a subject's presumed ideal point; here, the underlying continuum might be considered bipolar such as for the "Big Five" personality factors (for example, conscientious/nonconscientious; see Costa & McCrae, 1992). Whether such differently modeled items in a test would be useful in, say, employment selection, would seem ethically problematic. No longer would tests be used to choose the most able, but instead, to choose those who conform most closely to the needs of the workplace.

8.4 Psychometrics, Eugenics, and Immigration Restriction

Psychometrics and related statistical analyses and interpretation have played major parts in several ethically questionable episodes in United States history. This section discusses three such situations: (a) the eugenics movement from the first half of the 20th century and the enforced sterilization of those deemed "unfit" or "feebleminded," and who therefore had to be prevented from passing along such a trait and related ones to their children (for example, moral depravity, criminality, shiftlessness, insanity, poverty); (b) the use of psychological test data from Army draftees and recruits during World War I to justify racially restrictive United States immigration policies, particularly the Immigration Act of 1924. Here, the main purveyor of flawed statistical analyses was Carl Brigham's, *A Study of American Intelligence* (1923). This same Carl Brigham later developed the Scholastic Aptitude Test

(SAT) that helped launch the present-day Educational Testing Services; (c) the use of test data and associated analyses to justify the continuance of laws against interracial marriage (miscegenation); the prime example is the Racial Integrity Act of Virginia (1924).

8.4.1 Eugenics

The word *eugenics* was coined by Sir Francis Galton in 1883 using the Greek word "eu" (good or well) with the suffix "genēs" (born). Galton (1908) characterized eugenics as the "study of agencies under social control that may improve or impair the racial qualities of future generations" (p. 321). The search was for ways to improve the human gene pool, either through positive eugenics (by having the "best" marry and reproduce with the "best"), or negative eugenics (through, for example, segregation and colonization, sterilization, or euthanasia). The popular eugenics movement in the early decades of the 20th century, both in the United States and England, was heavily influenced by statisticians and biometricians ("Eugenics," 2011). In the United States, Charles Davenport was the most prominent as the director of Cold Spring Harbor Laboratory (1910) and founder of the Eugenics Record Office. In England, besides Galton there were the inaugural editors of *Biometrika*,[2] W.R.R. Weldon and Karl Pearson (a third inaugural editor was the U.S.-based Davenport).[3] A redaction follows from a Wikipedia entry on Karl Pearson and the subsection on "politics and eugenics" ("Karl Pearson," 2011):

> An aggressive eugenicist who applied his social Darwinism to entire nations, Pearson openly advocated "war" against "inferior races," and saw this as a logical implication of his scientific work on human measurement: "My view—and I think it may be called the scientific view of a nation," he wrote—"is that of an organized whole, kept up to a high pitch of internal efficiency by insuring that its numbers are substantially recruited from the better stocks, and kept up to a high pitch of external efficiency by contest, chiefly by way of war with inferior races." He reasoned that, if August Weismann's theory of germ plasm is correct, then the nation is wasting money when it tries to improve people who come from poor stock. Weismann claimed that acquired characteristics could not be inherited. Therefore, training benefits only the trained generation. Their children will not exhibit the learned improvements and, in turn, will need to be improved.

"No degenerate and feeble stock will ever be converted into healthy and sound stock by the accumulated effects of education, good laws, and sanitary surroundings. Such means may render the individual members of a stock passable if not strong members of society, but the same process will have to be gone through again and again with their offspring, and this in ever-widening circles, if the stock, owing to the conditions in which society has placed it, is able to increase its numbers." (Introduction, *The Grammar of Science*).

"History shows me one way, and one way only, in which a high state of civilization has been produced, namely, the struggle of race with race, and the survival of the physically and mentally fitter race. If you want to know whether the lower races of man can evolve a higher type, I fear the only course is to leave them to fight it out among themselves, and even then the struggle for existence between individual and individual, between tribe and tribe, may not be supported by that physical selection due to a particular climate on which probably so much of the Aryan's success depended ... " (Karl Pearson, *National Life from the Standpoint of Science* [London, 1905])

Other instances could be given for the role that some of our more famous psychometricians have played in justifying eugenic sterilization. We give one particularly late (and nasty) quotation from E. L. Thorndike's (1940), *Human Nature and the Social Order*:

By selective breeding supported by a suitable environment we can have a world in which all men will equal the top ten per cent of present men. One sure service of the able and good is to beget and rear offspring. One sure service (about the only one) which the inferior and vicious can perform is to prevent their genes from survival. (p. 957)

Henry Goddard and the Kallikak Family

Henry Goddard was a well-known (clinical) psychologist from the first half of the 20th century ("Henry H. Goddard," 2011). He was the first to distribute widely an English translation of the Binet intelligence test first developed in France and introduced the term "moron" to label people with IQs from 51 to 70. The range of 0 to 25 was reserved for "idiots," and 26 to 50 for "imbeciles." The more ambiguous and widely used term of "feebleminded" referred generally to those mental deficiencies that now might be considered various forms and grades of mental retardation or labeled as learning disabilities. Feeblemindedness might be assessed by a poor performance on a Binet test, or more commonly, by observation from a trained (always female) field worker, and often just by remembrances from others about individuals long dead. As noted in the introduction to Goddard's famous study, *The Kallikak Family: A Study in the Heredity of Feeble-mindedness* (1912), the field worker responsible for the assessment of feeblemindedness in the Kallikaks was Elizabeth Kite.

Goddard was best known for postulating that feeblemindedness was a hereditary trait, most likely caused by a single recessive gene, and thereby subject to the laws of Mendelian inheritance that had been rediscovered at the turn of the century. A main argument for the hereditary nature of feeblemindedness was the extensive case study of the Kallikak family, and the genealogy of the family's founder, Martin Kallikak. Martin, a Revolutionary War hero, was on his way home from battle, but stopped to dally once with a "feebleminded" barmaid. Martin went on to a morally upright life, marrying a Quaker woman and siring a large and prosperous New England family. The child who was the product of the single dalliance went on to establish another branch of the Kallikak family; this branch consisted mainly of those assessed as feebleminded (apparently, all by Elizabeth Kite). Goddard argued that the Kallikak study documented a natural (observational) experiment in the heritability of the lack of intelligence and its associated traits of morality and criminality.

The study of the Kallikak family was a popular and widely read cautionary tale about the perils of unfettered reproduction, and the importance of establishing a national eugenics policy so that feeblemindedness could not be passed on to future generations. Goddard himself argued for segregation into colonies; other eugenicists used the Kallikak study and other similar statistical arguments in the passage of laws involving forced sterilization and restricted immigration. Once again, we are reminded of the speciousness of naming something without really understanding it, and then being seduced by a proposed but unproven mechanism, this time that feeblemindedness was carried by a single recessive gene operating just like the smoothness or wrinkliness of the peas studied by the Austrian monk, Mendel. It might be prudent to keep in mind a quotation from John Stuart Mill (1869, vol. II, ch. xiv, p. 5, fn. 2):

> The tendency has always been strong to believe that whatever received a name must be an entity or being, having an independent existence of its own. And if no real entity answering to the name could be found, men did not for that reason suppose that none existed, but imagined that it was something peculiarly abstruse and mysterious.

Harry Laughlin and Buck v. Bell

The Eugenics Record Office founded by Charles Davenport was directed by Harry Laughlin from its inception in 1910 through to its closing in 1939. Laughlin figures prominently in the eugenics movement, particularly as an advocate of forced sterilization to eliminate the possibility of reproduction for "unfit" members of society. Laughlin constructed a "model law" for compulsory sterilization that he believed would surpass all constitutional challenges. Based on Laughlin's model, Virginia enacted such a sterilization law in 1924 that provided for the compulsory sterilization of persons deemed to be "feebleminded" including the "insane, idiotic, imbecilic, or epileptic" ("Harry H. Laughlin," 2011).

The first person ordered sterilized under Virginia law was Carrie Buck on the grounds that she was the "probable potential parent of socially inadequate offspring." This carefully chosen test case would go all the way to the Supreme Court as *Buck v. Bell* (1927), and result in one of the most notorious decisions ever handed down by the Court. The opinion in this 8 to 1 decision was written by the well-known jurist Oliver Wendell Holmes, Jr.

We give two items in appendices to this chapter: part of the deposition given by Laughlin about Carrie Buck's suitability for sterilization (given in a book by Harry Laughlin entitled *The Legal Status of Eugenical Sterilization* [1930]); and a redacted Supreme Court opinion in *Buck v. Bell* written by Holmes that contains the famous phrase, "three generations of imbeciles are enough."

8.4.2 Immigration Act of 1924

The general conclusion emphasized by nearly every investigator is that as regards "intelligence," the Germanic stock has on the average a marked advantage over the South European. And this result would seem to have had vitally important practical consequences in shaping the recent very stringent American laws as to admission of immigrants.

— Charles Spearman (*The Abilities of Man*, 1927)

Harry Laughlin, the head of the Eugenics Record Office met briefly in the discussion of *Buck v. Bell*, had another major success in 1924—the Johnson-Reed Immigration Act. Laughlin provided extensive statistical testimony to the United States Congress based primarily on the statistical analyses of Carl Brigham that we discuss below. Brigham's empirical interpretations were phrased within the racial ideology espoused by Madison Grant in *The Passing of the Great Race* (1916). Laughlin went on to be appointed as an "expert eugenics agent" to the Congressional Committee on Immigration and Naturalization.

The Immigration Act of 1924 set an initial yearly quota on immigration of 165,000, less than 20% of the pre-World War I average. Furthermore, it based ceilings on the number of immigrants from any particular area by the percentage of each such nationality recorded in the 1890 Census. Because most immigration from Southern and Eastern Europe occurred after 1890, it reduced to a trickle those people coming from two of the Caucasoid races (Alpine and Mediterranean according to the categories used by Madison Grant) and encouraged a Nordic influx. This type of quota system on immigration remained in effect until the 1960s.[4]

Carl Brigham

Carl Brigham was a psychologist at Princeton University in the 1920s and 1930s, but earlier had collaborated with Robert Yerkes on the development of the Army Alpha and (the nonverbal) Beta intelligence tests given to well over a million United States Army recruits during World War I ("Carl Brigham," 2010). Based on these data collected on recruits, Brigham published *A Study of American Intelligence* (1923), which quickly become an influential source for justifying the passage of the Immigration Act of 1924 and in popularizing and justifying the eugenics movement in the United States. The data and results were placed within the context of Nordic theory as espoused by Madison Grant and his hugely successful *The Passing of the Great Race* (1916). Nordic theory contends that the Caucasoid (European) race should be subdivided further into Nordic (Northern Europe), Alpine (Central Europe), and Mediterranean (Southern Europe).

Based on the Army tests, Brigham concluded that the Nordic race was intellectually superior and argued that immigration should be tightly controlled to prevent Alpine and Mediterranean immigration, and to thereby protect "American Intelligence." He was particularly concerned with miscegenation between blacks and whites, and viewed "Negroes" as the most inferior intellectually. The quality of the data with which Brigham had to work was extremely poor and subject to a variety of confoundings based on language and the reliance on current American cultural knowledge.[5] Also, the cavalier development of the combined indices based on combinations of Alpha, Beta, and Stanford–Binet scores for various subgroups seems particularly fraught with equating difficulties, so much so that many modern-day psychometricians would consider the use of such combined scores to be unethical. We give an extensive redaction in an appendix to this chapter of Brigham's introduction and conclusion chapters that give a sense of the psychometric sophistry of this study, and of the racially motivated interpretations in the flawed statistical analyses.

Carl Brigham in a 1930 *Psychological Review* article, "Intelligence Tests of Immigrant Groups," repudiated his whole study of American intelligence. Unfortunately, this came rather late; the damage had already been done in its justification for the restrictive Immigration Act of 1924 and other eugenic initiatives. A set of extractions from Brigham's article follows; these amount to a complete retraction of his earlier interpretations and conclusions:

> The question immediately arises as to whether or not the test score itself represents a single unitary thing. This question is crucial, for if a test score is not a unitary thing it should not be given one name. (p. 158)
>
> The present discussion in this field is no longer concerned so much with the validity of Spearman's theory as it is with the proper statement of the number of factors into which a test score may be resolved. (p. 159)
>
> Most psychologists working in the test field have been guilty of a

naming fallacy which easily enables them to slide mysteriously from the score in the test to the hypothetical faculty suggested by the name given to the test. Thus, they speak of sensory discrimination, perception, memory, intelligence, and the like while the reference is to a certain objective test situation. (pp. 159–160)

A far-reaching result of the recent investigations has been the discovery that test scores may not represent unitary things. It seems apparent that psychologists in adding scores in the sub-tests in some test batteries have been doing something akin to adding apples and oranges. A case in point is the army alpha test. (p. 160)

This fact together with the facts previously noted which suggest disparate group factors within the test show that the eight tests of army alpha should not have been added to obtain a total score, or, if added, similar total scores should not have been taken to represent similar performances in the test. If the army alpha test has thus been shown to be internally inconsistent to such a degree, then it is absurd to go beyond this point and combine alpha, beta, the Stanford–Binet and the individual performance tests in the so-called "combined scale," or to regard a combined scale score derived from one test or complex of tests as equivalent to that derived from another test or another complex of tests. As this method was used by the writer in his earlier analysis of the army tests as applied to samples of foreign born in the draft, that study with its entire hypothetical superstructure of racial differences collapses completely. (pp. 163–164)

These findings of Kelley as to the probable existence of independent traits rather effectively challenge current test procedure, as most tests combine verbal and mathematical materials indiscriminately. As most of the nonverbal tests deal with spatial relations, one should not attempt to find similarities between scores in verbal and nonverbal tests. The writer does not hold that studies of the independence of traits have advanced far enough to have final theoretical significance. It is, however, apparent that scores in tests found to be independent should not be added. (p. 165)

For purposes of comparing individuals or groups, it is apparent that tests in the vernacular must be used only with individuals having equal opportunities to acquire the vernacular of the test. This requirement precludes the use of such tests in making comparative studies of individuals brought up in homes in which the vernacular of the test is not used, or in which two vernaculars are used. The last condition is frequently violated here in studies of children born in this country whose parents speak another tongue. It is important, as the effects of bilingualism are not entirely known. (p. 165)

This review has summarized some of the more recent test findings which show that comparative studies of various national and racial groups may not be made with existing tests, and which show, in particular, that one of the most pretentious of these comparative racial studies—the writer's own—was without foundation. (p. 165)

An 1982 ETS research memorandum by Gary Saretzky, "Carl Campbell Brigham, the Native Intelligence Hypothesis, and the Scholastic Aptitude Test," gives a relevant quotation from one of Brigham's unpublished manuscripts (circa 1934) that reviewed the history of psychological measurement:

> The test movement came to this country some twenty-five or thirty years ago accompanied by one of the most glorious fallacies in the history of science, namely that the test measured *native intelligence* purely and simply without regard to training or schooling. I hope nobody believes that now. The test scores very definitely are a composite including schooling, family background, familiarity with English, and everything else, relevant and irrelevant. The *native intelligence* hypothesis is dead. (p. 11)

A second relevant quotation appeared in the *New York Times* (McDonald, December 4, 1938):

> The original and fallacious concept of the I.Q. ... was that it reported some mysterious attribute ... but now it is generally conceded that all tests are susceptible to training. ... it is ridiculous to claim that any test score is related to the germ plasm, and that alone.

8.4.3 Racial Purity Laws

Miscegenation laws enforce racial segregation at the level of marriage and intimate relations by criminal sanction, possibly including sex between members of two different races. In the United States, for example, such laws have been around since the late 17th century. The most famous was enacted more recently as the Racial Integrity Act of Virginia in 1924 (a particularly good year, it seems, for the passage of racial and eugenic laws). In *Loving v. Virginia* (1967), the Virginia law was declared unconstitutional by a 9 to 0 vote of the Supreme Court, thus ending all such race-based legal restriction in the United States.

We give excerpts from two items in an appendix to this chapter. The first is from the Virginia Racial Integrity Act itself that enacts the "one-drop rule," where any amount of African ancestry leads to a classification as "black." Note, however, the "Pocahontas exception" allowing up to one-sixteenth of American Indian blood with a permissable label of "white" (presumably, because of the genealogy of some prominent Virginians in 1924). The second set of excerpts provides part of the unanimous opinion in *Loving v. Virginia* written by then Chief Justice Earl Warren.

The type of miscegenation argument put forth depends on the particular "genetic theory" assumed. For example, we have Carl Brigham's assertion in 1923 that admixtures must result in a mean between two genetically determined traits:

> We may consider that the population of the United States is made up of four racial elements, the Nordic, Alpine, and Mediterranean races of Europe, and the Negro. If these four types blend in the future into one general American type, then it is a foregone conclusion that this future blended American will be less intelligent than the present native born American, for the general results of the admixture of higher and lower orders of intelligence must inevitably be a mean between the two. (p. 205)

Someone from Illinois, however (such as one of the authors (LH)), might listen instead to the corn breeders and the theory of heterosis or hybrid vigor, defined by the increased function of any biological quality in a hybrid offspring, and the occurrence of a genetically superior offspring from mixing the genes of its parents. Hybrid vigor was known before the rediscovery of Mendel's Laws at the turn-of-the-century, but obviously it has never been a preferred theory to use when races are mixed. We give a short excerpt from the Wikipedia article on "Heterosis," 2011 (notice the mention of Charles Darwin, and of Davenport and Holden, two faculty members from the University of Illinois in the late 1800s and early 1900s):

> Nearly all field corn (maize) grown in most developed nations exhibits heterosis. Modern corn hybrids substantially outyield conventional cultivars and respond better to fertilizer.
>
> Corn heterosis was famously demonstrated in the early 20th century by George H. Shull and Edward M. East after hybrid corn was invented by Dr. William James Beal of Michigan State University based on work begun in 1879 at the urging of Charles Darwin. Dr. Beal's work led to the first published account of a field experiment

demonstrating hybrid vigor in corn, by Eugene Davenport and Perry Holden, 1881. These various pioneers of botany and related fields showed that crosses of inbred lines made from a Southern dent and a Northern flint, respectively, showed substantial heterosis and out-yielded conventional cultivars of that era.

Appendix: Analysis of the Hereditary Nature of Carrie Buck

(From a Deposition of H. H. Laughlin in Circuit Court Proceedings)

1. Facts: Granting the truth of the following facts which were supplied by Superintendent A. S. Priddy of the State Colony for Epileptics and Feeble-Minded, Lynchburg, Va.:

(a) Propositus: "Carrie Buck: Mental defectiveness evidence by failure of mental development having a chronological age of 18 years, with a mental age of 9 years, according to Stanford Revision of Binet–Simon Test; and of social and economic inadequacy; has record during life of immorality, prostitution, and untruthfulness; has never been self-sustaining; has had one illegitimate child now about six months old and supposed to be mental defective. Carrie Buck has been duly and legally declared to be feeble-minded within the meaning of the laws of Virginia and was committed to the State Colony for Epileptics and Feeble-Minded, where she now is, on June 4, 1924. Date of birth July 2, 1906; place of birth, Charlottesville, Va.; present address, Colony, Va."

(b) Mother of Propositus: "Emma Buck, maiden name, Emma Harlow: Mental defectiveness evidence by failure of mental development, having a chronological age of 52 years, with a mental age, according to Stanford Revision of Binet-Simon Test, of seven years and eleven months; and of social and economic inadequacy. Has record during life of immorality, prostitution and untruthfulness; has never been self-sustaining, was maritally unworthy; having been divorced from her husband on account of infidelity; has had record of prostitution and syphilis; has had one illegitimate child and probably two other inclusive of Carrie Buck, a feeble-minded patient in the State Colony (Va.). Date of birth, November 18, 1872; place of birth, Charlottesville, Va."

(c) Family History: "These people belong to the shiftless, ignorant, and worthless class of anti-social whites of the south. She (the propositus) has a sister and two half-brothers, whose paternal parentage cannot be determined ... She has life-long record of moral delinquency and has borne one illegitimate child, considered feeble-minded." (According to depositions of the Red Cross nurse, Miss Caroline E. Wilhelm, of Charlottesville, Va., in the proceeding committing Carrie Buck to the State Colony. Carrie Buck's illegitimate baby gave evidence of mental defectiveness at an early age.) " ... this girl comes from a shiftless, ignorant, and moving class of people, and it is impossible to get intelligent and satisfactory data, though I have had Miss Wilhelm of the Red Cross of Charlottesville try to work out their line. We have several Bucks and Harlows, but on investigation it is denied that they are any kin to the Harlows; the maternal grandfather of Carrie Buck, and there is considerable doubt as to her being a 'Buck,' but the line of baneful heredity seems conclusive and unbroken on the side of her mother (Harlow), but all the Bucks and Harlows we have here, descend from the Bucks and Harlows of Albemarle County, in which the City of Charlottesville and the University of Virginia are located, and I believe they are of the same stock. She (the propositus) has two or three half-brothers and sisters, but at an early age they were taken from the custody of their mother and legally adopted by people not related to them. All I can learn about Emma Buck's father, Richard Harlow, the grandfather of Carrie, was (that he) died from spinal trouble. Carrie Buck, when four years old, was adopted by Mrs. J. T. Dobbs of Charlottesville, who kept her until her moral delinquencies culminated in the illegitimate birth of a child referred to. She attended school five years and attained the 6th grade; she was fairly helpful in the domestic work of the household under strict supervision; so far as I understand, no physical defect or mental trouble attended her early years. She is well grown, has rather badly formed face; of a sensual emotional reaction, with a mental age of nine years; is incapable of self-support and restraint except under strict supervision."

II. Analysis of Facts: Generally feeble-mindedness is caused by the inheritance of degenerate qualities; but sometimes it may be caused by environmental factors which are not hereditary. In the case given, the evidence points strongly toward the feeble-mindedness and moral delinquency of Carrie Buck being due, primarily, to inheritance and not to environment.

Appendix: Buck v. Bell (1927)

Mr. Justice Holmes delivered the opinion of the Court.

. . .

Carrie Buck is a feeble minded white woman who was committed to the State Colony above mentioned in due form. She is the daughter of a feeble minded mother in the same institution, and the mother of an illegitimate feeble minded child. She was eighteen years old at the time of the trial of her case in the Circuit Court, in the latter part of 1924. An Act of Virginia, approved March 20, 1924 recites that the health of the patient and the welfare of society may be promoted in certain cases by the sterilization of mental defectives, under careful safeguard, etc.; that the sterilization may be effected in males by vasectomy and in females by salpingectomy, without serious pain or substantial danger to life; that the Commonwealth is supporting in various institutions many defective persons who if now discharged would become a menace but if incapable of procreating might be discharged with safety and become self-supporting with benefit to themselves and to society; and that experience has shown that heredity plays an important part in the transmission of insanity, imbecility, etc. The statute then enacts that whenever the superintendent of certain institutions including the above named State Colony shall be of opinion that it is for the best interests of the patients and of society than an inmate under his care should be sexually sterilized, he may have the operation performed upon any patient afflicted with hereditary forms of insanity, imbecility, etc., on complying with the very careful provisions by which the act protects the patients from possible abuse.

. . .

The attack is not upon the procedure but upon the substantive law. It seems to be contended that in no circumstances could such an order be justified. It certainly is contended that the order cannot be justified upon the existing grounds. The judgment finds the facts that have been recited and that Carrie Buck "is the probable potential parent of socially inadequate offspring, likewise afflicted, that she may be sexually sterilized without detriment to her general health and that her welfare and that of society will be promoted by her sterilization," and thereupon makes the order. In view of the general declarations of the legislature and the specific findings of the Court, obviously we cannot say as matter of law that the grounds do not exist, and if they exist they justify the result. We have seen more than once that the public welfare may call upon the best citizens for their lives. It would be strange if it could not call upon those who already sap the strength of the State for these lesser sacrifices, often not felt to be such by those concerned, in order to prevent our being swamped with incompetence. It is better for all the world, if instead of waiting to execute degenerate offspring for crime, or to let them starve for their imbecility, society can prevent those who are manifestly unfit from continuing their kind. The principle that sustains compulsory vaccination is broad enough to cover cutting the Fallopian tubes. ... Three generations of imbeciles are enough.

But, it is said, however it might be if this reasoning were applied generally, it fails when it is confined to the small number who are in the institutions named and is not applied to the multitudes outside. It is the usual last resort of constitutional arguments to point out shortcomings of this sort. But the answer is that the law does all that is needed when it does all that it can, indicates a policy, applies it to all within the lines, and seeks to bring within the lines all similarly situated so far and so fast as its means allow. Of course so far as the operations enable those who otherwise must be kept confined to be returned to the world, and thus open the asylum to others, the equality aimed at will be more nearly reached.

Judgment affirmed.

Appendix: Excerpts From Brigham's (1923) *A Study of American Intelligence*

Given in the Appendix Supplements

Appendix: Racial Integrity Act of Virginia (1924); Loving v. Virginia (1967)

Given in the Appendix Supplements.

Notes

[1]In discussing possible negative consequences of high-stakes testing in the classroom (such as that mandated by the 2001 "No Child Left Behind" Act), we should keep Campbell's (1975) law in mind:

> The more any quantitative social indicator is used for social decision making, the more subject it will be to corruption pressures and the more apt it will be to distort and corrupt the social processes it is intended to monitor. (p. 35)

Campbell's law is generally relevant to any area where numbers drive some societal process and where "scamming the numbers" can lead to personal gain or glory. One such recent instance occurred at the University of Illinois College of Law. It involved law class profiles and how they were constructed and reported to various agencies by the Admissions Dean, Paul Pless (who resigned in November 2011). We give a redaction of the Investigative Report issued under the direction of the Office of University Counsel and the University Ethics Office (November 7, 2011):

> As detailed in this Report, the University of Illinois College of Law (COL) disseminated inaccurate class profile information in six out of the ten years reviewed. The inaccurate information related to undergraduate grade-point averages (GPA), Law School Admission Test (LSAT) scores, and acceptance rates. Underlying this information were unjustified changes to the GPAs and LSAT scores of individual students and other data manipulation. A single COL

employee, who resigned during the course of this investigation, was responsible for this data manipulation. At a minimum, this former employee exhibited gross incompetence in the performance of his job duties. Moreover, the investigative record supports the conclusion that with respect to certain COL classes, the former employee knowingly and intentionally changed and manipulated data in order to inflate GPA and LSAT statistics and decrease acceptance rates. (p. 1)

Class of 2014:

The selectivity data publicly reported for the Class of 2014 marks the culmination of a pattern of increasingly inaccurate data reporting that began at least as far back as 2008 (for the Class of 2011). Through an assortment of strategies, COL made a concerted effort to bring in its most highly credentialed class ever in 2011 (i.e., the Class of 2014), and in August 2011, it seemed as though it had succeeded in doing so. That month, COL announced that it had enrolled "the most academically distinguished ... class in [COL's] history."

COL touted the Class of 2014 as having a median LSAT score of 168 (96th percentile) and a median GPA of 3.81 (on a 4.0 scale). This LSAT median elevated COL into a "rarefied level" that it had considered profoundly unrealistic just three years earlier. And the GPA median exceeded, ahead of schedule, the ambitious five-year median GPA goal of 3.7 that COL had set in 2006 as part of a strategic plan.

On August 22, 2011, Pless gave a presentation to COL faculty that covered the academic credentials of the incoming class, in isolation and in relation to historical trends. Pless represented to those in attendance that the Class of 2014 had a median LSAT score of 168 and a median GPA of 3.81. As the COL official with principal responsibility for recruiting applicants and sole decision-making authority over individual admissions decisions, Pless was congratulated for what he appeared to have accomplished with the Class of 2014.

The investigation has determined that, in this instance (as with earlier classes), appearances were deceiving. Based on data maintained by the Law School Admission Council (LSAC), the actual median LSAT score for the class was 163, not 168, and the median GPA for the class was 3.67, not 3.81.

On an annual basis, COL set specific goals for the median LSAT score and the median GPA that it hoped to achieve with the upcoming J.D. class. The Class of 2014's median LSAT score not only failed to meet that year's goal of 168, but it dropped four points from the previous year's median of 167. The Class of 2014's median GPA (3.7) was one-tenth of a point below that year's goal of 3.8, which was the median GPA that COL had reported (inaccurately) for the previous year's class.

The investigation has revealed that underlying the inflated median figures that COL announced for the Class of 2014 were changes to the majority of the LSAT scores and to almost a third of the GPAs of the members of the class. These changes were made on spreadsheets that Pless maintained and used to calculate the class's median LSAT score and GPA. These spreadsheets demonstrate that:

— Changes were made to the LSAT scores of 109 students and to the GPAs of 58 students [note that the total class size was 184];

— Twenty-five students had both their LSAT score and GPA changed, while 42 students had neither changed;

— Every change was upward;

— The largest LSAT score change was 12 points, which occurred five times (all from 156 to 168);

— One hundred LSAT scores were increased by at least two points, while 64 scores were increased by at least seven points;

— The largest GPA change was 1.03 points (from 2.59 to 3.62);

— GPAs of 4.0 were ascribed to eight international students, even though under American Bar Association (ABA) reporting rules none of these students should have been treated as having had any GPA for purposes of computing the class's median GPA; and

— Thirty-six GPAs were increased by at least one-tenth of a point.

The vast majority of the changes "crossed" the announced, incorrect medians—that is, they raised scores that were below the announced medians to scores at or above the announced medians. Other changes were, in magnitude and number, sufficient to increase the 25th and 75th percentile figures to levels well beyond those supported by LSAC data. All of the GPA changes were to five specific end values: (i) eight changes to 3.61 (the incorrect 25th percentile); (ii) 12 changes to 3.62; (iii) 13 changes to 3.81 (the announced, incorrect median); (iv) four changes to 3.87 (the incorrect 75th percentile); and (v) 21 changes to 4.0 (for international students and students admitted through the Illinois Law Early Action Program (iLEAP)). (pp. 5–7)

Over Pless's seven-year tenure as Admissions Dean, COL showed steady, and occasionally dramatic, improvement in the main numeric factors used by *US News and World Report* (USNWR) to gauge the academic credentials of a law school class—the class's median LSAT score and median GPA. This improvement helped COL attain and hold a place among the ... top 25 United States law schools, according to USNWR. Pless's perennial success in recruiting highly credentialed classes into COL helped earn him praise from COL leadership and a series of pay raises that took his annual salary from $72,000 in 2004 (his first year as Admissions Dean) to $130,051 in 2011. (pp. 8–9)

The facts and circumstances uncovered by this investigation, as well as the extensive forensic analysis conducted as part of the investigation, however, support the conclusion that the changes to the LSAT scores and GPAs of individual students were not inadvertent, but made knowingly and intentionally by Pless, and Pless alone, to give the appearance that the classes at issue had higher marks than they had in reality. Similarly, the forensic analysis supports the inference that, with respect to the classes of 2012 through 2014, Pless also knowingly and intentionally overstated the total number of applications to COL, while simultaneously understating the number of offers of admission extended, so as to give the appearance that the acceptance rates in those years were lower (and that COL was therefore more selective) than was in fact the case. (p. 9)

The investigation has revealed, however, that COL's reporting and dissemination of inaccurate selectivity data persisted and went undiscovered for years, in part, because COL lacked effective internal controls and oversight to prevent, deter, and detect the miscalculation (inadvertent or intentional) of selectivity data. It is apparent that the COL administration, through the tenures of two Deans and two Interim Deans, never appreciated the compliance risks in this area and thus saw no need to establish such controls. This stemmed in large measure from long held and widely shared perceptions of Pless as an outstanding employee with a particular acumen for delivering highly credentialed classes, exceptional data management skills, and strong tendencies toward forthrightness and transparency. This led Dean after Dean to certify to the ABA that COL's selectivity statistics were accurate, when in fact they were not for six of the ten years reviewed.

As this investigation demonstrates, the heavy reliance that the COL administration placed on Pless was misguided. To be sure, the benefit of hindsight can easily obscure the practical difficulties of recognizing latent risk factors at the time those risks manifested in problematic conduct, and a compliance regime that reposes some trust and autonomy in employees is certainly not inherently flawed. Nevertheless, adequate controls and oversight would have been particularly useful and advisable here, given that (i) the data reporting and verification duties were consolidated in one person, Pless, who stood to benefit professionally and personally from positive outcomes; (ii) access to the LSAC database and the data relating to individual students contained therein was restricted to Admissions Office personnel, all of whom were in positions subordinate to Pless; and (iii) no persons or entities outside COL (e.g., a separate college or office within the University, the ABA, USNWR, LSAC) had a practice or requirement of performing, or did perform, audits or similar activities by which to verify the accuracy of the data. (p. 10)

Development and Implementation of Appropriate Controls, Including Effective Segregation of Duties:

We recommend that the COL Dean ensure that a comprehensive review of control procedures within COL is conducted to identify and implement best practices in the area of data compilation and reporting, in consultation with professionals and/or institutions having relevant experience and expertise (e.g., data verification processes, fraud detection). While we acknowledge the need, in an extraordinarily competitive and often time-sensitive environment, for the Admissions Office to be nimble and prompt in making admissions-related decisions (e.g., decisions to admit, scholarship awards) and otherwise responding to applicant inquiries and concerns, a revamped control regime should avoid excessive concentration of decision-making authority in any single individual, and responsibility for compiling, computing, and verifying data should be appropriately distributed across a sufficient number of individuals to ensure full compliance with applicable rules and guidelines. (p. 113)

[2]The subtitle for the journal was "a journal for the statistical study of biological problems," and reflected an obvious emphasis on biometry and the newly rediscovered Mendelian genetics.

[3]To give a sense of the seriousness of Davenport's eugenic views, we give part of a lecture he gave in 1916 at the Golden Jubilee of the Battle Creek Sanitarium; its title was "Eugenics as a Religion" (Davenport, 1916):

It is for men as a social species to develop a social order of the highest, most effective type—one in which each person born is physically fit, well endowed mentally for some kind of useful work, temperamentally calm and cheerful and with such inhibitions as will enable him to control his instinctive reactions so as to meet the mores of the community in which he lives.

Do you agree that this is the highest aim of the species? Have you the instinct of love of the race? If so then for you, eugenics may be vital and a religion that may determine your behavior.

Every religion, it appears, should have a creed. So I suggest a creed for the religion of eugenics:

I believe in striving to raise the human race, and more particularly our nation and community to the highest place of social organization, or cooperative work and of effective endeavor.

I believe that no merely palliative measures of treatment can ever take the place of good stock with *innate* excellence of physical and mental traits and moral control.

I believe that to secure to the next generation the smallest burden of defective development, or physical stigmata, or mental defect, of weak inhibitions, and the largest proportion of physical, mental

and moral fitness, it is necessary to make careful marriage selection—
not on the ground of the qualities of the individual, merely, but of
his or her family traits; and I believe that I can never realize the ideals I have for my children without this basis of appropriate *germinal*
factors.

I believe that I am the trustee of the germ plasm that I carry,
that I betray the trust if, (that germ plasm being good) I so act
as to jeopardize it, with its excellent possibilities, or, for motives of
personal convenience, to unduly limit offspring.

I believe that, having made our choice in marriage carefully, we,
the married pair, should seek to have 4 to 6 children ... that our
carefully selected germ plasms shall be reproduced in adequate degree and that this preferred stock shall not be swamped by that less
carefully selected.

I believe that, having children with the determiners of peculiarly
good traits it is the duty of parents to give peculiarly good training
and culture to such children to ensure the highest development and
effectiveness of such traits.

I believe in the maintenance of a high quality of hereditary traits
in the nation.

I believe in such a selection of immigrants as shall not tend to
adulterate our national germ plasm with socially unfit traits.

I believe in such sanitary measures as shall protect, as far as
possible, from accidental and unselective mortality, the offspring of
carefully selected matings.

I believe in repressing my instincts when to follow them would
injure the next generation.

I believe in doing it for the race.

[4]As apparent from the wide support for immigration restriction in the
period after World War I, the thought of a major influx in the Alpine and
Mediterrean races truly alarmed many Americans. This prejudice is manifest in one of the most famous trials of the 20th century, that of Sacco and
Vanzetti. Ferdinando Sacco and Bartolomeo Vanzetti were immigrant anar-
chists convicted of a 1920 murder of two men during an armed robbery in
South Braintree, Massachusetts. The trial and successive appeals were highly
politicized and controversial; Sacco and Vanzetti were finally executed on August 23, 1927. To give a sense of the continuing historical importance of these
trials, we give part of the Wikipedia entry on Sacco and Vanzetti that deals
with the proclamation of Massachusetts Governor Dukakis in 1977 ("Sacco
and Vanzetti," 2011):

In 1977, as the 50th anniversary of the executions approached, Massachusetts Governor Michael Dukakis asked the Office of the Governor's Legal Counsel to report on "whether there are substantial

grounds for believing—at least in the light of the legal standards of today—that Sacco and Vanzetti were unfairly convicted and executed" and to recommend appropriate action. The resulting "Report to the Governor in the Matter of Sacco and Vanzetti" detailed grounds for doubting that the trial was conducted fairly in the first instance and argued as well that such doubts were only reinforced by "later-discovered or later-disclosed evidence." The Report questioned prejudicial cross-examination that the trial judge allowed, the judge's hostility, the fragmentary nature of the evidence, and eyewitness testimony that came to light after the trial. It found the judge's charge to the jury troubling for the way it emphasized the defendants' behavior at the time of their arrest and highlighted certain physical evidence that was later called into question. The Report also dismissed the argument that the trial had been subject to judicial review, noting that "the system for reviewing murder cases at the time ... failed to provide the safeguards now present."

Based on recommendations of the Office of Legal Counsel, Dukakis declared August 23, 1977, the 50th anniversary of their execution, Nicola Sacco and Bartolomeo Vanzetti Memorial Day. His proclamation, issued in English and Italian, stated that Sacco and Vanzetti had been unfairly tried and convicted and that "any disgrace should be forever removed from their names." He did not pardon them, because that would imply they were guilty. Neither did he assert their innocence. A resolution to censure the Governor failed in the Massachusetts Senate by a vote of 23 to 12. Dukakis later expressed regret only for not reaching out to the families of the victims of the crime.

[5] As one particularly bald-faced equating of "general information" and "intelligence," consider the following paragraph from Brigham's study:

> After weighing all the evidence, it would seem that we are justified in ignoring most of the arm-chair criticisms of this test and in accepting the experimental evidence tending to show that the test was a fairly good one. The assumption underlying the use of a test of this type is that the more intelligent person has a broader range of general information than an unintelligent person. Our evidence shows that this assumption is, in the main, correct. (p. 31)

Part II

Data Presentation and Interpretation

Chapter 9

Background: Data Presentation and Interpretation

> We have the duty of formulating, of summarizing, and of communi-
> cating our conclusions, in intelligible form, in recognition of the right
> of other free minds to utilize them in making their own decisions.
> – R. Fisher ("Statistical Methods and Scientific Induction," 1955,
> *JRSS(B)*, p. 77))

The goal of statistics is to gain understanding from data. The
methods of presentation and analyses used should not only allow
us to "tell the story" in the clearest and fairest way possible, but
more primarily, to help uncover what the story is in the first place.
When results are presented, there is a need to be sensitive to the
common and perhaps not-so-common missteps that result from a
superficial understanding and application of the methods in statistics. It is insufficient just to "copy and paste" without providing
context for how good or bad the methods are that are being used,
and understanding what is behind the procedures producing the
numbers. We will present in this introductory section some of the
smaller pitfalls to be avoided; a number of larger areas of concern
will be treated in separate subsections.

Some of the more self-contained missteps:

(1) Even trivial differences between groups will be statistically
significant when sample sizes are large. Significance should never
be confused with importance; the current emphasis on the use of
confidence intervals and the reporting of effect sizes reflects this
point. Conversely, lack of statistical significance does not mean
that the effect is therefore zero.[1] Such a confusion was behind the
ongoing debacle for Vioxx, the now withdrawn pain killer believed
responsible for thousands of deaths by lethal heart attack (for
background, see the Wikipedia entries on "COX-2 inhibitor," 2011,
and "Rofecoxib," 2011). One of the original clinical trials (called

ADVANTAGE) for Vioxx against naproxen (for example, Aleve) showed eight heart attacks for Vioxx compared to one for naproxen (Lisse et al., 2003). Although not "significant," the difference was also apparently not zero. Many studies later have confirmed the wisdom of the decision to withdraw Vioxx from the market—but the eight to one incident numbers from the original clinical trial, although "nonsignificant," should have provided a much earlier warning.[2]

(2) As some current textbooks still report inappropriately, a significance test does not evaluate whether a null hypothesis is true. A p-value measures the "surprise value" of a particular observed result conditional on the null hypothesis being true.

(3) Degrees of freedom do not refer to the number of independent observations within a dataset. The term indicates how restricted the quantities are that are being averaged in computing various statistics, such as sums of squares between or within groups.

(4) Although the central limit theorem justifies assertions of robustness when dealing with means, the same is not true for variances. The common tests on variances are notoriously nonrobust and should not be used; robust alternatives are available in the form of sample reuse methods such as the jackknife and the bootstrap.

(5) Do not carry out a test for equality of variances before performing a two-independent samples t-test. A quotation from George Box (1953, p. 33) contrast the good robustness properties of the t-test with the nonrobustness of the usual tests for variances: "To make the preliminary test on variances is rather like putting to sea in a rowing boat to find out whether conditions are sufficiently calm for an ocean liner to leave port."[3]

(6) Measures of central tendency and dispersion, such as the mean and variance, are not *resistant* in that they are influenced greatly by extreme observations. The median and interquartile range, on the other hand, are resistant, and each observation counts the same in the calculation of the measure.[4]

(7) Do not ignore the repeated-measures nature of your data, and use methods appropriate for independent samples. For example, don't perform an independent samples t-test on "before" and "after" data in a time-series intervention study. Generally, the

standard error of a mean difference must include a correction for correlated observations, as is routinely done in a paired (matched samples) *t*-test.

(8) The level of measurement used for your observations limits the inferences that are meaningful. For example, interpreting the relative sizes of differences makes little sense on nominal or ordinal data. Also, performing arithmetic operations on data that are nominal, or at best ordinal, may produce inappropriate descriptive interpretations as well.

(9) Do not issue blanket statements as to the impossibility of carrying out reasonable testing, confidence interval construction, or cross-validation. It is almost always now possible to use resampling methods that do not rely on parametric models or restrictive assumptions, and which are computer-implemented for immediate application.

(10) Keep in mind the distinctions between fixed and random effects models and the differing test statistics they may necessitate. The output from some statistical package may use a default understanding of how the factors are to be interpreted. If your context is different, then appropriate calculations must be made, sometimes "by hand." To parody the Capital One Credit Card commercial: "What's in your denominator?"

(11) Do not report all of the eight or so decimal places given in typical computer output. Two decimal places are needed at most, and often, one is all that is really justified. As an example, consider how large a sample is required to support the reporting of a correlation to more than one decimal place (answer: given the approximate standard error of $\frac{1}{\sqrt{n}}$, a sample size greater than 400 would be needed to give a 95% confidence interval of $\pm.1$).[5]

(12) It is generally wise to avoid issuing statements that might appear to be right but with some deeper understanding are just misguided:

(a) "Given the size of a population, it is impossible to achieve accuracy with a sample"; this reappears regularly with the discussion of undercount and the census.

(b) "Always divide by $n-1$ when calculating a variance to give the 'best' estimator"; if you divide by n or $n+1$, the estimator has a smaller expected error of estimation, which to many is more important than being "unbiased." Also, we note that no one ever really worries that the usual correlation coefficient is a "biased" estimate of its population counterpart.

(c) "ANOVA is so robust that all of its assumptions can be violated at will"; although it is true that normality is not crucial if sample sizes are reasonable in size (and the central limit theorem is of assistance), and homogeneity of variances doesn't really matter as long as cell sizes are close, the independence of errors assumption is critical, and one can be led far astray when it doesn't hold; for example, in intact groups, spatial contexts, and repeated measures.

(d) Don't lament the dearth of one type of individual from the very upper scores on some test without first noting possible differences in variability. Even though mean scores may be the same for groups, those with even slightly larger variances will tend to have more representatives in both the upper and lower echelons.

(13) Unless a compelling reason exists, avoid using one-tailed tests. Even the mechanisms for carrying out traditional one-tailed hypothesis tests, the chi-square and F distributions, have two tails, and both ought to be considered. The logic of hypothesis testing is that if an event is sufficiently unlikely, we must reconsider the truth of the null hypothesis. Thus, for example, if an event falls in the lower tail of the chi-square distribution, it implies that the model fits too well. If investigators had used two-tailed tests, the data fabrications of Cyril Burt might have been uncovered much earlier (see Dorfman, 1978).

(14) A confidence level does not give the probability that repeated estimates would fall into that particular confidence interval. Instead, a confidence level is an indication of the proportion of time the intervals cover the true value of the parameter under consideration if we repeat the complete confidence interval construction process.

In concluding these introductory comments about the smaller missteps to be avoided, we note the observations of Edward Tufte on the ubiquity of PowerPoint (PP) for presenting quantitative data, and the degradation it produces in our ability to communicate (Tufte, 2006, p. 26; italics in the original):

The PP slide format has the worst signal/noise ratio of any known method of communication on paper or computer screen. Extending PowerPoint to embrace paper and internet screens pollutes those display methods.

Generally, PowerPoint is poor at presenting statistical evidence, and is not a good replacement for technical reports, numerical data presented in detailed handouts, and the like.[6] It is now part of our "pitch culture," where, for example, we are sold on what drugs to take, but are not provided with the type of detailed numerical evidence we should have for an informed decision about benefits and risks. In commenting on the incredible obscuration of important (numerical) data that surrounded the use of PowerPoint-type presentations in the crucial briefings on the first Shuttle accident of Challenger in 1986, Richard Feynman noted (Feynman, 1988):

Then we learned about 'bullets'—little black circles in front of phrases that were supposed to summarize things. There was one after another of these little goddamn bullets in our briefing books and on slides. (pp. 126–127)

9.1 Weight-of-the-Evidence Arguments in the Presentation and Interpretation of Data

Before discussing how weight-of-the-evidence (WOE) arguments might be framed, we start with several admonitions regarding what data should be presented in the first place. To begin, it is questionable professionally to engage in "salami science," where a single body of work is finely subdivided into "least publishable units," known by its acronym of LPUs. To avoid such salami science, an emphasis exists in the better publication outlets for the behavioral sciences on the combined reporting of multiple studies delineating a common area so that a compelling WOE argument might be constructed. Second, in medically related studies, the underreporting of research can be seen as scientific misconduct. As noted in the article by Iain Chalmers (1990, "Underreporting Research is Scientific Misconduct," *Journal of the American*

Medical Association, 263, 1405–1408), the results of many clinical trials never appear in print, and among many of those that do, there is insufficient detail to assess the validity of the study. The consequences of underreporting can be serious. It could compromise treatment decisions for patients; do injustice to those patients who participated in the trials; and waste scarce resources and funds available for medically relevant trials. As Chalmers remarks: "Studies should be accepted or rejected on the basis of whether they have been well conceptualized and competently executed, not on the basis of the direction or magnitude of any differences observed between comparison groups" (p. 1407).

A final admonition is noted in a 2011 article by Simmons, Nelson, and Simonsohn, "False-Positive Psychology: Undisclosed Flexibility in Data Collection and Analysis Allows Presenting Anything as Significant" (*Psychological Science, 22,* 1359–1366). Simmons et al. (2011) provide several suggestions for authors and reviewers to mitigate the problem of false-positive publication; we paraphrase these recommendations below:

For authors—

(1) before data collection begins, a rule must be decided on when data collection will be terminated; the rule must be stated in the article;

(2) because of the inherent variability in all behavioral data, at least twenty observations per cell should be obtained, absent a compelling cost-of-data-collection argument;

(3) all variables collected in a study must be listed;

(4) all experimental conditions must be identified, including the failed manipulations;

(5) whenever observations are excluded, the author(s) must provide the results obtained without exclusion, and justify the rule used to eliminate observations;

(6) whenever an analysis includes covariates, the results obtained without the covariates should be reported.

For reviewers—

(1) ensure that the author requirements are followed;

(2) be tolerant of imperfections in data and in the results;

(3) ask for evidence that the results do not depend on arbitrary analytic decisions;

(4) whenever data collection and analyses are questionable, reviewers should ask the author(s) to conduct an replication.

For those disciplines of primary relevance to this book—the social and behavioral sciences, and the health and medical fields—it is rare, if ever, to have a crucial experiment that would decisively resolve an issue at hand. Instead, we hear more about the "weight of the evidence" (WOE) being the eventual decider. In law, the WOE may merely refer to preponderance; that is, something that is more likely than not. In other areas, however, a WOE argument is usually advanced to make a (causal) claim, or to indicate that the risks far outweigh the benefits of a medical procedure, drug, supplement, pesticide, and so on.[7]

In the health and medical areas involving drugs or agents of some kind, there may be many separate sources of data available: randomized-controlled experiments, observational studies, meta-analyses, animal studies, biological mechanism plausibilities, structure-activity relationships (such as the comparison of chemical structures between suspected and known toxins), chromosomal damage in human cells *in vitro*, wildlife abnormalities, and so on. A WOE argument can depend on many types of data, but it usually reduces to an integrative but subjective determination by individuals in the form of a committee. Although an explicit use of the term "weight of the evidence" may not actually be present, surrogate statements probably are, such as "the benefits do not outweigh the risks involved." We give three recent examples for these kinds of decisions:

(1) In June 2010, a federal advisory panel for the FDA recommended against approving the drug, flibanserin, to treat "female sexual desire disorder." The panel said the drug's impact was "not robust enough to justify the risks," which can include nausea, dizziness, fatigue, anxiety, and insomnia. An earlier FDA staff report, relied on by the advisory panel, found that the drug had not been shown to increase women's desire (and, therefore, to combat a diagnosis of "hypoactive sexual desire disorder," which is included in the *Diagnostic and Statistical Manual of Mental Disorders*). For further discussion of this episode, see the *New York Times* article by Duff Wilson, "Drug for Sexual Desire Disorder Opposed by Panel" (June 18, 2010).[8]

(2) The EPA has begun yet another study of the herbicide at-

razine ("Atrazine," 2010); a WOE verdict was originally expected toward the end of 2010, but that date has come and gone without a ruling. One of the potentially damning pieces of data against atrazine is its effect on male amphibians, where male frogs become emasculated. As an apparent hormone disrupter in an animal species, it is unclear what the effects of atrazine are on humans. Given that millions of pounds are used regularly (on corn crops, in particular), which then leaches into the ground water and eventually into the drinking water supply, it is important for the EPA to get this right. To see how such a WOE process will unfold, the news release of October 7, 2009, from the EPA follows (entitled "EPA Begins New Scientific Evaluation of Atrazine"):

WASHINGTON – The United States Environmental Protection Agency is launching this year a comprehensive new evaluation of the pesticide atrazine to determine its effects on humans. At the end of this process, the agency will decide whether to revise its current risk assessment of the pesticide and whether new restrictions are necessary to better protect public health. One of the most widely used agricultural pesticides in the United States, atrazine can be applied before and after planting to control broadleaf and grassy weeds. EPA will evaluate the pesticide's potential cancer and noncancer effects on humans. Included in this new evaluation will be the most recent studies on atrazine and its potential association with birth defects, low birth weight, and premature births.

"One of Administrator Jackson's top priorities is to improve the way EPA manages and assesses the risk of chemicals, including pesticides, and as part of that effort, we are taking a hard look at the decision made by the previous administration on atrazine," said Steve Owens, assistant administrator for EPA's Office of Prevention, Pesticides and Toxic Substances. "Our examination of atrazine will be based on transparency and sound science, including independent scientific peer review, and will help determine whether a change in EPA's regulatory position on this pesticide is appropriate." During the new evaluation, EPA will consider the potential for atrazine cancer and noncancer effects, and will include data generated since 2003 from laboratory and population studies. To be certain that the best science possible is used in its atrazine human health risk assessment and ensure transparency, EPA will seek advice from the independent Scientific Advisory Panel (SAP) established under the Federal Insecticide, Fungicide and Rodenticide Act.

(3) The third example we give is of the FDA banning on

April 12, 2004, the herbal supplement Ephedra, which contains ephedrine alkaloids, a sympathomimetic drug; that is, one that increases blood pressure and heart rate by constricting blood vessels. We give the conclusions section for the Final Rule Summary from the FDA ("Final Rule Declaring Dietary Supplements Containing Ephedrine Alkaloids Adulterated Because They Present an Unreasonable Risk," 2004). It is a model of how a WOE decision can be presented convincingly (or alternatively, for a "not enough benefit for the risk incurred" conclusion):

> Multiple studies demonstrate that dietary supplements containing ephedrine alkaloids, like other sympathomimetics, raise blood pressure and increase heart rate. These products expose users to several risks, including the consequences of a sustained increase in blood pressure (e.g., serious illnesses or injuries that include stroke and heart attack that can result in death) and increased morbidity and mortality from worsened heart failure and proarrhythmic effects. Although the proarrhythmic effects of these products typically occur only in susceptible individuals, the long-term risks from elevated blood pressure can occur even in nonsusceptible, healthy individuals. These risks are neither outweighed by any known or reasonably likely benefits when dietary supplements containing ephedrine alkaloids are used under conditions suggested or recommended in their labeling, such as for weight loss, athletic performance, increased energy or alertness, or eased breathing. Nor do the benefits outweigh the risks under ordinary conditions of use, in the absence of suggested or recommended conditions of use in product labeling. As discussed above in section V.C of this document, the best scientific evidence of benefit is for modest short-term weight loss; however, such benefit would be insufficient to bring about an improvement in health that would outweigh the concomitant health risks. The other possible benefits discussed in section V.C of this document, have less scientific support. Even assuming that these possible benefits in fact occur, such temporary benefits are also insufficient to outweigh health risks that can lead to serious long-term or permanent consequences like heart attack, stroke, and death. On the other hand, we have determined that there are benefits from the use of OTC and prescription drug products containing ephedrine alkaloids in certain populations for certain disease indications that outweigh their risks. As with other sympathomimetics, the risks posed by dietary supplements containing ephedrine alkaloids for continuous, long-term use cannot be adequately mitigated without physician supervision. Temporary, episodic use can be justified only if a known or reasonably likely benefit outweighs the known and reasonably likely risks.

Similar to OTC single ingredient ephedrine products, dietary supplements containing ephedrine alkaloids could theoretically be marketed without physician supervision for a very temporary, episodic use if there were adequate evidence that the use resulted in a benefit sufficient to outweigh the risks of these products. However, we are currently unaware of any such use, and our experience with ephedrine and pseudoephedrine OTC drug products suggests that such benefits will be demonstrable only for disease uses. Therefore, we conclude that dietary supplements containing ephedrine alkaloids present an unreasonable risk of illness or injury under conditions of use recommended or suggested in labeling or under ordinary conditions of use, if the labeling does not suggest or recommend conditions of use. (p. 6827)

The National Research Council and its Committee on the Framework for Evaluating the Safety of Dietary Supplements, issued its report in 2005: *Dietary Supplements: A Framework for Evaluating Safety*. In it, the committee commented that because only limited data were usually available to assess supplement risk, the "appropriate scientific standard to be used to overturn this basic assumption of safety is to demonstrate significant or unreasonable risk, not *prove* [their italics] that an ingredient is unsafe " (p. 14). In contrast to the regulation of drugs, supplements have been "*assumed* [their italics] to be safe, but have not been required to be proven safe" (p. 14). We provide an information table from this report summarizing the principles for evaluating data to assess unreasonable risk. These principles have a much greater applicability than just to the regulation of supplements.

Guiding Principles for Evaluating Data to Determine Unreasonable Risk
General principles
– Absence of evidence of risk does not indicate that there is no risk.
– Proof of causality or proof of harm is not necessary to determine unreasonable or significant risk.
– Integration of data across different categories of information and types of study design can enhance biological plausibility and identify consistencies, leading to conclusions regarding levels of concern for an adverse event that may be associated with use of a dietary supplement.
Human data
– A credible report or study finding of a serious adverse event in

humans raises concern about the ingredient's safety and requires further information gathering and evaluation; final judgment, however, will require consideration of the totality of the evidence.

– Historical use should not be used as *prima facie* evidence that the ingredient does not cause harm.

– Considerable weight can be given to a lack of adverse events in large, high quality, randomized clinical trials or epidemiological studies that are adequately powered and designed to detect adverse effects.

Animal data

– Even in the absence of information on adverse events in humans, evidence of harm from animal studies is often indicative of potential harm to humans.

Related substances

– Scientific evidence for risk can be obtained by considering if the plant constituents are compounds with established toxicity, are closely related in structure to compounds with established toxicity, or the plant source of the botanical dietary supplement itself is a toxic plant or is taxonomically related to a known toxic plant.

– Supplement ingredients that are endogenous substances or that may be related to endogenous substances should be evaluated to determine if their activities are likely to lead to serious effects. Considerations should include the substances ability to raise the steady-state concentration of biologically active metabolites in tissues and whether the effect of such increases would be linked to a serious health effect.

In vitro data

– Validated *in vitro* studies can stand alone as independent indicators of risk to human health if a comparable exposure is attained in humans and the *in vitro* effects correlate with a specific adverse health effect in humans or animals.

[In this report, *in vitro* assays are considered validated when their results have been proven to predict a specific effect in animals and/or humans with reasonable certainty (not necessarily universally accepted or without detractors)]. (p. 12)

Historically, there have been many changed practices due to an eventual WOE argument, even though possibly no identifiable group of individuals ever made it explicit. We give several examples of this below:

(1) The Halsted radical mastectomy surgery is a procedure for breast cancer where the breast, underlying chest muscle, and lymph nodes of the axilla are removed ("Radical mastectomy," 2010). It was developed by William Halsted in 1882. About 90%

of women treated for breast cancer in the United States from 1895 to the mid-1970s had radical mastectomies. This morbid surgery is now rarely performed except in extreme cases.

(2) Phrenology is a system where the personality traits of a person can be inferred from the shape of the skull ("Phrenology," 2010).[9] It was developed by the German physician Franz Gall in 1796, and was popular throughout the 19th and early 20th century.[10]

(3) In the 1940s, William Herbert Sheldon pioneered the use of anthropometry to categorize people into somatotypes: endomorphic (soft and round), mesomorphic (stocky and muscular), and ectomorphic (thin and fragile). People could be graded on one-to-seven point scales as to the degree they exhibited each of the three somatotypes. For example, a "pure" mesomorph would be a 1-7-1, a "pure" ectomorph, a 1-1-7, and so on. Sheldon divided personality characteristics into three categories, where each body type had a corresponding personality profile: endotonia (physical comfort, food, and socializing); mesotonia (physical action and ambition); ectotonia (privacy and restraint). Sheldon saw a strong correlation between mesotonia and mesomorphs, and concluded that these mesotonic individuals would descend into criminality ("Somatotypes of Sheldon," 2011).[11]

Apart from the idea of connecting somatotypes and personality, Sheldon is best known for his involvement in the taking of nude posture photographs, both male and female, throughout the Ivy League from the 1940s to the 1960s. For an entertaining exposé of this whole episode, the reader is referred to an article from the *New York Times* by Ron Rosenbaum (January 15, 1995) entitled "The Great Ivy League Nude Posture Photo Scandal." The last line is telling: "In the Sheldon rituals, the student test subjects were naked—but it was the emperors of scientific certainty who had no clothes."

(4) Graphology is the study and analysis of handwriting in relation to human psychology and personality assessment ("Graphology," 2010). Given the paucity of empirical studies that show any validity, it is generally now considered a pseudoscience.

(5) Polygraph lie detection has much the same status as graphology. A perusal of the National Research Council report, *The Polygraph and Lie Detection* (2003), should be enough to

place polygraph examination into the same category as phrenology.

(6) The practice of bloodletting was the most common medical procedure performed by doctors from antiquity until the early 20th century ("Bloodletting," 2010). Barber poles, with the intermixture of white and red, indicate where bloodletting could be performed. Leeches also became popular in the early 19th century, with hundreds of millions being used by physicians throughout the century.

(7) Many health-related and other beliefs and practices have been debunked by WOE arguments. For example, it was once believed that listening to classical music could make you smarter ("Mozart effect," 2010).[12] A short section from the Wikipedia article on the political impact of the Mozart effect, follows:

> The popular impact of the theory was demonstrated on January 13, 1998, when Zell Miller, governor of Georgia, announced that his proposed state budget would include $105,000 a year to provide every child born in Georgia with a tape or CD of classical music. Miller stated "No one questions that listening to music at a very early age affects the spatial-temporal reasoning that underlies math and engineering and even chess." Miller played legislators some of Beethoven's "Ode to Joy" on a tape recorder and asked "Now, don't you feel smarter already?" Miller asked Yoel Levi, music director of the Atlanta Symphony, to compile a collection of classical pieces that should be included. State representative Homer M. DeLoach said "I asked about the possibility of including some Charlie Daniels or something like that, but they said they thought the classical music has a greater positive impact. Having never studied those impacts too much, I guess I'll just have to take their word for that."

Arguments based on appeals to a weight-of-the-evidence approach are similar to other activities that academics and professionals commonly engage in. One such familiar activity is the discursive literature review (subject to all the usual forms of confirmation bias, of course). Here, many studies are qualitatively integrated to come up with some type of WOE conclusion, and thereby plan for further studies. The thousands of literature reviews done for doctoral dissertations based on empirical studies in the social and behavioral sciences commonly reflect this genre. The process of construct validation in the social and behavioral sciences and psychometrics is another such activity. This refers to whether a scale (such as the Hare Psychopathy Checklist [2003]) measures the psychological construct it supposedly measures. It usually involves the correlation of the measure in question with other variables that are known to be related to the construct being assessed. We can easily disappear into the nomological net of Cronbach and Meehl (1955); that is, does the construct behave as it should within a system of related constructs (the nomological set or net). Finally, a WOE process that uses all available evidence relevant to the causal hypothesis under study is completely consistent with the type of argumentation guided by the Bradford–Hill criteria discussed in Chapter 11 on inferring causality.[13]

9.1.1   A Case Study in Data Interpretation: Brown v. Board of Education (1954)

Brown v. Board of Education (1954) is a landmark decision of the United States Supreme Court, declaring that state laws establishing separate public schools for black and white students were unconstitutional and violated the Equal Protection Clause of the Fourteenth Amendment. The Warren Court's unanimous (9–0) decision stated categorically that "separate educational facilities are inherently unequal," thus overturning the 1896 (7–1) decision in *Plessy v. Ferguson* (1896). This turn-of-the-century ruling held that the "separate but equal" provision of private services mandated by state government is constitutional under the Equal Protection Clause.

To give a flavor of the *Plessy v. Ferguson* decision, a short extract is given from the conclusion section of the majority opinion:

> We consider the underlying fallacy of the plaintiff's argument to consist in the assumption that the enforced separation of the two races stamps the colored race with a badge of inferiority. If this be so, it is not by reason of anything found in the act, but solely because the colored race chooses to put that construction upon it. ... The argument also assumes that social prejudices may be overcome by legislation, and that equal rights cannot be secured to the negro except by an enforced commingling of the two races. We cannot accept this proposition. If the two races are to meet upon terms of social equality, it must be the result of natural affinities, a mutual appreciation of each

other's merits, and a voluntary consent of individuals. As was said by the Court of Appeals of New York in People v. Gallagher,

"this end can neither be accomplished nor promoted by laws which conflict with the general sentiment of the community upon whom they are designed to operate. When the government, there-fore, has secured to each of its citizens equal rights before the law and equal opportunities for improvement and progress, it has accom-plished the end for which it was organized, and performed all of the functions respecting social advantages with which it is endowed."

Legislation is powerless to eradicate racial instincts or to abolish distinctions based upon physical differences, and the attempt to do so can only result in accentuating the difficulties of the present sit-uation. If the civil and political rights of both races be equal, one cannot be inferior to the other civilly or politically. If one race be inferior to the other socially, the Constitution of the United States cannot put them upon the same plane.

It is true that the question of the proportion of colored blood necessary to constitute a colored person, as distinguished from a white person, is one upon which there is a difference of opinion in the different States, some holding that any visible admixture of black blood stamps the person as belonging to the colored race ... ; others that it depends upon the preponderance of blood ... ; and still others that the predominance of white blood must only be in the proportion of three-fourths. ... But these are questions to be determined under the laws of each State, and are not properly put in issue in this case. Under the allegations of his petition, it may undoubtedly become a question of importance whether, under the laws of Louisiana, the petitioner belongs to the white or colored race.

The crucial difference in deciding *Brown v. Board of Education* in 1954 versus *Plessy v. Ferguson* in 1896 was the wealth of dis-positive behavioral science research and data then available on the deleterious effects on children of segregation and the philosophy of "separate but equal." As stated by Chief Justice Earl Warren in the unanimous opinion: "To separate them from others of simi-lar age and qualifications solely because of their race generates a feeling of inferiority as to their status in the community that may affect their hearts and minds in a way unlikely to ever be undone." The experimental data collected by Kenneth and Mamie Clark, in particular, were central to the Court's reasoning and decision; the Clark "doll test" studies were especially convincing as to the effects that segregation had on black school children's mental status. We quote a short summary of the doll experiments carried out from

the late 1930s to the early 1940s given in the Wikipedia article on
the Clarks ("Kenneth and Mamie Clark," 2011):

> The Clarks' doll experiments grew out of Mamie Clark's master's
> degree thesis. They published three major papers between 1939 and
> 1940 on children's self perception related to race. Their studies found
> contrasts among children attending segregated schools in Washing-
> ton, DC versus those in integrated schools in New York. They found
> that black children often preferred to play with white dolls over black;
> that, asked to fill in a human figure with the color of their own skin,
> they frequently chose a lighter shade than was accurate; and that the
> children gave the color "white" attributes such as good and pretty,
> but "black" was qualified as bad and ugly. They viewed the results as
> evidence that the children had internalized racism caused by being
> discriminated against and stigmatized by segregation.

A redaction of the *Brown v. Board of Education* opinion is given
in an appendix to this chapter (we omit the now famous Footnote
Eleven citing the dispositive mid-twentieth-century behavioral science research).

9.1.2   A Case Study in Data Interpretation: Matrixx Initiatives, Inc. v. Siracusano (2011)

We conclude this chapter with a lengthy redaction from the recent Supreme Court case of *Matrixx Initiatives, Inc. v. Siracusano*
(2011). It is a remarkable example of cogent causal and statistical reasoning. Although assisted by many "friend of the court"
briefs, and probably more than a few bright law clerks, this opinion delivered by the "wise Latina woman," Sonia Sotomayor, is an
exemplary presentation of a convincing WOE argument. It serves
as a model for causal reasoning in the presence of a "total mix" of
evidence.

The Matrixx case involved the nasal spray and gel Zicam manufactured by Matrixx Initiatives, Inc. From 1999 to 2004, the company received reports that Zicam might have led users to suffer the
"adverse event" of losing a sense of smell—anosmia. The company
did not disclose these adverse event reports to potential investors
under the excuse they were not "statistically significant." There
was no attempt to explain how "statistically significance" could
even be determined given that no "control group" was available

for comparison. The suit against Matrixx Initiatives was for securities fraud and for making statements that omitted material information, defined as information that reasonable investors would consider as substantially altering the "total mix" of available information. Here, "total mix" refers to all the information typically relied on in a WOE argument.

In arguing their position, Matrixx commented that in the several randomized clinical trials carried out for Zicam, adverse effects of anosmia did not appear (or, at least, the number of adverse events were not statistically different from those identified in the control groups). What Matrixx did not acknowledge is that the small sample sizes of the clinical trials may well have failed to identify any rare events. But once on the market and used by more individuals, rare events might well happen, and in relatively substantial numbers given the size of the treatment base. These post-marketing or phase IV trials are supposed to monitor the continued safety of over-the-counter and prescription products available to the public. The FDA has an automated reporting system in place (the Adverse Events Reporting System (AERS)), that tracks adverse events.[14] There is no need to invoke the idea of "statistical significance." Indeed, it is simply not applicable or even computable in this case. Once enough events get reported and a credible causal argument made, the product might be recalled or, at least, additional warning labels included—think of Avendia, Vioxx, or any of a number of prescription drugs now absent from the market.

Appendix: Brown v. Board of Education (1954)

Given in the Appendix Supplements.

Appendix: Matrixx Initiatives, Inc. v. Siracusano (2011)

It might be helpful to define a few of the legal terms used in the ruling:

To act with *scienter* implies "a mental state embracing the intent to deceive, manipulate, or defraud" ("Scienter law & legal definition," 2011);

A *reasonable person standard* refers to a hypothetical individual exercising average care, skill, and judgment in conduct; a reasonable person acts as a comparative standard for determining liability ("Reasonable/prudent man law & legal definition," 2011);

Something is *dispositive* when it is decisive or conclusive, and, for example, settles a dispute or question ("Dispositive," 2010);

When arguing the case of either party, one is said to *plead* the case ("Plead," 2010);

Something has *materiality* (or is *material*) if it is relevant and consequential to an issue under discussion. It does not have to be "statistically significant" to be material. In an earlier court ruling, the Supreme Court stated: "an omitted fact is material if there is a substantial likelihood that a reasonable shareholder would consider it important in deciding how to vote" ("Material," 2010);

A *bright-line* rule is an absolute criterion ("Bright-line rule," 2011); here, statistical significance is *not* such a bright-line rule when deciding about evidence disclosure to investors buying Matrixx stock.

The Securities Exchange Act of 1934, Section 10: Manipulative and Deceptive Devices, states:

> It shall be unlawful for any person, directly or indirectly, by the use of any means or instrumentality of interstate commerce or of the mails, or of any facility of any national securities exchange –
>
> . . .
>
> [and in section 10(b)]
>
> b. To use or employ, in connection with the purchase or sale of any security registered on a national securities exchange or any security not so registered, or any securities-based swap agreement (as defined in section 206B of the Gramm-Leach-Bliley Act), any manipulative or deceptive device or contrivance in contravention of such

rules and regulations as the Commission may prescribe as necessary or appropriate in the public interest or for the protection of investors.

The Securities Exchange Commission Rule 10b-5, Employment of Manipulative and Deceptive Practices:

It shall be unlawful for any person, directly or indirectly, by the use of any means or instrumentality of interstate commerce, or of the mails or of any facility of any national securities exchange,

(a) To employ any device, scheme, or artifice to defraud,

(b) To make any untrue statement of a material fact or to omit to state a material fact necessary ... to make the statements made, in the light of the circumstances under which they were made, not misleading, or

(c) To engage in any act, practice, or course of business which operates or would operate as a fraud or deceit upon any person, in connection with the purchase or sale of any security.

Given in the Appendix Supplements.

Notes

[1]Several other confusions about statistical significance should be pointed out. One is discussed in the Andrew Gelman and Hal Stern article, "The Difference Between 'Significant' and 'Not Significant' is Not Itself Statistically Significant" (*American Statistician*, 60, 2006, 328–331). The point here is that in making a comparison between two treatments, the statistical significance of the observed difference should be explicitly looked at, and not just the difference between their separate significance levels (based on, say, two one-sample tests). A similar confusion is presented by a 2×2 analysis-of-variance design when one simple main effect is significant and the other is not. This situation is not equivalent to finding a significant interaction. Apparently, this error is very prevalent in neuroscience; see Nieuwenhuis, Forstmann, and Wagenmakers, "Erroneous Analyses of Interaction in Neuroscience: A Problem of Significance" (*Nature Neuroscience*, 2011, *14*, 1105–1107).

[2]Some of the history behind the Vioxx incident is listed in the Suggested Reading (A.6.3): "Despite Warnings, Drug Giant Took Long Path to Vioxx Recall" (Alex Berenson et al., *New York Times*, November 14, 2004). The FDA has estimated that in the five years the drug was on the market, Vioxx may have caused anywhere from 88 to 139 thousand heart attacks, with 30 to 40% of these fatal.

[3]There are several versions of this quotation. The one following can be

considered a personal communication to LH taken from class notes (October 1973) while LH was auditing George Box's course on time series analysis: "to test for equality of variances before carrying out an independent samples t-test is like putting a rowboat out on the ocean to see if it is calm enough for the Queen Mary."

[4]In reporting an "average," it is important to distinguish between the types that could be given. For example, an arithmetic mean may provide one form for what is considered "average," but it may not be what is also "typical." For that, the mode or median could be a better means of communication. This can be especially true when dealing with money having underlying skewed distributions, such as for average tax cuts, tax increases, salaries, and home sale prices.

On a related interpretive matter, when reporting a number, no matter how large it may seem, you need to know the base to which it refers in deciding whether it is really "big." Thus, 500 million dollars may seem enormous when aggregated over all people in the United States, but it represents less than 2 dollars per person.

[5]Several quotations given below all pertain to the issue of approximation and accuracy:

> It is the mark of an educated man to look for precision in each class of things just so far as the nature of the subject admits; it is evidently equally foolish to accept probable reasoning from a mathematician and to demand from a rhetorician scientific proofs.
> – Aristotle, *Nicomachean Ethics*
> A little inaccuracy sometimes saves tons of explanation.
> – H. H. Munro (as Saki)
> Truth is much too complicated to allow anything but approximations.
> – John von Neumann
> Although this may seem a paradox, all exact science is dominated by the idea of approximation.
> – Bertrand Russell
> Far better an approximate answer to the *right* question, which is often vague, than an *exact* answer to the wrong question, which can always be made precise.
> – John W. Tukey

[6]Two items on PowerPoint are in the Suggested Reading, Section C.3, on how this method of conveying information is detrimental to the decision making processes within the ongoing war effort in Afghanistan: C.3.1, "We Have Met the Enemy and He Is PowerPoint" (Elisabeth Bumiller, *New York Times*, April 26, 2010); C.3.2, "Essay: Dumb-Dumb Bullets" (T. X. Hammes, *Armed Forces Journal*, 2009, *47*, 12, 13, 28).

[7]In making decisions it is important to consider how big the potential

benefits are, what the risks and costs might be, and what the weight of the evidence really shows. In issues of toxicology, for example, we have the famous quotation from Paracelsus (1493–1541):

German: Alle Ding' sind Gift, und nichts ohn' Gift; allein die Dosis macht, dass ein Ding kein Gift ist. (All things are poison and nothing is without poison, only the dose permits something not to be poisonous; "Paracelsus," 2010.)

So, in making decisions to take medicines, ingest supplements, spray insecticides, and so forth, is there evidence that no harm will result, that dosage information is understood, and have all the possible consequences been assessed (remember the "law of unintended consequences")? There are numerous examples where wrong decisions have resulted from incompleteness of data as to possible side effects. An example, labeled as one of the greatest science blunders in the last twenty years, is in the gasoline additive MTBE (Newman, October 1, 2000). MTBE came into use when lead additives were being phased out in the late 1970s. Unfortunately, MTBE is highly water soluble, and tends to leak from storage tanks at gasoline stations. Although it was banned by the Clinton administration in 1999, 14 percent of urban drinking water wells are contaminated with MTBE, with no cost-effective method of removal—and apparently, with no real knowledge of possible ill-effects at the dosage levels present.

[8]In the references, see Division of Reproductive and Urologic Products, Office of New Drugs, Center for Drug Evaluation and Research, Food and Drug Administration. (2010, May 20). *Background document for meeting of Advisory Committee for Reproductive Health Drugs (June 18, 2010): NDA 22-236, Flibanserin, Boehringer Ingelheim.*

[9]Until the depression era of the 1930s, phrenology was a widely held "theory" ("Phrenology," 2010). The phrenological argument was straightforward: people have diverse mental capacities or faculties, localized in the brain. The strength of faculty expression was related to the size of that part of the brain, in turn affecting the contours on the surface of the skull. The field of what is called vocational psychology today seemed particularly vulnerable to phrenological reasoning. One example is the 1897 text by Nelson Sizer with the following long title: *Choice of Pursuits; or, what to do, and why, describing seventy-five trades and professions, and the talents and temperaments required for each; Also, how to educate, on phrenological principles, each man for his proper work. Together with portraits and biographies of more than one hundred successful thinkers and workers.*

To assist in the implementation of a phrenological "reading," a machine named the "psycograph" was developed by Henry Lavery in the 1930s. It consisted of 1,954 parts in a metal carrier with a motor-driven belt, all housed inside a walnut case containing statements about thirty-two mental faculties. Each faculty was rated on a 1 (deficient) to 5 (superior) scale. A person's "score" was determined by the way the thirty-two probes made contact with the head once the headpiece was lowered. The operator used a lever to activate the motor; low-voltage signals from the headpiece resulted in the appropriate statement for each faculty being consecutively stamped out ("History of phrenology and the psycograph," 2011)

Some of the history of the psycograph (or "psychograph" as it is referred to below), and its relation to vocational psychology (and in particular, to the University of Minnesota) is given by Guenter Risse's entertaining article, "Vocational Guidance During the Depression: Phrenology Versus Applied Psychology" (*Journal of the History of the Behavioral Sciences*, 1976, *12*, 130–140). The abstract follows:

> The paper describes the design and use of a machine, the "Psychograph," which automatically measured the size and shape of the skull and provided evaluation of mental traits according to phrenological principles. Developed in 1930, the psychograph was billed as a diagnostic tool capable of providing suitable vocational guidance the thousands of unemployed as a result of the Depression. Its appearance prompted a vigorous opposition from the Psychology Department at the University of Minnesota, especially in the person of Donald L. Paterson. Subsequently, the psychograph was merely exploited for its entertainment value and disappeared after the 1933 World's Fair in Chicago. (p. 130)

[10]There is also the field of physiognomy, where it is believed that a person's character or personality is manifested in facial features ("Physiognomy," 2010). Lingering aspects of these ideas seem to popup every now and then in the popular media—but phrenology is a dead issue.

[11]Related to Sheldon's connection between somatotypes and personality, there is the much earlier (late 1800s) Cesare Lombrosco theory of anthropological criminality. This idea held that body characteristics and criminality are connected, and "born criminals" could be identified by physical defects (or stigmata), thus confirming a criminal as savage, or atavistic (that is, a throwback) ("Anthropological criminology," 2011).

[12]One practice no longer engaged in because of a WOE understanding of radiation risk was in buying new shoes for school during the 1950s. Typically, a shoe-fitting fluoroscope was available, where you literally stood on top of an x-ray tube (with all of your organs exposed), to see how your foot bones wiggled nicely in your new shoes. And possibly, once your shoes were bought, you could then play on the machine somewhat longer while your mother tried on shoes for herself.

[13]Although a WOE argument can be made convincing for climate change, there are still some troubling issues in how data are (mis)presented in making the case. To illustrate, we give a short quotation from a recent article from Justin Gillis of the *New York Times* (July 7, 2010): "British Panel Clears Scientists." If the "clever maneuver" mentioned below was not completely

and fully disclosed and discussed in the United Nations report, it is an obvious cause for concern:

> The issue involved a graphic for a 1999 United Nations report ... Dr. Jones wrote an e-mail message saying he had used a "trick" to "hide" a problem in the data. ... Dr. Jones said he had meant "trick" only in the sense of a clever maneuver.

[14]We give a description of the AERS from the FDA website:

> The Adverse Event Reporting System is a computerized information database designed to support the FDA's post-marketing safety surveillance program for all approved drug and therapeutic biologic products. The FDA uses AERS to monitor for new adverse events and medication errors that might occur with these marketed products.
>
> Reporting of adverse events from the point of care is voluntary in the United States. The FDA receives some adverse event and medication error reports directly from health care professionals (such as physicians, pharmacists, nurses, and others) and consumers (such as patients, family members, lawyers, and others). Healthcare professionals and consumers may also report these events to the products' manufacturers. If a manufacturer receives an adverse event report, it is required to send the report to the FDA as specified by regulations. The MedWatch site provides information about mandatory reporting.
>
> The structure of AERS is in compliance with the international safety reporting guidance issued by the International Conference on Harmonisation. Adverse events in AERS are coded to terms in the Medical Dictionary for Regulatory Activities terminology.
>
> AERS is a useful tool for the FDA, which uses it for activities such as looking for new safety concerns that might be related to a marketed product, evaluating a manufacturer's compliance to reporting regulations and responding to outside requests for information. The reports in AERS are evaluated by clinical reviewers in the Center for Drug Evaluation and Research and the Center for Biologics Evaluation and Research to monitor the safety of products after they are approved by the FDA. If a potential safety concern is identified in AERS, further evaluation might include epidemiological studies. Based on an evaluation of the potential safety concern, the FDA may take regulatory action(s) to improve product safety and protect the public health, such as updating a product's labeling information, restricting the use of the drug, communicating new safety information to the public, or, in rare cases, removing a product from the market.
>
> AERS data do have limitations. First, there is no certainty that

the reported event was actually due to the product. The FDA does not require that a causal relationship between a product and event be proven, and reports do not always contain enough detail to properly evaluate an event. Further, the FDA does not receive all adverse event reports that occur with a product. Many factors can influence whether or not an event will be reported, such as the time a product has been marketed and publicity about an event. Therefore, AERS cannot be used to calculate the incidence of an adverse event in the United States population.

Chapter 10

(Mis)reporting of Data

> The government are very keen on amassing statistics—they collect them, add them, raise them to the nth power, take the cube root and prepare wonderful diagrams. But what you must never forget is that every one of these figures comes in the first instance from the chowkydar [village watchman], who just puts down what he damn well pleases.
> — Josiah Stamp (*Some Economic Factors in Social Life*, 1929)

The Association for Psychological Science publishes a series of timely monographs on *Psychological Science in the Public Interest*. One recent issue was from Gerd Gigerenzer and colleagues, entitled "Helping Doctors and Patients Make Sense of Health Statistics" (Gigerenzer et al., 2007). It discusses aspects of statistical literacy as it concerns health, both our own individually as well as societal health policy more generally. Some parts of being statistically literate may be fairly obvious; we know that just making up data, or suppressing information even of supposed outliers without comment, is unethical. The topics touched upon by Gigerenzer et al. (2007), however, are more subtle. If an overall admonition is needed, it is that context is always important, and the way data and information are presented is absolutely crucial to an ability to reason appropriately and act accordingly. We review several of the major issues raised by Gigerenzer et al. in the discussion to follow.

We begin with a quotation from Rudy Guiliani from a New Hampshire radio advertisement that aired on October 29, 2007, during his run for the Republican presidential nomination:

> I had prostate cancer, five, six years ago. My chances of surviving prostate cancer and thank God I was cured of it—in the United States, 82 percent. My chances of surviving prostate cancer in England, only 44 percent under socialized medicine.

Not only did Guiliani not receive the Republican presidential nomination, he was just plain wrong on survival chances for prostate cancer. The problem is a confusion between survival and mortality rates. Basically, higher survival rates with cancer screening do not imply longer life.

To give a more detailed explanation, we define a five-year survival rate and an annual mortality rate:

five-year survival rate = (number of diagnosed patients alive after five years)/(number of diagnosed patients);
annual mortality rate = (number of people who die from a disease over one year)/(number in the group).

The inflation of a five-year survival rate is caused by a *lead-time bias*, where the time of diagnosis is advanced (through screening) even if the time of death is not changed. Moreover, such screening, particularly for cancers such as prostate, leads to an *overdiagnosis bias*, the detection of a pseudodisease that will never progress to cause symptoms in a patient's lifetime. Besides inflating five-year survival statistics over mortality rates, overdiagnosis leads more sinisterly to overtreatment that does more harm than good (for example, incontinence, impotence, and other health-related problems).

Screening does not "prevent cancer," and early detection does not prevent the risk of getting cancer. One can only hope that cancer is caught, either by screening or other symptoms, at an early enough stage to help. It is also relevant to remember that more invasive treatments are not automatically more effective. A recent and informative summary of the dismal state and circumstances surrounding cancer screening generally, appeared in the *New York Times* as a "page one and above the fold" article by Natasha Singer (July 16, 2009), "In Push for Cancer Screening, Limited Benefits."

A major area of concern in the clarity of reporting health statistics is in how the data are framed as relative risk reduction or as absolute risk reduction, with the former usually seeming much more important than the latter. We give examples that present the same information:

Relative risk reduction: If you have this test every two years, your chance of dying from the disease will be reduced by about one third over the next ten years.
Absolute risk reduction: If you have this test every two years, your

chance of dying from the disease will be reduced from 3 in 1000 to 2 in 1000, over the next ten years.

A useful variant on absolute risk reduction is given by its reciprocal, the *number needed to treat*; if 1000 people have this test every two years, one person will be saved from dying from the disease every ten years.[1]

Because bigger numbers garner better headlines and more media attention, it is expected that relative rather than absolute risks are the norm. It is especially disconcerting, however, to have potential benefits (of drugs, screening, treatments, and the like) given in relative terms, but harm in absolute terms that is typically much smaller numerically. The latter has been referred to as "mismatched framing" by Gigerenzer and colleagues.

An ethical presentation of information avoids nontransparent framing of information, whether intentional or unintentional. Intentional efforts to manipulate or persuade people are particularly destructive, and unethical by definition. As Tversky and Kahneman (see 1981) have noted many times, framing effects and context have major influences on a person's decision processes. Whenever possible, give measures that have operational meanings with respect to the sample at hand (for example, the Goodman–Kruskal γ, the median or the mode, the interquartile range) and avoid measures that do not, such as the odds ratio. This advice is not always followed (see, for example, the Agency for Healthcare Research and Quality's, *National Healthcare Disparities Report 2008*, in which the efficacy of medical care is compared across various groups in plots with the odds ratio as the dependent variable. As might be expected, this section's impact on the public consciousness was severely limited).

In a framework of misreporting data, we have the all-to-common occurrence of inflated and sensational statistics intended to have some type of dramatic effect. As noted succinctly by Joel Best in his 2005 *Statistical Science* article, "Lies, Calculations and Constructions," "Ridiculous statistics live on, long after they've been thoroughly debunked; they are harder to kill than vampires" (p. 211). We might see a three-stage process in the use of inflated statistics: first, there is a tale of atrocity (think Roman Polanski's 1968 movie, *Rosemary's Baby*); the problem is then given a name (for example, the presence of satanic cults in our midst); and finally, an inflated and, most likely, incorrect statistic is given that is intended to alarm (for example, there are well over 150,000 active satanic cults throughout the United States and Canada). Remember that when a statement seems counterintuitive and nonsensical, it probably is.[2]

Another issue in the reporting of data is when the context for a statement is important but is just not given (or is suppressed), resulting in a misinterpretation, or at least, an overinterpretation. These examples are legion and follow the types illustrated below:[3]

(a) The chances of a married man becoming an alcoholic are double those of a bachelor because 66% of souses are spouses. (This may not be so dramatic when we also note that 75% of all men over 20 are married.)

(b) Among 95% of couples seeking divorce, either one or both do not attend church regularly. (This example needs some baserate information to effect a comparison; for example, what is the proportion of couples generally, where one or both do not attend church regularly.)

(c) Over 65% of all accidents occur within 25 miles of home, and at a speed of 40 miles per hour or less. (An obvious question to ask is where most of one's driving is done.)

(d) Héctor Luna, who went 2-for-5 and raised his average to .432, had his fourth straight multi-hit game for the Cardinals, who have won six of seven overall (Associated Press; "Recap of St. Louis Cardinals vs. Pittsburgh Pirates," April 26, 2004). (Reporting of data should provide a context that is internally consistent; here, the word "raised" is odd.)

(e) A recent article posted on the MSNBC website had the title, "1 in 5 US Moms Have Kids With Multiple Dads, Study Says," with the teaser line: "Poverty, lack of education and divorce perpetuate lack of opportunities" (Carroll, April 1, 2011). The first short paragraph below leaves one with the question: can the child of a mother with only one child have multiple fathers?:

> [O]ne in five of all American moms have kids who have different birth fathers, a new study shows. And when researchers look only at moms with two or more kids, that figure is even higher: 28 percent have kids with at least two different men.

(f) In 2004, President Bush boasted that the average tax cut was $1,586; this was an arithmetical average so it was considerably inflated by many large numbers for the very wealthy. A more transparent measure would have been the median of $470, where half of individuals and families got more of a cut than this, but half got less.[4]

(g) A "page one" article in the *New York Times* by Sam Roberts (January 16, 2007), had the eye-catching title "51% of Women Are Now Living Without Spouse." Its first line reads: "For what experts say is probably the first time, more American women are living without a husband than with one, according to the *New York Times* analysis of census results." A month later (February 11, 2007), the Public Editor of the *Times*, Byron Calame, had a column entitled "Can a 15-Year-Old Be a 'Woman Without a Spouse?'" Part of the second paragraph reads: "But the new majority materialized only because *The Times* chose to use survey data that counted, as spouseless women, teenagers 15 through 17—almost 90 percent of whom were living with their parents."

In providing data intended to assist in assessing risk, it is important to present the information in a context that will help individuals make decisions that are best for their own circumstances. For example, in evaluating the risk of driving or flying, the statistics on fatalities/miles flown may not be the most useful number to consider. In a plane crash, it is generally an all-or-nothing proposition of life or death; there is no possibility of using one's own skills to avoid it. In a car, however, we may feel more in control, particularly if we don't engage in drinking and driving, are not cell-phone distractible, and so on.[5] Or, in deciding whether to get immunized against some disease purported to be leading to the next pandemic, the data should be presented as to the personal risk reduction of getting the disease in relation to any downside that conditions on who we are (for example, age, sex, present and previous diseases). The 1976 mass national inoculation program pushed by President Ford should serve as a cautionary reminder. The program was terminated after 40 million people were inoculated for a flu that never did come. The downside were the deaths of many elderly people whose vaccination touched off neurological problems, especially the rare Guillain–Barre syndrome. Although it may have been reasonable to err on the side of caution at the time and attempt a national inoculation program to generate the needed "herd immunity" and curtail any possibility of a pandemic, in the future it may be unreasonable to jeopardize the more vulnerable among us, for example, the very young and old, to attain this magical "herd immunity." For an evaluation of this episode, see Robert Mackey, "Lessons From the Non-Pandemic of 1976" (*New York Times*, April 30, 2009).

A good example of the power of data to influence personal behavior, such as immunization, was given to one of us recently (LH) on an annual visit to his doctor. She suggested strongly that LH receive a single dose of the shingles vaccine now that he was over sixty years old. Knowing there would be substantial push-back on the need to do this given a long-standing aversion to needles, she was ready with an information sheet from the Centers for Disease Control and Prevention; part of the information on this sheet is reproduced in an endnote.[6] After digesting the numerical data provided, and remembering several horribly suffering acquaintances who had gotten shingles, LH was first in line with a sleeve up and ready to receive whatever number of needle punctures would be necessary.

Because of the importance of context and framing in presenting and interpreting data, it is important to avoid falling prey to the cliché that "the data speak for themselves." This is never true; data are constructed by a process requiring many decisions as to what is observed and presented. This can be a benign procedure when the definitional boundaries are clear. Sometimes, however, data collection is used with a more nefarious intent, with two of the most blatant named "astroturfing" and "push polling."

As defined in an OpEd by Ryan Sager in the *New York Times* (August 18, 2009), "Keep Off the Astroturf," "astroturfing" refers to the simulation of a (large) "grass-roots" movement by manufacturing a large number of responses. Thus, whenever there are data given in the form of the number of, say, "concerned citizens," this can be highly inflated. Senator Lloyd Bentsen of Texas coined the term in 1985 to describe the enormous amount of mail he received promoting the interests of insurance companies. Ryan quotes Bentsen as saying, "a fellow from Texas can tell the difference between grass-roots and Astroturf ... this is generated mail." A much older example might be recalled by those who read Shakespeare's *Julius Caesar* in high school. Cassius astroturfed Brutus to convince him to assassinate Caesar. He constructed letters of public support "in several hands" to appear "as if they came from several citizens":

> I will this night,
> In several hands, in at his windows throw,
> As if they came from several citizens,
> Writings all tending to the great opinion
> That Rome holds of his name, wherein obscurely
> Caesar's ambition shall be glanced at.
> And after this let Caesar seat him sure,
> For we will shake him, or worse days endure.
> – From Act 1, Scene 2
> (THE ARDEN SHAKESPEARE, edited by D. Daniell, 1998, p. 183)

One can only imagine what electronic e-mail is now doing to this practice.

A "push poll" is a survey in which the data (that is, the responses) are not of any real interest. Questions are posed to "push" the respondent in a particular direction. A good example of a push poll is given by Richard H. Davis in a *Boston Globe* article (March 21, 2004), entitled "The Anatomy of a Smear Campaign." Davis was Senator John McCain's presidential campaign manager in 2000 when McCain was running against Bush in the Republican primaries. During the contentious (and decisive) South Carolina contest, push polling was used to imply that McCain's adopted Bangladeshi-born daughter was his own illegitimate black child. The question posed to McCain supporters was whether they would be more or less likely to vote for McCain if they knew he had fathered an illegitimate black child. As we know, Bush went on to win the South Carolina primary, and ultimately the Presidency.

The two final topics in this chapter on the (mis)reporting of data are developed in separate subsections: one deals generally with the social construction of statistics, and then more specifically with the use of false precision present in a dataset to argue for undeserved superiority; the other involves adjusting for variables such as age, sex, and race in the reporting of statistics common to public health.

10.1   The Social Construction of Statistics

> If you want to get people to believe something really, really stupid, just stick a number on it.
>> Charles Seife (*Proofiness*, 2010)

In a series of three entertaining books, Joel Best has made the point that most statistics presented to the public are socially constructed, and therefore, possibly not a perfect reflection of reality. A critical consumer has to know why and by whom the numbers were generated and in what context they are being used. So yet again, context and framing are crucial to how numbers should be interpreted. The titles of these three books are given below along with their complete subtitles:

Damned Lies and Statistics: Untangling Numbers From the Media, Politicians, and Activists (2001)

More Damned Lies and Statistics: How Numbers Confuse Public Issues (2004)

Stat-Spotting: A Field Guide to Identifying Dubious Data (2008)

Best offers a variety of useful observations as to how we might go about being intelligent consumers of the numerical information that is continually provided by the media. Criminologists, for example, have used a phrase, "the dark figure," to refer to the number of crimes that don't appear in official statistics. More generally, every social problem has a dark figure of some sort defined by the difference between officially recorded incidents and the true number. It may be big (as in reporting prostitution), or small (as in reporting homicides), but it exists for most social problems that come to mind (Best, 2001, pp. 33–34).

Once a dark figure, such as declaring the number of homeless in the United States to be so many million, has been estimated and made public, that number goes through a "number laundering" process where the best guess source is now forgotten, and the number is treated as a fact. People who repeat or create the "fact" may even come to have a stake in defending it. Also, any estimate can be defended just by impugning the motives of someone disputing the figure.

Besides giving high-end guesses of a dark figure to promote attention to whatever social problems are being addressed, the use of especially broad definitions supports much larger estimates of a problem's size. Depending on how instances of such things as "bullying," "hunger," or "mental illness" are defined, very different estimates of a dark figure could be generated. A recent article by Jenny Anderson in the *New York Times* (November 7, 2011) has the title "National Study Finds Widespread Sexual Harassment of Students in Grades 7 to 12." It first line reads

> Nearly half of 7th to 12th graders experienced sexual harassment in the last school year, according to a study scheduled for release on Monday, with 87 percent of those who have been harassed reporting negative effects such as absenteeism, poor sleep and stomachaches.

An obvious question that arises is the broadness of the definition used for the term "sexual harassment." Depending on how the boundaries of a vague term are delineated, the numbers then attached are open to differing interpretations as to severity. Other instances of equivocation in definition are generally problematic as well. For example, questions have been raised about a current epidemic of autism among children, and whether this might be attributable to environmental contaminants or vaccination. It may be that the definition of what is considered autistic behavior has just been broadened to include an "Autism Spectrum," a term now used to describe pervasive developmental disorders, which include among others, Autistic Disorder, Asperger's Disorder, Childhood Disintegrative Disorder, Rett's Disorder, and the catch-all term, Pervasive Development Disorder Not Otherwise Specified.

Attaching pejorative terms without precise definition to characterize some circumstance has the potential of being misleading if not ethically questionable. Another concern is in the use of surrogate variables to evaluate some social or educational program that may be only peripherally related to what should be a primary standard—a statistical "bait-and-switch" if you will. An all-too-common practice in evaluation contexts is to change the target whenever you can't reach the original goal. An article by Sam Dillon in the *New York Times* (April 25, 2011) illustrates the problem. It has the self-explanatory title, "High School Classes May Be Advanced in Name Only." The article discusses Advanced

Placement classes and other such offerings in terms of burgeoning participation rates. Unfortunately, even through students are getting more credits in more advanced classes, they are not scoring any higher on standardized tests. As noted by Dillon, the number of Advanced Placement exams taken by American high school students has more than doubled over the last decade to 3.1 million in 2010, but the failure rate is also much higher. Evaluation by participation rate is not the same as evaluation by exam competency. There are no A's for effort in these evaluative circumstances.

When examples are used to substitute for explicit definition, there is an additional increased risk of distortion in our understanding of a problem's magnitude. In the language of decision theory, an activist trying to argue for the importance of a new social ill would consider false negatives (that is, not counting someone when they should be counted) to be more troubling than false positives (that is, counting someone when they shouldn't be counted). Public opinion and attitudes toward most social issues are too complicated to fall within simple pro or con categories. Thus, someone advocating for a particular cause and then conducting a survey can decide on how to word questions to slant responses in particular directions that allow different interpretations of results. These biases and a recognition that the measurement itself may have been devised to minimize false negatives may be hidden.

It is important to know the population surveyed if appropriate interpretations are to be made of the responses. The randomness of the chosen sample is a hoped-for ideal that is rarely achieved. Even when we try to approximate randomness by, say, the random-digit dialing of landlines, the influence of high cell-phone usage in an area is unknown. Asking about homelessness at bus stations or estimating one's electoral chances based only on the enthusiasm of the crowds at political rallies may suggest obvious biases. But subtler things can happen that remain hidden. Generally, the pseudo-randomness of a sample is more important than sample size, and the use of Internet-based surveys would seem dubious indeed. Also, a large sample size may give a false impression of the accuracy of the data. Who cares about a margin of error that is plus or minus three percentage points if you are surveying the wrong crowd? In short, good statistics are based on more than guessing (for example, of the dark figure). Clear and reasonable definitions that are not overly broad are necessary, as well as clear and reasonable measures with appropriate wording of questions that doesn't direct answers in a particular way. Finally, good representative samples are crucial to extrapolation and interpretation.

In Joel Best's discussion of statistical information of which one should be wary, he provides a number of clever names. A *mutant* statistic is one mangled beyond recognition from its original form (2001, pp. 1–2). Best recalls a dissertation proposal reporting that "every year since 1950, the number of American children gunned down has doubled" (p. 3). Thus, if one child were gunned down in 1950, we would be up to 35 million in 1995 and one trillion by 2014—not very plausible given the earth's population. Possibly the following statement might have been intended: "the number of American children killed each year by guns has doubled since 1950." Generally, complex statistics are particularly prone to mutation, especially in the hands of a statistically innumerate public. As Best (2004) summarizes:

> Today, virtually anyone can produce—if not necessarily understand—highly sophisticated statistics. This ability has created a continual escalation in the complexity of statistical analyses, in an effort to specify increasingly complicated relationships among ever more variables, by using measures that ever fewer people can hope to understand. (p. 142)

Another category of bad statistics are "missing numbers," referring to what has not been counted (Best, 2004, pp. 1–25). For example, there may be a declaration of an epidemic of school gun violence because of several salient occurrences (remember, the plural of "anecdote" is not "data");[7] or definitions that specify what is to be counted may be problematic to implement (for example, what characterizes a "missing child," or how does one count an occurrence of "child abuse"). The uncounted may result from the census (the proverbial "undercount"), or be more hidden, such as the practice of religion or the specification of race. There are also the forgotten numbers that could provide context when comparing current health scares; for example, how many died in the flu epidemic of 1918, or what is the historical number of deaths from tuberculosis or the plague.[8]

One particularly fraught name that Best uses is that of a "scary number" (Best, 2004, pp. 63–90). Given the incredible needs of the cable and broadcast news media, numbers are continually sought that will command our attention, and thereby allow the industry to sell us more things we probably don't really need. Risk numbers are given in bigger scary relative values rather than in smaller absolute terms; social problems are given big estimates (for example, inflated dark figures); trends always seem to be troubling; the various scenarios envisioned for whatever is being followed are invariably apocalyptic; and so on. When we get frightened, the focus is on what scares us rather than on the actual risk. This type of reaction is named "probability neglect" in the judgment and decision-making literature (Sunstein, 2002).[9]

As an example of the type of scary number and scenario that might be encountered (but in a more or less unknown context), a report from the Environmental Working Group is given in an endnote.[10] It has the title "Canaries in the Kitchen: Teflon Toxicosis," with a subtitle of "EWG finds heated Teflon pans can turn toxic faster than DuPont claims." The estimate is that perhaps thousands of pet bird deaths each year are linked to Teflon.

Based on the social construction of the numbers we are bombarded with continually, the question to ask is not necessarily "is it true?", but rather, "how and by whom was it produced?" This admonition applies to simple number misuse (for example, when colleges scam the numbers in the annual "Best Colleges" issue of *US News & World Report*), and to those that have serious national military implications (for example, the bodycount and pacification figures reported in the Vietnam war, or more recently, the counts of weapons of mass destruction in justifying the Iraq War).

Charles Seife in his 2010 book, *Proofiness: The Dark Arts of Mathematical Deception*, provides several additional memorable names in pointing out various acts of numerical mendacity. Seife defines "proofiness," a numerical version of Stephen Colbert's notion of "truthiness" ("Truthiness," 2010), as "the art of using bogus mathematical arguments to prove something that you know in your heart to be true—even when it's not" (p. 14). One of the first terms Seife introduces is used for fabricated statistics—"Potemkin numbers"—the numerical equivalent of Potemkin villages. To explain the historical allusion, we quote from Seife:

According to legend, Prince Grigory Potemkin didn't want the empress of Russia to know that a region in the Crimea was a barren wasteland. Potemkin felt he had to convince the empress that the area was thriving and full of life, so he constructed elaborate facades along her route—crudely panted wooden frameworks meant to look like villages and towns from afar. Even though these "Potemkin villages" were completely empty—a closer inspection would reveal them to be mere imitations of villages rather than real ones—they were good enough to fool the empress, who breezed by them without alighting from her carriage. (p. 215)

Potemkin numbers seem to be everywhere once we look: we have the thousands of pet birds dying from Teflon toxicosis; Joe McCarthy's 205 known communists in the State Department (made all the more impressive by the exactness of the given number); or numerable examples that arise from the Minnesota Republican, Michele Bachman (Obama's trip to India cost 200 million dollars per day, the number of Tea Party protesters numbered 20 to 45 thousand, and so on). Irrespective of political leanings, when things are made up as one goes along, Potemkin numbers are sure to arise.

Another useful term introduced in *Proofiness* is *disestimation*, or the underestimating of the uncertainties associated with most numbers, and generally, taking a number too seriously. We give three examples of disestimation: the first, which immediately follows, is a humorous anecdote that Seife presents about a guide at a national history museum; the second but not so humorous tale, is about the "smoothness" of Old Gold cigarettes; the third concerns a belief that by continual recounts in contested elections, the "true winner" will eventually be identified.[11]

There's an anecdote about an aging guide at a natural history museum. Every day, the guide gives tours of the exhibits, without fail ending with the most spectacular sight in the museum. It's a skeleton of a fearsome dinosaur—a *tyrannosaurus rex*—that towers high over the wide-eyed tour group. One day, a teenager gestures at the skeleton and asks the guide, "How old is it?"

"Sixty-five million and thirty-eight years old," the guide responds proudly.

"How could you possibly know that?" the teenager shoots back.

"Simple! On the very first day that I started working at the museum, I asked a scientist the very same question. He told me that the skeleton was sixty-five million years old. That was thirty-eight years ago." (p. 18)

The second example that exploits false precision for nefarious purposes occurred in the 1950s with the P. Lorillard Company and its marketing of Old Gold cigarettes. Two excerpts are given in an appendix that relate to the history of this case. The first item is a brief summary taken from a book by Susan Wagner, *Cigarette Country: Tobacco in American History and Politics* (1971); the second is more lengthly and taken directly from the opinion in *P. Lorillard Co. v. Federal Trade Commission* (1950).

The third example of disestimation is the notion of *electile dysfunction* defined as a collective belief that a "true" election winner will finally emerge if only enough recounts are carried out (2010, pp. 125–166). Recent examples of this are the 2000 Florida vote in *Bush v. Gore* or the Minnesota Senate election between Al Franken and Norm Coleman. Any true (but razor-thin) margin of winning will be swamped by errors in absentee ballots; write-ins; machine ((un)readable) failures; and so on. In short, there is no way to determine a "true" winner in these closely contested elections; possibly, the fairest mechanism to both candidates would be a simple coin toss.

10.2   Adjustments for Groups Not Comparable On a Variable, Such As Age

Data may require preliminary adjustment when presented for groups of individuals on events related to a variable such as age (for example, the presence or absence of disease, death, injury, and suicide). If the groups differ in their distributions on this variable, two types of adjustment are generally possible: direct and indirect. Relying on age as the variable of interest throughout this discussion, direct age adjustment adopts a standard population to eliminate the effects of age differences between the groups to be compared. This direct adjustment is eventually implemented using a weighted average. Indirect adjustment compares the mortality rates for a standard population classified by age to those for

TABLE 10.1: United States 2000 standard population weights for direct age adjustment; adapted from the State of New Mexico Indicator-Based Information System.

	Age group	United States 2000 population projection (in thousands)	Weight
1	Under 1 year	3,795	0.01
2	1 – 4 years	15,192	0.06
3	5 – 14 years	39,977	0.15
4	15 – 24 years	38,077	0.14
5	25 – 34 years	37,233	0.14
6	35 – 44 years	44,659	0.16
7	45 – 54 years	37,030	0.13
8	55 – 64 years	23,961	0.09
9	65 – 74 years	18,136	0.07
10	75 – 84 years	12,315	0.04
11	85 years plus	4,259	0.02
	Total	274,634	1.00 (to rounding)

the single group being studied. The final device applied in indirect adjustment is standardized mortality ratios (SMRs), defined as the ratio of the incidences of death observed in the study group to the incidences of death expected if the study group had the same incidences of death as the selected standard population. If the events observed are for disease occurrence rather than death, the ratios are standard morbidity ratios, but the same SMR acronym applies.[12]

To give examples of how direct and indirect age adjustment would work, we rely on the State of New Mexico Indicator-Based Information System (NM-IBIS) and the data provided on their website:

`http://ibis.health.state.nm.us`

Three tables follow: Table 10.1 is the standard population used to directly adjust rates for age, and is based on the age distribution for the whole of the United States in 2000; Table 10.2 provides the death rates for diabetes mellitus for the state of New Mexico (2003–2005), plus the age-adjusted rates based on the United States age distribution; the death rates in Table 10.3 are specific to Sierra County, New Mexico (2003–2005); again, the given age-adjusted rates are based on the United States population.

The standard population weights of Table 10.1 are calculated

298   *A Statistical Guide for the Ethically Perplexed*

TABLE 10.2: Age-adjusted death rate for diabetes mellitus, state of New Mexico, 2003–2005; adapted from the State of New Mexico Indicator-Based Information System.

Age group	Number of deaths (3-year sum)	Population counts (3-year sum)	Age-specific rate (per 100,000)	Cross products with weight
1	0	84,952	0	0
2	0	325,508	0	0
3	2	828,663	0.24	0.03
4	2	893,809	0.22	0.03
5	19	718,484	2.64	0.36
6	61	810,632	7.52	1.22
7	160	833,948	19.19	2.59
8	297	602,768	49.27	4.30
9	443	381,451	116.14	7.67
10	546	235,030	232.31	10.41
11	369	82,660	446.41	6.92
All	1,899	5,797,906	32.75	33.54

by dividing the number of individuals in a certain age category by the total population—for example, in the "25 – 34 years" group, we have 37,233 people [given in thousands], divided by the total population (that is, 274,634 [again, given in thousands]), to produce the weight of 0.14 (= 37,233/274,634). Obviously, the sum of all these weights over the eleven age groups must be 1.00 (to rounding error).

Table 10.2 provides the age-adjusted death rate for the complete state of New Mexico (over a 3-year period). The age-specific rate for a given age category is the number of deaths divided by the number in that category (per 100,000 individuals). For example, in the "25 – 34 years" group, the Age-Specific Rate is 2.64 (= (19/718,484) × 100,000). The age-adjusted overall death rate of 33.54 is then the sum of the cross product values (between the United States Standard Population Weights and the age-specific rates), which gives a weighted average of the age-specific rates. In this case, this age-adjusted value of 33.53 is not much different from the unadjusted overall age rate of 32.75.

The situation for Sierra County in Table 10.3 is somewhat different. Here, the unadjusted overall age rate is 53.72, but when age adjusted, it drops to 27.01. Thus, what might have caused a

TABLE 10.3: Age-adjusted death rate for diabetes mellitus, Sierra County, New Mexico, 2003–2005; adapted from the State of New Mexico Indicator-Based Information System.

Age group	Number of deaths (3-year sum)	Population counts (3-year sum)	Age-Specific rate (per 100,000)	Cross products with weight
1	0	350	0	0
2	0	1,266	0	0
3	0	4,384	0	0
4	0	4,526	0	0
5	0	2,977	0	0
6	1	4,269	23.43	3.81
7	0	5,581	0	0
8	1	5,985	16.71	1.46
9	11	5,946	185.01	12.22
10	6	4,086	146.85	6.58
11	3	1,584	189.45	2.94
All	22	40,952	53.72	27.01

health alert if the unadjusted statistic were reported is mitigated dramatically when the age-adjusted statistic is given. In fact, the 27.01 value is even below the state of New Mexico age-adjusted figure of 33.54.

The calculation of an SMR is immediate. For example, there is an age-specific death rate of 232.31 (per 100,000) for the "75 – 84 years" group for the whole state of New Mexico. For Sierra County, the number of individuals in this particular age group is 4,086, with 6 deaths observed due to diabetes mellitus. The number expected based on the New Mexico rate of 232.31 per 100,000 is 9.5, giving an SMR (ratio of observed to expected) of $6/9.5 = .63$. The ratio is less than 1.0, indicating fewer deaths than what would be expected based on the age-specific rate for the state of New Mexico as a whole. If desired, various types of confidence intervals for an SMR are available, such as those based on Poisson models for rare events. The computational formulas and underlying theory are given by Kahn and Sempos (1989; *Statistical Methods in Epidemiology*).

Appendix: P. Lorillard Co. v. Federal Trade Commission (1950)

In *P. Lorillard Co. v. FTC*, the company was charged by the FTC with making a distorted use of a *Reader's Digest* article that discussed the harmful effects of various brands of cigarettes. A laboratory had concluded that no particular brand of cigarettes was substantially more harmful than any other. A table of variations in brand characteristics was inserted in the article to show the insignificance of the differences that existed in the tar and nicotine content of the smoke produced by the various brands. The table indicated that Old Golds had less nicotine and tars, although the difference was so small as to be insignificant. Lorillard launched a national advertising campaign stressing that the *Reader's Digest* test proved that its brand was "lowest in nicotine and tars," and defended its advertising before the FTC on the ground that it had truthfully reported what had been stated in the article. In a 1950 decision, the Fourth Circuit Court of Appeals, upholding the commission's cease-and-desist order, declared that Lorillard's advertising violated the FTC Act because, by printing only a small part of the article, it created an entirely false and misleading impression. "To tell less than the whole truth is a well-known method of deception," the court ruled (Source: Susan Wagner, 1971, pp. 72–73).

The Court opinion is given in the Appendix Supplements.

Notes

[1]In addition to the use of relative and absolute risk, or the number needed to treat, a fourth way of presenting benefit would be as an increase in life expectancy. For example, one might say that women who participate in screening from the ages of 50 to 69 increase their life expectancy by an average of 12 days. This is misleading in terms of a benefit to any one particular individual; it is much more of an all-or-nothing situation, like a lottery. Nobody who plays a lottery gains the expected payout; you either win it all or not.

[2]A particularly egregious example of data misinterpretation (or, at least, data taken out of context), occurred in 1986 when *Newsweek* reported that an unmarried 40-year-old woman had "as much chance of marrying as being killed by a terrorist" (Salholz, June 2, 1986). This quickly was treated as a fact in the popular culture, and caused many women substantial consternation. The observation was based on a Harvard-Yale study that had stated that a 40-year-old unmarried woman had a 1.3% chance of marrying (Bennett, Bloom, & Craig, 1989; Bloom & Bennett, September, 1985). True, but what was lost was the definition of the class of 40-year-old women being referred to; it was

those who never married, and did not include anyone divorced or widowed; also, it did not differentiate between women who wanted to marry and those who chose not to for whatever reason.

[3]As another instance where context is crucial, it can be considered a fallacy to compare the average age at which, say, death occurs rather than comparing the risk of death between groups of the same age. For example, suppose there are two groups of individuals: A: all elementary students in the United States; B: all members of special forces units in the United States military. The average age of death is much lower for group A than B, but that does not imply it is riskier to be an elementary student than a military commando. Generally, the average age at death doesn't reflect the risk of death but only a characteristic of those who die. It only looks at those who die and ignores all those who survive. A similar reasoning anomaly happens when the age at onset of a particular kind of disease (for example, lung cancer in chemical factory workers) is compared to that for the general population. An average age being lower only implies that factory workers may be younger to begin with. The comparison needed is of the risk for lung cancer occurrence in groups having the same age.

[4]This episode and several others using an "average" defined by the arithmetical mean to confuse, are discussed further in Jackson and Jamieson's 2007 book, *unSpun: Finding Facts in a World of Disinformation* (p. 55).

[5]A similar story might be told of requiring separate infant child seats in planes, and the additional payments such seats generate. If this policy induced families to drive rather than fly, it could also have the unintended consequence of more transportation-related child deaths.

[6]Fact sheet from the Centers for Disease Control and Prevention (October 6, 2009):

SHINGLES VACCINE: WHAT YOU NEED TO KNOW
What is shingles?
Shingles is a painful skin rash, often with blisters. It is also called Herpes Zoster, or just Zoster.

A shingles rash usually appears on one side of the face or body and lasts from 2 to 4 weeks. Its main symptom is pain, which can be quite severe. Other symptoms of shingles can include fever, headache, chills and upset stomach. Very rarely, a shingles infection can lead to pneumonia, hearing problems, blindness, brain inflammation (encephalitis), or death.

For about 1 person in 5, severe pain can continue even long after the rash clears up. This is called post-herpetic neuralgia.

Shingles is caused by the Varicella Zoster virus, the same virus that causes chickenpox.

Only someone who has had chickenpox—or, rarely, has gotten chickenpox vaccine—can get shingles. The virus stays in your body, and can cause shingles many years later.

You can't catch shingles from another person with shingles. However, a person who has never had chickenpox (or chickenpox vaccine) could get chickenpox from someone with shingles. This is not very common.

Shingles is far more common in people 50 years of age and older than in younger people. It is also more common in people whose immune systems are weakened because of a disease such as cancer, or drugs such as steroids or chemotherapy.

At least 1 million people a year in the United States get shingles. Shingles vaccine:

A vaccine for shingles was licensed in 2006. In clinical trials, the vaccine reduced the risk of shingles by 50%. It can also reduce pain in people who still get shingles after being vaccinated.

A single dose of shingles vaccine is recommended for adults 60 years of age and older.

[7]There is now the coined word *anecdata* which refers to using anecdotal information as if it were data. A definition from the online Urban Dictionary follows (BabbleOn5, 2010): "The usage of multiple points of anecdotal data to confirm any stipulation, often used to 'prove' pseudo-scientific claims of illnesses and treatments." (As an example, Jenny McCarthy uses anecdata collected from various sources to support her pseudo-scientific treatment for childhood autism.)

[8]We are reminded of two famous Donald Rumsfeld quotations:

[A]s we know, there are known knowns; there are things we know that we know. We also know there are known unknowns; that is to say, we know there are some things we do not know. But there are also unknown unknowns—here are things we don't know we don't know (DoD news briefing, 2002).

Look for what's missing. Many advisors can tell a President how to improve what's proposed or what's gone amiss. Few are able to see what isn't there (Teany, 2003).

[9]A set of particularly insidious numbers used to scare for political effects are those predicting that Europe will soon become a majority-Muslim "Eurabia," with widespread demographic domination by Muslims. An article in the *New York Times* by Laurie Goodstein (January 27, 2011) has the title, "Forecast Sees Muslim Population Leveling Off." The gist of the argument is that although the number of Muslims around the world will increase at twice the rate of non-Muslims, the rapid growth will level off. As Muslim women receive education and jobs, and migration occurs to cities along with an improvement in living standards, the birth rate will come to resemble that of other nations, as it always has throughout history.

[10]Report from the Environmental Working Group (Houlihan, Thayer, & Klein, 2003):

> In two to five minutes on a conventional stovetop, cookware coated with Teflon and other nonstick surfaces can exceed temperatures at which the coating breaks apart and emits toxic particles and gases linked to hundreds, perhaps thousands, of pet bird deaths and an unknown number of human illnesses each year, according to tests commissioned by Environmental Working Group (EWG).
>
> In new tests conducted by a university food safety professor, a generic nonstick frying pan preheated on a conventional, electric stovetop burner reached 736 °F in three minutes and 20 seconds, with temperatures still rising when the tests were terminated. A Teflon pan reached 721 °F in just five minutes under the same test conditions, as measured by a commercially available infrared thermometer. DuPont studies show that the Teflon offgases toxic particulates at 464 °F. At 680 °F Teflon pans release at least six toxic gases, including two carcinogens, two global pollutants, and MFA, a chemical lethal to humans at low doses. At temperatures that DuPont scientists claim are reached on stovetop drip pans (1000 °F), nonstick coatings break down to a chemical warfare agent known as PFIB, and a chemical analog of the WWII nerve gas phosgene.
>
> For the past fifty years DuPont has claimed that their Teflon coatings do not emit hazardous chemicals through normal use. In a recent press release, DuPont wrote that "significant decomposition of the coating will occur only when temperatures exceed about 660 degrees F (340 degrees C). These temperatures alone are well above the normal cooking range."
>
> These new tests show that cookware exceeds these temperatures and turns toxic through the common act of preheating a pan, on a burner set on high.
>
> In cases of "Teflon toxicosis," as the bird poisonings are called, the lungs of exposed birds hemorrhage and fill with fluid, leading to suffocation. DuPont acknowledges that the fumes can also sicken people, a condition called "polymer fume fever." DuPont has never studied the incidence of the fever among users of the billions of nonstick pots and pans sold around the world. Neither has the company studied the long-term effects from the sickness, or the extent to which Teflon exposures lead to human illnesses believed erroneously to be the common flu.
>
> The government has not assessed the safety of nonstick cookware. According to a Food and Drug Administration (FDA) food safety scientist: "You won't find a regulation anywhere on the books that specifically addresses cookwares," although the FDA approved Teflon for contact with food in 1960 based on a food frying study

that found higher levels of Teflon chemicals in hamburger cooked on heat-aged and old pans. At the time, FDA judged these levels to be of little health significance.

Of the 6.9 million bird-owning households in the US that claim an estimated 19 million pet birds, many don't know know that Teflon poses an acute hazard to birds. Most nonstick cookware carries no warning label. DuPont publicly acknowledges that Teflon can kill birds, but the company-produced public service brochure on bird safety discusses the hazards of ceiling fans, mirrors, toilets, and cats before mentioning the dangers of Teflon fumes.

As a result of the new data showing that nonstick surfaces reach toxic temperatures in a matter of minutes, EWG has petitioned the Consumer Product Safety Commission (CPSC) to require that cookware and heated appliances bearing nonstick coatings must carry a label warning of the acute hazard the coating poses to pet birds. Additionally, we recommend that bird owners completely avoid cookware and heated appliances with nonstick coatings. Alternative cookware includes stainless steel and cast iron, neither of which offgases persistent pollutants that kill birds.

[11] The pedantry of too much precision leads to an amusing story about the nineteenth-century inventor Charles Babbage, considered the "father of the computer." In a 1961 book edited by Philip and Emily Morrison, *Charles Babbage and His Calculating Engines: Selected Writings by Charles Babbage and Others*, a letter is quoted that Babbage supposedly wrote to the poet Alfred Lord Tennyson regarding the poem, "The Vision of Sin":

> "Every minute dies a man, / Every minute one is born": I need hardly point out to you that this calculation would tend to keep the sum total of the world's population in a state of perpetual equipoise, whereas it is a well-known fact that the said sum total is constantly on the increase. I would therefore take the liberty of suggesting that in the next edition of your excellent poem the erroneous calculation to which I refer should be corrected as follows: "Every moment dies a man, / And one and a sixteenth is born." I may add that the exact figures are 1.167, but something must, of course, be conceded to the laws of metre. (p. xxiii)

[12] Another well-known and extensively developed example of age adjustment is given in Mosteller and Tukey's 1977 textbook, *Data Analysis and Regression: A Second Course in Statistics* (pp. 225–229). It concerns death rates in Maine and South Carolina in 1930 and is a good example where the comparison of overall crude death rates produces misleading results. South Carolina had higher death rates than Maine in all but one age class (and even here they were close). Nonetheless, the crude death rate of South Carolina (at 1288.8 per 100,000) made it look better than Maine (at 1390.8 per 100,000). The

reason for this misleading benefit of living in South Carolina (pun intended) is that Maine's population was generally much older than South Carolina's.

Chapter 11

Inferring Causality

> If a person a) is sick, b) receives treatment intended to make him better, and c) gets better, then no power of reasoning known to medical science can convince him that it may not have been the treatment that restored his health.
> — Sir Peter Brian Medawar (*The Art of the Soluble*, 1967)

The aim of any well-designed experimental study is to make a causal claim, such as "the difference observed between two groups is *caused* by the different treatments administered." To make such a claim we need to know the counterfactual: what would have happened if this group had not received the treatment? This counterfactual is answered most credibly when subjects are assigned to the treatment and control groups at random. In this instance, there is no reason to believe that the group receiving the treatment condition would have reacted any differently (than the control condition) had it received the control condition. If there is no differential experimental mortality to obscure this initial randomness, one can even justify the analyses used by how the groups were formed (for example, by randomization tests, or their approximations defined by the usual analysis methods based on normal theory assumptions). As noted by R. A. Fisher (1971, p. 34), "the actual and physical conduct of an experiment must govern the statistical procedure of its interpretation." When the gold standard of inferring causality is not met, however, we are in the realm of quasi-experimentation, where causality must be approached differently.

When data from observational/quasi-experimental studies are to be interpreted, the question of causality often gets rephrased as one of whether the experiment is "internally valid." Internal validity addresses the question of what manifestly appears to distinguish the groups of respondents, and whether this could also be the reason for the observed differences. Such a reinterpretation has been a part of the graduate education of psychologists for over forty years, primarily through Campbell and Stanley (1966) (*Experimental and Quasi-Experimental Designs for Research*), and later through all of its many successors (e.g., Shadish, Cook, & Campbell, 2002). Basically, one tries to argue for causality by eliminating various threats to the internal validity of the quasi-experiment. We list eight threats to internal validity reviewed by Winch and Campbell (1969):[1]

(1) *History*: Events, other than the experimental treatment occurring between pre-test and post-test and thus providing alternate explanations of effects.
(2) *Maturation*: Processes within the respondents or observed social units producing changes as a function of the passage of time per se, such as growth, fatigue, secular trends, etc.
(3) *Testing*: The effect of taking a test on the scores of a second testing. The effect of publication of a social indicator on subsequent readings of that indicator.
(4) *Instrumentation*: Changes in the calibration of a measuring instrument or changes in the observers or scores used producing changes in the obtained measurements.
(5) *Statistical regression artifacts*: Pseudo-shifts occurring when persons or treatment units have been selected upon the basis of their extreme scores.
(6) *Selection*: Biases resulting from differential recruitment of comparison groups, producing different mean levels on the measure of effects.
(7) *Experimental mortality*: The differential loss of respondents from comparison groups.
(8) *Selection-maturation interaction*: Selection biases resulting in differential rates of "maturation" or autonomous change (p. 141).

In addition to eliminating threats to internal validity in arguing for plausible causal explanations when true experiments are not possible (for example, as in establishing the links between smoking and lung cancer in humans), it is important to verify the tell-tale signs of a potentially causal connection: the association needs to be consistent; higher "doses" should be connected to stronger responses; alleged causes should be plausible and temporally appropriate by preceding the effect in question. The most authoritative source for methodologies in observational studies are the two volumes by Rosenbaum: *Observational Studies* (2002) and *The Design of Observational Studies* (2009).

A potential statistical solution to the equating of intact groups is sometimes offered in the form of analysis of covariance, where adjustments in treatment effects are made based on existing group differences for a collection of covariates. Unfortunately, and contrary to the dreams of a quasi-experimenter, this is just not possible. We quote from Fredric Lord's (1967) classic article, "A Paradox in the Interpretation of Group Comparisons":

> [T]here simply is no logical or statistical procedure that can be counted on to make proper allowances for uncontrolled preexisting differences between groups. The researcher wants to know how the groups would have compared if there had been no preexisting uncontrolled differences. The usual research study of this type is attempting to answer a question that simply cannot be answered in any rigorous way on the basis of available data. (p. 305)

In contexts where group differences on other variables may influence the response variable of interest, it is best to block on these variables and randomly assign subjects within blocks to the treatments. Given the horrendous difficulties encountered in treatment-covariate confounding in ANCOVA, this type of statistical control by blocking is just not legitimately done after the fact. The analysis of covariance cannot be relied on to adjust properly for or to control the differences between naturally occurring groups. A particularly informative and cogent review of the difficulties involved with unraveling Lord's Paradox is given by Holland and Rubin (1983) and Wainer and Brown (2004). Typically, the goal of ANCOVA is causal inference—showing that the treatment *caused* an effect after adjusting for preexisting differences. Thus, an unraveling of Lord's Paradox requires a model for causal inference. These two articles use Rubin's model for causal inference to show that the choice of which untestable assumption is made determines the validity of the subsequent analysis. A narrower view, focusing just on the limitations of ANCOVA, is found in Miller and Chapman (2001).

Recalling from an earlier chapter the traditional adage that "correlation does not imply causation," we might add a corollary: merely identifying a plausible causal mechanism doesn't change the import of this caveat. The plausible mechanism may result from a third confounding or "lurking" variable just as the original correlation is spurious and due to the influence of a third variable. A salient illustration of this faulty reasoning was the initial observation that women with herpes are more likely than other women to develop cervical cancer. Because of general beliefs about biomedical activity, some investigators concluded that the relationship was causal and cancer must result from the herpes. Later research, however, showed that the primary cause of cervical cancer was human papilloma virus (HPV), and herpes was just a marker for sexual activity. Women with multiple sexual partners were more likely to be exposed not just to herpes but also to HPV. The apparent association between herpes and cervical cancer was the result of other variables. For a further discussion of this and similar examples of faulty causal reasoning, see Kaye and Freedman (2011).[2]

11.1   Causuistry

> How easy it is to work over an undigested mass of data and emerge with a pattern, which at first glance, is so intricately put together that it is difficult to believe it is nothing more than the product of a man's brain.
> — Martin Gardner (*Fads and Fallacies in the Name of Science*, 1957)

The legitimate word *casuistry* has several meanings (e.g., it can refer to specious reasoning intended to mislead, or to the evaluation of right and wrong by analyzing specific cases that illustrate general ethical rules) (*American Heritage Dictionary of the English Language*, 2000/2006). In his 2010 book, *Proofiness*, Charles Seife adds another "u" to coin a new word, "causuistry," the ascribing of cause to what are just accidental associations (p. 44). As Seife points out, sometimes things just happen for no apparent reason. In medicine, for example, the term *idiopathic* is commonly applied when some condition appears to arise spontaneously or for an unknown reason.

People generally have a difficult time accepting that an event or situation is of unknown or idiopathic origin (think of Job: "Why me, Lord?"). Seife coins another word that helps explain this human failing, "randumbness," the insistence on order in the presence of what is more often just chaos, or in creating pattern when this is none to see (p. 55). In a *ScienceNews* article by Bruce Bower (February 12, 2011), "In the Zone," an argument is reviewed (though it is unprovable) that evolutionary pressures may be partially responsible for our minds seeing streaks, such as the "hot hand" in basketball.

One particularly troubling result of the need to have causal explanations is when we observe some grouping of unusual medical cases, such as that for cancer, birth defects, or similar bad things. Even though such clusters most often happen just randomly, we still search for cause, and perhaps, for someone or something to blame. A good and insightful review of this kind of specious "culling for clusters" phenomenon, and the subsequent "search for causes" is given in a *New Yorker* article by Atul Gawande, "The Cancer-Cluster Myth" (February 8, 1999). Generally, such anomalies appear regularly in the popular media, which is always on the lookout for compelling stories to hold our attention. One such recent example was given in the *New York Times* (November 22, 2010) by Jesse McKinley, "In California, Birth Defects Show No Link."

Assertions of specious causation never seem to disappear for events that are merely identified (i.e., "cherry-picked") as a result of their salient extremeness. Because of the implicit search carried out over a large database, what we try to explain is likely a result of mere chance and the operation of various random processes. Currently, we have the earnest news stories about master teachers who supply "added value" to their students, and are identified by an amazing change seen in test scores over time. Such a teacher's behavior is then scrutinized, hoping to add to the collection of "best practices" that could be passed on to others with less or negative "value added." Most (or all) of this is just chasing moonbeams. It is best to keep in mind that subjective belief is insufficient as evidence of causation, as is inferring causality not from a controlled study but from a few selected anecdotal observations.[3]

One insidious public health instance of such search for causation and a good exemplar for the fallacy of *post hoc ergo propter hoc* (after this, therefore because of this) is the anti-vaccine movement. This campaign is fueled by a false belief that a causal link exists between the administration of childhood vaccines and illnesses such as autism. As a direct result of not immunizing children for the various childhood illnesses that were once fairly common (for example, polio, MMR [measles, mumps, and rubella], DPT [diphtheria, pertussis (whooping cough), and tetanus]), the protective "herd immunity" levels are dropping, possibly allowing a resurgency for these diseases. Two books by Paul Offit lay out the issues well for this false ascription of causality between the administering of vaccines and the subsequent onset of an illness: *Autism's False Prophets: Bad Science, Risky Medicine, and the Search for a Cure* (2008) and *Deadly Choices: How the Anti-Vaccine Movement Threatens Us All* (2011).[4]

The anti-vaccine movement seems to be kept alive by a combination of proposing faulty but conceivable causal mechanisms (for example, mercury used in the manufacture of vaccine, or immune system compromises because of the large number of vaccinations given simultaneously at an early age), and by those few instances where vaccines may have actually caused problems. For example, there was the mass national flu inoculation fiasco under President Ford in 1976, where a number of elderly people apparently developed the rare Guillain–Barre syndrome (Mackey, 2009); or some children with a preexisting compromised immune system becoming paralyzed when given the live Sabin polio vaccine as opposed to the dead Salk alternative (Offit, 2011, p. 58).[5]

One explanatory causal mechanism that people act on at times is superstition. The field of sports seems particularly prone to postulating superstitious causal mechanisms—the pitcher who jumps over a chalk line rather than stepping on it, or someone like the now-retired baseball player called the "human rain delay," Mike Hargrove, and his very elaborate routine at the plate after each pitch.

Once people believe there is an epidemic afoot, there is an associated drive to look for someone or something to blame, even though the epidemic may have resulted just from a broadening of a diagnostic criterion, such as what is now labeled by a more inclusive phrase of an "autism spectrum." Causuistry results from this need to link every effect with some sort of cause, and generally to search for an explanatory causal mechanism. In the clinical psychology arena, for example, there is the following phenomenon: people brought to emergency rooms as young children for suspected sexual abuse are interviewed twenty years later but can't remember the incident. The conclusion is that this must be amnesia, with the repression of memory being the operative causal mechanisms. An alternative explanation is not considered that the sexual abuse never occurred or was so "mild" in form that it was forgotten and not just repressed.

Several causation fallacies are important to avoid. One has to do with the effect of small numbers and when random variability produces wide shifts that are then interpreted causally. If states, for example, award special recognition to schools showing marked improvement in test scores, they should not be overly surprised when only small schools are the ones garnering attention. Bigger changes and smaller sample sizes are intimately tied. A second trap is to be seduced by the lure of a coherent narrative. Thus, we might argue that heroin addiction is primed by the use of a "gateway" drug such as tobacco or marijuana; here, the word "gateway" lends credence to a causal assertion even though it isn't true. Just the naming or postulating of some mechanism doesn't provide proof that a causal link is present. A well-publicized fiasco to keep in mind that is discussed elsewhere (Section 15.1) is on the "heart protective" value of hormone replacement therapy, or the now debunked presence of a "timing hypothesis" as to when hormone replacement therapy should begin for it to be protective.

11.2 The Bradford–Hill Criteria for Determining a Causal Connection

Austin Bradford Hill (Fellow of the Royal Statistical Society; 1897–1991) is considered by many to have been the world's foremost medical statistician of the twentieth century ("Austin Bradford Hill," 2010). He pioneered the randomized clinical trial, working with streptomycin in treating tuberculosis in the 1940s. Hill's most famous contributions, however, were with Richard Doll. Hill and Doll (pun intended) established the link between lung cancer and smoking in the 1950s. Interestingly, R. A. Fisher was one of Hill's endorsers for Fellow of the Royal Society in 1954, but strongly disagreed with the conclusions and procedures used in making the smoking–cancer link, and criticized the work in both the popular press and academic publications. In a letter to the *NewScientist* (2008, *197*(2646)), one of Fisher's former students, Jean Marston, wrote that although Fisher was personally skeptical of a link between smoking and cancer, he encouraged her work with mice on the inheritance of a susceptibility to lung tumors induced by the chemicals in tobacco smoke. Marston also noted that Fisher had prepared an elegant proof of a direct correlation between the import of apples and the increased divorce rate. However, because this work was funded by the British Tobacco Research Council, Fisher was considered biased and the work was suspect.

The nine Bradford–Hill criteria for establishing causality were published by Hill in 1965 (*Proceedings of the Royal Society of Medicine*, *58*, 295–300). They are not meant as a checklist, and Hill himself noted: "None of my nine viewpoints can bring indisputable evidence for or against the cause-and-effect hypothesis and none can be required *sine qua non*" (p. 299). In any case, here they are:

Strength: A weak association does not imply that there is no causal effect; however, the stronger the association, the more likely it is causal.

Consistency: Consistent findings observed by different people in different places with different samples strengthens the likelihood of a causal effect.

Specificity: Causation is likely if a very specific population at a specific site has a disease with no other likely explanation. The more specific an association between a factor and an effect, the larger the probability of a causal relationship.

Temporality: The effect has to occur after the cause (and if there is an expected delay between the cause and expected effect, the effect must occur after that delay).

Biological gradient: Greater exposure should generally lead to greater incidence of the effect (that is, a dose-response relationship). In some cases, the mere presence of the factor (above some

threshold, say) can trigger the effect; in others, an inverse proportion might even be observed, where greater exposure leads to lower
incidence.

Plausibility: A plausible mechanism between cause and effect is
helpful (but knowledge of the mechanism may be limited by current knowledge).

Coherence: Coherence between epidemiological and laboratory
findings increases the likelihood of an effect. But as Hill noted,
"lack of such [laboratory] evidence cannot nullify the epidemiological effect on associations" (p. 298).

Experiment: Occasionally, it is possible to appeal to experimental evidence, such as when an agent is removed, a reduction in a
particular disease is observed.

Analogy: The effect of similar factors may be considered.

We suggest three additional criteria:

(a) Have alternative explanations been considered thoroughly?
possibilities include confounding with the presence of other variables; bias operating with respect to the subjects being studied;
or by how exposure or disease status is assessed.

(b) Are there effects of ceasing or altering exposure?

(c) Are the findings consistent with other relevant knowledge?
For example, this may occur when a lessened (national) use of some
agent (for example, smoking) appears associated with a reduced
occurrence of some disease (for example, the prevalence of lung
cancer).

11.3 Some Historical Health and Medical Conceptions of Disease Causality

A later section of this book concerns the collection of data and
experimental design and introduces the term *differential etiology*
(Henifin, Kipen, & Poulter, 2000, p. 481). Simply put, this involves
the investigation and reasoning that lies behind the determination
of external causation, usually for a disease. It typically proceeds by
a process of elimination, and is directed toward identifying disease
causes, but not to determining treatment. A treatment empha-

sis falls under the notion of *differential diagnosis* (Goldstein & Henifin, 2000, p. 433).[6]

In the early to late 1800s, the dominant idea used to explain infectious diseases was miasma theory. Although diseases such as smallpox, syphilis, and measles were believed to be contagious, it was unclear what to think about the epidemic diseases such as cholera, typhoid, and typhus. The editor of the *Lancet*, Thomas Wakley, for example, wrote an 1853 editorial that asked the question: "What is cholera?":

> All is darkness and confusion, vague theory, and a vain speculation. Is it a fungus, an insect, a miasm, an electrical disturbance, a deficiency of ozone, a morbid off-scouring from the intestinal canal? We know nothing; we are at sea in a whirlpool of conjecture. (p. 393)

Miasma theory held that certain diseases, such as cholera, were caused by a noxious form of "bad air" (a miasma), filled with particles from decomposed matter (called miasmata). It could be identified by its foul smell.

The miasma theory of disease became a very appealing causal explanation in the middle 1800s for English public health reformers who could now focus on environmental problems as opposed to those tied only to infection and personal health. We quote part of the Wikipedia article on "Miasma theory" (2010) that mentions the work of two influential 19th century English statisticians, William Farr and Florence Nightingale:

> In the 1850s, miasma was used to explain the spread of cholera in London and in Paris, partly justifying Haussmann's latter renovation of the French capital. The disease was said to be preventable by cleansing and scouring of the body and items. Dr. William Farr, the assistant commissioner for the 1851 London Census, was an important supporter of the miasma theory. He believed that cholera was transmitted by air, and that there was a deadly concentration of *miasmata* near the River Thames' banks. The wide acceptance of miasma theory during the cholera outbreaks overshadowed the partially correct theory brought forth by John Snow that cholera was spread through water. This slowed the response to the major outbreaks in the Soho district of London and other areas. Another proponent of the miasmatic theory was Crimean War nurse Florence Nightingale (1820–1910), who became famous for her work in making hospitals sanitary and fresh-smelling.

The theory of miasma disease made sense to the English sanitary reformers of the mid-nineteenth century. Miasma explained why cholera and other diseases were epidemic in places where the water was undrained and very foul-smelling. The theory led to improvements in the sanitation systems, which coincidentally led to decreased episodes of cholera, thus helping to support the theory.

Even though the miasmatic theory has been disproven by the discovery of viruses and bacteria, it helped make the connection between poor sanitation and disease. This caused public health reforms and encouraged cleanliness, even though some doctors still did not wash their hands between patients. They believed that the miasmata were only airborne, and would not be stuck on the doctors' hands.

The miasmatic theory was consistent with the observations that disease was associated with poor sanitation (and hence foul odours) and that sanitary improvements reduced disease, but not with the observations of microbiology that led to the germ theory of disease.

So far as cholera is concerned, the miasmatic theory was disproved by John Snow following an epidemic in Soho, central London in 1854. Because of the miasmatic theory's predominance among Italian scientists, the 1854 discovery by Filippo Pacini of the bacillus that caused the disease was completely ignored, and the bacteria had to be rediscovered thirty years later by Robert Koch.

A remnant of this theory is the name of malaria, from Italian *mala aria* ("bad air").

There have been several historical instances where causes have been asserted for disease occurrence that later were proved wrong (usually, this happened after the germ theory of disease was developed in the late 1800s). One particularly interesting historical tale involves Ignaz Semmelweis, who proposed a causal theory of puerperal (childbed) fever, which was rampant in the maternity wards of Europe in the middle 1800s ("Ignaz Semmelweis," 2010). Semmelweis believed that childbed fever was the result of doctors dissecting cadavers and then delivering babies without first washing their hands. Arguing that cadaverous particles, similar to those in miasma theory, were being absorbed by the patients, Semmelweis institutionalized hand washing with a lime and chlorine solution (a disinfectant that removed the cadaverous odor), before reentering the maternity wards. The results were dramatic in terms of the tenfold reduction in mortality from childbed fever. The story behind Semmelweis is informative enough that we give a fairly extensive redaction in an appendix to this section from the Wikipedia article devoted to him.

The story of Ignaz Semmelweis highlights the importance of observation in generating working hypotheses, and then in the collection of data to verify possible causal mechanisms. He had four separate pieces of information that led him to try the use of disinfectants: (1) two maternity wards with dramatically different rates of childbed fever, with the only difference between them being the doctors present in one who also worked with cadavers, and midwives in the second who did not help in autopsies; (2) the death of Semmelweis' friend, Jakob Kolletschka, who was cut with a student scalpel while doing a postmortem exam; his own autopsy showed the same pathology as the women dying of puerperal fever; (3) because the chlorine and lime solution covered up the cadaver smells, Semmelweis conjectured that it might also stop what he thought was being passed to the women giving birth; and finally, (4) there were unexplained lower mortality rates for women actually having "street births."

The careful collection of time series mortality statistics before and after the introduction of the use of hand disinfectant lead Semmelweiss to conclude that indeed it was the cadaverous material that was the cause of the childbed fever. Although the effect of mortality reduction was present with the use of disinfectants, Semmelweis got the specific causal mechanism wrong; it wasn't necessarily cadavers, but most often *Streptococcus pyogenes* present in unclean areas generally. We should remember that this was before the germ theory of disease was established in the later 1800s, and therefore, we shouldn't blame Semmelweis for taking advantage of the "law of unintended consequences," where eliminating the cadaver odor also killed the *Streptococcus pyogenes*.[7]

One could argue that Semmelweis was seduced by his strong belief in the biological causal mechanism of cadaver particles being resorbed by women giving birth. And to some extent, this insistence on a single cause created difficulty in the adoption of the Semmelweis disinfectant theory. Seduction by biological mechanism occurs to this day. One example will be discussed in detail later involving hormone replacement therapy for women. Just because a plausible biochemical mechanism could be advanced for why hormone therapy might be heart-protective doesn't necessarily mean that it has to be true. In fact, hormone therapy is generally the opposite of heart-protective. The reason the supposed benefits were observed in the women studied was likely due to the "healthy-user bias" (Taubes, September 16, 2007). Women who took the hormone replacements regularly were generally different (for example, more health-conscious and diligent) than those who didn't, and this masked the severe downside of hormone therapy.[8]

Ignaz Semmelweis

Ignaz Philipp Semmelweis (July 1, 1818 – August 13, 1865), was a Hungarian physician described as the "savior of mothers," who discovered by 1847 that the incidence of puerperal fever could be drastically cut by the use of hand disinfection (by means of hand washing with chlorinated lime solution) in obstetrical clinics. Puerperal fever (or *childbed fever*) was common in mid-19th-century hospitals and often fatal, with mortality at 10%–35%. Semmelweis postulated the theory of washing with "chlorinated lime solutions" in 1847 while working in Vienna General Hospital's First Obstetrical Clinic, where doctors' wards had three times the mortality of midwives' wards. He published a book of his findings in childbed fever in *Etiology, Concept and Prophylaxis of Childbed Fever.*

Despite various publications of results where hand-washing reduced mortality to below 1%, Semmelweis's practice earned widespread acceptance only years after his death, when Louis Pasteur confirmed the germ theory. In 1865, a nervous breakdown (or possibly Alzheimer's) landed him in an asylum, where Semmelweis died of injuries, at age 47.

...

Discovery of cadaverous poisoning

Semmelweis was appointed assistant to Professor Klein in the *First Obstetrical Clinic* of the Vienna General Hospital on July 1, 1846. A comparable position today would be head resident. His duties were, amongst others, to examine patients each morning in preparation for the professor's rounds, supervise difficult deliveries and teach obstetrical students. He was also responsible for the clerical records.

Maternity institutions were set up all over Europe to address problems of infanticide of "illegitimate" children. They were set up as gratis institutions and offered to care for the infants, which made them attractive to underprivileged women, including prostitutes. In return for the free services, the women would subject themselves to the training of doctors and midwives. There were two maternity clinics at the Viennese hospital. The *First Clinic* had an average maternal mortality rate due to puerperal fever of about 10% (actual rates fluctuated wildly). The _Second Clinic's_ rate was considerably lower, averaging less than 4%. This fact was known outside the hospital. The two clinics admitted on alternate days but women begged to be admitted to the Second Clinic due to the bad reputation of the First Clinic. Semmelweis described desperate women begging on their knees not to be admitted to the First Clinic. Some women even preferred to give birth in the streets, pretending to have given sudden birth _en route_ to the hospital (a practice known as _street births_), which meant they would still qualify for the child care benefits without having been admitted to the clinic. Semmelweis was puzzled that puerperal fever was rare amongst women giving street births. "To me, it appeared logical that patients who experienced street births would become ill at least as frequently as those who delivered in the clinic. ... What protected those who delivered outside the clinic from these destructive unknown endemic influences?"

Semmelweis was severely troubled and literally sickened that his First Clinic had a much higher mortality rate due to puerperal fever than the Second Clinic. It "made me so miserable that life seemed worthless." The two clinics used almost the same techniques, and Semmelweis started a meticulous process of eliminating all possible differences, including even religious practices. The only major difference was the individuals who worked there. The First Clinic was the teaching service for medical students, while the Second Clinic had been selected in 1841 for the instruction of midwives only.

He excluded "overcrowding" as a cause, since the Second Clinic was always more crowded and yet the mortality was lower. He eliminated climate as a cause because the climate was the same. The breakthrough occurred in 1847, following the death of his good friend Jakob Kolletschka, who had been accidentally poked with a student's scalpel while performing a postmortem examination. Kolletschka's own autopsy showed a pathology similar to that of the women who were dying from puerperal fever. Semmelweis immediately proposed a connection between cadaveric contamination and puerperal fever.

He concluded that he and the medical students carried "cadaverous particles" on their hands from the autopsy room to the patients they examined in the First Obstetrical Clinic. This explained why the student midwives in the Second Clinic, who were not engaged in autopsies and had no contact with corpses, saw a much lower mortality rate.

The germ theory of disease had not yet been developed at the time. Thus, Semmelweis concluded that some unknown "cadaverous material" caused childbed fever. He instituted a policy of using a solution of chlorinated lime (modern calcium hypochlorite, the compound used in today's common household chlorine bleach solution) for washing hands between autopsy work and the examination of patients. He did this because he found that this chlorinated solution worked best to remove the putrid smell of infected autopsy tissue, and thus perhaps destroying the causal "poisonous" or contaminating "cadaveric" agent hypothetically being transmitted by this material.

The result was that the mortality rate dropped tenfold, comparable to the Second Clinic's. The mortality rate in April 1847 was 18.3%; after handwashing was instituted in mid-May, the rates in June were 2.2%, July 1.2%, August 1.9%, and, for the first time since the introduction of anatomical orientation, the death rate was zero in two months in the year following this discovery.

Efforts to reduce childbed fever

. ... Semmelweis's hypothesis, that there was only one cause, that all that mattered was cleanliness, was extreme at the time, and was largely ignored, rejected or ridiculed. He was dismissed from the hospital for political reasons and harassed by the medical community in Vienna, being eventually forced to move to Pest.

Semmelweis was outraged by the indifference of the medical profession and began writing open and increasingly angry letters to prominent European obstetricians, at times denouncing them as irresponsible murderers. His contemporaries, including his wife, believed he was losing his mind, and in 1865 he was committed to an asylum. In an ironic twist of fate, he died there of septicaemia only 14 days later, possibly after being severely beaten by guards. Semmelweis's practice earned widespread acceptance only years after his death, when Louis Pasteur developed the germ theory of disease, offering a theoretical explanation for Semmelweis's findings. He is considered a pioneer of antiseptic procedures.

Conflict with established medical opinions

Semmelweis's observations conflicted with the established scientific and medical opinions of the time. The theory of diseases was highly influenced by ideas of an imbalance of the basic "four humours" in the body, a theory known as dyscrasia, for which the main treatment was bloodlettings. Medical texts at the time emphasized that each case of disease was unique, the result of a personal imbalance, and the main difficulty of the medical profession was to establish precisely each patient's unique situation, case by case.

The findings from autopsies of deceased women also showed a confusing multitude of various physical signs, which emphasized the belief that puerperal fever was not one, but many different, yet unidentified, diseases. Semmelweis's main finding—that all instances of puerperal fever could be traced back to only one single cause: lack of cleanliness—was simply unacceptable. His findings also ran against the conventional wisdom that diseases spread in the form of "bad air," also known as miasmas or vaguely as "unfavourable atmospheric-cosmic-terrestrial influences." Semmelweis's groundbreaking idea was contrary to all established medical understanding.

As a result, his ideas were rejected by the medical community. Other more subtle factors may also have played a role. Some doctors, for instance, were offended at the suggestion that they should wash their hands; they felt that their social status as gentlemen was inconsistent with the idea that their hands could be unclean.

Specifically, Semmelweis's claims were thought to lack scientific basis, since he could offer no acceptable explanation for his findings. Such a scientific explanation was made possible only some decades later, when the germ theory of disease was developed by Louis Pasteur, Joseph Lister, and others.

During 1848, Semmelweis widened the scope of his washing protocol to include all instruments coming in contact with patients in labor, and used mortality rate time series to document his success in virtually eliminating puerperal fever from the hospital ward.

Hesitant publication of results and first signs of trouble

Toward the end of 1847, accounts of Semmelweis's work began to spread around Europe. Semmelweis and his students wrote letters to the directors of several prominent maternity clinics describing their recent observations. Ferdinand von Hebra, the editor of a leading Austrian medical journal, announced Semmelweis's discovery in the December 1847 and April 1848 issues of the medical journal. Hebra claimed that Semmelweis's work had a practical significance comparable to that of Edward Jenner's introduction of cowpox inoculations to prevent smallpox.

In late 1848, one of Semmelweis's former students wrote a lecture explaining Semmelweis's work. The lecture was presented before the Royal Medical and Surgical Society in London and a review published in the *Lancet*, a prominent medical journal. A few months later, another of Semmelweis's former students published a similar essay in a French periodical.

As accounts of the dramatic reduction in mortality rates in Vienna were being circulated throughout Europe, Semmelweis had reason to expect that the chlorine washings would be widely adopted, saving tens of thousands of lives. Early responses to his work also gave clear signs of coming trouble, however. Some physicians had clearly misinterpreted his claims. James Young Simpson, for instance, saw no difference between Semmelweis's groundbreaking findings and the British idea suggested by Oliver Wendell Holmes in 1843 that childbed fever was contagious (i.e., that infected persons could pass the infection to others). Indeed, initial responses to Semmelweis's findings were that *he had said nothing new.*

In fact, Semmelweis was warning against all decaying organic matter, not just against a specific contagion that originated from victims of childbed fever themselves. This misunderstanding, and others like it, occurred partly because Semmelweis's work was known only through secondhand reports written by his colleagues and students. At this crucial stage, Semmelweis himself had published nothing. These and similar misinterpretations would continue to cloud discussions of his work throughout the century.

Some accounts emphasize that Semmelweis refused to communicate his method officially to the learned circles of Vienna, nor was he eager to explain it on paper.

. . .

Legacy

Semmelweis's advice on chlorine washings was probably more influential than he realized himself. Many doctors, particularly in Germany, appeared quite willing to experiment with the practical handwashing measures that he proposed, but virtually everyone rejected his basic and groundbreaking theoretical innovation—that the disease had only one cause, lack of cleanliness. Professor Gustav Adolf Michaelis from a maternity institution in Kiel replied positively to Semmelweis's suggestions—eventually he committed suicide, however, because he felt responsible for the death of his own cousin, whom he had examined after she gave birth.

Only belatedly did his observational evidence gain wide acceptance; more than twenty years later, Louis Pasteur's work offered a *theoretical* explanation for Semmelweis's observations—the germ theory of disease. As such, the Semmelweis story is often used in university courses with epistemology content, for example, philosophy of science courses—demonstrating the virtues of empiricism or positivism and providing a historical account of which types of knowledge count as scientific (and thus accepted) knowledge, and which do not. It is an irony that Semmelweis's critics considered themselves positivists. They could not accept his ideas of minuscule and largely invisible amounts of decaying organic matter as a cause of every case of childbed fever. To them, Semmelweis seemed to be reverting to the speculative theories of earlier decades that were so repugnant to his positivist contemporaries.

The so-called Semmelweis reflex—a metaphor for a certain type of human behavior characterized by reflex-like rejection of new knowledge because it contradicts entrenched norms, beliefs or paradigms—is named after Semmelweis, whose perfectly reasonable hand-washing suggestions were ridiculed and rejected by his contemporaries.

11.4 Medical Error as a (the) Causative Factor

The term *iatrogenesis* refers to the inadvertent adverse effects or complications that result from medical treatment ("Iatrogenesis," 2010). These errors can be widely defined: surgical complications, drug interactions, negligence, and so on. As just discussed, the major iatrogenic catastrophe of the 19th century was puerperal fever, resulting from pathogen transfer from autopsy room to maternity patients. More recently, major iatrogenic errors have been in the use of radiation, the antibiotic resistance in bacteria, and the various hospital anomalies against which the Institute for Healthcare Improvement have been campaigning for some time. A former President and CEO of the Institute for Healthcare Improvement was Donald Berwick, then Professor of Pediatrics and Health Care Policy in the Department of Pediatrics at the Harvard Medical School. He was nominated on April 19, 2010, to be the Administrator of the Centers for Medicare and Medicaid Services but faced a formidable Senate confirmation fight up to July of 2010. At that point, he was just named to the position on a recess appointment by President Obama. As noted by Robert Pear in his article "Obama's Pick to Head Medicare and Medicaid Resigns Post" (November 23, 2011), Berwick resigned his position at the end of November 2011. Pear comments that Berwick had "become a symbol of all that Republicans dislike in President Obama's health care policies."[9]

Dr. Berwick's views on reducing iatrogenic error in hospitals have been well known for some time. He has been successful in his various campaigns to save lives (the Institute for Healthcare Improvement has completed a 100K Lives Campaign and is now working on a 5 Million Lives Campaign). We give a *Newsweek* article below (and extending into the endnotes) that Berwick wrote in December 2005, "Keys to Safer Hospitals," laying out what needs to be done in the nation's hospitals to prevent needless life-threatening errors:[10]

Sometime soon, I will need a new right knee. If all goes well, it will be quite a relief. An artificial joint can be a modern miracle, the alternative to decades of pain and hobbling. Here's the problem. Instead of helping me, health care might kill me. In 1999, the Institute of Medicine shocked the nation with an authoritative report on hospital errors. The report concluded that up to 98,000 Americans each year die in hospitals, not from the diseases that brought them there but from injuries caused by their medical care: preventable bleeding or infections, a medication mix-up, a respirator tube put in the wrong way and a lot more. I have climbed Mount Rainier five times. Each time I made that tough trek, my risk of dying was about 100 times smaller than the risk I will face on the operating table.[11]

We end this section on iatrogenic errors by mentioning two items. One is a short article from *ScienceNews* entitled "July: When Not to Go to the Hospital" (Janet Raloff; June 2, 2010); this discusses the theory that "green doctors may account for a one-month national blip in death rates from a particular type of medical error." The second is to refer the reader to the Internet Movie Database for the 1971 movie *The Hospital*, starring George C. Scott.

Notes

[1]We are reminded of the quotation from Arthur Conan Doyle (and Sherlock Holmes) (*The Sign of Four*): "Eliminate all other factors, and the one which remains must be the truth."

[2]A successful causal argument in a health-related area may lead to changed clinical practice that by itself has some unintended consequences. A good example of this is reported in the recent *New York Times* article by Christie Aschwanden (December 19, 2011) with the title "Studies Suggest an Acetaminophen-Asthma Link." There has been a dramatic and perplexing increase in childhood asthma over the last thirty years. As researchers have now noted, this asthma epidemic accelerated from the 1980s onward, about the same time that giving aspirin to children was stopped in favor of acetaminophen (e.g., Tylenol) because of aspirin's link to Reye's syndrome. If you read the warnings on your aspirin bottle, you will see something like the following: "Reyes Syndrome: children and teenagers who have or are recovering from chicken pox or flu-like symptoms should not use this product. When using this product, if changes in behavior with nausea and vomiting occur, consult a doctor because these symptoms could be an early sign of Reye's Syndrome, a rare but serious illness." There are no similar warnings as yet on Tylenol (acetaminophen) and asthma, but one pediatrician who studies the possible acetaminophen–asthma link, John McBride, provided an apt precautionary quote in the Aschwanden article: "I cannot say with 100 percent certainty that acetaminophen makes asthma worse, but I can say that if I had a child with asthma, I would give him or her ibuprofen for the time being ... I think the burden of proof is now to show that it's safe."

[3]Claims of causation are often made not on the basis of any data but from the construction of some plausible conspiracy theory. Women's Right to Know Acts, for example, may require doctors to point out a specious connection between abortion and breast cancer; here, the postulated conspiracy theory involves doctors and institutions, such as Planned Parenthood, all getting "filthy rich" from the abortion industry they work in. (For a discussion of the Woman's Right to Know Act passed by the Minnesota Legislature in 2003, see http://www.health.state.mn.us/wrtk.) Another well-known example is the supposed connection between brain tumors and the consumption of the artificial sweetener aspartame. The collusion here involves those wishing to make millions from the manufacture of aspartame and the FDA Center for Food Safety and Applied Nutrition, which obviously gets its "take" in this endeavor ("Aspartame," 2012).

[4]The 1905 Supreme Court case of *Jacobson v. Massachusetts* upheld the authority of states to pass compulsory vaccination laws. The Court argued the view that an individual's freedom must at times be subordinated to the common welfare and subject to the police power of the state. We give the

operative part of the decision below (and note that it was cited in the famous forced sterilization case of *Buck v. Bell* (1927), discussed in Section 8.4.1. This latter opinion had the oft-quoted line from Oliver Wendell Holmes, Jr.: "It is better for all the world, if instead of waiting to execute degenerate offspring for crime, or to let them starve for their imbecility, society can prevent those who are manifestly unfit from continuing their kind. The principle that sustains compulsory vaccination is broad enough to cover cutting the Fallopian tubes (Jacobson v. Massachusetts). Three generations of imbeciles are enough."
From *Jacobson v. Massachusetts*:

There is, of course, a sphere within which the individual may assert the supremacy of his own will, and rightfully dispute the authority of any human government—especially of any free government existing under a written constitution, to interfere with the exercise of that will. But it is equally true that in every well-ordered society charged with the duty of conserving the safety of its members the rights of the individual in respect of his liberty may at times, under the pressure of great dangers, be subjected to such restraint, to be enforced by reasonable regulations, as the safety of the general public may demand. An American citizen arriving at an American port on a vessel in which, during the voyage, there had been cases of yellow fever or Asiatic cholera, he, although apparently free from disease himself, may yet, in some circumstances, be held in quarantine against his will on board of such vessel or in a quarantine station, until it be ascertained by inspection, conducted with due diligence, that the danger of the spread of the disease among the community at large has disappeared. The liberty secured by the 14th Amendment, this court has said, consists, in part, in the right of a person 'to live and work where he will'; and yet he may be compelled, by force if need be, against his will and without regard to his personal wishes or his pecuniary interests, or even his religious or political convictions, to take his place in the ranks of the army of his country, and risk the chance of being shot down in its defense. It is not, therefore, true that the power of the public to guard itself against imminent danger depends in every case involving the control of one's body upon his willingness to submit to reasonable regulations established by the constituted authorities, under the sanction of the state, for the purpose of protecting the public collectively against such danger.

[5]One of us (LH) had had some experiences with forced vaccination and testing that reinforces the notion that individuals should be allowed to "opt out" if there is a verifiable reason, but probably not just from the causal hysteria generated by the anti-vaccine movement and its celebrities: Don Imus, Jenny McCarthy, Jim Carry, Bill Maher, among others.

Growing up in the 1940s and 1950s, LH was never given a smallpox vaccination because of a severe case of childhood eczema. Apparently, the way a smallpox vaccine was administered by scratching the upper arm and producing a rather large lesion and eventual scab would have done the same to every place there was an eczema outbreak, producing an extremely sick child. The second incident that wasn't so understanding was the time LH started his first academic job in 1970 at the University of Wisconsin, Madison. Two preemployment items needed to be dealt with: the signing of a loyalty oath ("I am not now, nor have I ever been, a member of the Communist Party"), and the taking of a Tine test for tuberculosis. The answer to a question of consequence for refusal to sign the oath or take the Tine test was simple—no oath, no test, no job! (This is akin to a more recent experience with Comcast in Santa Fe, when it demanded a Social Security Number before providing service—no number, no Internet.) The Tine test was administered at the old University Hospital by a nurse, having the name of Ratched if I recall correctly. She requested my return in a few days; I complied along with an arm that was completely red, raised, and swollen. I was immediately quarantined until a disconfirming chest X-ray was performed. Nurse Ratched failed to ask whether my mother ever had TB, which she did in the 1920s.

[6]Think of the television series, *House, MD*, and the cry of "give me the differentials."

[7]Another prominent figure in the middle 19th century who argued that puerperal fever may be passed from patient to patient by doctors and nurses, was Oliver Wendell Holmes (1809–1894). Although regarded as one of the best writers of the 19th century, Holmes was also an academic and practicing physician. His major medical work was published in 1843, and entitled *The Contagiousness of Puerperal Fever*; this tract was apparently unknown to Semmelweis until at least he had formulated his own cadaverous theory for the cause of puerperal fever. We give a short section from Wikipedia on Holmes that is embedded in the article on "Puerperal fever" (2010). As you can see, Holmes did not fare much better than Semmelweis in terms of acceptance:

> In 1843, Oliver Wendell Holmes published *The Contagiousness of Puerperal Fever* and controversially concluded that puerperal fever was frequently carried from patient to patient by physicians and nurses; he suggested that clean clothing and avoidance of autopsies by those aiding birth would prevent the spread of puerperal fever. Holmes stated that " ... in my own family, I had rather that those I esteemed the most should be delivered unaided, in a stable, by the mangerside, than that they should receive the best help, in the fairest apartment, but exposed to the vapors of this pitiless disease."
>
> Holmes' conclusions were ridiculed by many contemporaries, including Charles Delucena Meigs, a well-known obstetrician, who stated "Doctors are gentlemen, and gentlemen's hands are clean." Richard Gordon states that Holmes' exhortations "outraged obstetricians, particularly in Philadelphia." In those days, "surgeons oper-

ated in blood-stiffened frock coats—the stiffer the coat, the prouder the busy surgeon," "pus was as inseparable from surgery as blood," and "cleanliness was next to prudishness." He quotes Sir Frederick Treves on that era: "There was no object in being clean ... Indeed, cleanliness was out of place. It was considered to be finicking and affected. An executioner might as well manicure his nails before chopping off a head."

[8]To establish a causal link between a microbe and a disease, a set of postulates were formulated by Robert Koch and Fredrich Loeffler (1884), and published by Koch in 1890. Koch himself applied (Koch's) postulates to the etiology of tuberculosis and anthrax. Part of the Wikipedia entry on "Koch's postulates" (2010) is given below:

Koch's postulates are:
1) The microorganism must be found in abundance in all organisms suffering from the disease, but should not be found in healthy animals.
2) The microorganism must be isolated from a diseased organism and grown in pure culture.
3) The cultured microorganism should cause disease when introduced into a healthy organism.
4) The microorganism must be re-isolated from the inoculated, diseased experimental host and identified as being identical to the original specific causative agent.

However, Koch abandoned the universalist requirement of the first postulate altogether when he discovered asymptomatic carriers of cholera and, later, of typhoid fever. Asymptomatic or subclinical infection carriers are now known to be a common feature of many infectious diseases, especially viruses such as polio, herpes simplex, HIV, and hepatitis C. As a specific example, all doctors and virologists agree that poliovirus causes paralysis in just a few infected subjects, and the success of the polio vaccine in preventing disease supports the conviction that the poliovirus is the causative agent.

The third postulate specifies "should," not "must," because as Koch himself proved in regard to both tuberculosis and cholera, not all organisms exposed to an infectious agent will acquire the infection. Noninfection may be due to such factors as general health and proper immune functioning; acquired immunity from previous exposure or vaccination; or genetic immunity, as with the resistance to malaria conferred by possessing at least one sickle cell allele.

The second postulate may also be suspended for certain microorganisms which we cannot (at the present time) grow in pure culture, such as some viruses. In summary, a body of evidence that satisfies

Koch's postulates is sufficient but not necessary to establish causation.

History:

Koch's postulates were developed in the 19th century as general guidelines to identify pathogens that could be isolated with the techniques of the day. Even in Koch's time, it was recognized that some infectious agents were clearly responsible for disease even though they did not fulfill all of the postulates. Attempts to rigidly apply Koch's postulates to the diagnosis of viral diseases in the late 19th century, at a time when viruses could not be seen or isolated in culture, may have impeded the early development of the field of virology. Currently, a number of infectious agents are accepted as the cause of disease despite their not fulfilling all of Koch's postulates. Therefore, while Koch's postulates retain historical importance and continue to inform the approach to microbiologic diagnosis, fulfillment of all four postulates is not required to demonstrate causality.

[9]Robert Pear has several articles in the *New York Times* on the nomination of Donald Berwick: "President Nominates Professor to Health Job," (April 19, 2009); "Confirmation Fight on Health Chief," (June 21, 2010); "Obama to Bypass Senate to Name Health Official," (July 6, 2010).

[10]The figure given by Berwick of 98,000 Americans dying each year from iatrogenic error is a rate of about 270 per day, or the equivalent of a jumbo jet crash per day. One would expect such a high frequency of crashes to generate immediate and serious investigation, with all jumbo jet flights probably canceled until the problem was resolved.

[11]Continued from the main text:

Even if the surgery doesn't kill me, it may still cause needless harm. The reason I need a new knee is that I have osteoarthritis—the result of a botched and unnecessary knee operation 30 years ago, when I was a naive and trusting medical student. What could go wrong this time? My postoperative pain may not be adequately controlled. I may receive the wrong dose of blood thinner, causing bleeding in my stomach. Someone may overlook the little patch of pneumonia on my routine postoperative chest X-ray, causing me to remain on a respirator in the intensive-care unit for several days. Or the hospital may fail to take steps that could prevent the pneumonia in the first place.

Fortunately, hospitals are beginning to realize that it doesn't have to be this way. On Dec. 14, 2004, the Institute for Healthcare Improvement, a nonprofit organization headquartered in Cambridge, Mass., launched the 100,000 Lives Campaign, a broad national effort to achieve the most urgent reforms. Mainstream leadership groups like the American Medical Association, the American Nurses Association and the Joint Commission on Accreditation of Healthcare Organizations immediately signed on to the campaign. Several federal agencies—including the Centers for Disease Control and Prevention, the Centers for Medicare and Medicaid Services, the Veterans Health Administration and the Agency for Healthcare Research and Quality—pledged support as well.

We have identified six basic measures that could save as many as 100,000 lives a year if even 2,000 hospitals adopted them. It's surprising to learn that these standards aren't already the norm—but the norms may finally be changing. Nearly 3,000 American hospitals have enrolled in our 100,000 Lives Campaign over the past year, and more than half are reporting their monthly death rates so that we (and they) can track progress. That takes courage in a world where hospitals, fearing blame and lawsuits, too often feel the need to hide their mistakes.

What exactly will it take to improve the quality of care? Here are the prescriptions that we and our partners are advancing. You don't have to be a doctor to understand them.

PREVENT RESPIRATOR PNEUMONIA: VAPs, or ventilator-associated pneumonias, are often deadly lung infections that people on respirators can get (after surgery, for example). A few simple maneuvers, like elevating the head of the hospital bed and frequently cleaning the patient's mouth, can eliminate them. Dominican Hospital in Santa Cruz, Calif., just celebrated one full year without a single VAP—a result most doctors would have thought impossible.

PREVENT IV-CATHETER INFECTIONS: Central-line infections occur when bacteria contaminate catheters that deliver food and medicine intravenously. Dr. Peter Pronovost of Johns Hopkins University recently reported that 70 hospitals in Michigan, California, Iowa and Indiana cut their central-line infections by half, saving an estimated $165 million from complications to boot. How did they do it? They made it easy for doctors and nurses to wash their hands between patients, adopted simple procedures for changing the bandages around the catheters and made absolutely sure that no catheter remained in a vein even one hour longer than needed.

STOP SURGICAL-SITE INFECTIONS: Surgical-site infections are a major cause of complications and deaths after operations. Last year Mercy Health Center in Oklahoma City operated on 1,200 consecutive patients without a single wound infection—by adopting a series of simple preventive measures. These include giving the right antibiotics at the right time during surgery, enforcing strict hand-washing and avoiding shaving the surgery site before the operation (clipping hair avoids nicking the skin and is safer).

RESPOND RAPIDLY TO EARLY-WARNING SIGNALS: A nurse or visitor is often the first person to notice that a patient is in trouble. By setting up special rapid-response teams, hospitals can ensure that these critical warnings are never missed or ignored. Busy physicians may resent the false alarms, but lives are saved when hospitals take nurses' concerns seriously and respond within minutes. Australian researchers have found that rapid-response teams may be able to cut hospital death rates by 20 percent or more. The University of Pittsburgh Medical Center is testing an even more innovative way to use rapid-response teams. The staff trains patients' visiting family members to call for assistance whenever they sense trouble. The new protocol, dubbed Condition H (for "Help"), has already saved lives.

MAKE HEART-ATTACK CARE ABSOLUTELY RELIABLE: The scientifically correct treatments for heart attacks could save far more lives if we used them reliably. The 100,000 Lives Campaign simply asks hospitals to ensure that every patient gets every medication—and treatment recommended by the American College of Cardiology and other expert bodies. These measures include aspirin and a beta blocker on arrival and a stent or clot buster promptly after admission. McLeod Regional Medical Center in Florence, S.C., has cut the death rate among its heart-attack patients from 10 percent (the United States average) to about 4 percent. All the hospital had to do was ensure 100 percent reliability.

STOP MEDICATION ERRORS: Medication errors kill tens of thousands of patients a year, yet many are easily prevented. One secret is to "reconcile" medications whenever patients move from one care setting to another—from hospital to home, or even from one place to another within a hospital. The reconciliation protocol assigns a doctor or nurse at every step to check and recheck: are the medicines the patient gets after the transfer exactly the ones planned before the transfer? If not, the mistake gets corrected right away.

How much difference are we making through these efforts? We don't yet know whether the campaign will save 100,000 lives in its first year. Talk is cheap; changing the culture of a hospital is hard. But I've got a stake in it. When I close my eyes on the operating table so that a surgical team can implant my shiny new pain-free titanium knee, I know exactly what I want: safe, effective care, without a single complication.

Chapter 12

Simpson's Paradox

> Stigler's Law of Eponymy: Every scientific discovery is named after the last individual too ungenerous to give due credit to his predecessors.
> — Stephen Stigler (*Statistics on the Table*, 2002)

An unusual phenomenon occurs so frequently in the analysis of multiway contingency tables that it has been given the label of "Simpson's Paradox" (Simpson, 1951; Yule, 1903). Basically, various relations that appear to be present when data are conditioned on the levels of one variable, either disappear or change "direction" when aggregation occurs over the levels of the conditioning variable. A well-known real-life example is the Berkeley sex bias case applicable to graduate school (Bickel, Hammel, & O'Connell, 1975). The table below shows the aggregate admission figures for the fall of 1973:

	Number of applicants	Percent admitted
Men	8442	44
Women	4321	35

Given these data, there appears to be a *primae facie* case for bias because a lower percentage of women than men is admitted.

Although a bias seems to be present against women at the aggregate level, the situation becomes less clear when the data are broken down by major. No department is significantly biased against women; in fact, most have a small bias against men: Simpson's paradox has occurred! Apparently, women tend to apply to competitive departments with lower rates of admission among qualified applicants (for example, English); men tend to apply to departments with generally higher rates of admission (for example, Engineering).[1]

A different example showing a similar point can be given using data on the differential imposition of a death sentence depending on the race of the defendant and the victim. These data are from twenty Florida counties during 1976-1977 (Radelet, 1981):

	Death Penalty	
Defendant	Yes	No
White	19 (12%)	141
Black	17 (10%)	149

Because 12% of white defendants receive the Death penalty and only 10% of blacks, at this aggregate level there appears to be no bias against blacks. But when the data are disaggregated, the situation appears to change:

		Death Penalty	
Victim	Defendant	Yes	No
White	White	19 (13%)	132
White	Black	11 (17%)	52
Black	White	0 (0%)	9
Black	Black	6 (6%)	97

When aggregated over victim race, there is a higher percentage of white defendants (12%) receiving the death penalty than black defendants (10%), so apparently, there is a slight race bias against whites. But when looking within the race of the victim, black defendants have the higher percentages of receiving the death sentence compared to white defendants (17% to 13% for white victims; 6% to 0% for black victims). The conclusion is disconcerting: the value of a victim is worth more if white than if black, and because more whites kill whites, there appears to be a slight bias against whites at the aggregate level. But for both types of victims, blacks are more likely to receive the death penalty.[2]

Simpson's Paradox is part of a larger class of reversal paradoxes (Messick & van de Geer, 1981). Several of these possibilities have been mentioned briefly in Chapter 5 on correlation:

(1) Based on the algebraic constraints for correlations given in Chapter 5, suppose that performance on each of two developmental tasks has a positive correlation with age; an observed positive correlation between the two tasks could conceivably reverse when

a partial correlation is computed between the two tasks that "holds age constant."

(2) Illusory (positive) correlations that result from a "lurking" or confounding variable might be reversed when that variable is controlled. One particularly memorable example given below is from Messick and van de Geer (1981):

> A well-known example is the apparent paradox that the larger the number of firemen involved in extinguishing a fire, the larger the damage. Here the crucial third variable, of course, is the "severity of the fire"; for fires of equal severity, one would hope that the correlation would have a reversed sign. (p. 584)

A common way to explain what occurs in Simpson's Paradox is to use contingency tables. For convenience, we restrict discussion to the simple $2 \times 2 \times 2$ case, and use the "death penalty" data as an illustration. There are two general approaches based on conditional probabilities. One involves weighted averages; the second relies on the language of events being conditionally positively correlated, but unconditionally negatively correlated (or the reverse). To set up the numerical example, define three events: A, B, and C:

A: the death penalty is imposed;
B: the defendant is black;
C: the victim is white.

For reference later, we give a collection of conditional probabilities based on frequencies in the $2 \times 2 \times 2$ contingency table:

$P(A|B) = .10; P(A|\bar{B}) = .12; P(A|B \cap C) = .17;$
$P(A|\bar{B} \cap C) = .13; P(A|\bar{B} \cap \bar{C}) = .00;$
$P(C|B) = .38; P(\bar{C}|B) = .62; P(C|\bar{B}) = .94;$
$P(\bar{C}|\bar{B}) = .38; P(C) = .66; P(\bar{C}) = .34.$

The explanation for Simpson's Paradox based on a weighted average begins by formally stating the paradox through conditional probabilities: It is possible to have

$$P(A|B) < P(A|\bar{B}) ,$$

but

$$P(A|B \cap C) \geq P(A|\bar{B} \cap C) ;$$

$$P(A|B \cap \bar{C}) \geq P(A|\bar{B} \cap \bar{C}) \ .$$

So, conditioning on the C and $\bar{C}$ events, the relation reverses.

In labeling this reversal as anomalous, people reason that the conditional probability, $P(A|B)$, should be an average of

$$P(A|B \cap C) \text{ and } P(A|B \cap \bar{C}) \ ,$$

and similarly, that $P(A|\bar{B})$ should be an average of

$$P(A|\bar{B} \cap C) \text{ and } P(A|\bar{B} \cap \bar{C}) \ .$$

Although this is true, it is not a simple average but one that is weighted:

$$P(A|B) = P(C|B)P(A|B \cap C) + P(\bar{C}|B)P(A|B \cap \bar{C}) \ ;$$

$$P(A|\bar{B}) = P(C|\bar{B})P(A|\bar{B} \cap C) + P(\bar{C}|\bar{B})P(A|\bar{B} \cap \bar{C}) \ .$$

If B and C are independent, $P(C|B) = P(C|\bar{B}) = P(C)$ and $P(\bar{C}|B) = P(\bar{C}|\bar{B}) = P(\bar{C})$. Also, under such independence, $P(C)$ and $P(\bar{C})$ $(= 1 - P(C))$ would be the weights for constructing the average, and no reversal would occur. If B and C are not independent, however, a reversal can happen, as it does for our "death penalty" example:

$.10 = P(A|B) = (.38)(.17) + (.62)(.06);$
$.12 = P(A|\bar{B}) = (.94)(.13) + (.62)(.00).$

So, instead of the weights of .66 $(= P(C))$ and .34 $(= P(\bar{C}))$, we use .38 $(= P(C|B))$ and .62 $(= P(\bar{C}|B))$; and .94 $(= P(C|\bar{B}))$ and .06 $(= P(\bar{C}|\bar{B}))$.

The second approach characterizes a reversal through events being positively correlated conditionally but negatively correlated unconditionally (or, the words "positively" and "negatively" could be interchanged). Defining two events A and B to be negatively correlated when $P(A \cap B) < P(A)P(B)$, it is possible to have

$$P(A \cap B|C) \geq P(A|C)P(B|C) \ ;$$

$$P(A \cap B|\bar{C}) \geq P(A|\bar{C})P(B|\bar{C}) ,$$

but

$$P(A \cap B) < P(A)P(B) .$$

Again, conditioning on the C and $\bar{C}$ events reverses the relation.

The "death penalty" example shows the reversal defined in this way, based on the following (conditional) probabilities:

$P(A) = .11$; $P(B) = .51$; $P(A \cap B) = .05$;
$P(A \cap B|C) = .05$; $P(A \cap B|\bar{C}) = .05$;
$P(A|C) = .14$; $P(B|C) = .24$; $P(A|\bar{C}) = .05$; $P(B|\bar{C}) = .92$.

Thus, we have:

$$P(A \cap B|C) = .05 \geq P(A|C)P(B|C) = (.14)(.29) = .04 ;$$

$$P(A \cap B|\bar{C}) = .05 \geq P(A|\bar{C})P(B|\bar{C}) = (.05)(.92) = .046 ,$$

but yet:

$$P(A|B) = .05 < P(A)P(B) = (.11)(.51) = .056 .$$

An obvious advantage to using the weighted average perspective over that focusing on conditional/unconditional events is the more dramatic numerical size discrepancies.

Figure 12.1 provides a convenient graphical representation for the reversal paradox in our "death penalty" illustration. This representation generalizes to any $2 \times 2 \times 2$ contingency table. The x-axis is labeled as percentage of victims who are white; the y-axis has a label indicating the probability of death penalty imposition. This probability generally increases along with the percentage of victims that are white. Two separate lines are given in the graph reflecting this increase, one for Black defendants and one for white defendants. Note that the line for the black defendant lies wholly above that for the white defendant, implying that irrespective of the percentage of victims that may be white, the imposition of the death penalty has a greater probability for a black defendant compared to a white defendant.

FIGURE 12.1: A graph demonstrating the reversal paradox with the Florida death-penalty data.

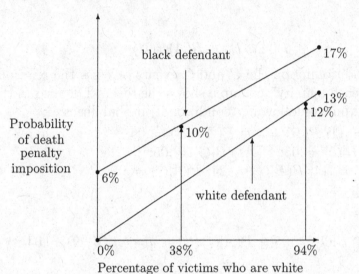

The reversal paradox of having a higher death penalty imposition for whites (of 12%) compared to blacks (of 10%) in the 2×2 contingency table aggregated over the race of the victim, is represented by two vertical lines in the graphs. Because black defendants have 38% of their victims being white, the vertical line from the x-axis value of 38% intersects the black defendant line at 10%; similarly, because white defendants have 94% of their victims being white, the vertical line from the x-axis value of 94% intersects the white defendant line at (a higher value of) 12%. The reversal occurs because there is a much greater percentage of white victims for white defendants than for black defendants. (The two lines in the graph can be constructed readily by noting how the endpoints were obtained of 0% and 6%, and of 13% and 17%. When the percentage of white victims along the x-axis is 0%, that is the same as having a black victim [which immediately generates the graph values of 0% and 6%]; if the percentage of white victims is 100%, this is equivalent to the victim being white [and again, immediately provides the other two endpoints of 13% and 17%]).

We conclude with yet another example of Simpson's Paradox (taken from Wainer, 2005, pp. 63–67) and a solution, standardization, that makes the paradox disappear. Consider the results from the National Assessment of Educational Progress (NAEP) shown in Table 12.1 (Mullis et al., 1993). The 8th grade students in Nebraska scored 6 points higher in mathematics than their counterparts in New Jersey. White students do better in New Jersey, and so do black students; in fact, all students do better in New Jersey. How is this possible? Again, this is an example of Simpson's Paradox. Because a much greater proportion of Nebraska's 8th grade students (87%) are from the higher scoring white population than in New Jersey (66%), their scores contribute more to the total.

Is ranking states on such an overall score sensible? It depends on the question that these scores are being used to answer. If the question is "I want to open a business. In which state will I find a higher proportion of high-scoring math students to hire?", the unadjusted score is sensible. If, however, the question of interest is: "I want to enroll my children in school. In which state are they likely to do better in math?", a different answer is required. Irrespective of race, children are more likely to do better in New Jersey. When questions of this latter type are asked more frequently, it makes sense to adjust the total to reflect the correct answer. One way to do this is through the method of standardization, where each state's score is based upon a common demographic mixture. In this instance, a sensible mixture to use is that of the nation overall. After standardization, the result obtained is the score we would expect each state to have if it had the same demographic mix as the nation. When this is done, New Jersey's score is not affected much (273 instead of 271), but Nebraska's score shrinks substantially (271 instead of 277).[3]

Although Simpson's Paradox is subtle, experience has taught us that a graphic depiction often aids understanding. A graphic representation of Simpson's Paradox was provided by three Korean statisticians (Jeon, Chung, & Bae, 1987), and independently by Baker and Kramer in 2001. Consider the graphic representation of the results from this table shown in Figure 12.2. A solid diagonal line shows the average NAEP math score for various proportions of white examinees in Nebraska. At the extreme left, if no whites took the test, the mean score would be that for nonwhites, 236. At the extreme right is what the mean score would be if only whites took the test, 281. The large black dot labeled "277" represents the observed score for the mixture that includes 87% whites. A

FIGURE 12.2: A Baker–Kramer plot of the New Jersey–Nebraska average 8th grade National Assessment of Educational Progress (NAEP) mathematics scores.

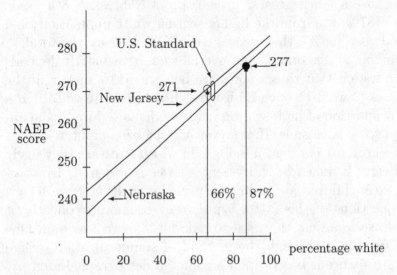

second solid line above the one for Nebraska shows the same thing for New Jersey; the large open dot labeled "271" denotes the score for a mixture in which 66% of those tested were white.

We see that for any fixed percentage of whites on the horizontal axis, the advantage of New Jersey over Nebraska is the same, two NAEP points. But because Nebraska has a much larger proportion of higher scoring white examinees, its mean score is higher than that of New Jersey. The small vertical box marks the percentage mixture representing the United States as a whole, and hence, encloses the standardized values. The graph makes clear how and why standardization works; it uses the same location on the horizontal axis for all groups being compared.

Simpson's Paradox generally occurs when data are aggregated. If data are collapsed across a subclassification (such as grades, race, or age), the overall difference observed might not represent what is really occurring. Standardization can help correct this, but nothing will prevent the possibility of yet another subclassification, as yet

TABLE 12.1: National Assessment of Educational Progress (NAEP) 1992 8th grade mathematics scores.

	State	White	Black	Other Non-White	Stand-ardized
Nebraska	277	281	236	259	271
New Jersey	271	283	242	260	273
		% Population			
Nebraska		87%	5%	8%	
New Jersey		66%	15%	19%	
Nation		69%	16%	15%	

unidentified, from changing things around. We believe, however, that knowing of the possibility helps contain the enthusiasm for what may be overly impulsive first inferences.[4]

The only way to rule out the possibility of a third factor reversing the observed ordering is by randomizing subjects to groups. Randomization will automatically adjust (in the limit) for any "missing third factors" that could reverse the direction of the observed effect. This emphasizes a key point, the ignorance of which has yielded ethically dubious conclusions in the past. When we are making comparisons among preexisting groups (by sex or race, for example), and because we cannot therefore credibly believe the groups were formed at random, we cannot rule out the possibility of a missing third factor reversing the observed differences. An understanding of this fact should help dampen any propensity for making invidious comparisons when random assignment is not possible. It parallels Paul Holland's quip: "no causation without manipulation" (Holland, 1986, p. 959).

Although Simpson's Paradox has been known by this name only rather recently (as coined by Colin Blyth in 1972), the phenomenon has been recognized and discussed for well over a hundred years; in fact, it has a complete textbook development in Yule's *An Introduction to the Theory of Statistics*, first published in 1911.[5] In honor of Yule's early contribution (Yule, 1903), we sometimes see the title of the Yule–Simpson effect. But most often, Stigler's Law of Eponymy, quoted at the beginning of this chapter, is operative, and Simpson is given sole naming credit for the phenomenon.[6]

Notes

[1]A question exists as to whether an argument for bias "falls apart" because of Simpson's paradox? Interesting, in many cases the authors have seen like this, there is a variable that if interpreted in a slightly different way would make a case for bias even at the disaggregated level. Here, why do the differential admission quotas interact with sex? In other words, is it inherently discriminatory to women if the majors to which they apply most heavily are also those with the most limiting admission quotas?

[2]Simpson's Paradox is a very common occurrence, and even through it can be "explained away" by the influence of differential marginal frequencies, the question remains as to why the differential marginal frequencies are present in the first place. Generally, a case can be made that gives an argument for bias or discrimination in an alternative framework, for example, differential admission quotas or differing values on a life. A more recent study similar to Radelet (1981) is from the *New York Times*, April 20, 2001, reported in a short article by Fox Butterfield, "Victims' Race Affects Decisions on Killers' Sentence, Study Finds."

Although not explicitly a Simpson's Paradox context, there are similar situations that appear in various forms of multifactor analysis of variance that raise cautions about aggregation phenomena. The simplest dictum is that "you cannot interpret main effects in the presence of interaction." This admonition is usually softened when the interaction is not disordinal, and the graphs of means don't actually cross. In these instances it may be possible to eliminate the interaction by some relatively simple transformation of the data, and produce an "additive" model. Because of this, noncrossing interactions might be considered "unimportant." Similarly, the absence of parallel profiles (that is, when interaction is present) may hinder the other tests for the main effects of coincident and horizontal profiles. Possibly, if the profiles show only an "unimportant" interaction, such evaluations could proceed.

[3]Standardization, by normalizing to a given mixture on a variable (for example, race), appears in several different guises throughout this book. For example, in the chapter on (mis)reporting of data, there is a subsection on adjusting groups that were not comparable on some variable (such as age). Also, and although the mechanics of analysis of covariance have not been discussed explicitly, the method can be seen as equating groups to have the "same average value" on the covariate. This equating is carried out by first assuming that all within-group slopes of the regressions of the dependent measure on the covariate are the same. Intuitively, groups are then equated by "sliding" them along their regression lines to achieve a common covariate value.

[4]Fienberg (1988, p. 40) discusses an interesting example of Simpson's Paradox as it occurred in a court case involving alleged racial employment dis-

crimination in the receipt of promotions. In this instance, blacks were being "under-promoted" in virtually every pay grade, but because of the differing numbers of blacks and whites in the various grades, blacks appeared to be "over-promoted" in the aggregate. As always, before an overall conclusion is reached based on data that have been aggregated over a variable (such as pay grade), it is always wise to "look under the hood."

[5]We give the section of Yule's text that discusses Simpson's Paradox, but obviously without the name:

> Misleading associations ... may easily arise through the mingling of records ... which a careful worker would keep distinct.
>
> Take the following case, for example. Suppose there have been 200 patients in a hospital, 100 males and 100 females, suffering from some disease. Suppose further, that the death-rate for males (the case mortality) has been 30 per cent, for females 60 per cent. A new treatment is tried on 80 per cent of the males and 40 per cent of the females, and the results published without distinction of sex. The three attributes, with the relations of which we are here concerned, are *death*, *treatment*, and *male sex*. The data show that more males were treated than females, and more females died than males; there-fore the first attribute is associated negatively, the second positively, with the third. It follows that there will be an illusory negative association between the first two—*death* and *treatment*. If the treatment were completely inefficient we should, in fact, have the following results:

	Males	Females	Total
Treated and died	24	24	48
Treated and did not die	56	16	72
Not treated and died	6	36	42
Not treated and did not die	14	24	38

> i.e., of the treated, only $48/12 = 40$ per cent died, while of those not treated $42/80 = 52.5$ per cent died. If this result were stated without any reference to the fact of the mixture of the sexes, to the different proportions of the two that were treated and to the different death-rates under normal treatment, then some value in the new treatment would appear to be suggested. To make a fair return, either the results for the two sexes should be stated separately, or the same proportion of the two sexes must receive the experimental treatment. Further, care would have to be taken in such a case to see that there was no selection (perhaps unconscious) of the less severe cases for treatment, thus introducing another source of fallacy (*death* positively associated with *severity*, *treatment* negatively associated with *severity*, giving rise to illusory negative association between *treatment* and *death*). (pp. 49–50)

[6]See C.4.1 on the ubiquity of Simpson's Paradox in day-to-day reporting of economic statistics; this article is by Cari Tuna, *Wall Street Journal* (December 2, 2009), "When Combined Data Reveal the Flaw of Averages."

Chapter 13

Meta-Analysis

> Now that we have all this useful information, it would be nice to do
> something with it. (Actually, it can be emotionally fulfilling just to
> get the information. This is usually only true, however, if you have
> the social life of a kumquat.)
> — *UNIX Programmer's Manual*

In his Presidential Address to the American Educational Research
Association, Glass (1976) introduced the idea of "meta-analysis," referring to a statistical integration of a set of studies
that ask a common research question. The title of his address, "Primary, Secondary, and Meta-Analysis of Research," distinguishes
three types of data analysis.[1] A *primary analysis* is the initial data
analysis for an original research study. A *secondary analysis* is a
reexamination of an existing dataset, possibly with different statistical and/or interpretative tools than originally used or available.
Mosteller and Moynihan's (1972) reanalysis of the Coleman (1966)
data on equality of educational opportunity is a famous example of
a secondary data analysis. Finally, a *meta-analysis* combines the
analyses (both primary and secondary) for a number of studies
into a coherent statistical review. This is in contrast to the more
usual discursive literature review that was common up to the time
of Glass's address.

The three decades that followed Glass's introduction of meta-analysis has seen an explosion of such studies published in journals in the behavioral and social sciences, education, health, and
medicine. In the behavioral sciences, meta-analyses appear regularly in the field's premier journals (e.g., *Psychological Bulletin*);
for medical- and health-related topics, we now have the extensive internationally organized Cochrane Collaboration, founded
in 1993. The handbook produced by this latter consortium, the
Cochrane Handbook for Systematic Reviews of Interventions (Higgins & Green, 2008), although designed for researchers in medicine,

is also useful in other areas. The Campbell Collaboration, an organization similar to the Cochrane Collaboration, was founded in 1999 and dedicated to Donald Campbell. It is devoted to systematic reviews of interventions in the social, behavioral, and educational areas. Many of these reviews are now published electronically in the free online journal, *Campbell Systematic Reviews*.[2]

A meta-analysis involves the use of a common measure of "effect size" that is then aggregated over studies to give an "average" effect. The heterogeneity in the individual effects can be used to generate confidence intervals for an assumed population effect parameter, or related to various study characteristics to assess why effect sizes might systematically vary apart from the inherent random variation present within any single study. The common effect measures for dichotomous outcomes are odds ratios, risk ratios, or risk differences. For continuous data, there are Pearson correlation coefficients or Cohen effect sizes defined by between-group mean differences divided by within-group standard deviations.

The use of meta-analysis has been involved in several publicly controversial cases in the recent past. Probably the most sensationalized was a meta-analysis published in *Psychological Bulletin* (1998, *124*, 22–53) by Rind, Tromovitch, and Bauserman entitled "A Meta-Analytic Examination of Assumed Properties of Child Sexual Abuse Using College Samples." The abstract follows:

Many lay persons and professionals believe that child sexual abuse (CSA) causes intense harm, regardless of gender, pervasively in the general population. The authors examined this belief by reviewing 59 studies based on college samples. Meta-analyses revealed that students with CSA were, on average, slightly less well adjusted than controls. However this poorer adjustment could not be attributed to CSA because family environment (FE) was consistently confounded with CSA. FE explained considerably more adjustment variance than CSA, and CSA-adjustment relations generally became nonsignificant when studies controlled for FE. Self-reported reactions to and effects from CSA indicated that negative effects were neither pervasive nor typically intense, and that men reacted much less negatively than women. The college data were completely consistent with data from national samples. Basic beliefs about CSA in the general population were not supported. (p. 22)

As might have been expected, this meta-analysis caused quite an uproar, including a unanimous condemnation resolution from Congress. We give an OpEd item from the *Los Angeles Times* by Carol Tavris ("The Politics of Sex Abuse," July 19, 1999):[3]

I guess I should be reassured to know that Congress disapproves of pedophilia and the sexual abuse of children. On July 12, the House voted unanimously to denounce a study that the resolution's sponsor, Matt Salmon (R-Ariz.), called "the emancipation proclamation of pedophiles." In a stunning display of scientific illiteracy and moral posturing, Congress misunderstood the message, so they condemned the messenger.

What got Congress riled was an article published last year in the journal *Psychological Bulletin*, which is to behavioral science what the *Journal of the American Medical Association* is to medicine. Articles must pass rigorous peer review, during which they are scrutinized for their methods, statistics and conclusions. The authors of the article—Bruce Rind, Philip Tromovitch and Robert Bauserman— statistically analyzed 59 studies, involving more than 37,000 men and women, on the effects of childhood sexual abuse on college students. (A previous paper reviewed studies of more than 12,000 adults in the general population.)

The findings, reported with meticulous detail and caution, are astonishing. The researchers found no overall link between childhood sexual abuse and later emotional disorders or unusual psychological problems in adulthood. Of course, some experiences, such as rape by a father, are more devastating than others, such as seeing a flasher in an alley. But the children most harmed by sexual abuse are those from terrible family environments, where abuse is one of many awful things they have to endure.

Perhaps the researchers' most inflammatory finding, however, was that not all experiences of child-adult sexual contact have equally emotional consequences nor can they be lumped together as "abuse." Being molested at the age of 5 is not comparable to choosing to have sex at 15. Indeed, the researchers found that two thirds of males who, as children or teenagers, had had sexual experiences with adults did not react negatively.

Shouldn't this be good news? Shouldn't we be glad to know which experiences are in fact traumatic for children, and which are not upsetting to them? Shouldn't we be pleased to get more evidence of the heartening resilience of children? And "more" evidence it is, for abundant research now shows that most people, over time, cope successfully with adversity—even war. Many not only survive, but find meaning and strength in the experience, discovering psychological resources they did not know they had.

But the fact that many people survive life's losses and cruelties is surely no endorsement of child abuse, rape, or war. A criminal act is still a criminal act, even if the victim eventually recovers. If I get over having been mugged, it's still illegal for someone to mug me, and if I recover from rape, my recovery should offer no mercy for rapists. If a child eventually recovers from molestation by an adult, pedophilia is still illegal and wrong. Moreover, the fact that many people recover on their own says nothing about the importance of promoting interventions that help those who cannot.

To conclude our discussion of this particular meta-analysis episode, we point the reader to a somewhat later article by Benedict Carey (*New York Times*, November 9, 2004), entitled "Long After Kinsey, Only the Brave Study Sex";[4] note the differing characterization of *Psychological Bulletin* as "an arcane journal" as compared to Tavris. If there is a summary observation to be made here, it might be that even meta-analyses cannot prevent moral outrage when the topics studied strike a particularly sensitive chord in some individuals.

One of the first demands in carrying out any meta-analysis is a preliminary selection of the studies to be included, and in turn, who exactly are the individuals to be studied. Usually, this is implemented by an explicit set of exclusionary and inclusionary criteria. Given this selection, it is then obviously important to temper one's overall conclusion to what studies were actually integrated. Some difficulty with carrying out this admonition occurs whenever the cable news networks demand fodder to fill their airtime. Witness the recent controversy about the effectiveness of antidepressants as judged by a particular meta-analysis published in the *Journal of the American Medical Association* (Fournier, et al., 2010). We give a beginning excerpt from an OpEd item by Judith Warner (*New York Times*, January 8, 2010), entitled "The Wrong Story About Depression":[5]

> "STARTLING results," promised the CNN teasers, building anticipation for a segment on this week's big mental health news: a study led by researchers at the University of Pennsylvania indicating that the antidepressants Paxil and imipramine work no better than placebos ("than sugar pills," said CNN) for people with mild to moderate depression.
>
> Happy pills don't work, the story quickly became, even though, boiled down to that headline, it was neither startling nor particularly true.

It sounded true. After all, any number of experts have argued that antidepressants—and selective serotonin reuptake inhibitors like Paxil in particular—are overhyped and oversold. And after years of hearing about shady practices within the pharmaceutical industry, and of psychiatrists who enrich themselves in the shadows by helping the industry market its drugs, we are primed to believe stories of psychiatric trickery.

Yet in all the excitement about "startling" news and "sugar pills," a more nuanced and truer story about mental health care in America was all but lost.

That story begins to take shape when you consider what the new study actually said: Antidepressants do work for very severely depressed people, as well as for those whose mild depression is chronic. However, the researchers found, the pills don't work for people who aren't really depressed—people with short-term, minor depression whose problems tend to get better on their own. For many of them, it's often been observed, merely participating in a drug trial (with its accompanying conversation, education and emphasis on self-care) can be anti-depressant enough.

To ignore the acute or chronic nature of the depression for those individuals included in the studies evaluated, and more generally, to ignore who the subjects actually are in a statistical integration, can turn meta-analysis into an unethically motivated strategy of data analysis and persuasion.

The practice of meta-analysis brings with it the possibility of imposing some order in a possibly less than completely organized research area. First, apparent conflicts may be resolvable; then, variation in our measures of effect may be related to the characteristics of the amassed studies (though what is referred to as meta-regression, or better, meta-multiple-regression). In analogy to the distinction between fixed and random effects analysis of variance, it is even possible to change the inferential goal from a specific set of studies (that is, the fixed-effects notion), to the universe of possible studies (that is, the random-effects notion). In all of this, and because of the increase in overall sample size by the process of aggregating studies, we have greater precision of estimation for the overall effect.

As might be expected with any useful statistical method, meta-analysis also has some problems. Several of these are noted below:

Publication bias: In conducting a meta-analysis, the goal is to include all relevant studies that pass the set of inclusionary criteria, whatever they may be. This can involve finding or considering studies that may, for whatever reason, remain unpublished. Folk wisdom and/or experience tells us that negative results and those that are just "nonsignificant" might not be publishable because of suppression, say, by Big Pharma (or Big Tobacco), or the inherent reluctance of editors to use valuable pages to publish noninteresting (that is, statistically nonsignificant) results. The tendency to see a larger than what might be expected proportion of significant results in the published literature is *prima facie* evidence for a publication bias toward statistically significant results ("Publication bias," 2010; Rosenthal, 1979).

Both the file-drawer problem (characterized by negative or nonsignificant results being tucked away in a cabinet), and bias resulting from explicit decisions not to place studies in the public domain are problematic. If present, these tendencies would falsely inflate the effect estimates; thus, various ways of detection (and hopefully, correction) have been proposed. One relatively simple diagnostic is labeled a "funnel plot," where effect magnitudes (on the horizontal axis) are plotted against some measure of sample size (or an estimate of precision that depends on sample size) on the vertical axis. If sample size is unrelated to the true magnitude of effect, as it should be, and there is no publication bias, the plot should resemble a funnel where estimates associated with the smaller sample sizes spread out over a wider range at the bottom of the funnel. The degree to which symmetry is not present around some vertical line through the funnel may indicate that publication bias has occurred (that is, for the smaller sample sizes, there is an asymmetry in that there are more published results than might be expected). Again, one can be luckier for small samples in getting a spuriously significant and larger effect estimate than for larger sample sizes (this is sometimes discussed under the rubric of "small-study effects").

The type of unpublished literature alluded to here has led to another issue in meta-analysis and elsewhere, in how to deal generally with "gray literature," or material that has not been subjected to the usual standards of peer review. The major issue with gray literature is a lack of vetting, which can be a major problem when biased, wrong, fraudulent, and so on. Witness the recent case of the

IPCC Climate Change Report (2007) merely lifting the Himalayan glacier melt claim from an unsubstantiated news source, and then one editor's decision not to remove it because of its potential for dramatic persuasion (Raloff, January 19, 2010).[6]

Study inclusion: Any meta-analysis requires a decision as to what studies to include. Gene Glass's original idea was to be very inclusive and to "bring on all comers." If effects varied widely, these could then be related to study characteristics (for example, randomized or not, study settings, general age and sex distribution of the subjects). Other entities (such as the Cochrane Collaboration) emphasize the need to be very selective, and to include only randomized clinical trials. The inclusion of nonrandomized observational studies, they might argue, just moves the bias up to a different level when subjects partially self-select into the various treatment groups. But even when restricted to randomized clinical trials, bias can creep in by some of the mechanisms we mention later in Chapter 15, Experimental Design and the Collection of Data. In general, we must be ever vigilant to the effects of confirmation bias, where we decide, possibly without really knowing that we are doing so, on what the "truth" should be before we begin to amass our studies. If the inclusionary criteria are set to show what we know should be there (or, to set the exclusionary criteria to eliminate any inconvenient nonconforming studies), the "truth" is being constructed and not discovered. For one of the better recent discussions of the operation of confirmation bias in its many guises, we recommend the literate discussion in Tavris and Aronson's *Mistakes Were Made (But Not By Me)* (2007). The subtitle gives a sense of the book's message: *Why We Justify Foolish Beliefs, Bad Decisions, and Hurtful Acts.*

The need to be explicit about the inclusionary criteria for studies in a meta-analysis may be obvious, but some of the resulting interpretative differences can run very deep indeed. There are problems inherent in the variation of who the subjects are in the various studies, how the treatments might vary, what type and form the measures take that are used to evaluate the treatments, and so on. As an example of what difficulties can happen, a study reported in the *New England Journal of Medicine* some years ago raised quite a sizable kerfuffle about these issues: "Discrepancies Between Meta-Analyses and Subsequent Large, Randomized Controlled Trials" (LeLorier, Grégoire, Benhaddad, Lapierre, & Derderian; *New England Journal of Medicine*, 1997, *337*, 536–542).[7] As another example, the meta-analysis study on depression referred to earlier is a good illustration of the need to always interpret one's results in terms of who exactly is being treated and what the results really then do say.

The "apples and oranges" criticism: One commonly heard criticism of meta-analysis is that you can't meaningfully mix "apples and oranges" together, but that's exactly what a meta-analysis tries to do. On the face of it such a complaint may sound reasonable, but more often it is a red herring. Think about how the practice of modern statistics proceeds. In a multiple regression that attempts to predict some output measure from a collection of predictor variables, the latter can be on any scale whatsoever (that is, differing means and variances, for example). If that bothers the sensibilities of the analyst, everything can be reduced to z-scores, and the results (re)interpreted in terms of all variables now being forced to be commensurable (with means of zero and variances of 1).

Typically, it is not necessary in a meta-analysis to move to this level of enforced z-score commensurability. If we are measuring behavior change, for example, with a number of manifestly different observed measures that may vary from study to study, then as long as a strong common factor underlies the various measures (in the traditional Spearman sense), it is reasonable to normalize the measures and include them in a meta-analysis. If the observed measures are more factorially complex, a "multivariate meta-analysis" might make sense using several normalized measures that tap the distinct domains (in the tradition of Thurstone [1935] group factors).

Individual differences: When our cognitively oriented colleagues recruit subjects for their studies, they are usually not very picky. Subjects are more or less interchangeable for the processes they are interested in, and all are equally representative of what is being studied. Some of our other colleagues with more of an individual differences emphasis (for example, those in the industrial/organization, social/personality, and developmental fields), are typically more concerned with variety and diversity because what is being studied is probably related to who the subjects are, such as age, education, political leanings, attitudes more generally. And commonly, that is part of what is being studied.

Meta-analysis in its usual vanilla form is not concerned with individual differences to any great extent. What is analyzed are averages at a group level. But averages cannot do justice to what occurs internally within an intervention or study. Who benefits or doesn't? A zero overall effect could result from and mask the situation in which some do great, some do badly, and many just don't change at all.[8] We give a quotation from Lee Cronbach (1982) to this effect:[9]

> [S]ome of our colleagues are beginning to sound like a kind of Flat Earth Society. They tell us that the world is essentially simple: most social phenomena are adequately described by linear relations; one-parameter scaling can discover coherent variables independent of culture and population; and inconsistencies among studies of the same kind will vanish if we but amalgamate a sufficient number of studies. ... The Flat Earth folk seek to bury any complex hypothesis with an empirical bulldozer." (p. 70)

Each of us wants decisions made by doctors to be based on what is best for the unique "me." It is somewhat irrelevant that such and such of an average change occurred for a group administered this particular regimen if such a change did not occur for the likes of "us" in this group. What would seem to be required are data on a gamut of individuals varying over a variety of conditions and demographics, and with reasonably good sample sizes for those subgroups that would be our particular niches. To paraphrase an old political observation: "what is good for the country may not be good for me." (In fact, given the ubiquity of Simpson's Paradox, it might not be good for anyone.)

One of the reasons that meta-analysis was necessary in the first place was the burgeoning number of studies available on any number of topics but where the summary data were available (if at all) in only extremely cryptic forms. In fact, at one time, only p-values were given (and possibly without even a mention of what tests were actually done); any descriptive summary information was suppressed by editors looking for some economy in the number of pages published. There was never the possibility of giving both a table and a figure for the same data, even though they might be extremely valuable for subsequent secondary or meta-analyses. But now, given the Internet and cheap data-storage availabilities for supplementary and primary information, it may be time to impose the gold standard in terms of data availability—individual subject (or patient) data. Also, we should make sure that the subject or patient characteristics relevant to outcomes are available within the same archive.

In an interview with Daniel Robinson in the *Educational Researcher* (*33*(3), 2004), Gene Glass provided the following pertinent comment:

> That's why our biggest challenge is to tame the wild variation in our findings not by decreeing this or that set of standard protocols but by describing and accounting for the variability in our findings. The result of a meta-analysis should never be an average; it should be a graph. (p. 29)

Vote-counting meta-analysis: One strategy suggested for reconciling possibly conflicting studies is to count the number of times that led to a rejection/acceptance of the null hypothesis (at say, the .05 level) over all available studies. Then, according to some variation on majority rule, a conclusion of "effect" or "no effect" is made. There are several problematic aspects to this strategy: the sample sizes for the individual studies are a primary determinant of significance; the actual size of an effect plays a more secondary role when it should be utmost; even if we want to rely on p-values to make a conclusion, there are much better ways of aggregating the p-values over studies to get one such overall p-value. For a cautionary tale about the inadvisability of vote-counting methods, we suggest the article by Hedges, Laine, and Greenwald in the *Educational Researcher* (1994, *23*(3), 5–14): "An Exchange: Part I: Does Money Matter? A Meta-Analysis of Studies of the Effects of Differential School Inputs on Student Outcomes." An earlier article by Hanushek (1989) had used vote counting to conclude that "there is no strong or systematic relationship between school expenditures and student performance" (p. 45). Hedges et al. come to a very different conclusion using better meta-analysis techniques on exactly the same set of studies.

Notes

[1]The Gene Glass Presidential Address is listed in the Suggested Reading, C.5.4.

[2]A particularly good recent and brief survey of the plethora of meta-analysis resources oriented toward psychology is the review chapter (aptly entitled "Meta-Analysis") by A. P. Field (2009), in *The Sage Handbook of Quantitative Methods in Psychology*. A new journal, *Research Synthesis Methods*, sponsored by the Society for Research Synthesis Methodology, was introduced by Wiley in 2010. This outlet has the interdisciplinary goal of following work on all facets of research synthesis of the type represented by both the Cochrane and Campbell Collaborations.

[3]The Carol Tavris OpEd article is also listed in the Suggested Reading, A.4.7; it is given here in its entirety with the latter part given immediately below:

> The article by Rind and his colleagues, however, has upset two powerful constituencies: religious fundamentalists and other conservatives who think this research endorses pedophilia and homosexuality, and psychotherapists who believe that all sexual experiences in childhood inevitably cause lifelong psychological harm. These groups learned about the research last December, when the National Association for the Research and Therapy of Homosexuality, or NARTH, posted an attack on the paper on their Web site.
>
> NARTH endorses the long-discredited psychoanalytic notion that homosexuality is a mental disorder and that it is a result of seduction in childhood by an adult. Thus NARTH was exercised by the study's findings that most boys are not traumatized for life by experiences with older men (or women) and that these experiences do not "turn them" into homosexuals.
>
> NARTH's indictment of the article was picked up by right-wing magazines, organizations and radio talk-show hosts, notably Laura Schlessinger. They in turn contacted allies in Congress, and soon the study was being used as evidence of the liberal agenda to put a pedophile in every home, promote homosexuality and undermine "family values."
>
> The conservatives found further support from a group of clinicians who still maintain that childhood sexual abuse causes "multiple personality disorder" and "repressed memories." These ideas have been as discredited by research as the belief that homosexuality is a mental illness or a chosen "lifestyle," but their promulgators cannot let them go. These clinicians want to kill the Rind study because they fear that it will be used to support malpractice claims against

their fellow therapists. And, like their right-wing allies, they claim the article will be used to protect pedophiles in court.

But all scientific research, on any subject, can be used wisely or stupidly. For clinicians to use the "exoneration of pedophiles" argument to try to suppress this article's important findings, and to smear the article's authors by impugning their scholarship and motives, is particularly reprehensible. They should know better. The Bible can be used wisely or stupidly, too.

And so the American Psychological Association (the journal's publisher) has been under constant attack by the Christian Coalition, Republican congressmen, panicked citizens, radio talk-show hosts and a consortium of clinicians that reads like a "Who's Who" in the multiple personality disorder and repressed-memories business. The APA has responded that future articles on sensitive subjects will be more carefully considered for their "public policy implications" and that the article would be re-reviewed by independent scholars. It assured Congress that "the sexual abuse of children is a criminal act that is reprehensible in any context."

These placatory gestures are understandable given the ferocity of the attacks. But the APA missed its chance to educate the public and Congress about the scientific method, the purpose of peer review, and the absolute necessity of protecting the right of its scientists to publish unpopular findings. Researchers cannot function if they have to censor themselves according to potential public outcry or are silenced by social pressure, harassment, or political posturing from those who misunderstand or disapprove of their results.

On emotionally sensitive topics such as sex, children and trauma, we need all the clear-headed information we can get. We need to understand what makes most people resilient, and how to help those who are not. We need to understand a whole lot more about sexuality, including children's sexuality. Congress and clinicians may feel a spasm of righteousness by condemning scientific findings they dislike. But their actions will do no more to reduce the actual abuse of children than posting the Ten Commandments in schools will improve children's morality.

[4]The Benedict Carey article is listed in the Suggested Reading, Section A.4.10.

[5]Section C.5.6 lists three articles on this antidepressant episode, including the complete Judith Warner OpEd contribution. Of the other two, one is by Benedict Carey reporting on the original meta-analysis ("Popular Drugs May Benefit Only Severe Depression," *New York Times*, January 6, 2010); the third is by a medical doctor, Richard A. Friedman, who, on the basis of the results of the meta-analysis, cautions individuals against discontinuing use of

antidepressants ("Before You Quit Antidepressants ... " *New York Times*, January 11, 2010).

[6]The problems generated from various forms of publication bias are legion. One recent and readable exposé appeared in the *New Yorker* by Jonah Lehrer (December 13, 2010), "The Truth Wears Off: Is There Something Wrong With the Scientific Method?" The basic issue is that some supposed "big result," upon replication, generally declines. In fact, this phenomenon appears to be so universal it has now been labeled the "decline effect" (by Lehrer and others). One possible explanation (among several) goes something like this: the "true effect" of whatever is under study is only modest, so various studies of efficacy show effects too small to be acceptable for publication. Then one researcher happens, by chance, to obtain a big enough result to get published in a top-tier journal. All the previous studies now come out of the file drawers are submitted for publication and are successful because they represent replications of the "big observed effect." The dates of the accepted replications actually predate the focal study and usually appear in second-tier journals. Assiduous reviewers of the literature soon discover that the size of the effect has declined.

Such a plausible explanation for the decline effect, however, does not apply in all instances where publication bias appears to be present. A *PloS Medicine* article (2005) by John Ioannidis and colleagues (Pan et al., 2005), "Local Literature Bias in Genetic Epidemiology: An Empirical Evaluation of the Chinese Literature," reports on a meta-analysis of the replications for a number of such "big effect" studies appearing in the Chinese literature. The Chinese medical literature is essentially invisible to Western researchers but not conversely. Chinese researchers are well aware of the Western literature, at least that published in the top-tier journals. So, one might expect the same decline effect regularly seen in the Western literature. But no! The same (or larger) effects are found as in the original "big effect" studies. Something very odd appears to be going on, to say the least

[7]Part of the controversy surrounding this article is reviewed in the Suggested Reading—there is a precipitating *New England Journal of Medicine* editorial (C.5.2) by John C. Bailar III, "The Promise and Problems of Meta-Analysis" (1997); several letters to the editor and rejoinders are given in C.5.3 that were published at various times throughout 1998.

[8]Also, see Sam Savage (2009) *The Flaw of Averages* for many business examples of dubious inference based on just averages. Savage suggests a reliance of distributions instead of just means, and provides software to put this recommendation into effect.

[9]One is also reminded of the quotation from Ernest Rutherford: "The only possible interpretation of any research whatever in the 'social sciences' is: some do, some don't." The entertaining online Gene Glass article ("Meta-Analysis at 25") is listed in the Suggested Reading, C.5.5.

Chapter 14

Statistical Sleuthing and Explanation

> My mother made me a scientist without ever intending to. Every Jewish mother in Brooklyn would ask her child after school: 'So? Did you learn anything today?' But not my mother. She always asked me a different question. 'Izzy,' she would say, 'did you ask a good question today?'
> — Isidor Tabi (Nobel Prize in Physics, 1944; quotation given by John Barell in *Developing More Curious Minds*, 2003)

Modern statistics is often divided into two parts: exploratory and confirmatory. Confirmatory methods were developed over the first half of the 20th century, principally by Karl Pearson and Ronald Fisher. This was, and remains, a remarkable intellectual accomplishment. The goal of confirmatory methods is largely judicial: they are used to weigh evidence and make decisions. The aim of exploratory methods is different. They are useful in what could be seen as detective work; data are gathered and clues are sought to enable us to learn what might have happened. Exploratory analysis generates the hypotheses that are tested by the confirmatory methods. Surprisingly, the codification, and indeed the naming of exploratory data analysis, came after the principal work on the development of confirmatory methods was complete. John Tukey's (1977) influential book changed everything. He taught us that we should understand what might be true before we learn how well we have measured it.

The focus so far in this book has been mainly on confirmatory methods, and how they can be used ethically. This chapter shifts to the direction of exploratory methods. It is well beyond our scope to provide even a summary of all the methods that are available. Instead, we shall choose a very few to discuss briefly and point the way to some secondary sources. Tukey's opus is the place to start, but there are many other authors whose prose might be less idiosyncratic. A selection of sources is listed in the endnote to this paragraph.[1]

Last, a brief warning. Exploration is where the action is. It is more exciting than confirmation and much more fun. More is written about the lives of detectives than of judges: we hear more often of the exploits of Sherlock Holmes than of Oliver Wendell. It is easy to be seduced by the fun side—to look at a large dataset and, by using the powerful sets of exploratory tools now available, generate convincing arguments about what happened. But here is where the discipline of confirmation is critical. And nothing is more important in trying to decide about whether what has been uncovered is real than that touchstone of scientific validity, reproducibility. If, after sifting through a huge morass of data, you uncover a delicate jewel, you must now look in a previously unexplored dataset and see if your finding is confirmed. If it is not, what you found is likely due to chance.

14.1   Sleuthing Interests and Basic Tools

Some of the more enjoyable intellectual activities statisticians engage in might be called *statistical sleuthing*—the use of various statistical techniques and methods to help explain or "tell the story" about some given situation. We first give a flavor of several areas where such sleuthing has been of explanatory assistance:

(a) The irregularities encountered in Florida during the 2000 Presidential election and why; see, for example, Alan Agresti and Brett Presnell, "Misvotes, Undervotes, and Overvotes: The 2000 Presidential Election in Florida" (*Statistical Science*, *17*, 2002, 436–440).

(b) The attribution of authorship for various primary sources; for example, we have the seminal work by Mosteller and Wallace (1964) on the disputed authorship of some of the Federalist Papers.[2]

(c) Searching for causal factors and situations that might influence disease onset; for example, "Statistical Sleuthing During

Epidemics: Maternal Influenza and Schizophrenia" (Nicholas J. Horton & Emily C. Shapiro, *Chance*, *18*(1), 2005, 11–18);

(d) Evidence of cheating and corruption, such as the Justin Wolfers (2006) article on point shaving in NCAA basketball as it pertains to the use of Las Vegas point spreads in betting (but, also see the more recent article by Bernhardt and Heston [2010] disputing Wolfers' conclusions);

(e) The observations of Quetelet's from the middle 1800s that based on the very close normal distribution approximations for human characteristics, there were systematic understatements of height (to below 5 feet, 2 inches) for French conscripts wishing to avoid the minimum height requirement needed to be drafted (Stigler, 1986, pp. 215–216);

(f) Defending someone against an accusation of cheating on a high-stakes exam when the "cheating" was identified by a "cold-hit" process of culling for coincidences, and with subsequent evidence provided by a selective search (that is, a confirmation bias). A defense that a false positive has probably occurred requires a little knowledge of Bayes' theorem and the positive predictive value.[3]

(g) Demonstrating the reasonableness of results that seem "too good to be true" without needing an explanation of fraud or misconduct. An exemplar of this kind of argumentation is in the article, "A Little Ignorance: How Statistics Rescued a Damsel in Distress" (Peter Baldwin and Howard Wainer, *Chance*, 2009, *22*(3), 51–55).[4]

A variety of sleuthing approaches are available to help explain what might be occurring over a variety of different contexts. Some of these have been introduced already: Simpson's Paradox, Bayes' rule and baserates, bounds provided by corrections for attenuation, regression toward the mean, the effects of culling on the identification of false positives and the subsequent inability to cross-validate, the ecological fallacy, the operation of randomness and the difficulty in "faking" such a process, confusions caused by misinterpreting conditional probabilities, illusory correlations, restrictions of range for correlations, and so on. We mention a few other tools below that may provide some additional assistance: the use of various discrete probability distributions, such as the binomial, Poisson, or those for runs, in constructing convincing explanations for some phenomena; the digit regularities suggested by Benford's law (Benford, 1938); a reconception of some odd probability problems by considering pairs (what might be labeled as the "the birthday probability model" [Mindinow, 2009, p. 64]); and the use of the statistical techniques in survival analysis to model time-to-event processes.[5]

The simplest probability distribution has only two event classes (for example, success/fail, live/die, head/tail, 1/0). A process that follows such a distribution is called Bernoulli; typically, our concern is with repeated and independent Bernoulli trials. Using an interpretation of the two event classes of heads (H) and tails (T), assume $P(H) = p$ and $P(T) = 1 - p$, with p being invariant over repeated trials (that is, the process is stationary). The probability of any sequence of size n that contains k heads and $n - k$ tails is $p^k(1 - p)^{n-k}$. Commonly, our interest is in the distribution of the number of heads (say, X) seen in the n independent trials. This random variable follows the binomial distribution:

$$P(X = r) = \binom{n}{r} p^r (1 - p)^{n-r} \,,$$

where $0 \leq r \leq n$, and $\binom{n}{r}$ is the binomial coefficient:

$$\binom{n}{r} = \frac{n!}{(n - r)! r!} \,,$$

using the standard factorial notation.

Both the binomial distribution and the underlying repeated Bernoulli process offer useful background models against which to compare observed data, and to evaluate whether a stationary Bernoulli process could have been responsible for its generation. For example, suppose a Bernoulli process produces a sequence of size n with r heads and $n - r$ tails. All arrangements of the r Hs and $n - r$ Ts should be equally likely (cutting, say, various sequences of size n all having r Hs and $n - r$ Ts from a much longer process); if not, possibly the process is not stationary or the assumption of independence is inappropriate. A similar use of the binomial would first estimate p from the long sequence, and then use this value to find the expected number of heads in sequences of a smaller size n; a long sequence could be partitioned into segments of this size and the observed number of heads compared to what would be expected. Again, a lack of fit between the observed and expected might suggest lack of stationarity or trial dependence (a more formal assessment of fit could be based on the usual chi-square goodness-of-fit test).

A number of different discrete distributions prove useful in statistical sleuthing. We mention two others here, the Poisson and a distribution for the number of runs in a sequence. A discrete random variable, X, that can take on values 0, 1, 2, 3, ... , follows a Poisson distribution if

$$P(X = r) = \frac{e^{-\lambda}\lambda^r}{r!} ,$$

where λ is an intensity parameter, and r can take on any integer value from 0 onward. Although a Poisson distribution is usually considered a good way to model the number of occurrences for rare events, it also provides a model for spatial randomness as the example adapted from Feller (1968, Vol. 1, pp. 160–161) illustrates:

Flying-bomb hits on London. As an example of a spatial distribution of random points, consider the statistics of flying-bomb hits in the south of London during World War II. The entire area is divided into 576 small areas of 1/4 square kilometers each. Table 14.1 records the number of areas with exactly k hits. The total number of hits is 537, so the average is .93 (giving an estimate for the intensity parameter, λ). The fit of the Poisson distribution is surprisingly good. As judged by the χ^2-criterion, under ideal conditions, some 88 per cent of comparable observations should show a worse agreement. It is interesting to note that most people believed in a tendency of the points of impact to cluster. If this were true, there would be a higher frequency of areas with either many hits or no hits and a deficiency in the intermediate classes. Table 14.1 indicates a randomness and homogeneity of the area, and therefore, we have an instructive illustration of the established fact that to the untrained eye, randomness appears as regularity or tendency to cluster (the appearance of this regularity in such a random process is sometimes referred to as "Poisson clumping").[6]

To develop a distribution for the number of runs in a sequence, suppose we begin with two different kinds of objects (say, white (W) and black (B) balls) arranged randomly in a line. We count the number of runs, R, defined by consecutive sequences of all Ws

TABLE 14.1: Flying-bomb hits on London.

Number of hits	0	1	2	3	4	5 or more
Number of areas	229	211	93	35	7	1
Expected number	226.74	211.39	98.54	30.62	7.14	1.57

or all Bs (including sequences of size 1). If there are n_1 W balls and n_2 B balls, the distribution for R under randomness can be constructed (e.g., Mood and Graybill, 1963, pp. 410–412). We note the expectation and variance of R, and the normal approximation:

$$E(R) = \frac{2n_1 n_2}{n_1 + n_2} + 1 \; ;$$

$$V(R) = \frac{2n_1 n_2 (2n_1 n_2 - n_1 - n_2)}{(n_1 + n_2)^2 (n_1 + n_2 - 1)} \; ;$$

and

$$\frac{R - E(R)}{\sqrt{V(R)}}$$

is approximately (standard) normal with mean zero and variance one. Based on this latter distributional approximation, an assessment can be made as to the randomness of the process that produced the sequence, and whether there are too many or too few runs for the continued credibility that the process is random. Run statistics have proved especially important in monitoring quality control in manufacturing, but these same ideas could be useful in a variety of statistical sleuthing tasks.

Besides the use of formal probability distributions, there are other related ideas that might be of value in the detection of fraud or other anomalies. One such notion, called Benford's law, has captured some popular attention; for example, see the article by Malcolm W. Browne, "Following Benford's Law, or Looking Out for No. 1" (*New York Times*, August 4, 1998).[7] Benford's law gives a "probability distribution" for the first digits (1 to 9) found for many (naturally) occurring sets of numbers. If the digits in some collection (such as tax returns, campaign finances, (Iranian) election results, or company audits) do not follow this distribution, there is a *prima facie* indication of fraud.[8]

Benford's law gives a discrete probability distribution over the digits 1 to 9 according to:

$$P(X = r) = \log_{10}(1 + \frac{1}{r}) \, ,$$

for $1 \leq r \leq 9$. Numerically, we have the following:

r	Probability	r	Probability
1	.301	6	.067
2	.176	7	.058
3	.125	8	.051
4	.097	9	.046
5	.079		

Although there may be many examples of using Benford's law for detecting various monetary irregularities, one of the most recent applications is to election fraud, such as in the 2009 Iranian Presidential decision. A recent popular account of this type of sleuthing is Carl Bialik's article, "Rise and Flaw of Internet Election-Fraud Hunters" (*Wall Street Journal*, July 1, 2009).[9] It is always prudent to remember, however, that heuristics, such as Benford's law and other digit regularities, might point to a potentially anomalous situation that should be studied further, but violations of these presumed regularities should never be considered definitive "proof."

Another helpful explanatory probability result is commonly referred to as the "birthday problem" (Mlodinow, 2009, p. 64): what is the probability that in a room of n people, at least one pair of individuals will have the same birthday. As an approximation, we have $1 - e^{-n^2/(2 \times 365)}$; for example, when $k = 23$, the probability is .507; when $k = 30$, it is .706. These surprisingly large probability values result from the need to consider matchings over all pairs of individuals in the room; that is, there are $\binom{n}{2}$ chances to consider for a matching, and these inflate the probability beyond what we might intuitively expect. We give an example from Leonard Mlodinow's book, *The Drunkard's Walk* (2009):

> Another lottery mystery that raised many eyebrows occurred in Germany on June 21, 1995. The freak event happened in a lottery named Lotto 6/49, which means that the winning six numbers are drawn

from the numbers 1 to 49. On the day in question the winning numbers were 15-25-27-30-42-48. The very same sequence had been drawn previously, on December 20, 1986. It was the first time in 3,016 drawings that a winning sequence had been repeated. What were the chances of that? Not as bad as you'd think. When you do the math, the chance of a repeat at some point over the years comes out to around 28 per cent. (p. 65)

14.2 Survival Analysis

The reason that the term "censored" is used is that in the pessimistic vocabulary of survival-analysis, life is a temporary phenomenon and someone who is alive is simply not dead yet. What the statistician would like to know is how long he or she lived but this information is not (yet) available and so is censored.
— Stephen Senn (*Dicing with Death*, 2003)

The area of statistics that models the time to the occurrence of an event, such as death or failure, is called *survival analysis*. Some of the questions survival analysis is concerned with include: what is the proportion of a population that will survive beyond a particular time; among the survivors, at what (hazard) rate will they die (or fail); how do the circumstances and characteristics of the population change the odds of survival; can multiple causes of death (or failure) be taken into account. The primary object of interest is the survival function, specifying the probability that time of death (the term to be used generically from now on), is later than some specified time. Formally, we define the survival function as: $S(t) = P(T > t)$, where t is some time, and T is a random variable denoting the time of death. The function must be nonincreasing, so: $S(u) \leq S(v)$, when $v \leq u$. This reflects the idea that survival to some later time requires survival at all earlier times as well.

The most common way to estimate $S(t)$ is through the now ubiquitous Kaplan–Meier (1958) estimator, which allows a certain (important) type of right-censoring of the data. This censoring is where the corresponding objects have either been lost to observation or their lifetimes are still ongoing when the data were analyzed. Explicitly, let the observed times of death for the N members under study be $t_1 \leq t_2 \leq \cdots \leq t_N$. Corresponding to each t_i is the number of members, n_i, "at risk" just prior to t_i; d_i is the number of deaths at time t_i. The Kaplan–Meier nonparametric maximum likelihood estimator, $\widehat{S(t)}$, is a product:

$$\widehat{S(t)} = \prod_{t_i \leq t}(1 - \frac{d_i}{n_i}) \ .$$

When there is no right-censoring, n_i is just the number of survivors prior to time t_i; otherwise, n_i is the number of survivors minus the number of censored cases (by that time t_i). Only those surviving cases are still being observed (that is, not yet censored), and thus at risk of death. The function $\widehat{S(t)}$ is a nonincreasing step function, with steps at $t_i, 1 \leq i \leq N$; it is also usual to indicate the censored observations with tick marks on the graph of $\widehat{S(t)}$.

The original Kaplan and Meir article that appeared in 1958 (Kaplan, E. L., & Meier, P., "Nonparametric Estimation From Incomplete Observations," *Journal of the American Statistical Association*, *53*, 457–481), is one of the most heavily cited articles in all of the sciences. It was featured as a "Citation Classic" in the June 13, 1983 issue of *Current Contents: Life Sciences*. As part of this recognition, Edward Kaplan wrote a short retrospective that we excerpt below:

> This paper began in 1952 when Paul Meier at Johns Hopkins University (now at the University of Chicago) encountered Greenwood's paper on the duration of cancer. A year later at Bell Telephone Laboratories I became interested in the lifetimes of vacuum tubes in the repeaters in telephone cables buried in the ocean. When I showed my manuscript to John W. Tukey, he informed me of Meier's work, which already was circulating among some of our colleagues. Both manuscripts were submitted to the *Journal of the American Statistical Association*, which recommended a joint paper. Much correspondence over four years was required to reconcile our differing approaches, and we were concerned that meanwhile someone else might publish the idea.
>
> The nonparametric estimate specifies a discrete distribution, with all the probability concentrated at a finite number of points, or else (for a large sample) an actuarial approximation thereto, giving the probability in each of a number of successive intervals. This paper

considers how such estimates are affected when some of the lifetimes are unavailable (censored) because the corresponding items have been lost to observation, or their lifetimes are still in progress when the data are analyzed. Such items cannot simply be ignored because they may tend to be longer-lived than the average. (p. 14)

To indicate the importance of the Kaplan–Meier estimator in sleuthing within the medical/pharmaceutical areas and elsewhere, we give the two opening paragraphs of Malcolm Gladwell's *New Yorker* article (May 17, 2010) entitled "The Treatment: Why Is It So Difficult to Develop Drugs for Cancer?":[10]

In the world of cancer research, there is something called a Kaplan–Meier curve, which tracks the health of patients in the trial of an experimental drug. In its simplest version, it consists of two lines. The first follows the patients in the "control arm," the second the patients in the "treatment arm." In most cases, those two lines are virtually identical. That is the sad fact of cancer research: nine times out of ten, there is no difference in survival between those who were given the new drug and those who were not. But every now and again—after millions of dollars have been spent, and tens of thousands of pages of data collected, and patients followed, and toxicological issues examined, and safety issues resolved, and manufacturing processes fine-tuned—the patients in the treatment arm will live longer than the patients in the control arm, and the two lines on the Kaplan–Meier will start to diverge.

Seven years ago, for example, a team from Genentech presented the results of a colorectal-cancer drug trial at the annual meeting of the American Society of Clinical Oncology—a conference attended by virtually every major cancer researcher in the world. The lead Genentech researcher took the audience through one slide after another—click, click, click—laying out the design and scope of the study, until he came to the crucial moment: the Kaplan–Meier. At that point, what he said became irrelevant. The members of the audience saw daylight between the two lines, for a patient population in which that almost never happened, and they leaped to their feet and gave him an ovation. Every drug researcher in the world dreams of standing in front of thousands of people at ASCO and clicking on a Kaplan–Meier like that. "It is why we are in this business," Safi Bahcall says. Once he thought that this dream would come true for him. It was in the late summer of 2006, and is among the greatest moments of his life. (p. 69)

A great deal of additional statistical material involving survival functions can be helpful in our sleuthing endeavors. Survival functions may be compared over samples (for example, the log-rank test), and generalized to accommodate different forms of censoring; the Kaplan–Meier estimator has a closed-form variance estimator (for example, the Greenwood formula); various survival models can incorporate a mechanism for including covariates (for example, the proportional hazard models introduced by Sir David Cox; see Cox and Oakes, 1984: *Analysis of Survival Data*). All of the usual commercial software (SAS, SPSS, SYSTAT) include modules for survival analysis. And, as might be expected, a plethora of cutting edge routines are in R, as well as in the Statistics Toolbox in MATLAB.

14.3 Statistical Sleuthing and the Imposition of the Death Penalty: McCleskey v. Kemp (1987)

> Those whom we would banish from society or from the human community itself often speak in too faint a voice to be heard above society's demand for punishment. It is the particular role of courts to hear these voices, for the Constitution declares that the majoritarian chorus may not alone dictate the conditions of social life.
>
> — Supreme Court Justice Brennan (dissenting in *McCleskey v. Kemp*)

The United States has had a troubled history with the imposition of the death penalty. Two amendments to the Constitution, the Eighth and the Fourteenth, operate as controlling guidelines for how death penalties are to be decided on and administered (if at all). The Eighth Amendment prevents "cruel and unusual punishment"; the Fourteenth Amendment contains the famous "equal protection" clause:

> No State shall make or enforce any law which shall abridge the privileges or immunities of citizens of the United States; nor shall any State deprive any person of life, liberty, or property, without due process of law; nor deny to any person within its jurisdiction the equal protection of the laws.

Various Supreme Court rulings over the years have relied on the Eighth Amendment to forbid some punishments entirely and to exclude others that are excessive in relation to the crime or the competence of the defendant. One of the more famous such rulings was in *Furman v. Georgia* (1972), which held that an arbitrary and inconsistent imposition of the death penalty violates both the Eighth and Fourteenth Amendments, and constitutes cruel and unusual punishment. This ruling lead to a moratorium on capital punishment throughout the United States that extended to 1976 when another Georgia case was decided in *Gregg v. Georgia* (1976).

Although no majority opinion was actually written in the 5 to 4 decision in *Furman v. Georgia*, Justice Brennan writing separately in concurrence noted that

> There are, then, four principles by which we may determine whether a particular punishment is 'cruel and unusual' ... [the] essential predicate [is] that a punishment must not by its severity be degrading to human dignity ... a severe punishment that is obviously inflicted in wholly arbitrary fashion ... a severe punishment that is clearly and totally rejected throughout society ... a severe punishment that is patently unnecessary.

Brennan went on to write that he expected that no state would pass laws obviously violating any one of these principles; and that court decisions involving the Eighth Amendment would use a "cumulative" analysis of the implication of each of the four principles.

The Supreme Court case of *Gregg v. Georgia* reaffirmed the use of the death penalty in the United States. It held that the imposition of the death penalty does not automatically violate the Eighth and Fourteenth Amendments. If the jury is furnished with standards to direct and limit the sentencing discretion, and the jury's decision is subjected to meaningful appellate review, the death sentence may be constitutional. If, however, the death penalty is mandatory, so there is no provision for mercy based on the characteristics of the offender, then it is unconstitutional.

This short background on *Furman v. Georgia* and *Gregg v. Georgia* brings us to the case of *McCleskey v. Kemp* (1987), of primary interest in this section. For us, the main importance of *McCleskey v. Kemp* is the use and subsequent complete disregard of a monumental statistical study by David C. Baldus, Charles Pulaski, and George G. Woodworth, "Comparative Review of Death Sentences: An Empirical Study of the Georgia Experience" (*Journal of Criminal Law and Criminology*, 1983, *74*, 661–753).[11] In *McCleskey v. Kemp*, the Court held that despite statistical evidence of a profound racial disparity in application of the death penalty, such evidence is insufficient to invalidate a defendant's death sentence. The syllabus of this ruling is given below. To see additional contemporary commentary, an article by Anthony Lewis lamenting this ruling appeared in the *New York Times* (April 28, 1987), entitled "Bowing To Racism."

United States Supreme Court, McCleskey v. Kemp (1987): Syllabus

In 1978, petitioner, a black man, was convicted in a Georgia trial court of armed robbery and murder, arising from the killing of a white police officer during the robbery of a store. Pursuant to Georgia statutes, the jury at the penalty hearing considered the mitigating and aggravating circumstances of petitioner's conduct, and recommended the death penalty on the murder charge. The trial court followed the recommendation, and the Georgia Supreme Court affirmed. After unsuccessfully seeking post-conviction relief in state courts, petitioner sought habeas corpus relief in Federal District Court. His petition included a claim that the Georgia capital sentencing process was administered in a racially discriminatory manner in violation of the Eighth and Fourteenth Amendments. In support of the claim, petitioner proffered a statistical study (the Baldus study) that purports to show a disparity in the imposition of the death sentence in Georgia based on the murder victim's race and, to a lesser extent, the defendant's race. The study is based on over 2,000 murder cases that occurred in Georgia during the 1970's, and involves data relating to the victim's race, the defendant's race, and the various combinations of such persons' races. The study indicates that black defendants who killed white victims have the greatest likelihood of receiving the death penalty. Rejecting petitioner's constitutional claims, the court denied his petition insofar as it was based on the Baldus study, and the Court of Appeals affirmed the District Court's decision on this issue. It assumed the validity of the Baldus study, but found the statistics insufficient to demonstrate unconstitutional discrimination in the Fourteenth Amendment context or to show irrationality, arbitrariness, and capriciousness under Eighth Amendment analysis.

Held:

1. The Baldus study does not establish that the administration of the Georgia capital punishment system violates the Equal Protection Clause.

(a) To prevail under that Clause, petitioner must prove that the decision makers in his case acted with discriminatory purpose. Petitioner offered no evidence specific to his own case that would support an inference that racial considerations played a part in his sentence, and the Baldus study is insufficient to support an inference that any of the decision makers in his case acted with discriminatory purpose. This Court has accepted statistics as proof of intent to discriminate in the context of a State's selection of the jury venire, and in the context of statutory violations under Title VII of the Civil Rights Act of 1964. However, the nature of the capital sentencing decision and the relationship of the statistics to that decision are fundamentally different from the corresponding elements in the venire selection or Title VII cases. Petitioner's statistical proffer must be viewed in the context of his challenge to decisions at the heart of the State's criminal justice system. Because discretion is essential to the criminal justice process, exceptionally clear proof is required before this Court will infer that the discretion has been abused.

(b) There is no merit to petitioner's argument that the Baldus study proves that the State has violated the Equal Protection Clause by adopting the capital punishment statute and allowing it to remain in force despite its allegedly discriminatory application. For this claim to prevail, petitioner would have to prove that the Georgia Legislature enacted or maintained the death penalty statute because of an anticipated racially discriminatory effect. There is no evidence that the legislature either enacted the statute to further a racially discriminatory purpose or maintained the statute because of the racially disproportionate impact suggested by the Baldus study.

2. Petitioner's argument that the Baldus study demonstrates that the Georgia capital sentencing system violates the Eighth Amendment's prohibition of cruel and unusual punishment must be analyzed in the light of this Court's prior decisions under that Amendment. Decisions since Furman v. Georgia, have identified a constitutionally permissible range of discretion in imposing the death penalty. First, there is a required threshold below which the death penalty cannot be imposed, and the State must establish rational criteria that narrow the decision-maker's judgment as to whether the circumstances of a particular defendant's case meet the threshold. Second, States cannot limit the sentencer's consideration of any relevant circumstance that could cause it to decline to impose the death penalty. In this respect, the State cannot channel the sentencer's discretion, but must allow it to consider any relevant information offered by the defendant.

3. The Baldus study does not demonstrate that the Georgia capital sentencing system violates the Eighth Amendment.

(a) Petitioner cannot successfully argue that the sentence in his case is disproportionate to the sentences in other murder cases. On the one hand, he cannot base a constitutional claim on an argument that his case differs from other cases in which defendants did receive the death penalty. The Georgia Supreme Court found that his death sentence was not disproportionate to other death sentences imposed in the State. On the other hand, absent a showing that the Georgia capital punishment system operates in an arbitrary and capricious manner, petitioner cannot prove a constitutional violation by demonstrating that other defendants who may be similarly situated did not receive the death penalty. The opportunities for discretionary leniency under state law do not render the capital sentences imposed arbitrary and capricious. Because petitioner's sentence was imposed under Georgia sentencing procedures that focus discretion "on the particularized nature of the crime and the particularized characteristics of the individual defendant," it may be presumed that his death sentence was not "wantonly and freakishly" imposed, and thus that the sentence is not disproportionate within any recognized meaning under the Eighth Amendment.

(b) There is no merit to the contention that the Baldus study shows that Georgia's capital punishment system is arbitrary and capricious in application. The statistics do not prove that race enters into any capital sentencing decisions or that race was a factor in petitioner's case. The likelihood of racial prejudice allegedly shown by the study does not constitute the constitutional measure of an unacceptable risk of racial prejudice. The inherent lack of predictability of jury decisions does not justify their condemnation. On the contrary, it is the jury's function to make the difficult and uniquely human judgments that defy codification and that build discretion, equity, and flexibility into the legal system.

(c) At most, the Baldus study indicates a discrepancy that appears to correlate with race, but this discrepancy does not constitute a major systemic defect. Any mode for determining guilt or punishment has its weaknesses and the potential for misuse. Despite such imperfections, constitutional guarantees are met when the mode for determining guilt or punishment has been surrounded with safeguards to make it as fair as possible.

4. Petitioner's claim, taken to its logical conclusion, throws into serious question the principles that underlie the entire criminal justice system. His claim easily could be extended to apply to other types of penalties and to claims based on unexplained discrepancies correlating to membership in other minority groups and even to gender. The Constitution does not require that a State eliminate any

demonstrable disparity that correlates with a potentially irrelevant factor in order to operate a criminal justice system that includes capital punishment. Petitioner's arguments are best presented to the legislative bodies, not the courts.

We make a number of comments about the majority opinion in *McCleskey v. Kemp* just summarized in the syllabus and noted in the article by Anthony Lewis. First, it is rarely the case that a policy could be identified as the cause for an occurrence in one specific individual. The legal system in its dealings with epidemiology and toxicology has generally recognized that an agent can never be said to have been the specific cause of, say, a disease in a particular individual. This is the notion of specific causation, which is typically unprovable. As an alternative approach to causation, courts have commonly adopted a criterion of general causation defined by relative risk being greater than 2.0 (as discussed in Section 4.2 on probability and litigation) to infer that a toxic agent was more likely than not the cause of a specific person's disease (and thus open to compensation).[12] To require that a defendant prove that the decision makers in his particular case acted with discriminatory malice is to set an unreachable standard. So is an expectation that statistics could ever absolutely prove "that race enters into any capital sentencing decisions or that race was a factor in petitioner's case." Statistical sleuthing can at best identify anomalies that need further study, for example, when Benford's law is used to identify possible fraud, or the Poisson model is used to suggest a lack of spatial clustering. But, irrespective, the anomalies cannot be just willed away as if they never existed.

The statement that "petitioner cannot successfully argue that the sentence in his case is disproportionate to the sentences in other murder cases" again assigns an impossible personal standard. It will always be impossible to define unequivocally what the "comparables" are that might be used in such comparisons. The operation of confirmation biases would soon overwhelm any attempt to define a set of comparables. Even realtors have huge difficulties in assigning comparable sales to a given property when deciding on an asking or selling price. Usually, realtors just fall back on a simple linear rule of dollars per square foot. But unfortunately, nothing so simple exists in defining comparables in imposing (or not) death sentences in Georgia.

If it can be shown that an enacted (legislative or legal) policy has the effect of denying constitutional rights for an identifiable group of individuals, then that policy should be declared discriminatory and changed. It should never be necessary to show that the enactors of such a policy consciously meant for that effect to occur—the law of unintended consequences again rears its ugly head—or that in one specific case it was operative. When policies must be carried out through human judgment, any number of subjective biases may be present at any given moment, and without any possibility of identifying which ones are at work and which ones are not.

In various places throughout the majority opinion, there appears to be argument by sheer assertion with no other supporting evidence at all. We all need to repeat to ourselves the admonition that just saying so doesn't necessarily make it so. Thus, we have the admission that there appears to be discriminatory effects correlated with race, with the empty assertion that "this discrepancy does not constitute a major systemic weakness" or "despite such imperfection, constitutional guarantees are met." This is just nonsense, pure and simple.

The final point in the syllabus is that "if the Petitioner's claim is taken to its logical conclusion, questions arise about the principles underlying the entire criminal justice system." Or in Justice Brennan's dissent, the majority opinion is worried about "too much justice." God forbid that other anomalies be identified that correlate with membership in other groups (for example, sex, age, other minorities) that would then have to be dealt with.

The *New York Review of Books* in its December 23, 2010 issue scored a coup by having a lead article entitled "On the Death Sentence," by retired Supreme Court Justice John Paul Stevens. Stevens was reviewing the book, *Peculiar Institution: America's Death Penalty in an Age of Abolition* (by David Garland). In the course of his essay, Stevens comments on *McCleskey v. Kemp* and notes that Justice Powell (who wrote the majority opinion) in remarks he made to his biographer, said that he should have voted the other way in the *McCleskey* 5 to 4 decision. It's too bad we cannot retroactively reverse Supreme Court rulings, particularly given the doctrine of *stare decisis* ("Stare decisis," 2011), according to which judges are obliged to respect the precedents set by prior decisions. The doctrine of *stare decisis* suggests that no amount of statistical evidence will ever be sufficient to declare the death penalty in violation of the "equal protection" clause of the Fourteenth Amendment. The relevant quotation from the Stevens review follows:

> In 1987, the Court held in McCleskey v. Kemp that it did not violate the Constitution for a state to administer a criminal justice system under which murderers of victims of one race received death sentences much more frequently than murderers of victims of another race. The case involved a study by Iowa law professor David Baldus and his colleagues demonstrating that in Georgia murderers of white victims were eleven times more likely to be sentenced to death than were murderers of black victims. Controlling for race-neutral factors and focusing solely on decisions by prosecutors about whether to seek the death penalty, Justice Blackmun observed in dissent, the effect of race remained "readily identifiable" and "statistically significant" across a sample of 2,484 cases.
>
> That the murder of black victims is treated as less culpable than the murder of white victims provides a haunting reminder of once-prevalent Southern lynchings. Justice Stewart, had he remained on the Court, surely would have voted with the four dissenters. That conclusion is reinforced by Justice Powell's second thoughts; he later told his biographer that he regretted his vote in McCleskey.

We give redactions of the majority opinion and dissent in an appendix (by Justice Brennan) for *McCleskey. v. Kemp*. It is a pity that Brennan's dissent did not form the majority opinion as it would have but for Justice Powell's vote that in hindsight he wished he could change. It also would have given greater legitimacy and importance to such landmark statistical studies as done by Baldus, et al. (1983). We will leave readers to peruse the majority and dissenting opinions and arriving at their own identification of outrageous argumentation on either side. There are several howlers, such as the statement that an r-squared of .48 is not very good because then "the model does not predict the outcomes in half the cases." In reading the majority and dissenting opinions, it is best to keep in mind the word "opinion." Such opinions include disregarding incontrovertible statistical evidence that something is amiss in the administration of the Georgia death penalty, wherever that may arise from. Although the cause may be ambiguous, there is no doubt that it results from all the various actors in the legal system who make the series of decisions necessary in determining who lives and who dies.

Appendix: United States Supreme Court, McCleskey v. Kemp (1987): Majority Opinion and Dissent

Given in the Appendix Supplements.

Appendix: Cheating? by Howard Wainer

My annual retreat in the Adirondack Mountains of upstate New York was interrupted by a phone call from an old friend. It seemed that he was hired as a statistical consultant for an unfortunate young man who had been accused of cheating on a licensing exam. My friend, a fine statistician, felt that he could use a little help on the psychometric aspects of the case. After hearing the details I agreed to participate.

The Details of the Case: The client had taken, and passed, a licensing test. It was the third time he took the exam—the first two times he failed by a small amount, but this time he passed, also by a small amount. The licensing agency, as part of their program to root out cheating, did a pro forma analysis in which they calculated the distance between all pairs of examinees based on the number of incorrect item responses those examinees had in common. There were 11,000 examinees and so they calculated the distances for all 65 million pairs. After completing this analysis they concluded that 46 examinees (23 pairs) were much closer together than one would expect. Of these 23 pairs, only one took the exam at the same time in the same room. They sat one row apart in the same room. At this point a more detailed investigation was done in which their actual test booklets were examined. The test booklets were the one place in which examinees could do scratch work before deciding on the correct answer. The investigator concluded on the basis of this examination that there was not enough work in the test booklets to allow him to conclude that the examinee had actually done the work, and so reached the decision that the examinee had copied and his score was not earned. His passing score was then disallowed, and he was forbidden from applying to take this exam again for ten years. The second person of the questioned pair sat in front, and so it was decided that she could not have copied and hence she faced no disciplinary action. The examinee who was to be punished brought suit.

Industry Standards: There are many testing companies. Most of them are special purpose organizations that deal with one specific exam. Typically, their resources are too limited to allow the kinds of careful, rigorous, research required to justify the serious consequences that the investigation of cheating might lead to. The Educational Testing Service (ETS), whose annual income of almost a billion dollars, makes it the six hundred pound gorilla. ETS has carefully studied methods for the detection of cheating, and has led the way in the specification of procedures that are detailed in a collaborative effort of the three major professional organizations concerned with testing. Most smaller testing organizations simply adopt the methods laid out by ETS and the other large organizations. To deviate from established practices would require extensive (and expensive) justification. The key standards for this situation are:

1. Statistical evidence of cheating is almost never used as the primary motivator of an investigation. Usually, there is an instigating event, for example, a proctor reporting seeing something odd, reports of examinees boasting of having cheated, other examinees reporting knowledge of collusion, an extraordinarily large increase in score from a previous time. Only when such reports are received and documented is a statistical analysis of the sort done in this case prepared—but as confirmation. They are not done as an exploratory procedure.

2. The subject of the investigation is the score, not the examinee. After completing an examination in which all evidence converges to support the hypothesis that cheating might have occurred, the examinee receives a letter that states that the testing company cannot stand behind the validity of the test score and hence will not report it to whoever required the result.

3. The finding is tentative. Typically, the examinee is provided with five options (taken directly from ETS, 1993):

(i) The test taker may provide information that might explain the questioned circumstances (for example, after an unusually large jump in score, the test taker might bring a physician's note confirming severe illness when the test was originally taken). If accepted, the score obtained is immediately reported.

(ii) The test taker may elect to take an equivalent retest privately, at no cost, and at a convenient time to confirm the score being reviewed. If the retest score is within a specified range of the questioned score, the original score is confirmed. If it is not, the test taker is offered a choice of other options.

(iii) The test taker may choose independent arbitration of the issue by the American Arbitration Association. The testing organization pays the arbitration fee and agrees to be bound by the arbitrator's decision.

(iv) The test taker may opt to have the score reported accompanied by the testing organization's reason for questioning it and the test taker's explanation.

(v) The test taker may request cancelation of the score. The test fee is returned, and the test taker may attend any regularly scheduled future administration.

Option (ii) is, by far, the most common one chosen.

Why are these standards important?: Obviously, all three of these standards were violated in this summer's investigation. And, I will conclude, the testing organization ought to both change their methods in the future as well as revise the outcome of the current case. To support my conclusion, let me illustrate, as least partially, why time, effort, and the wisdom borne of experience has led serious testing organizations to scrupulously follow these guides. The example, using mammograms for the early detection of breast cancer among women (given in Section 3.3), illustrates how the problem of false positives is the culprit that bedevils attempts to find rare events.

How accurate is the testing program in question?: The short answer is, "I don't know." To calculate this we would need to know (i) how many cheaters there are—analogous to how many cancer tumors, and (ii) how accurate is the cheating detection methodology. Neither of these numbers is known; they are both missing. Let us treat them as we would any missing data and, through multiple imputation, develop a range of plausible values. I will impute a single value for these unknowns but invite readers to stick in as many others as you wish to span the space of plausibility. Let us assume that there are 100 cheaters out of the 11,000 examinees. If this number seems unreasonable to you, substitute in another value that you find more credible. And, let us assume that the detection scheme is, like a well-done mammogram, 90% accurate (although since the reliability of the test is about 0.80, this is a remarkably Pollyannaish assumption). Let us further assume that the errors are symmetric—it is as accurate at picking out a cheater as it is at identifying an honest examinee. This too is unrealistic, since the method of detection has almost no power at detecting someone who only copied a very few items, or who, by choosing a partner wisely, only copied correct answers. Nevertheless, we must start someplace, so let's see where these assumptions carry us.

We are trying to estimate the probability of being a cheater given that the detection method singles us out as one. The numerator of this probability is 90; there are 100 honest-to-goodness cheaters, which are identified with 90% accuracy. The denominator is the true cheaters identified minus 90, plus the 1,100 false positives (10% of the 11,000 honest examinees). So the fraction is $90/1,190 = 7.6\%$. Or, more than 92% of those identified as cheaters were falsely accused, given these assumptions! If you believe that these assumptions are wrong, change them and see where it leads. The only assumption that must be adhered to is that no detection scheme is 100% accurate. As Damon Runyon pointed out, "nothing in life is more than 4 to 1." In the current situation, the pool of positive results was limited to 46 individuals. How can we adapt this result to the model illustrated by the mammogram example? One approach might be to treat those 46 as being drawn at random from the population of the 1,190 yielded by our example, thus maintaining the finding that only 7.6% of those identified as cheaters were correctly so identified. But a different interpretation might be to believe that the testing organization recognized their detection methodology had little power in finding an examinee that only copied a few answers from a neighbor. This reality might have suggested that the cost of missing such a cheater was minor. Since all they could detect were those who had not only copied a substantial number of items, but were unfortunate enough to have copied a substantial number of wrong answers. With this limited goal in mind, they set a dividing line for cheating so that only 46 people out of 11,000 were snared as possible. Then, they were forced to conclude that 22 of the 23 pairs of examinees were false positives, because of the physical impossibility of copying. Thus, despite the uncontroversial evidence that at least 22 of 23 were false positives they concluded that if it could be than it was. It is certainly plausible that all 23 pairs were false positives and that one of them happened to occur in the same testing center. This conclusion is bolstered by noting that the validity of only one section of the test was being questioned and that the test taker passed the other sections.

The scanning of the test booklet for corroborating evidence is a red herring, for although we know that the one candidate in question had what were deemed insufficient supporting calculations, we don't know what the test booklets of other examinees looked like—they were all destroyed. It seems clear from these analyses that there is insufficient evidence for actions as draconian as those imposed.

Conclusions: Of course it is important for any testing organization to impose standards of proper behavior upon its examinees. A certified test score loses value if it can be obtained illegally. And yet any detection scheme is imperfect. And the costs of its imperfections are borne by the innocents accused. With due deliberation, we have determined that the inconvenience and expense involved for the vast majority of women who have an unnecessary mammogram is offset by the value of early detection for the small percentage of women whose lives are lengthened because of the early detection. The arguments made by those opposed to capital punishment have been strengthened enormously by the more than 200 innocent men released from long prison terms in the 20 years since DNA evidence has become more widely used. This result dramatically makes the point that the imperfections of our detection methods argues against draconian measures when other alternatives are available.

References:

American Educational Research Association, the American Psychological Association and the National Council on Measurement in Education (1999). *Standards for Educational and Psychological Testing*. Washington, DC: American Psychological Association.

Educational Testing Service (1993). *Test Security: Assuring Fairness for All.* Princeton, NJ: Educational Testing Service.

Notes

[1]References on exploratory data analysis:

Behrens, J. T., & Yu, C. H. (2003). Exploratory data analysis. In J. Schinka & W. F. Velicer (Eds.), *Handbook of psychology, Vol. 2: Research methods in psychology* (pp. 33–64). New York: Wiley.

Hoaglin, D. C., Mosteller, F., & Tukey, J. W. (1983). *Understanding robust and exploratory data analysis.* New York: Wiley.

Hoaglin, D. C, Mosteller, F., & Tukey, J. W. (1985). *Exploring data tables, trends, and shapes.* New York: Wiley.

[2]The entry C.6.4 in the Suggested Reading provides an entertaining article from the *Atlantic* on authorship issues in works supposedly by Shakespeare: "The Ghost's Vocabulary" (Edward Dolnick, October, 1991).

[3]We give an item in an appendix to this chapter written by Howard Wainer, entitled "Cheating?". This short article provides exactly the type of cautionary tale alluded to here. Also, see Chapter 8 in Wainer's *Uneducated Guesses*, 2011, Princeton University Press.

[4]This is listed in the Suggested Reading, C.6.3.

[5]There are several quantitative phenomena useful in sleuthing but which are less than transparent to understand. One particularly bedeviling result is called the Inspection Paradox (Aczel, 2004, pp. 63–68). Suppose a light bulb now burning above your desk (with an average rated life of, say, 2000 hours), has been in operation for a year. It now has an expected life longer than 2000 hours because it has already been on for a while, and therefore cannot burn out at any earlier time than right now. The same is true for life spans in general. Because we have not, as they say, "crapped out" as yet, and we cannot die at any earlier time than right now, our lifespans have an expectancy longer than what they were when we were born. This is good news brought to you by Probability and Statistics!

[6]A plot of the logarithm of the "number of areas" against the "number of hits" produces a negatively associated and near to a linear fit. This is akin to Zipf's law, where the frequency of, say, a word is inversely proportional to its rank in the frequency table ("Zipf's law," 2012).

[7]This piece is listed in the Suggested Reading, C.6.1.

[8]The International Society for Clinical Biostatistics through its Subcommittee on Fraud published a position paper entitled "The Role of Biostatistics in the Prevention, Detection, and Treatment of Fraud in Clinical Trials" (Buyse et al., *Statistics in Medicine*, 1999, *18*, 3435–3451). Its purpose was to

point out some of the ethical responsibilities the statistical community has in helping monitor clinical studies with public or personal health implications. The abstract is given below, but we still refer the reader directly to the article for more detail on a range of available statistical sleuthing tools (including Benford's law) that can assist in uncovering data fabrication and falsification:

Recent cases of fraud in clinical trials have attracted considerable media attention, but relatively little reaction from the biostatistical community. In this paper we argue that biostatisticians should be involved in preventing fraud (as well as unintentional errors), detecting it, and quantifying its impact on the outcome of clinical trials. We use the term "fraud" specifically to refer to data fabrication (making up data values) and falsification (changing data values). Reported cases of such fraud involve cheating on inclusion criteria so that ineligible patients can enter the trial, and fabricating data so that no requested data are missing. Such types of fraud are partially preventable through a simplification of the eligibility criteria and through a reduction in the amount of data requested. These two measures are feasible and desirable in a surprisingly large number of clinical trials, and neither of them in any way jeopardizes the validity of the trial results. With regards to detection of fraud, a brute force approach has traditionally been used, whereby the participating centres undergo extensive monitoring involving up to 100 per cent verification of their case records. The cost-effectiveness of this approach seems highly debatable, since one could implement quality control through random sampling schemes, as is done in fields other than clinical medicine. Moreover, there are statistical techniques available (but insufficiently used) to detect "strange" patterns in the data including, but no limited to, techniques for studying outliers, inliers, overdispersion, underdispersion and correlations or lack thereof. These techniques all rest upon the premise that it is quite difficult to invent plausible data, particularly highly dimensional multivariate data. The multicentric nature of clinical trials also offers an opportunity to check the plausibility of the data submitted by one centre by comparing them with the data from all other centres. Finally, with fraud detected, it is essential to quantify its likely impact upon the outcome of the clinical trial. Many instances of fraud in clinical trials, although morally reprehensible, have a negligible impact on the trial's scientific conclusions. (pp. 3435–3436)

[9] This article is listed in the Suggested Reading, C.6.2.

[10] The article is listed in the Suggested Reading, B.2.4.

[11] For an extended, book length version of this article and an explicit discussion of *McCleskey v. Kemp*, see *Equal Justice and the Death Penalty: A*

Legal and Empirical Analysis, by David C. Baldus, George C. Woodworth, and Charles A. Pulaski, Jr., Boston: Northeastern University Press, 1990.

[12]In his dissent, Justice Brennan makes this exact point when he states: "For this reason, we have demanded a uniquely high degree of rationality in imposing the death penalty. A capital sentencing system in which race more likely than not plays a role does not meet this standard."

Part III

Experimental Design and the Collection of Data

Chapter 15

Background: Experimental Design and the Collection of Data

> If we take in our hand any volume; of divinity or school metaphysics, for instance; let us ask, *Does it contain any experimental reasoning concerning matter of fact and existence?* No. Commit it then to the flames: for it can contain nothing but sophistry and illusion.
> — David Hume (*An Enquiry Concerning Human Understanding*, 1748)

Any discussion of data collection, surveys, and experimental design will involve many words and phrases that may not have an immediately obvious interpretation. We first characterize some of these below, in no particular order:[1]

Controlled experiment: The results from a group where a treatment was imposed are compared to the results from a second (control) group that is identical except for the treatment administration.[2]

Observational study: (sometimes referred to as a natural or quasi-experiment, or a "status" study) A study of group(s) that precludes actual treatment manipulation; in other words, one has to "play the hand that is dealt."

Retrospective study: Subjects are first identified by some procedure; subsequent to this identification, historical data are then collected. The often used "case-control" methodology, to be discussed shortly, is perhaps the best-known example of a retrospective study.

Prospective study: Subjects are identified and new data are then obtained.

Counterbalance: Attempts to even out various effects, such as priming or carry-over, by systematic or random presentation strategies.

Placebo: A typically harmless treatment administered as a control in comparison to a real treatment.

Placebo effect: An improvement in a person's condition regarded as the effect of the person's belief in the utility of the treatment used.

Voluntary response: Individuals who volunteer their responses or participation may be inherently different than nonvolunteers.

Nonresponse bias: Those individuals responding to questions are inherently different in unknown ways compared to those who do not respond.[3]

Double-blind: Neither subjects nor experimenters know who is getting what treatment.

Wait list: When it is unethical to deny treatment, it still may be possible to have a randomly constructed group just wait some period of time for treatment. In the meantime, responses for the "wait" group could still be obtained and compared to those from subjects immediately given the treatment.

Framing (wording) of questions: As always, context is crucial and different responses may be given depending on how a question is asked and by whom. To emphasize the importance of wording and context in asking questions of an individual, we refer the reader to an article written by Dalia Sussman, "Opinion Polling: A Question of What to Ask" (*New York Times*, February 27, 2010).

Chapter 11 of this book discussed the problem of inferring causality and listed eight threats to the internal validity of a quasi-experiment (Winch & Campbell, 1969). A second type of validity is also of general interest in both true and quasi-experiments, "external validity" (or possibly, "ecological validity"): to what extent do the results provide a correct basis for generalizations to other populations, settings, or treatment and measurement variables. We list six threats to external validity that should be noted whenever data are to be collected and interpreted (Winch & Campbell, 1969):

1. *Interaction effects of testing*: the effect of a pretest in increasing or decreasing the respondent's sensitivity or responsiveness to the experimental variable, thus making the results obtained for a pretested population unrepresentative of the effects of the experimental variable for the unpretested universe from which the experimental respondents were selected.

2. *Interaction of selection and experimental treatment*: unrepresentative responsiveness of the treated population.

3. *Reactive effects of experimental arrangements*: "artificiality"; con-

ditions making the experimental setting atypical of conditions of regular application of the treatment;

4. *Multiple-treatment interference*: where multiple treatments are jointly applied, effects atypical of the separate application of the treatments.

5. *Irrelevant responsiveness of measures*: apparent effects produced by inclusion of irrelevant components in complex measures.

6. *Irrelevant replicability of treatments*: failure of replications of complex treatments to include those components actually responsible for the effects. (p. 141)

15.1   Observational Studies: Interpretation

Caution is needed whenever the results of an observational study are used to inform decisions regarding health practice, social policy, or other similar choices. A good illustration is the Nurses' Health Study, started by Frank Speizer at Harvard in 1976 to assess the long-term consequences of oral contraceptive use. This prospective cohort study of about 122,000 nurses came to a dramatic conclusion in 1985—women taking estrogen had only a third as many heart attacks as women who had never taken the drug (Stampfer et al., 1985). The inference was made that estrogen was protective against heart attacks until women passed through menopause. This belief provided the foundation for therapeutic practice for the next two decades, at least until the results of two clinical trials were announced. The two trials, HERS (Hulley et al., 1998; Heart and Estrogen-progestin Replacement Study) and WHI (Rossouw et al., 2002; Women's Health Initiative) came to conclusions opposite that for a protective effect of hormone replacement therapy (or, more commonly, HRT); in fact, HRT constituted a potential health risk for all postmenopausal women, particularly for heart attacks, strokes, blood clots, breast cancer, and possibly even dementia.

Discrepancies between the results of observational studies and randomized clinical trials appear so frequently that some epidemiologists question the entire viability of the field. A 2001 editorial in the *International Journal of Epidemiology* by George Davey Smith and Shah Ebrahim was aptly entitled "Epidemiology—Is It Time to Call It a Day?". This echoes a quotation from John Bailar III at the National Academy of Sciences (as reported in a Gary Taubes article in the *New York Times* [September 16, 2007] referenced below):

> The appropriate question is not whether there are uncertainties about epidemiologic data, rather, it is whether the uncertainties are so great that one cannot draw useful conclusions from the data.

The differences between what was seen in the Nurses' observational study in comparison to the clinical trials is likely due to the *healthy-user bias* (Taubes, 2007). Basically, women who were on HRT are different than those who are not, both in engaging in activities that are good for them—taking a prescribed drug, eating a healthy diet, exercise—and in their demographics—thinner, better educated, wealthier, more health conscious generally. Some cautions about hasty causal conclusions should have been raised earlier by the results from another observational study, named the Walnut Creek Study (Petitti, Perlman, & Sidney, 1987). An epidemiologist, Diana Petitti, overseeing this cohort of 16,500, reported an apparent heart protective effect of HRT in 1987. But, she also found a more sizable reduction in death from homicide, accidents, and suicide. It is difficult to maintain HRT as a causative factor for heart protection in the face of such anomalous evidence.

There are several effects similar to or part of the healthy-user bias that should be noted as possible explanatory mechanisms for associations seen in observational data. One that has the potential to be particularly insidious is called the *compliance* or *adherer effect*. To illustrate, we give an excerpt from a Gary Taubes article in the *New York Times* (September 16, 2007), "Do We Really Know What Makes Us Healthy":[4]

> The lesson comes from an ambitious clinical trial called the Coronary Drug Project that set out in the 1970s to test whether any of five different drugs might prevent heart attacks. The subjects were some 8,500 middle-aged men with established heart problems. Two-thirds of them were randomly assigned to take one of the five drugs and the other third a placebo. Because one of the drugs, clofibrate, lowered cholesterol levels, the researchers had high hopes that it would ward off heart disease. But when the results were tabulated after five years,

clofibrate showed no beneficial effect. The researchers then considered the possibility that clofibrate appeared to fail only because the subjects failed to faithfully take their prescriptions.

As it turned out, those men who said they took more than 80 percent of the pills prescribed fared substantially better than those who didn't. Only 15 percent of these faithful "adherers" died, compared with almost 25 percent of what the project researchers called "poor adherers." This might have been taken as reason to believe that clofibrate actually did cut heart-disease deaths almost by half, but then the researchers looked at those men who faithfully took their placebos. And those men, too, seemed to benefit from adhering closely to their prescription: only 15 percent of them died compared with 28 percent who were less conscientious. "So faithfully taking the placebo cuts the death rate by a factor of two," says David Freedman, a professor of statistics at the University of California, Berkeley. "How can this be? Well, people who take their placebo regularly are just different than the others. The rest is a little speculative. Maybe they take better care of themselves in general. But this compliance effect is quite a big effect."

The moral of the story, says Freedman, is that whenever epidemiologists compare people who faithfully engage in some activity with those who don't—whether taking prescription pills or vitamins or exercising regularly or eating what they consider a healthful diet—the researchers need to account for this compliance effect or they will most likely infer the wrong answer. They'll conclude that this behavior, whatever it is, prevents disease and saves lives, when all they're really doing is comparing two different types of people who are, in effect, incomparable.

This phenomenon is a particularly compelling explanation for why the Nurses' Health Study and other cohort studies saw a benefit of H.R.T. in current users of the drugs, but not necessarily in past users. By distinguishing among women who never used H.R.T., those who used it but then stopped and current users (who were the only ones for which a consistent benefit appeared), these observational studies may have inadvertently focused their attention specifically on, as Jerry Avorn says, the "Girl Scouts in the group, the compliant ongoing users, who are probably doing a lot of other preventive things as well."

Two other possible explanatory mechanisms for what we might see in an observational study are the *prescriber effect* and the *eager-patient effect*. Again, we give a short excerpt from the article by Gary Taubes (September 16, 2007):

If we think like physicians, Avorn explains, then we get a plausible explanation: "A physician is not going to take somebody either dying of metastatic cancer or in a persistent vegetative state or with end-stage neurologic disease and say, 'Let's get that cholesterol down, Mrs. Jones.' The consequence of that, multiplied over tens of thousands of physicians, is that many people who end up on statins are a lot healthier than the people to whom these doctors do not give statins. Then add into that the people who come to the doctor and say, 'My brother-in-law is on this drug,' or, 'I saw it in a commercial,' or, 'I want to do everything I can to prevent heart disease, can I now have a statin, please?' Those kinds of patients are very different from the patients who don't come in. The *coup de grâce* then comes from the patients who consistently take their medications on an ongoing basis, and who are still taking them two or three years later. Those people are special and unusual and, as we know from clinical trials, even if they're taking a sugar pill they will have better outcomes."

Apart from the situation introduced earlier of asking questions of an individual regarding matters of opinion, the appropriate framing of questions is also crucial to the collection of valid data in all health-related studies. The Gary Taubes (September 16, 2007) article is explicit about how the framing of questions concerning HRT may have lead to the discrepancies between the observational studies and the subsequent clinical trials. We present the relevant excerpt:

Even the way epidemiologists frame the questions they ask can bias a measurement and produce an association that may be particularly misleading. If researchers believe that physical activity protects against chronic disease and they ask their subjects how much leisure-time physical activity they do each week, those who do more will tend to be wealthier and healthier, and so the result the researchers get will support their preconceptions. If the questionnaire asks how much physical activity a subject's job entails, the researchers might discover that the poor tend to be more physically active, because their jobs entail more manual labor, and they tend to have more chronic diseases. That would appear to refute the hypothesis.

The simpler the question or the more objective the measurement the more likely it is that an association may stand in the causal pathway, as these researchers put it. This is why the question of whether hormone-replacement therapy affects heart-disease risk, for instance, should be significantly easier to nail down than whether any aspect of diet does. For a measurement "as easy as this," says Jamie Robins, a Harvard epidemiologist, "where maybe the confounding is not horrible, maybe you can get it right." It's simply easier to imagine that women who have taken estrogen therapy will remember and report that correctly—it's yes or no, after all—than that they will recall and report accurately what they ate and how much of it over the last week or the last year.

But as the H.R.T. experience demonstrates, even the timing of a yes-or-no question can introduce problems. The subjects of the Nurses' Health Study were asked if they were taking H.R.T. every two years, which is how often the nurses were mailed new questionnaires about their diets, prescription drug use and whatever other factors the investigators deemed potentially relevant to health. If a nurse fills out her questionnaire a few months before she begins taking H.R.T., as Colditz explains, and she then has a heart attack, say, six months later, the Nurses' study will classify that nurse as "not using" H.R.T. when she had the heart attack.

As it turns out, 40 percent of women who try H.R.T. stay on it for less than a year, and most of the heart attacks recorded in the W.H.I. and HERS trials occurred during the first few years that the women were prescribed the therapy. So it's a reasonable possibility that the Nurses' Health Study and other observational studies misclassified many of the heart attacks that occurred among users of hormone therapy as occurring among nonusers. This is the second plausible explanation for why these epidemiologic studies may have erroneously perceived a beneficial association of hormone use with heart disease and the clinical trials did not.

15.2   Observational Studies: Types

The field of epidemiology is concerned with diseases and injuries, and how they might be caused and/or prevented. Because it is typically unethical to do a randomized clinical trial with the type of agents of interest to epidemiologists, observational data may be the only information available. In an observational framework, four types of design are typically identified: cohort, case-control, cross-section, ecological. We give the definitions for each of these below, taken from Green, Freedman, and Gordis (2000), "Reference Guide on Epidemiology," in the *Reference Manual on Scientific Evidence*. The specific definition for a cohort study was also given earlier in our discussion of probability of causation, but for completeness we repeat it here as well:[5]

cohort study. The method of epidemiologic study in which groups of individuals can be identified who are, have been, or in the future may be differentially exposed to an agent or agents hypothesized to influence the probability of occurrence of a disease or other outcome. The groups are observed to find out if the exposed group is more likely to develop disease. The alternative terms for a cohort study (concurrent study, follow-up study, incidence study, longitudinal study, prospective study) describe an essential feature of the method, which is observation of the population for a sufficient number of person-years to generate reliable incidence or mortality rates in the population subsets. This generally implies study of a large population, study for a prolonged period (years), or both. (p. 389)

case-control study. Also, case-comparison study, case history study, case referent study, retrospective study. A study that starts with the identification of persons with a disease (or other outcome variable) and a suitable control (comparison, reference) group of persons without the disease. Such a study is often referred to as retrospective because it starts after the onset of disease and looks back to the postulated causal factors. (p. 388)

cross-sectional study.[6] A study that examines the relationship between disease and variables of interest as they exist in a population at a given time. A cross-sectional study measures the presence or absence of disease and other variables in each member of the study population. The data are analyzed to determine if there is a relationship between the existence of the variables and disease. Because cross-sectional studies examine only a particular moment in time, they reflect the prevalence (existence) rather than the incidence (rate) of disease and can offer only a limited view of the causal association between the variables and disease. Because exposures to toxic agents often change over time, cross-sectional studies are rarely used to assess the toxicity of exogenous agents. (p. 390)

ecological study. Also, demographic study. A study of the occurrence of disease based on data from populations, rather than from individuals. An ecological study searches for associations between the incidence of disease and suspected disease-causing agents in the studied populations. Researchers often conduct ecological studies by examining easily available health statistics, making these studies relatively inexpensive in comparison with studies that measure disease and exposure to agents on an individual basis. (p. 391)

Cohort studies are usually prospective and compare the incidence of a disease in exposed and unexposed groups. A temporal ordering is present in the relationship between the agent and the onset of the disease, so it is possible to follow a cohort to assess whether the disease occurs after exposure to the agent (a necessary consideration for any causal interpretation). Here, the independent variable is one of exposure/nonexposure; the dependent variable is disease condition (present/absent). Within a familiar 2×2 contingency table framework, the relative risk, defined earlier, is the ratio of proportions for those having the disease within the exposed and unexposed groups.

A case-control study is well exemplified by the smoking/lung cancer investigations.[7] Individuals who have a disease are first identified. A "comparable" group without the disease is then constructed, and the exposure to the agent of interest compared for the "cases" versus the "controls." Retrospective case-control studies can generally be completed more quickly than a cohort study that requires tracking over time. Thus, they are suited for the study of rare diseases that would require the recruitment of a prohibitive number of subjects for a comparable cohort study. In a case-control context, disease status is now the independent variable and exposure is the dependent. There is a comparison of the exposure of those with the disease (the "cases") to those without the disease (the "controls"). Unfortunately, a calculation of relative risk is no longer meaningful. We give a quotation from the "Reference Guide on Epidemiology" (Green et al., 2000) in the *Reference Manual on Scientific Evidence*:

> A relative risk cannot be calculated for a case-control study, because a case-control study begins by examining a group of persons who already have the disease. That aspect of the study design prevents a researcher from determining the rate at which individuals develop the disease. Without a rate or incidence of disease, a researcher cannot calculate a relative risk. (p. 350, footnote 46)

The Wikipedia entry on case-control studies notes the difficulties that confounding creates in generating valid interpretations. A memorable phrase is used for this: "it is difficult, often impossible, to separate the chooser from the choice." We give the paragraph below from Wikipedia that discusses problems with case-control studies, and which includes the phrase just noted:

> One problem is that of confounding. The nature of case-control studies is such that it is difficult, often impossible, to separate the chooser from the choice. For example, studies of road accident victims found

that those wearing seat belts were 80% less likely to suffer serious injury or death in a collision, but data comparing rates for those collisions involving two front-seat occupants of a vehicle, one belted and one unbelted, show a measured efficacy only around half that. Several case-control studies have shown a link between bicycle helmet use and reductions in head injury, but long-term trends—including from countries which have substantially increased helmet use through compulsion—show no such benefit. Analysis of the studies shows substantial differences between the 'case' and 'control' populations, with much of the measured benefit being due to fundamental differences between those who choose to wear helmets voluntarily and those who do not. ("Case-control study," 2010)

Another manifestation of case-control confounding occurs when the two groups are formed by a lottery involving people hoping to share in a scarce resource. Consider an example where the scarce resource is admission to a superior publicly-supported high school in Chicago, which can be offered to only a limited number of individuals. Here, a random lottery is used to decide between those who get the experience (the "cases") and those who must remain at a more traditional institution (the "controls"). A separate assessment of the "cases" and their presumed increased performance is insufficient to argue for the "value added" by the enhanced experience. At the least, a comparison must be made to the performance of the "controls" not chosen through the lottery. It may well be that those individuals voluntarily entering a lottery are inherently different than those who choose not to so engage. In these kinds of situations, there is always a need to separate the effects of an intervention from those willing to undergo an intervention; this is just another version of distinguishing between the chooser and the choice.

An ecological study is carried out at the level of groups. The overall rate of a disease in groups is compared to other differences that might be present for these same groups. One has to be careful not to commit the ecological fallacy and attribute what might be present at a group level to what is also true at an individual level. It would be necessary to follow up any group-level associations with a study at the individual level to make the type of individual level conclusions one would usually wish to have.

In addition to the four common types of observational study, we might add a fifth that is characterized by the general form of the intervention. An *encouragement design* is one in which the active treatment is just the encouragement to do something, for example, take a drug to reduce blood pressure, change a diet to be more healthy, exercise regularly (Bradlow, 1998). Instead of trying to evaluate an actual treatment regime when the compliance with it is unknown or difficult to control, the target of inference changes to one of encouragement having an effect or not. In some cases where the choice of not giving encouragement would not be considered unethical, it even may be possible to effect some approximation to a randomized study where what is being randomly assigned is encouragement.

The concept of an encouragement design fits well with a recent and popular book in behavioral economics by Richard Thaler and Cass Sunstein entitled *Nudge: Improving Decisions About Health, Wealth, and Happiness* (2008). The central suggestion is to "nudge" (or "encourage") people toward healthier, safer, and generally better lives while addressing large social issues such as environmental damage, climate change, and rising health care costs. Given Susstein's current position in the Obama White House, possibly a lot more nudging and encouragement lie ahead for the American people.

Irrespective of the type of study adopted, it is important to consider how bias may affect the conclusions. Bias refers to anything, other than sampling error, that results in a specious association, and which thereby compromises validity. Generally, there are two broad categories of bias.[8] The first, *selection bias*, is about who gets into a study. Are there systematic differences in characteristics between those who get in and those who don't? Because of some reason that relates to what is being studied, people may be unwilling to be part of a study or they drop out after entering. Obviously, this affects the inferences that can be legitimately drawn. The second category of bias is *information bias*, a flaw in the measurement of either or both the exposure and the disease. As part of this, there is *recall bias* where people with the disease may remember past exposures better. This bias may be particularly problematic in a case-control study where the control group may not remember their exposures correctly. In general, *misclassifica-*

tion bias is when people are misclassified according to exposure or disease status. For disease, the diagnostic criteria should be good. For example, a home pregnancy kit may be too unreliable a test for studying spontaneous abortions and their relation to other factors. Confounding (and "lurking") third variables seem to exist everywhere, making it incumbent on all who report observational results to be ever vigilant, lest misinformation does more harm than good as it possibly did in the case of HRT.

15.3   Observational Studies: Additional Cautions

Although we have given a number of cautions in this chapter as to how observational data should be considered with care, and there are several more in Chapter 11 on inferring causality (for example, the Bradford–Hill criteria), a number of other sources provide additional wise guidance. One that we excerpt next is a short section on observational studies from the "Reference Guide on Statistics" in the *Reference Manual on Scientific Evidence* (authored by David H. Kaye and David A. Freedman, 2000):

> The bulk of the statistical studies seen in court are observational, not experimental. Take the question of whether capital punishment deters murder. To do a randomized controlled experiment, people would have to be assigned randomly to a control group and a treatment group. The controls would know that they could not receive the death penalty for murder, while those in the treatment group would know they could be executed. The rate of subsequent murders by the subjects in these groups would be observed. Such an experiment is unacceptable—politically, ethically, and legally.
>
> Nevertheless, many studies of the deterrent effect of the death penalty have been conducted, all observational, and some have attracted judicial attention. Researchers have catalogued differences in the incidence of murder in states with and without the death penalty, and they have analyzed changes in homicide rates and execution rates over the years. In such observational studies, investigators may speak of control groups (such as the states without capital punishment) and of controlling for potentially confounding variables (for example, worsening economic conditions). However, association is not causation, and the causal inferences that can be drawn from

such analyses rest on a less secure foundation than that provided by a randomized controlled experiment.

Of course, observational studies can be very useful. The evidence that smoking causes lung cancer in humans, although largely observational, is compelling. In general, observational studies provide powerful evidence in the following circumstances:

The association is seen in studies of different types among different groups. This reduces the chance that the observed association is due to a defect in one type of study or a peculiarity in one group of subjects.

The association holds when the effects of plausible confounding variables are taken into account by appropriate statistical techniques, such as comparing smaller groups that are relatively homogeneous with respect to the factor.[9]

There is a plausible explanation for the effect of the independent variables; thus, the causal link does not depend on the observed association alone. Other explanations linking the response to confounding variables should be less plausible.

When these criteria are not fulfilled, observational studies may produce legitimate disagreement among experts, and there is no mechanical procedure for ascertaining who is correct. In the end, deciding whether associations are causal is not a matter of statistics, but a matter of good scientific judgment, and the questions that should be asked with respect to data offered on the question of causation can be summarized as follows:

Was there a control group? If not, the study has little to say about causation.

If there was a control group, how were subjects assigned to treatment or control: through a process under the control of the investigator (a controlled experiment) or a process outside the control of the investigator (an observational study)?

If the study was a controlled experiment, was the assignment made using a chance mechanism (randomization), or did it depend on the judgment of the investigator?

If the data came from an observational study or a nonrandomized controlled experiment, how did the subjects come to be in treatment or in control groups? Are the groups comparable? What factors are confounded with treatment? What adjustments were made to take care of confounding? Were they sensible? (pp. 94–96)

A particularly influential but problematic meta-analysis appeared in 1991 written by Meir Stampfer and Graham Colditz and entitled "Estrogen Replacement Therapy and Coronary Heart Disease: A Quantitative Assessment of the Epidemiologic Evidence"

(*Preventive Medicine, 20,* 47–63). We give the short abstract below in its entirety:

> Considerable epidemiological evidence has accumulated regarding the effect of post-menopausal estrogens on coronary heart disease risk. Five hospital-based case-control studies yielded inconsistent but generally null results; however, these are difficult to interpret due to the problems in selecting appropriate controls. Six population-based case-control studies found decreased relative risks among estrogen users, though only 1 was statistically significant. Three cross-sectional studies of women with or without stenosis on coronary angiography each showed markedly less atherosclerosis among estrogen users. Of 16 prospective studies, 15 found decreased relative risks, in most instances, statistically significant. The Framingham study alone observed an elevated risk, which was not statistically significant when angina was omitted. A reanalysis of the data showed a nonsignificant protective effect among younger women and a nonsignificant increase in risk among older women. Overall, the bulk of the evidence strongly supports a protective effect of estrogens that is unlikely to be explained by confounding factors. This benefit is consistent with the effect of estrogens on lipoprotein subfractions (decreasing low-density lipoprotein levels and elevating high-density lipoprotein levels). A quantitative overview of all studies taken together yielded a relative risk of 0.56 (95% confidence interval: 0.50-0.61), and taking only the internally controlled prospective and angiographic studies, the relative risk was 0.50 (95% confidence interval: 0.43-0.56). (p. 47)

A reprinting of this article occurred in 2004 (*International Journal of Epidemiology*) along with a collection of commentaries that tried to deal with the discrepancy between the results of this meta-analysis and subsequent clinical trials. One particularly good commentary was from Diana Petitti, "Hormone Replacement Therapy and Coronary Heart Disease: Four Lessons" (*International Journal of Epidemiology, 33,* 461–463). Paraphrasing, the four lessons referred to in this article are:

(1) Do not turn a blind eye to contradiction and ignore contradictory evidence; instead, try to understand the reasons behind the contradictions. At the time of the Stampfer and Colditz (1991) meta-analysis, there were data available from two major sources that contradicted the meta-analysis—The Coronary Drug Project and many studies on the use of oral contraceptives—which would

be difficult to reconcile with the Stampfer and Colditz conclusions. As noted by Petitti:

> If the epidemiological studies were true and oestrogen decreased the risk of CHD [Coronary Heart Disease] in women, it would mean that the effect of the oestrogen is 'crossed' by sex or by dose. That is, it would mean that oestrogen increases (or does not affect) the risk of CHD in men (Coronary Drug Project) but it decreases risk in women (epidemiological studies) or that oestrogen use increases (or does not affect the risk) of CHD at a high dose (Coronary Drug Project) but it decreases risk at a low dose (epidemiological studies).
>
> There are no examples where the effects of either exogenous or endogenous factors on CHD are crossed—the factor increases the risk of CHD in people of one sex and decreases risk in those of the opposite sex. There are no examples where the effect of exogenous administration of a drug on CHD is different between high and low doses of the drug. (p. 461)
>
> If there truly were an effect of oestrogen/progestin combinations in decreasing the risk of CHD in postmenopausal women (as hormone replacement therapy), it would mean that the effect of administration of combinations of oestrogen and progestin in women is crossed by age. That is, the drugs are a hazard to the heart in younger women (oral contraception) and a benefit in older women (hormone replacement therapy).
>
> There are no examples of where the effect of a drug on the risk of CHD at one age is the opposite of its effect at another. (p. 462)
>
> Crossed effects are contradictions. Contradiction must be identified and explained. There is much to learn from contradiction. (p. 462)

(2) Do not be seduced by mechanism. Even where a plausible mechanism exists, do not assume that we know everything about that mechanism and how it might interact with other factors. Apparently, Stampfer and Colditz (1991) were so seduced—(from their abstract, p. 47): "This benefit is consistent with the effect of estrogens on lipoprotein subfractions (decreasing low-density lipoprotein levels and elevating high-density lipoprotein levels)."[10]

(3) Suspend belief. In commenting on researchers defending observational studies, Pettiti (2004) notes: "belief caused them to be unstrenuous in considering confounding as an explanation for the studies" (p. 462). Don't be seduced by your desire to prove your case. Stampfer and Colditz ignored the well-known (in 1991) effects

of social class, education, and socioeconomic status on coronary heart disease.[11]

(4) Maintain scepticism. Question whether the factor under investigation can really be that important; consider what other differences might characterize the case and control groups. Be wary of extrapolating results beyond the limits of reasonable certainty (for example, with grandiose forecasts of number of "lives saved").[12]

15.4   Controlled Studies

The two most common controlled studies are named (for purposes of PubMed publication), a "randomized controlled trial" and a "controlled clinical trial." We give definitions from the National Library of Medicine:[13]

> Controlled Clinical Trial: A Controlled Clinical Trial (CCT) is a work consisting of a clinical trial involving one or more test treatments, at least one control treatment, specified outcome measures for evaluating the studied intervention, and a bias-free method for assigning patients to the test treatment. The treatment may be drugs, devices, or procedures studied for diagnostic, therapeutic, or prophylactic effectiveness.

> Randomized Controlled Trial: A Randomized Controlled Trial (RCT) is a work consisting of a clinical trial that involves at least one test treatment and one control treatment, concurrent enrollment and followup of the test- and control-treated groups, and in which the treatments to be administered are selected by a random process, such as the use of a random-numbers table.

An RCT is usually considered the gold standard for providing valid evidence of a (causal) effect in whatever area it is used (with a CCT coming in at a more distant second). These controlled studies, however, are subject to some of the same interpretive anomalies that plague observational studies. First, in considering how subjects are recruited, there is a need to follow whatever rules of informed consent are imposed by the institution overseeing the study. This implies that subjects must be told about the possible downsides of what may be administered, that they might be allocated to, say, a placebo or alternative condition, and they can terminate their participation at any time they might wish. Generally, informed consent is a voluntary and documented confirmation of a subject's willingness to participate in a trial, and before any protocol-related procedures or treatments are performed.[14]

The process of obtaining informed consent gives a potential subject the chance to just say "no" to entering the trial at the outset, or permission to stop participation at any time. Based on the difficulties seen in observational studies, it is clear that individuals who opt to begin participation and/or who continue are basically different from those who don't start or those who drop out. We have all the usual biases to worry about, such as the healthy-user, eager-patient and adherer. There is even a prescriber effect to consider—who are the individuals that were asked to join the study in the first place (Taubes, 2007). For example, some of the current thinking about HRT centers on the "timing hypothesis," where the age at which HRT starts is crucial for any protective effects to emerge. Obviously, the type of subject recruited for any study will influence greatly the type of question that can be asked and the conclusions that can be legitimately reached. It is generally true that determining a best treatment for an individual participant is fundamentally different from determining which treatment is best on average. The healthy-user bias acts as a general factor underlying many potential effects, complicating the interpretation for even the gold standard of an RCT. As Gary Taubes (2007) notes from the same *New York Times* article quoted earlier:

> Clinical trials invariably enroll subjects who are relatively healthy, who are motivated to volunteer and will show up regularly for treatments and checkups. As a result, randomized trials "are very good for showing that a drug does what the pharmaceutical company says it does," David Atkins, a preventive-medicine specialist at the Agency for Healthcare Research and Quality, says, "but not very good for telling you how big the benefit really is and what are the harms in typical people. Because they don't enroll typical people."

Even in the most rigorously conducted double-blind and placebo-controlled RCT, subjects must be informed they could be put into the placebo (or an alternative) condition. This knowledge might lead to several things. First, it may be possible for subjects who want the "real stuff" to pay for treatment by themselves, with no risk of getting placed into the placebo condition. Or, if the nonplacebo treatment has a possibly unknown downside, and more well-known older treatments already exist, why be part of the trial to begin with (think of Vioxx and naproxen [for example, Aleve]). Second, if a subject notices "no effect" once in a trial, the reasonable inference would be of being placed in the placebo condition. So why continue? Alternatively, if one has some adverse reaction to the assigned treatment condition, whatever it is, why not stop participation?

Once the data from a controlled study are available, it is tempting to engage in a process of data dredging (the older version of modern data mining), to see if various effects can be teased apart for subgroups. It deserves reminding that sample size issues and the culling of chance occurrences must always be accounted for. A particularly contentious part of this process arises when an "intention-to-treat" analysis is performed, alternatively labeled as "analyze as randomized" or "as randomized, so analyzed" ("Intention to treat analysis," 2012). An intention-to-treat analysis in an RCT uses *all* patients randomly assigned to the treatments, irrespective of whether they completed or even received the designated or intended treatment(s).[15]

The purpose of an intention-to-treat analysis is to avoid interpreting misleading artifacts that can arise in intervention research generally and to sidestep the problems of dropout and crossovers where the "wrong" treatments were actually administered. Intention-to-treat analyses provide information about the potential effects of a treatment policy rather than the possible effect of a specific treatment. The term "efficacy subset analysis" (or "treatment-received analysis") refers to the group of patients who received the designated treatments, regardless of initial randomization, and who have not dropped out. Critics of efficacy subset analysis argue that various unknown biases are introduced, with a possible inflation of the type-one error. Obviously, a full application of an intention-to-treat analysis can be done only when there are complete data for all randomized subjects. The use of an alternative missing data imputation method is generally problematic because data are typically not "missing at random," with "missingness" most likely related to the specific treatments imposed.

This is also true for the imputation method that uses the "last observation carried forward."

15.5 Controlled Studies: Additional Sources of Bias

The *Cochrane Handbook for Systematic Reviews of Interventions* includes a chapter entitled "Assessing Risk of Bias in Included Studies." Although directed toward an eventual meta-analysis, it includes a valuable discussion of bias sources that should be assessed in considering just a single controlled study. We begin with a classification scheme into five broad areas of bias (Higgins & Altman, 2008; adapted from Table 8.4.a, p. 195):

Selection bias: Systematic differences between the baseline characteristics of the groups to be compared;

Performance bias: Systematic differences between groups in the care provided, or in exposure to factors other than the interventions of interest;

Attrition bias: Systematic differences between groups in the number of withdrawals from a study;

Detection bias: Systematic differences between groups in how outcomes are determined;

Reporting bias: Systematic differences between reported and unreported findings.

Several additional terms introduced in this *Cochrane Handbook* chapter (pp. 193–194) pertain to the classification of bias just given: *sequence generation* and *allocation concealment* concern selection bias; *blinding* is relevant to performance, attrition, and detection bias; *incomplete outcome data* is pertinent to attrition bias; and *selective reporting* is obviously connected to reporting bias. Brief characterizations follow:

Sequence generation refers to the rule for subject allocation to treatment by some chance mechanism;

Allocation concealment concerns the steps taken to insure implementation of the sequence generation by preventing prior knowledge of the ensuing allocations;

Blinding reduces the risk of study participants or personnel knowing the intervention received;

Incomplete outcome data points to possibly biased outcomes because of attrition or study participant exclusion;

Selective reporting is the (generally unethical) censoring of data on study outcomes.

15.5.1   The Special Case of Medical Trials

Controlled studies carried out to obtain FDA (Food and Drug Administration) approval for some medical intervention, such as a drug, implant, or other device, have their own set of biases, with some unique to medical trials carried out to seek regulatory approvals. Obviously, all biases should be of concern in the interpretation of clinical trial data and how these might skew the outcome of a regulatory argument. Even though a study may begin as a well-designed randomized controlled trial, because of differential dropout and other forms of experimental mortality, that is not where one usually ends up. As Paul Holland has noted: "All randomized experiments are observational studies waiting to happen" (personal communication to HW, October 26, 1986.).[16] Because of this, biases discussed for observational studies are just as pertinent for those randomized clinical trials provided as evidence to obtain FDA approvals. In addition to recognizing possible observational study biases, it would seem prudent for the FDA to probe further and ask pointed questions as to how the studies were conducted. This level of added detail could help assess the veracity of the statements being made regarding the effectiveness and safety of the medical intervention.

It is important to know from where and how the subjects were recruited. For foreign trials, in particular, the kind of inducements offered should be known, and then, exactly who was so enticed to participate. Follow-up information should be available on the characteristics of the subject population (for example, general health and nutrition, age, sex, social class, education), and whether these may interact with the medical procedure being assessed to slant the results in particular ways. For nonforeign trials, the bonuses doctors receive for recruiting to the clinical trial could be seen as problematic inducements that might lead to registering patients who are not really eligible for the trial. The general question is always the following: were the correct patients enrolled so the trial provides information on safety and effectiveness directly relevant for the target group expected to receive the medical intervention?

A second central question would be a careful characterization of the treatments administered, both as to type and dosage. For instance, if a "me-too" drug, defined as one similar to others already available, is being evaluated, did the comparison involve an inert placebo or an inappropriate dose level of a competitor? Also, was the length of the trial too short to show the adverse events that might be apparent only in Phase IV postmarketing monitoring (think of Vioxx, Avendia, and the host of other withdrawn products)? Are all the data from the trial reported and not just selected portions that demonstrate what the petitioners wish to put forward? Were treatments compared to alternatives no longer under patent, and was the new treatment substantially better than generic alternatives? A third question involves what is being measured. Many times only surrogate outcomes are available (for example, lowered cholesterol), because the relevant clinical outcome is not available (for example, death from heart failure). Has the definition been changed, possibly arbitrarily, for the condition being treated—for obesity, hypertension, high cholesterol levels—and what effect does this have on the surrogate endpoint being assessed? Also, what is the strength of connection to the ultimate clinical endpoint? Is the treatment and trial for an ethically dubious disorder? For example, there are now drugs to treat the normal consequences of aging (menopause, osteopenia), or relatively common chronic conditions, such as GERD (gastroesophageal reflux disease), PMDD (premenstrual dysphoric disorder), PE (premature ejaculation), social anxiety disorder (treated with antidepressants), or low T (low testosterone). A final set of questions has to do with the trial sponsor(s). Strong effects seem to be obtained more often when drug companies conduct the trials involving patentable products. Are all trials being reported and registered before they begin? Are the trials being conducted outside the reach of the Declaration of Helsinki? Do the sponsors have complete control over what gets reported in the open literature? What is the fate of private Contract Research Organizations

(CROs) running clinical trials that don't obtain the results a sponsor would like?

Drug companies cannot generally promote their products for conditions not FDA approved, but doctors can engage in such off-label uses, possibly suggested through drug company sponsored "educational trips" or seminars. This violates at least the spirit of any FDA approval process for an intervention designed for a specific medical purpose. Drug companies can search around for other uses for their products without a stringent process of showing effectiveness (assuming that safety was established in the original approval). This practice gets compounded by the availability, through pharmacy chains, of a physician's prescribing history. Armed with this kind of information, drug representatives can make cases for all sorts of off-label applications of their proprietary products. A most public and ongoing off-label marketing scandal involves the drug Neurontin, originally approved for epilepsy but now marketed (off-label) as an almost general elixir. A lawsuit brought by Kaiser Foundation Health Plan (first decided in early 2010), found Pfizer, the maker of Neurontin, guilty of violating antiracketeering laws in promoting Neurotin for unapproved uses. Kaiser Foundation claimed it was misled into believing Neurontin was effective for off-label treatment of migraines, bipolar disorder, and many other conditions. A Bloomberg news article from January 28, 2011, provides an update on this whole sordid saga ("Pfizer Told to Pay $142.1 Million Over Marketing of Epilepsy Drug").

Marcia Angell, a physician and the first woman to serve as editor-in-chief of the *New England Journal of Medicine*, published a book in 2004 about how Big Pharma operates; it has the self-explanatory title, *The Truth About the Drug Companies: How They Deceive Us and What to Do about It*. In commenting on the conduct of medical trials, Angell argues that drug companies should not be allowed to control the testing of their own medical products. Moreover, clinical trial data should be the joint property of, say, NIH and the researchers who carried out the trials, and not in any way under the control of the sponsoring drug company. We give a short quotation from Angell's book that summarizes this position:

To ensure that clinical trials serve a genuine medical need and to see that they are properly designed, conducted, and reported, I propose that an Institute for Prescription Drug Trials be established within the National Institutes of Health to administer clinical trials of prescription drugs. (p. 245; original given in italics)

In an afterword to her book, Angell gives a few suggestions about what individuals can do to protect their interests when it comes to the pharmaceutical industry. First, when your doctor prescribes a new drug, ask for evidence that this is better than alternative treatments. Also, has the evidence been published in a peer-reviewed journal, or is it just from drug company representatives? To our members of Congress, ask about financial ties to the pharmaceutical industry. And finally, ignore all direct-to-consumer (DTC) drug advertising. These ads are meant to sell drugs and not to educate consumers in any altruistic manner. They serve to raise the prices of the medical products sold. As of now, only the United States and New Zealand allow DTC marketing of prescription drugs. The rest of the world seems to have noticed the ethical and regulatory questions raised by allowing DTC advertising. Specifically, to what extent do these ads unduly influence the prescribing of prescription medication, irrespective of any medical necessity, and based only on consumer demand (Angell, 2004, pp. 1261–1263).

15.5.2 An Introductory Oddity: The 1954 Salk Polio Vaccine Trials

The 1954 Salk polio vaccine trial was the biggest public health experiment ever conducted. One field trial, labeled an observed control experiment, was carried out by the National Foundation for Infantile Paralysis. It involved the vaccination, with parental consent, of second graders at selected schools in selected parts of the country. A control group would be the first and third graders at these same schools, and indirectly those second graders for whom parental consent was not obtained. The rates for polio contraction (per 100,000) are given below for the three groups (see Francis et al., 1955, for the definitive report on the Salk vaccine trials).[17]

Grade 2 (Vaccine): 25/100,000;
Grade 2 (No consent): 44/100,000;
Grades 1 and 3 (Controls): 54/100,000.

The interesting observation we will return to below is that the Grade 2 (No consent) group is between the other two in the probability of polio contraction. Counterintuitively, the refusal to give consent seems to be partially protective.

The second field trial was a (double-blind) randomized controlled experiment. A sample of children were chosen, all of whose parents consented to vaccination. The sample was randomly divided into two, with half receiving the Salk vaccine and the other half a placebo of inert salt water. There is a third group formed from those children with no parental consent and who therefore were not vaccinated. We give the rates of polio contraction (per 100,000) for the three groups:

Vaccinated: 28/100,000;

Control: 71/100,000;

No consent: 46/100,000.

Again, not giving consent appears to confer some type of immunity; the probability for contracting polio for the "no consent" group is between the other two.

The seeming oddity in the ordering of probabilities, where "no consent" seems to confer some advantage, is commonly explained by two "facts": (a) children from higher-income families are more vulnerable to polio; children raised in less hygienic surroundings tend to contract mild polio and immunity early in childhood while still under protection from their mother's antibodies; (b) parental consent to vaccination appears to increase as a function of education and income, where the better-off parents are much more likely to give consent. The "no consent" groups appear to have more natural immunity to polio than children from the better-off families. This may be one of the only situations we know of where children growing up in more resource-constrained contexts are conferred some type of advantage.

15.6   The Randomized Response Method

As noted earlier, how questions are framed and the context in which they are asked are crucial for understanding the meaning of the given responses. This is true both in matters of opinion polling and for collecting data on, say, the health practices of subjects. In these situations, the questions asked are usually not sensitive, and when framed correctly, honest answers are expected. For more sensitive questions about illegal behavior, (reprehensible) personal habits, suspect health-related behaviors, questionable attitudes, and so on, asking a question outright may not garner a truthful answer.

The randomized response method is one mechanism for obtaining "accurate" data for a sensitive matter at a group level (but not at the individual level). It was first proposed in 1965 by Stanley Warner in the *Journal of the American Statistical Association*, "Randomized Response: A Survey Technique for Eliminating Evasive Answer Bias" (*60*, 63–69). A modified strategy was proposed by Bernard Greenberg and colleagues in 1969, again in *JASA*: "The Unrelated Question Randomized Response Model: Theoretical Framework" (*64*, 520–539). We first illustrate Warner's method and then Greenberg's with an example.

Let Q be the question: "Have you ever smoked pot (and inhaled)?"; and $\bar{Q}$ the complement: "Have you never smoked pot (and inhaled)?" With some known probability, θ, the subject is asked Q; and with probability $(1 - \theta)$, is given $\bar{Q}$ to answer. The respondent determines which question is posed by means of a probability mechanism under his or her control. For example, if the respondent rolls a single die and a 1 or 2 appears, question Q is given; if 3, 4, 5, or 6 occurs, $\bar{Q}$ is given. So, in this case, $\theta = 1/3$.

As notation, let p be the proportion in the population for which the true response to Q is "yes"; $1 - p$ is then the proportion giving a "yes" to $\bar{Q}$. Letting P_{yes} denote the observed proportion of "yes" responses generally, its expected value is $\theta p + (1 - \theta)(1 - p)$; thus, p can be estimated as

$$\hat{p}_w = \frac{P_{yes} - (1 - \theta)}{2\theta - 1} \ ,$$

where the subscript w is used to denote Warner's method of estimation. Obviously, θ cannot be $1/2$ because the denominator would then be zero; but all other values are legitimate. The extremes of θ being 0 or 1, however, do not insure the "privacy" of a subject's response because the question actually answered would then be known.

The Greenberg method is referred to as the unrelated (or innocuous) question technique. The complement question $\bar{Q}$ is replaced with an unrelated question, say, Q_U, with a known probability of giving a "yes" response, say γ. For example, Q_U could be "Flip a coin. Did you get a head?" Here, $\gamma = 1/2$ for a "yes" response; the expected value of P_{yes} is $\theta p + (1 - \theta)\gamma$, leading to

$$\hat{p}_g = \frac{P_{yes} - (1 - \theta)\gamma}{\theta} ,$$

where the subscript g now refers to Greenberg's method of estimation.

To decide which strategy might be the better, the variances of the two estimates can be compared though closed-form formulas:

$$\text{Var}(\hat{p}_w) = \frac{p(1 - p)}{n} + \frac{\theta(1 - \theta)}{n(2\theta - 1)^2} ;$$

$$\text{Var}(\hat{p}_g) =$$
$$\frac{p(1 - p)}{n} + \frac{(1 - \theta)^2\gamma(1 - \gamma) + \theta(1 - \theta)(p(1 - \gamma) + \gamma(1 - p))}{n\theta^2} ,$$

where the number of respondents is denoted by n. As an example, suppose θ is .6; the coin flip defines Q_U so γ is .5; and let the true proportion p be .3. Using the variance formulas above: $\text{Var}(\hat{p}_w) = 6.21/n$ and $\text{Var}(\hat{p}_g) = .654/n$. Here, the Greenberg "innocuous question" variance is only about a tenth of that for the Warner estimate, making the Greenberg method much more efficient in this instance (that is, the sampling variance for the Greenberg estimate is much less than that for the Warner estimate).

Notes

[1]We give an amusing story, "Catapoultry," about the need to provide all relevant material to aid in an experimental replication. The November 1995 issue of *Feathers: A Newsletter of the California Poultry Industry Federation* tells the following story (called "The Dangers of Frozen Chickens, Part Two"):

> Seems the FAA has a device for testing the strength of windshields on airplanes. They point this thing at the plane's windshield and shoot a dead chicken at it at about the normal airplane speed. The theory is that if the windshield doesn't crack, it'll survive a real collision with a bird during flight. Now, the British were very impressed by this test and wanted to test a windshield on a brand new speedy locomotive. They borrowed the FAA's testing device, loaded in the chicken and fired away. The bird went through the windshield, broke the engineer's chair and made a major dent in the back wall of the engine cab. The British were quite shocked at this result and asked the FAA to re-check the test to see if everything was done correctly. The FAA went over the test thoroughly and had one recommendation: They suggested the Brits repeat the test—this time using a thawed chicken instead of a frozen one. (p. 7)

[2]The first person to conduct a controlled experiment was Francesco Redi (1626–1697), an Italian physician, naturalist, and poet ("Francesco Redi," 2011). In a series of experiments published in 1668 (*Experiments on the Generation of Insects*), Redi took the first steps toward refuting the Aristotelian theory of abiogenesis, or spontaneous generation. The prevailing wisdom of the time was that maggots formed naturally in rotting meat. We give an extract from the Mab Bigelow translation published in 1909. Note that besides introducing a controlled experiment, Redi also understood the importance of replication across a variety of conditions:

> Having considered these things, I began to believe that all worms found in meat were derived directly from the droppings of flies, and not from the putrefaction of the meat, and I was still more confirmed in this belief by having observed that, before the meat grew wormy, flies had hovered over it, of the same kind as those that later bred in it. Belief would be vain without the confirmation of experiment, hence in the middle of July I put a snake, some fish, some eels of the Amo, and a slice of milk-fed veal in four large, wide-mouthed flasks; having well closed and sealed them, I then filled the same number of flasks in the same way, only leaving these open. It was not long before the meat and the fish, in these second vessels, became wormy and

flies were seen entering and leaving at will; but in the closed flasks I did not see a worm, though many days had passed since the dead flesh had been put in them. Outside on the paper cover there was now and then a deposit, or a maggot that eagerly sought some crevice by which to enter and obtain nourishment. Meanwhile the different things placed in the flasks had become putrid and stinking; the fish, their bones excepted, had all been dissolved into a thick, turbid fluid, which on settling became clear, with a drop or so of liquid grease floating on the surface; but the snake kept its form intact, with the same color, as if it had been put in but yesterday; the eels, on the contrary, produced little liquid, though they had become very much swollen, and losing all shape, looked like a viscous mass of glue; the veal, after many weeks, became hard and dry.

Not content with these experiments, I tried many others at different seasons, using different vessels. . . . to leave nothing undone, I even had pieces of meat put under ground, but though remaining buried for weeks, they never bred worms, as was always the case when flies had been allowed to light on the meat. One day a large number of worms, which had bred in some buffalo meat, were killed by my order; having placed part in a closed dish, and part in an open one, nothing appeared in the first dish, but in the second worms had hatched, which changing as usual into egg-shape balls [pupae], finally became flies of the common kind. In the same experiment tried with dead flies, I never saw anything breed in the closed vessel. (pp. 34–35)

[3]The problems generated by nonresponse bias can appear in several forms. For example, nonresponse bias may be related to the size of an observed placebo effect. Consider a drug trial where the active treatment has a small positive effect, and the placebo has none. People who experience a negative or zero effect may be more likely to drop out of the study, with those remaining then generating an apparent positive placebo effect. In other words, differential dropout may help explain any manifest placebo effect.

[4]The complete Gary Taubes article is listed in the Suggested Reading, D.3.2. It is a readable and informative article explaining the discrepancies between what may be present in observational studies and what is then found in subsequent clinical trials.

[5]Terms commonly used in epidemiology: These are grouped below according to the chapters of the *Reference Manual on Scientific Evidence* in which they appeared.

"Reference Guide on Epidemiology" (Michael D. Green, D. Mical Freedman, & Leon Gordis):
agent: Also, risk factor. A factor, such as a drug, microorganism, chemical substance, or form of radiation, whose presence or absence

can result in the occurrence of a disease. A disease may be caused by a single agent or a number of independent alternative agents, or the combined presence of a complex of two or more factors may be necessary for the development of the disease. (p. 387)

bias: Any effect at any stage of investigation or inference tending to produce results that depart systematically from the true values. In epidemiology, the term bias does not necessarily carry an imputation of prejudice or other subjective factor, such as the experimenter's desire for a particular outcome. This differs from conventional usage, in which bias refers to a partisan point of view. (p. 388)

differential misclassification: A form of bias that is due to the misclassification of individuals or a variable of interest when the misclassification varies among study groups. This type of bias occurs when, for example, individuals in a study are incorrectly determined to be unexposed to the agent being studied when in fact they are exposed. (p. 390)

etiology: The cause of disease or other outcome of interest (p. 391). [The phrase "etiology unknown" simply means that it is of an unknown cause.]

misclassification bias. The erroneous classification of an individual in a study as exposed to the agent when the individual was not, or incorrectly classifying a study individual with regard to disease. Misclassification bias may exist in all study groups (nondifferential misclassification) or may vary among groups (differential misclassification). (p. 393)

pathognomonic: An agent is pathognomonic when it must be present for a disease to occur. Thus, asbestos is a pathognomonic agent for asbestosis. See signature disease. (p. 394)

secular-trend study: Also, time-line study. A study that examines changes over a period of time, generally years or decades. Examples include the decline of tuberculosis mortality and the rise, followed by a decline, in coronary heart disease mortality in the United States in the past fifty years. (p. 395)

signature disease. A disease that is associated uniquely with exposure to an agent (for example, asbestosis and exposure to asbestos). See pathognomonic. (p. 396)

teratogen: An agent that produces abnormalities in the embryo or fetus by disturbing maternal health or by acting directly on the fetus *in utero*. (p, 397)

"Reference Guide on Toxicology" (Bernard D. Goldstein & Mary Sue Henifin):

epigenetic: Pertaining to nongenetic mechanisms by which certain agents cause diseases, such as cancer. (p. 433)

mutagen: A substance that causes physical changes in chromosomes or biochemical changes in genes. (p. 435)

"Reference Guide on Medical Testimony" (Mary Sue Henifin, Howard M. Kipen, & Susan R. Poulter):

differential diagnosis: The term used by physicians to refer to the process of determining which of two or more diseases with similar symptoms and signs the patient is suffering from, by means of comparing the various competing diagnostic hypotheses with the clinical findings (p. 481). [The aim is to identify the disease to determine the treatment.]

differential etiology: A term used on occasion by expert witnesses or courts to describe the investigation and reasoning that leads to a determination of external causation, sometimes more specifically described by the witness or court as a process of identifying external causes by a process of elimination (p. 481). [Here, the goal is to identify the cause(s) of the disease but not to determine treatment.]

pathogenesis. The mode of origin or development of any disease or morbid process. (p. 482)

From the Cochrane Collaboration glossary:

phase I, II, III, IV trials: A series of levels of trials required of drugs before (and after) they are routinely used in clinical practice: Phase I trials assess toxic effects on humans (not many people participate in them, and usually without controls); Phase ll trials assess therapeutic benefit (usually involving a few hundred people, usually with controls, but not always); Phase III trials compare the new treatment against standard (or placebo) treatment (usually a full randomized controlled trial). At this point, a drug can be approved for community use. Phase IV monitors a new treatment in the community, often to evaluate long-term safety and effectiveness.

From www.MedicineNet.com:

idiopathic: of unknown cause. Any disease that is of uncertain or unknown origin may be termed "idiopathic."

[6] Although it may be tempting to make longitudinal inferences from cross-sectional data, this can produce incorrect conclusions. For example, consider a researcher of language development who spent a weekend in North Miami Beach; he observed that when people were young they spoke Spanish but when they were old they spoke Yiddish. He viewed this as suggesting an interesting developmental hypothesis, confirmed through the observation of adolescents working in local shops—they spoke mostly Spanish but also a little Yiddish.

[7] For another case-control success story, there is the identification of the synthetic estrogen DES given to mothers for pregnancy complications (from about 1940 to 1970), being the cause of a subsequent vaginal *adenocarcinoma* among those women exposed to DES *in utero* (Herbst, Ulfelder, & Poskanzer, 1971).

[8] For a more in depth discussion of the four bias terms given in this para-

graph, see Green et al. (2000); succinct definitions are given in the glossary to this later chapter.

[9]The idea is to control for the influence of a confounder by making comparisons separately within groups for which the confounding variable is nearly constant and therefore has little influence over the variables of primary interest. For example, smokers are more likely to get lung cancer than nonsmokers. Age, gender, social class, and region of residence are all confounders, but controlling for such variables does not really change the relationship between smoking and cancer rates. Furthermore, many different studies—of different types and on different populations—confirm the causal link. That is why most experts believe that smoking causes lung cancer and many other diseases.

[10]This is good advice more generally. Just because one can conjure up a reason for explaining why a particular result may have been observed, doesn't automatically mean that it is therefore true. For example, Caspi and colleagues (2003) identified a particular serotonin gene as possibly being related to depression, and we know that the common antidepressants all act on serotonin as re-uptake inhibitors. So, here is a reasonable mechanism to account for the relationship they apparently saw. The failure to replicate the Caspi results mentioned in Chapter 18, should be read against the seduction of an unproven serotonin mechanism.

[11]We are reminded of a quip generally attributed to George Box: "Don't emulate Pygmalion and fall in love with your model." Box provided a more detailed version of this sentiment in his R. A. Fisher Memorial Lecture, "Science and Statistics" (*Journal of the American Statistical Association*, 1976, *71*, pp. 791–799):

> The good scientist must have the flexibility and courage to seek out, recognize, and exploit such errors—especially his own. In particular, using Bacon's analogy, he must not be like Pygmalion and fall in love with his model. (pp. 791–792)

[12]As a general admonishment that might be best to keep in mind, remember the *law of unintended consequences*; any intervention in a complex system invariably creates unanticipated and often undesirable outcomes.

[13]An interesting historical use of controlled experimentation for scurvy is given in the Wikipedia article on the "Design of experiments" (2010):

> In 1747, while serving as surgeon on HM Bark Salisbury, James Lind carried out a controlled experiment to develop a cure for scurvy.
>
> Lind selected 12 men from the ship, all suffering from scurvy, and divided them into six pairs, giving each group different additions to their basic diet for a period of two weeks. The treatments were all remedies that had been proposed at one time or another. They were:
> – A quart of cider every day
> – Twenty five gutts (drops) of *elixir vitriol* (sulphuric acid) three times a day upon an empty stomach

– One half-pint of seawater every day

– A mixture of garlic, mustard, and horseradish in a lump the size of a nutmeg

– Two spoonfuls of vinegar three times a day

– Two oranges and one lemon every day

The men who had been given citrus fruits recovered dramatically within a week. One of them returned to duty after [six] days and the other became nurse to the rest. The others experienced some improvement, but nothing was comparable to the citrus fruits, which were proved to be substantially superior to the other treatments.

In this study, his subjects' cases "were as similar as I could have them," that is, he provided strict entry requirements to reduce extraneous variation. The men were paired, which provided replication. From a modern perspective, the main thing that is missing is randomized allocation of subjects to treatments.

[14]Informed consent should also mean just that, "informed." This implies that when drugs/screenings/treatments are involved, all the relevant probabilities are provided—positive predictive values, specificities, sensitivities, the likelihood of adverse events, and so on.

[15]Two terms defined in the Cochrane Collaboration Glossary (2005) are useful when discussing issues raised by an "intention-to-treat" analysis:

per protocol analysis: An analysis of the subset of participants from a randomized controlled trial who complied with the protocol sufficiently to ensure that their data would be likely to exhibit the effect of treatment. This subset may be defined after considering exposure to treatment, availability of measurements and absence of major protocol violations. The per protocol analysis strategy may be subject to bias as the reasons for noncompliance may be related to treatment.

performance bias: Systematic differences between intervention groups in care provided apart from the intervention being evaluated. For example, if participants know they are in the control group, they may be more likely to use other forms of care. If care providers are aware of the group a particular participant is in, they might act differently. Blinding of study participants (both the recipients and providers of care) is used to protect against performance bias.

[16]An implication of this quotation is the need for obtaining added detail even in randomized medical trials and the explicit collection of the same kind of covariate information done routinely in observational studies. The availability of this information offers some hope of post hoc adjustment and explanation of the trial results.

[17]The interpretation of results and the source of the information given in

this section, *An Evaluation of the 1954 Poliomyelitis Vaccine Trials*, is by Thomas Francis, Robert Korns, and colleagues (1955) (in particular, see Table 2b: Summary of Study of Cases by Diagnostic Class and Vaccination Status; p. 35).

Chapter 16

Ethical Considerations in Data Collection and Analysis Involving Human Experimentation

Statisticians and other quantitatively oriented behavioral and medical scientists who do analyses and interpretations of data obtained from human experimentation are expected to follow the established ethical guidelines that control such experimentation. The American Statistical Association, for example, in its *Ethical Guidelines* (1999), has an explicit section entitled "Responsibilities to Research Subjects (including census or survey respondents and persons and organizations supplying data from administrative records, as well as subjects of physically or psychologically invasive research)." We give four of the more germane points from this particular section (and reproduce the complete ASA Ethical Guidelines in an appendix to this chapter):

1. Know about and adhere to appropriate rules for the protection of human subjects, including particularly vulnerable or other special populations that may be subject to special risks or may not be fully able to protect their own interests. Ensure adequate planning to support the practical value of the research, validity of expected results, ability to provide the protection promised, and consideration of all other ethical issues involved.

6. Before participating in a study involving human beings or organizations, analyzing data from such a study, or accepting resulting manuscripts for review, consider whether appropriate research subject approvals were obtained. (This safeguard will lower your risk of learning only after the fact that you have collaborated on an unethical study.) Consider also what assurances of privacy and confidentiality were given and abide by those assurances.

7. Avoid or minimize the use of deception. Where it is necessary and provides significant knowledge—as in some psychological, sociological, and other research—ensure prior independent ethical review of the protocol and continued monitoring of the research.

8. Where full disclosure of study parameters to subjects or other investigators is not advisable, as in some randomized clinical trials, generally inform them of the nature of the information withheld and the reason for withholding it. As with deception, ensure independent ethical review of the protocol and continued monitoring of the research.

This chapter discusses three landmarks in the development of ethical guidelines for human experimentation: the Nuremberg Code resulting from the war crimes trial of Nazi doctors after the close of World War II, the passage of the National Research Act in 1974 partly because of public exposure of the Tuskegee syphilis experiment that ran from 1932 to 1972, and the Declaration of Helsinki first adopted in 1964 by the World Medical Association (and revised many times since) that until recently has been the guiding document internationally for all medically related experimental trials of drugs, medical products, vaccines, and similar health-related interventions.

16.1   The Nazi Doctors' Trial and the Nuremberg Code

A momentous event in the ethics of human experimentation occurred with the Nazi Doctors' Trial in Nuremberg in 1946. Formally known as *United States of America v. Karl Brandt et al.*, it produced the Nuremberg Code within the final ruling given by Justice Walter Beals. This short statement of ten principles has formed the basis for all later codifications of ethical principles governing human experimentation. The two United States doctors attached to the trial as advisers, Andrew Ivy and Leo Alexander, are believed jointly responsible for the wording of the Code in the form used by Justice Beals. The ten short principles are reproduced in an appendix to this chapter, with the first rule of informed consent being the longest and most important (see Temme, 2003).

As noted, all later incarnations of ethical guidelines for human experimentation include variations on these summary statements issued at Nuremberg, except for the fifth principle. Here, it is argued that experimentation, unethical to begin with, doesn't somehow become ethical merely by having the experimental researchers or physicians also willing to take part. Also, a study can be evaluated as being unethical at its outset, irrespective of the value of the data that might be obtained. No matter how laudable the goal, it can't justify an unethical mechanism for reaching it. Or, to put a Latin phrase to good use, this is not a situation of *exitus acta probat* (the outcome justifies the deed).

The person named in the Nuremberg Doctors' Trial, Karl Brandt, was Adolf Hitler's personal physician, and head of the administration for the Nazi euthanasia program from 1939 onward. In his position as Major General Reich Commissioner for Health and Sanitation, he was involved in incredibly brutal human experimentation. Brandt and six of the other named defendants were convicted of medical war crimes, many carried out at various Third Reich concentration camps, and were hanged at Landsberg Prison ("Karl Brandt," 2011).

The attorney for Karl Brandt raised several points of defense suggesting there were no real differences between what the Nazi doctors had done and the type of human experimentation performed in the United States. One major instance cited was the study of malaria vaccine on prisoners at Stateville Prison in Joliet, Illinois ("Stateville Penitentiary Malaria Study," 2011). This story was prominently featured in *LIFE* magazine (June 4, 1945), under the title, "Prisoner Malaria: Convicts Expose Themselves to Disease So Doctors Can Study It." To rebut this evidence, Andrew Ivy was summoned by the prosecution to testify at Nuremburg, which he did. But before going, Ivy asked the Illinois Governor, Dwight Green, to form an ad hoc committee to advise on ethical considerations in medical experimentation. Although the committee never met, Ivy testified at Nuremberg that the committee had issued a report, known as the Green report ("Green report," 2011). In fact, Ivy had written the document himself, justifying the ethicality of prison research and refusing to make any parallels to the Nazi medical experimentation. The Green report was later published in the *Journal of the American Medical Association* (1948), after minor belated input and editing from the Governor's ad hoc committee. For some decades thereafter, it served as a justification for continued medical research on prisoners.

A second defense argument raised by Brandt's attorney was

that the view of undesirable races and the resulting population policies of the Third Reich were not unusual or even unique to Nazi Germany. As documentation of this, excerpts from Madison Grant's, *The Passing of the Great Race* (1916), were introduced as evidence.[1] Supposedly, Grant's popular book was Adolf Hilter's favorite; he even wrote Grant a fan letter applauding it and commenting that the book was "his Bible." We redact in an appendix to this chapter some of Grant's fourth chapter, *The Competition of Races*, with parts italicized that were used as explicit defense evidence for Brandt ("Madison Grant," 2011).[2]

16.2   The National Research Act of 1974

The Tuskegee syphilis study is arguably the most infamous and unethical biomedical study ever performed in the United States. It was conducted by the United States Public Health Service from 1932 until its exposure in the national press in 1972. For some historical background, we redact below the introduction to the Wikipedia article on the "Tuskegee syphilis experiment" (2011):[3]

> The Public Health Service, working with the Tuskegee Institute, began the study in 1932. Investigators enrolled in the study 399 impoverished African-American sharecroppers from Macon County, Alabama, infected with syphilis. For participating in the study, the men were given free medical exams, free meals and free burial insurance. They were never told they had syphilis, nor were they ever treated for it. According to the Centers for Disease Control, the men were told they were being treated for "bad blood," a local term used to describe several illnesses, including syphilis, anemia and fatigue.
>
> The 40-year study was controversial for reasons related to ethical standards, primarily because researchers failed to treat patients appropriately after the 1940s validation of penicillin as an effective cure for the disease. Revelation of study failures led to major changes in United States law and regulation on the protection of participants in clinical studies. Now studies require informed consent (with exceptions possible for United States Federal agencies which can be kept secret by Executive Order), communication of diagnosis, and accurate reporting of test results.
>
> By 1947 penicillin had become the standard treatment for

syphilis. Choices might have included treating all syphilitic subjects and closing the study, or splitting off a control group for testing with penicillin. Instead, the Tuskegee scientists continued the study, withholding penicillin and information about it from the patients. In addition, scientists prevented participants from accessing syphilis treatment programs available to others in the area. The study continued, under numerous supervisors, until 1972, when a leak to the press resulted in its termination. Victims included numerous men who died of syphilis, wives who contracted the disease, and children born with congenital syphilis.[4]

The *National Research Act* (of 1974) was passed partly because of the Tuskegee study. It created the National Commission for the Protection of Human Subjects of Biomedical and Behavioral Research to oversee and regulate human experimentation. In turn, this Act lead to the 1979 Belmont Report, named for the Smithsonian Institution's Belmont Conference Center. The Report laid out the basic ethical principles identified by the Commission over some four years of deliberation. It lead to the formation of the Office for Human Research Protection (OHRP) within the United States Department of Heath and Human Services, and to the establishment of the now ubiquitous Institutional Review Boards for the protection of human subjects in all forms of medical and behavioral experimentation. The main body of the Belmont Report is given in an appendix to this chapter. Explicit attention should be focused on the three general and controlling ethical principles: respect for persons, beneficence, and justice.

16.3   The Declaration of Helsinki

From the late 1970s and continuing to the present, the International Committee of Medical Journal Editors has regularly updated a set of guidelines for writing and editing in biomedical publication. This document is entitled *Uniform Requirements for Manuscripts Submitted to Biomedical Journals: Writing and Editing for Biomedical Publications*. It includes the following section on the Protection of Human Subjects and Animals in Research:

> When reporting experiments on human subjects, authors should in-
> dicate whether the procedures followed were in accordance with the
> ethical standards of the responsible committee on human experimen-
> tation (institutional and national) and with the Helsinki Declaration
> of 1975, as revised in 2008. If doubt exists whether the research was
> conducted in accordance with the Helsinki Declaration, the authors
> must explain the rationale for their approach and demonstrate that
> the institutional review body explicitly approved the doubtful as-
> pects of the study. When reporting experiments on animals, authors
> should indicate whether the institutional and national guide for the
> care and use of laboratory animals was followed. (p. 6)

As this paragraph indicates, the Declaration of Helsinki adopted
by the World Medical Association is to be the controlling set of
ethical guidelines for human (medical) experimentation. The 2008
revision is given in an appendix to this chapter (with some parts
italicized that will be commented on later).

As of October 27, 2008, the United States Food and Drug Administration (FDA) discontinued its reliance on the Declaration
of Helsinki (DOH) in favor of an alternative—Guideline for Good
Clinical Practice (GCP)—developed with significant input from
the large international drug companies (see International Conference on Harmonisation of Technical Requirements for Registration
of Pharmaceuticals for Human Use). Part of a short item that appeared in the *Lancet* (2009, *373*, 13–14) is given below ("Helsinki
Discords: FDA, Ethics, and International Drug Trials"; Jonathan
Kimmelman, Charles Weijer, and Eric Meslin):

> Since 1964, the Declaration of Helsinki has stood as one of the world's
> most authoritative statements on ethical standards for human re-
> search. Drafted by the World Medical Association to provide med-
> ical researchers with ethical guidance, the Declaration has under-
> gone six major revisions, most recently in October, 2008. For many
> years the US Food and Drug Administration (FDA) has required
> that foreign clinical studies supporting applications for drug licen-
> sure comply with the Declaration. However, on Oct 27, 2008, the
> FDA formally discontinued its reliance on the Declaration and sub-
> stituted the International Conference on Harmonization's Guideline
> for Good Clinical Practice (GCP).
>
> The rationale behind the FDA's action is complex, and no doubt
> reflects an effort to balance important interests and public-policy
> goals. Among the FDA's reasons are the need to assure the quality
> of foreign data submitted to the agency, a wish to prevent confu-

sion among researchers when the Declaration of Helsinki undergoes revision, and a worry that future modifications could "contain provisions that are inconsistent with US laws and regulations". The FDA's latest action completes a process begun in 2001 when the agency declined to recognise the 2000 revision, in part due to the Declaration's restrictive stance on placebo-controlled trials in economically developing countries. The practical consequences of the FDA's current action are unclear because the ruling applies to only a subset of clinical trials—i.e., international trials. Moreover, several countries that host such research have regulations that endorse or emulate the Declaration of Helsinki.

Nevertheless, at a time when the volume of overseas trials is increasing, the FDA's new policy is troubling. First, the Declaration of Helsinki has a moral authority that GCP lacks. The Declaration has long been recognised as a leading international ethical standard for research. Whereas the World Medical Association includes 85 national medical societies from every part of the globe, the International Conference on Harmonization consists of only voting members from the USA, the European Union, and Japan. Indeed, the authors of GCP acknowledge the authority of the Declaration of Helsinki when they state that a goal of GCP is "consisten[cy] with the principles that have their origin in the Declaration of Helsinki". The FDA regulates the largest drug market in the world and we worry that its replacement of the Declaration of Helsinki with a less morally authoritative document may cause others to follow suit, thereby undermining international ethical standards for research. Second, the Declaration of Helsinki has a breadth and depth that GCP lacks. For sure, GCP covers similar topics to the Declaration, but the focus of GCP is regulatory harmonisation, not the articulation of ethical commitments. Careful examination of the two documents reveals several important ethical issues that are addressed in the Declaration about which GCP is silent (see the inset panel below).

Thus reliance on GCP rather than on the Declaration of Helsinki may result in less protection for research participants. If so, the FDA's action might lower the bar for international research under its purview—a scenario that has worried previous commentators.

Third, the FDA's departure from the Declaration of Helsinki could undermine its stated goals of clarity and regulatory harmonisation. For example, if many countries continue to use the Declaration, US researchers will encounter the same "confusion" that the FDA is attempting to prevent with its new rule. Similarly, should other countries follow the FDA's lead and abandon the Declaration of Helsinki, the result could be the balkanisation of ethical standards in international research.

In view of these concerns, we suggest the new US administration suspend this rule pending a review of the implications for US-sponsored research overseas. If such review confirms our concerns, the FDA should be directed to rejoin the international community in requiring that studies be done in accordance with the Declaration of Helsinki. We also see an important role for major medical societies—though they lack regulatory authority, collectively these organisations can give voice to the commitment of medical researchers to the Declaration's high ethical standards. The American Society of Gene Therapy is considering policy on this issue and others should follow suit.

Panel: Requirements in latest revision of Declaration of Helsinki but absent in GCP

— Investigators to disclose funding, sponsors, and other potential conflicts of interest to both research ethics committees and study participants [DOH para. 14]

— Study design to be disclosed publicly (e.g., in clinical trial registries) [DOH para. 19]

— Research, notably that in developing countries, to benefit and be responsive to health needs of populations in which it is done [DOH para. 17]

— Restricted use of placebo controls in approval process for new drugs and in research done in developing countries [DOH para. 32]

— Post-trial access to treatment [DOH para. 14 and 33]

— Authors to report results accurately, and publish or make public negative findings [DOH para. 30]

Appendix: American Statistical Association Ethical Guidelines for Statistical Practice

Prepared by the Committee on Professional Ethics
Approved by the Board of Directors, August 7, 1999
Executive Summary:
This document contains two parts: I. Preamble and II. Ethical Guidelines. The Preamble addresses A. Purpose of the Guidelines, B. Statistics and Society, and C. Shared Values. The purpose of the document is to encourage ethical and effective statistical work in morally conducive working environments. It is also intended to assist students in learning to perform statistical work responsibly. Statistics plays a vital role in many aspects of science, the economy, governance, and even entertainment. It is important that all statis-

tical practitioners recognize their potential impact on the broader society and the attendant ethical obligations to perform their work responsibly. Furthermore, practitioners are encouraged to exercise "good professional citizenship" in order to improve the public climate for, understanding of, and respect for the use of statistics throughout its range of applications.

The Ethical Guidelines address eight general topic areas and specify important ethical considerations under each topic.

A. Professionalism points out the need for competence, judgment, diligence, self-respect, and worthiness of the respect of other people.

B. Responsibilities to Funders, Clients, and Employers discusses the practitioner's responsibility for assuring that statistical work is suitable to the needs and resources of those who are paying for it, that funders understand the capabilities and limitations of statistics in addressing their problem, and that the funder's confidential information is protected.

C. Responsibilities in Publications and Testimony addresses the need to report sufficient information to give readers, including other practitioners, a clear understanding of the intent of the work, how and by whom it was performed, and any limitations on its validity.

D. Responsibilities to Research Subjects describes requirements for protecting the interests of human and animal subjects of research-not only during data collection but also in the analysis, interpretation, and publication of the resulting findings.

E. Responsibilities to Research Team Colleagues addresses the mutual responsibilities of professionals participating in multidisciplinary research teams.

F. Responsibilities to Other Statisticians or Statistical Practitioners notes the interdependence of professionals doing similar work, whether in the same or different organizations. Basically, they must contribute to the strength of their professions overall by sharing nonproprietary data and methods, participating in peer review, and respecting differing professional opinions.

G. Responsibilities Regarding Allegations of Misconduct addresses the sometimes painful process of investigating potential ethical violations and treating those involved with both justice and respect.

H. Responsibilities of Employers, Including Organizations, Individuals, Attorneys, or Other Clients Employing Statistical Practitioners encourages employers and clients to recognize the highly interdependent nature of statistical ethics and statistical validity. Employers and clients must not pressure practitioners to produce a particular "result," regardless of its statistical validity. They must avoid the potential social harm that can result from the dissemination of false or misleading statistical work.

I. PREAMBLE

A. Purpose of the Guidelines

The American Statistical Association's Ethical Guidelines for Statistical Practice are intended to help statistics practitioners make and communicate ethical decisions. Clients, employers, researchers, policymakers, journalists, and the public should be urged to expect statistical practice to be conducted in accordance with these guidelines and to object when it is not. While learning how to apply statistical theory to problems, students should be encouraged to use these guidelines, regardless of whether their target professional specialty will be "statistician." Employers, attorneys, and other clients of statistics practitioners have a responsibility to provide a moral environment that fosters the use of these ethical guidelines.

Application of these or any other ethical guidelines generally requires good judgment and common sense. The guidelines may be partially conflicting in specific cases. The application of these guidelines in any given case can depend on issues of law and shared values; work-group politics; the status and power of the individuals involved; and the extent to which the ethical lapses pose a threat to the public, to one's profession, or to one's organization. The individuals and institutions responsible for making such ethical decisions can receive valuable assistance by discussion and consultation with others, particularly persons with divergent interests with respect to the ethical issues under consideration.

B. Statistics and Society

The professional performance of statistical analyses is essential to many aspects of society. The use of statistics in medical diagnoses and biomedical research may affect whether individuals live or die, whether their health is protected or jeopardized, and whether medical science advances or gets sidetracked. Life, death, and health, as well as efficiency, may be at stake in statistical analyses of occupational, environmental, or transportation safety. Early detection and control of new or recurrent infectious diseases depend on sound epidemiological statistics. Mental and social health may be at stake in psychological and sociological applications of statistical analysis.

. Effective functioning of the economy depends on the availability of reliable, timely, and properly interpreted economic data. The profitability of individual firms depends in part on their quality control and market research, both of which should rely on statistical methods. Agricultural productivity benefits greatly from statistically sound applications to research and output reporting. Governmental policy decisions regarding public health, criminal justice, social equity, education, the environment, the citing of critical facilities, and other matters depend in part on sound statistics.

Scientific and engineering research in all disciplines requires the careful design and analysis of experiments and observations. To the extent that uncertainty and measurement error are involved-as they are in most research-research design, data quality management, analysis, and interpretation are all crucially dependent on statistical concepts and methods. Even in theory, much of science and engineering involves natural variability. Variability, whether great or small, must be carefully examined for both random error and possible researcher bias or wishful thinking.

Statistical tools and methods, as with many other technologies, can be employed either for social good or evil. The professionalism encouraged by these guidelines is predicated on their use in socially responsible pursuits by morally responsible societies, governments, and employers. Where the end purpose of a statistical application is itself morally reprehensible, statistical professionalism ceases to have ethical worth.

C. Shared Values

Because society depends on sound statistical practice, all practitioners of statistics, whatever their training and occupation, have social obligations to perform their work in a professional, competent, and ethical manner. This document is directed to those whose primary occupation is statistics. Still, the principles expressed here should also guide the statistical work of professionals in all other disciplines that use statistical methods. All statistical practitioners are obliged to conduct their professional activities with responsible attention to the following:

1. The social value of their work and the consequences of how well or poorly it is performed. This includes respect for the life, liberty, dignity, and property of other people.

2. The avoidance of any tendency to slant statistical work toward predetermined outcomes. (It is acceptable to advocate a position; it is not acceptable to misapply statistical methods in doing so.)

3. Statistics as a science. (As in any science, understanding evolves. Statisticians have a body of established knowledge, but also many unresolved issues that deserve frank discussion.)

4. The maintenance and upgrading of competence in their work.

5. Adherence to all applicable laws and regulations, as well as applicable international covenants, while also seeking to change any of those that are ethically inappropriate.

6. Preservation of data archives in a manner consistent with responsible protection of the safety and confidentiality of any human being or organization involved.

In addition to ethical obligations, good professional citizenship encourages the following:

7. Collegiality and civility with fellow professionals.

8. Support for improved public understanding of and respect for statistics.

9. Support for sound statistical practice, especially when it is unfairly criticized.

10. Exposure of dishonest or incompetent uses of statistics.

11. Service to one's profession as a statistical editor, reviewer, or association official and service as an active participant in (formal or informal) ethical review panels.

II. ETHICAL GUIDELINES

A. Professionalism

1. Strive for relevance in statistical analyses. Typically, each study should be based on a competent understanding of the subject-matter issues, statistical protocols that are clearly defined for the stage (exploratory, intermediate, or final) of analysis before looking at those data that will be decisive for that

stage, and technical criteria to justify both the practical relevance of the study and the amount of data to be used.

2. Guard against the possibility that a predisposition by investigators or data providers might predetermine the analytic result. Employ data selection or sampling methods and analytic approaches that are designed to ensure valid analyses in either frequentist or Bayesian approaches.

3. Remain current in dynamically evolving statistical methodology; yesterday's preferred methods may be barely acceptable today and totally obsolete tomorrow.

4. Ensure that adequate statistical and subject-matter expertise is both applied to any planned study. If this criterion is not met initially, it is important to add the missing expertise before completing the study design.

5. Use only statistical methodologies suitable to the data and to obtaining valid results. For example, address the multiple potentially confounding factors in observational studies and use due caution in drawing causal inferences.

6. Do not join a research project unless you can expect to achieve valid results and you are confident that your name will not be associated with the project or resulting publications without your explicit consent.

7. The fact that a procedure is automated does not ensure its correctness or appropriateness; it is also necessary to understand the theory, data, and methods used in each statistical study. This goal is served best when a competent statistical practitioner is included early in the research design, preferably in the planning stage.

8. Recognize that any frequentist statistical test has a random chance of indicating significance when it is not really present. Running multiple tests on the same dataset at the same stage of an analysis increases the chance of obtaining at least one invalid result. Selecting the one "significant" result from a multiplicity of parallel tests poses a grave risk of an incorrect conclusion. Failure to disclose the full extent of tests and their results in such a case would be highly misleading.

9. Respect and acknowledge the contributions and intellectual property of others.

10. Disclose conflicts of interest, financial and otherwise, and resolve them. This may sometimes require divestiture of the conflicting personal interest or withdrawal from the professional activity. Examples where conflict of interest may be problematic include grant reviews, other peer reviews, and tensions between scholarship and personal or family financial interests.

11. Provide only such expert testimony as you would be willing to have peer reviewed.

B. Responsibilities to Funders, Clients, and Employers

1. Where appropriate, present a client or employer with choices among valid alternative statistical approaches that may vary in scope, cost, or precision.

2. Clearly state your statistical qualifications and experience relevant to your work.

3. Clarify the respective roles of different participants in studies to be undertaken.

4. Explain any expected adverse consequences of failure to follow through on an agreed-upon sampling or analytic plan.

5. Apply statistical sampling and analysis procedures scientifically, without predetermining the outcome.

6. Make new statistical knowledge widely available to provide benefits to society at large and beyond your own scope of applications. Statistical methods may be broadly applicable to many classes of problem or application. (Statistical innovators may well be entitled to monetary or other rewards for their writings, software, or research results.)

7. Guard privileged information of the employer, client, or funder.

8. Fulfill all commitments.

9. Accept full responsibility for your professional performance.

C. Responsibilities in Publications and Testimony

1. Maintain personal responsibility for all work bearing your name; avoid undertaking work or coauthoring publications for which you would not want to acknowledge responsibility. Conversely, accept (or insist upon) appropriate authorship or acknowledgment for professional statistical contributions to research and the resulting publications or testimony.

2. Report statistical and substantive assumptions made in the study.

3. In publications or testimony, identify who is responsible for the statistical work if it would not otherwise be apparent.

4. Make clear the basis for authorship order, if determined on grounds other than intellectual contribution. Preferably, authorship order in statistical publications should be by degree of intellectual contribution to the study and material to be published, to the extent that such ordering can feasibly be determined. When some other rule of authorship order is used in a statistical publication, the rule should be disclosed in a footnote or endnote. (Where authorship order by contribution is assumed by those making decisions about hiring, promotion, or tenure, for example, failure to disclose an alternative rule may improperly damage or advance careers.)

5. Account for all data considered in a study and explain the sample(s) actually used.

6. Report the sources and assessed adequacy of the data.

7. Report the data cleaning and screening procedures used, including any imputation.

8. Clearly and fully report the steps taken to guard validity. Address the suitability of the analytic methods and their inherent assumptions relative to the circumstances of the specific study. Identify the computer routines used to implement the analytic methods.

9. Where appropriate, address potential confounding variables not included in the study.

10. In publications or testimony, identify the ultimate financial sponsor of the study, the stated purpose, and the intended use of the study results.

11. When reporting analyses of volunteer data or other data not representative of a defined population, include appropriate disclaimers.

12. Report the limits of statistical inference of the study and possible sources of error. For example, disclose any significant failure to follow through fully on an agreed sampling or analytic plan and explain any resulting adverse consequences.

13. Share data used in published studies to aid peer review and replication, but exercise due caution to protect proprietary and confidential data, including all data that might inappropriately reveal respondent identities.

14. As appropriate, promptly and publicly correct any errors discovered after publication.

15. Write with consideration of the intended audience. (For the general public, convey the scope, relevance, and conclusions of a study without technical distractions. For the professional literature, strive to answer the questions likely to occur to your peers.)

D. Responsibilities to Research Subjects (including census or survey respondents and persons and organizations supplying data from administrative records, as well as subjects of physically or psychologically invasive research)

1. Know about and adhere to appropriate rules for the protection of human subjects, including particularly vulnerable or other special populations that may be subject to special risks or may not be fully able to protect their own interests. Ensure adequate planning to support the practical value of the research, validity of expected results, ability to provide the protection promised, and consideration of all other ethical issues involved.

2. Avoid the use of excessive or inadequate numbers of research subjects by making informed recommendations for study size. These recommendations may be based on prospective power analysis, the planned precision of the study endpoint(s), or other methods to ensure appropriate scope to either frequentist or Bayesian approaches. Study scope also should take into consideration the feasibility of obtaining research subjects and the value of the data elements to be collected.[5]

3. Avoid excessive risk to research subjects and excessive imposition on their time and privacy.

4. Protect the privacy and confidentiality of research subjects and data concerning them, whether obtained directly from the subjects, other persons, or administrative records. Anticipate secondary and indirect uses of the data when obtaining approvals from research subjects; obtain approvals appropriate for peer review and independent replication of analyses.

5. Be aware of legal limitations on privacy and confidentiality assurances. Do not, for example, imply protection of privacy and confidentiality from legal processes of discovery unless explicitly authorized to do so.

6. Before participating in a study involving human beings or organizations, analyzing data from such a study, or accepting resulting manuscripts for review, consider whether appropriate research subject approvals were obtained. (This safeguard will lower your risk of learning only after the fact that you

have collaborated on an unethical study.) Consider also what assurances of privacy and confidentiality were given and abide by those assurances.

7. Avoid or minimize the use of deception. Where it is necessary and provides significant knowledge-as in some psychological, sociological, and other research-ensure prior independent ethical review of the protocol and continued monitoring of the research.

8. Where full disclosure of study parameters to subjects or other investigators is not advisable, as in some randomized clinical trials, generally inform them of the nature of the information withheld and the reason for withholding it. As with deception, ensure independent ethical review of the protocol and continued monitoring of the research.

9. Know about and adhere to appropriate animal welfare guidelines in research involving animals. Ensure that a competent understanding of the subject matter is combined with credible statistical validity.

E. Responsibilities to Research Team Colleagues

1. Inform colleagues from other disciplines about relevant aspects of statistical ethics.

2. Promote effective and efficient use of statistics by the research team.

3. Respect the ethical obligations of members of other disciplines, as well as your own.

4. Ensure professional reporting of the statistical design and analysis.

5. Avoid compromising statistical validity for expediency, but use reasonable approximations as appropriate.

F. Responsibilities to Other Statisticians or Statistics Practitioners

1. Promote sharing of (nonproprietary) data and methods. As appropriate, make suitably documented data available for replicate analyses, meta-data studies, and other suitable research by qualified investigators.

2. Be willing to help strengthen the work of others through appropriate peer review. When doing so, complete the review promptly and well.

3. Assess methods, not individuals.

4. Respect differences of opinion.

5. Instill in students an appreciation for the practical value of the concepts and methods they are learning.

6. Use professional qualifications and the contributions of the individual as an important basis for decisions regarding statistical practitioners' hiring, firing, promotion, work assignments, publications and presentations, candidacy for offices and awards, funding or approval of research, and other professional matters. Avoid as best you can harassment of or discrimination against statistical practitioners (or anyone else) on professionally irrelevant bases such as race, color, ethnicity, sex, sexual orientation, national origin, age, religion, nationality, or disability.

G. Responsibilities Regarding Allegations of Misconduct

1. Avoid condoning or appearing to condone careless, incompetent, or unethical practices in statistical studies conducted in your working environment or elsewhere.

2. Deplore all types of professional misconduct, not just plagiarism and data fabrication or falsification. Misconduct more broadly includes all professional dishonesty, by commission or omission, and, within the realm of professional activities and expression, all harmful disrespect for people, unauthorized use of their intellectual and physical property, and unjustified detraction from their reputations.

3. Recognize that differences of opinion and honest error do not constitute misconduct; they warrant discussion, but not accusation. Questionable scientific practices may or may not constitute misconduct, depending on their nature and the definition of misconduct used.

4. If involved in a misconduct investigation, know and follow prescribed procedures. Maintain confidentiality during an investigation, but disclose the results honestly after the investigation has been completed.

5. Following a misconduct investigation, support the appropriate efforts of the accused, the witnesses, and those reporting the possible scientific error or misconduct to resume their careers in as normal a manner as possible.

6. Do not condone retaliation against or damage to the employability of those who responsibly call attention to possible scientific error or misconduct.

H. Responsibilities of Employers, Including Organizations, Individuals, Attorneys, or Other Clients Employing Statistical Practitioners

1. Recognize that the results of valid statistical studies cannot be guaranteed to conform to the expectations or desires of those commissioning the study or the statistical practitioner(s). Any measures taken to ensure a particular outcome will lessen the validity of the analysis.

2. Valid findings result from competent work in a moral environment. Pressure on a statistical practitioner to deviate from these guidelines is likely to damage both the validity of study results and the professional credibility of the practitioner.

3. Make new statistical knowledge widely available in order to benefit society at large. (Those who have funded the development of statistical innovations are entitled to monetary and other rewards for their resulting products, software, or research results.)[6]

4. Support sound statistical analysis and expose incompetent or corrupt statistical practice. In cases of conflict, statistical practitioners and those employing them are encouraged to resolve issues of ethical practice privately. If private resolution is not possible, recognize that statistical practitioners have an ethical obligation to expose incompetent or corrupt practice before it can cause harm to research subjects or society at large.

5. Recognize that within organizations and within professions using statistical methods generally, statistics practitioners with greater prestige, power, or status have a responsibility to protect the professional freedom and responsibility of more subordinate statistical practitioners who comply with these guidelines.

6. Do not include statistical practitioners in authorship or acknowledge

their contributions to projects or publications without their explicit permission.

Appendix: The Nuremberg Code (1949)

1. The voluntary consent of the human subject is absolutely essential. This means that the person involved should have legal capacity to give consent; should be so situated as to be able to exercise free power of choice, without the intervention of any element of force, fraud, deceit, duress, overreaching, or other ulterior form of constraint or coercion; and should have sufficient knowledge and comprehension of the elements of the subject matter involved as to enable him/her to make an understanding and enlightened decision. This latter element requires that before the acceptance of an affirmative decision by the experimental subject there should be made known to him the nature, duration, and purpose of the experiment; the method and means by which it is to be conducted; all inconveniences and hazards reasonable to be expected; and the effects upon his health or person which may possibly come from his participation in the experiment.

The duty and responsibility for ascertaining the quality of the consent rests upon each individual who initiates, directs or engages in the experiment. It is a personal duty and responsibility which may not be delegated to another with impunity.

2. The experiment should be such as to yield fruitful results for the good of society, unprocurable by other methods or means of study, and not random and unnecessary in nature.

3. The experiment should be so designed and based on the results of animal experimentation and a knowledge of the natural history of the disease or other problem under study that the anticipated results will justify the performance of the experiment.

4. The experiment should be so conducted as to avoid all unnecessary physical and mental suffering and injury.

5. No experiment should be conducted where there is a prior reason to believe that death or disabling injury will occur; except, perhaps, in those experiments where the experimental physicians also serve as subjects.

6. The degree of risk to be taken should never exceed that determined by the humanitarian importance of the problem to be solved by the experiment.

7. Proper preparations should be made and adequate facilities provided to protect the experimental subject against even remote possibilities of injury, disability, or death.

8. The experiment should be conducted only by scientifically qualified persons. The highest degree of skill and care should be required through all stages of the experiment of those who conduct or engage in the experiment.

9. During the course of the experiment the human subject should be at liberty to bring the experiment to an end if he has reached the physical or mental state where continuation of the experiment seems to him to be impossible.

10. During the course of the experiment the scientist in charge must be prepared to terminate the experiment at any stage, if he has probable cause to believe, in the exercise of the good faith, superior skill and careful judgment required of him that a continuation of the experiment is likely to result in injury, disability, or death to the experimental subject.

Appendix: Excerpts from Madison Grant's *The Passing of the Great Race*(1916); the material in italics was offered as evidence in Karl Brandt's trial

Where two races occupy a country side by side, it is not correct to speak of one type as changing into the other. Even if present in equal numbers one of the two contrasted types will have some small advantage or capacity which the other lacks toward a perfect adjustment to surroundings. Those possessing these favorable variations will flourish at the expense of their rivals and their offspring will not only be more numerous, but will also tend to inherit such variations. In this way one type gradually breeds the other out. In this sense, and in this sense only, do races change.

Man continuously undergoes selection through the operation of the forces of social environment. (p. 46)

The lowering of the birth rate among the most valuable classes, while the birth rate of the lower classes remains unaffected, is a frequent phenomenon of prosperity. Such a change becomes extremely injurious to the race if unchecked, unless nature is allowed to maintain by her own cruel devices the relative numbers of the different classes in their due proportions. To attack race suicide by encouraging indiscriminate reproduction is not only futile but is dangerous if it leads to an increase in the undesirable elements. *What is needed in the community most of all is an increase in the desirable classes, which are of superior type physically, intellectually and morally and not merely an increase in the absolute numbers of the population. The value and efficiency of a population are not numbered by what the newspapers call souls, but by the proportion of men of physical and intellectual vigor.* (p. 47–48)

Where altruism, philanthropy or sentimentalism intervene with the noblest purpose and forbid nature to penalize the unfortunate victims of reckless breeding, the multiplication of inferior types is encouraged and fostered. Indiscriminate efforts to preserve babies among the lower classes often result in serious injury to the race. At the existing stage of civilization, the legalizing of birth control would probably be of benefit by reducing the number of offspring in the undesirable classes. Regulation of the number of children is, for good or evil, in full operation among the better classes and its recognition by the state would result in no further harm among them.

Mistaken regard for what are believed to be divine laws and a sentimental belief in the sanctity of human life tend to prevent both the elimination of defective infants and the sterilization of such adults as are themselves of no value to the community. The laws of nature require the obliteration of the unfit and human life is valuable only when it is of use to the community or race.

It is highly unjust that a minute minority should be called upon to supply brains for the unthinking mass of the community, but it is even worse to burden the responsible and larger but still overworked elements in the community with an ever increasing number of moral perverts, mental defectives and hereditary cripples. As the percentage of incompetents increases, the burden of their support will become ever more onerous until, at no distant date, society will in self-defense put a stop to the supply of feebleminded and criminal children of weaklings.

The church assumes a serious responsibility toward the future of the race whenever it steps in and preserves a defective strain. The marriage of deaf mutes was hailed a generation ago as a triumph of humanity. Now it is recognized as an absolute crime against the race. A great injury is done to the community by the perpetuation of worthless types. These strains are apt to be meek and lowly and as such make a strong appeal to the sympathies of the successful. Before eugenics were understood much could be said from a Christian and humane viewpoint in favor of indiscriminate charity for the benefit of the individual. The societies for charity, altruism or extension of rights, should have in these days, however, in their management some small modicum of brains, otherwise they may continue to do, as they have sometimes done in the past, more injury to the race than black death or smallpox.

As long as such charitable organizations confine themselves to the relief of suffering individuals, no matter how criminal or diseased they may be, no harm is done except to our own generation and if modern society recognizes a duty to the humblest malefactors or imbeciles that duty can be harmlessly performed in full, provided they be deprived of the capacity to procreate their defective strain. Those who read these pages will feel that there is little hope for humanity, but the remedy has been found, and can be quickly and mercifully applied. *A rigid system of selection through the elimination of those who are weak or unfit—in other words, social failures—would solve the whole question in a century, as well as enable us to get rid of the undesirables who crowd our jails, hospitals and insane asylums. The individual himself can be nourished, educated and protected by the community during his lifetime, but the state through sterilization must see to it that his line stops with him or else future generations will be cursed with an ever increasing load of victims of misguided sentimentalism.* This is a practical, merciful and inevitable solution of the whole problem and can be applied to an ever widening circle of social discards, beginning always with the criminal, the diseased and the insane and extending gradually to types which may be called weaklings rather than defectives and perhaps ultimately to worthless race types. (pp. 48–51)

Under existing conditions the most practical and hopeful method of race improvement is through the elimination of the least desirable elements in the nation by depriving them of the power to contribute to future generations. It is well known to stock breeders that the color of a herd of cattle can be modified by continuous destruction of worthless shades and of course this is true of other characters. Black sheep, for instance, have been practically obliterated by cutting out generation after generation all animals that show this color phase, until in carefully maintained flocks a black individual only appears as a rare sport.

In mankind it would not be a matter of great difficulty to secure a general consensus of public opinion as to the least desirable, let us say, ten per cent of the community. When this unemployed and unemployable human residuum has been eliminated together with the great mass of crime, poverty, alcoholism and feeblemindedness associated therewith it would be easy to consider the advisability of further restricting the perpetuation of the then remaining least valuable types. By this method mankind might ultimately become sufficiently intelligent to choose deliberately the most vital and intellectual strains to carry on the race. (pp. 53–54)

At the present time the Nordic race is undergoing selection through alcoholism, a peculiarly Nordic vice, and through consumption. Both these dread scourges unfortunately attack those members of the race that are otherwise most desirable, differing in this respect from filth diseases like typhus, typhoid or smallpox. One has only to look among the more desirable classes for the victims of rum and tubercule to realize that death or mental and physical impairment through these two causes have cost the race many of its most brilliant and attractive members. (p. 55)

Appendix: The Belmont Report

Given in the Appendix Supplements.

Appendix: The Declaration of Helsinki

Ethical Principles for Medical Research Involving Human Subjects
A. INTRODUCTION
1. The World Medical Association (WMA) has developed the Declaration

of Helsinki as a statement of ethical principles for medical research involving human subjects, including research on identifiable human material and data. The Declaration is intended to be read as a whole and each of its constituent paragraphs should not be applied without consideration of all other relevant paragraphs.

2. Although the Declaration is addressed primarily to physicians, the WMA encourages other participants in medical research involving human subjects to adopt these principles.

3. It is the duty of the physician to promote and safeguard the health of patients, including those who are involved in medical research. The physician's knowledge and conscience are dedicated to the fulfilment of this duty.

4. The Declaration of Geneva of the WMA binds the physician with the words, "The health of my patient will be my first consideration," and the International Code of Medical Ethics declares that, "A physician shall act in the patient's best interest when providing medical care."

5. Medical progress is based on research that ultimately must include studies involving human subjects. Populations that are underrepresented in medical research should be provided appropriate access to participation in research.

6. In medical research involving human subjects, the well-being of the individual research subject must take precedence over all other interests.

7. The primary purpose of medical research involving human subjects is to understand the causes, development and effects of diseases and improve preventive, diagnostic and therapeutic interventions (methods, procedures and treatments). Even the best current interventions must be evaluated continually through research for their safety, effectiveness, efficiency, accessibility and quality.

8. In medical practice and in medical research, most interventions involve risks and burdens.

9. Medical research is subject to ethical standards that promote respect for all human subjects and protect their health and rights. Some research populations are particularly vulnerable and need special protection. These include those who cannot give or refuse consent for themselves and those who may be vulnerable to coercion or undue influence.

10. Physicians should consider the ethical, legal and regulatory norms and standards for research involving human subjects in their own countries as well as applicable international norms and standards. No national or international ethical, legal or regulatory requirement should reduce or eliminate any of the protections for research subjects set forth in this Declaration.

B. PRINCIPLES FOR ALL MEDICAL RESEARCH

11. It is the duty of physicians who participate in medical research to protect the life, health, dignity, integrity, right to self-determination, privacy, and confidentiality of personal information of research subjects.

12. Medical research involving human subjects must conform to generally accepted scientific principles, be based on a thorough knowledge of the scientific literature, other relevant sources of information, and adequate laboratory

and, as appropriate, animal experimentation. The welfare of animals used for research must be respected.

13. Appropriate caution must be exercised in the conduct of medical research that may harm the environment.

14. *The design and performance of each research study involving human subjects must be clearly described in a research protocol. The protocol should contain a statement of the ethical considerations involved and should indicate how the principles in this Declaration have been addressed. The protocol should include information regarding funding, sponsors, institutional affiliations, other potential conflicts of interest, incentives for subjects and provisions for treating and/or compensating subjects who are harmed as a consequence of participation in the research study. The protocol should describe arrangements for post-study access by study subjects to interventions identified as beneficial in the study or access to other appropriate care or benefits.*

15. The research protocol must be submitted for consideration, comment, guidance and approval to a research ethics committee before the study begins. This committee must be independent of the researcher, the sponsor and any other undue influence. It must take into consideration the laws and regulations of the country or countries in which the research is to be performed as well as applicable international norms and standards but these must not be allowed to reduce or eliminate any of the protections for research subjects set forth in this Declaration. The committee must have the right to monitor ongoing studies. The researcher must provide monitoring information to the committee, especially information about any serious adverse events. No change to the protocol may be made without consideration and approval by the committee.

16. Medical research involving human subjects must be conducted only by individuals with the appropriate scientific training and qualifications. Research on patients or healthy volunteers requires the supervision of a competent and appropriately qualified physician or other health care professional. The responsibility for the protection of research subjects must always rest with the physician or other health care professional and never the research subjects, even though they have given consent.

17. *Medical research involving a disadvantaged or vulnerable population or community is only justified if the research is responsive to the health needs and priorities of this population or community and if there is a reasonable likelihood that this population or community stands to benefit from the results of the research.*

18. Every medical research study involving human subjects must be preceded by careful assessment of predictable risks and burdens to the individuals and communities involved in the research in comparison with foreseeable benefits to them and to other individuals or communities affected by the condition under investigation.

19. *Every clinical trial must be registered in a publicly accessible database before recruitment of the first subject.*

20. Physicians may not participate in a research study involving human

subjects unless they are confident that the risks involved have been adequately assessed and can be satisfactorily managed. Physicians must immediately stop a study when the risks are found to outweigh the potential benefits or when there is conclusive proof of positive and beneficial results.

21. Medical research involving human subjects may only be conducted if the importance of the objective outweighs the inherent risks and burdens to the research subjects.

22. Participation by competent individuals as subjects in medical research must be voluntary. Although it may be appropriate to consult family members or community leaders, no competent individual may be enrolled in a research study unless he or she freely agrees.

23. Every precaution must be taken to protect the privacy of research subjects and the confidentiality of their personal information and to minimize the impact of the study on their physical, mental and social integrity.

24. In medical research involving competent human subjects, each potential subject must be adequately informed of the aims, methods, sources of funding, any possible conflicts of interest, institutional affiliations of the researcher, the anticipated benefits and potential risks of the study and the discomfort it may entail, and any other relevant aspects of the study. The potential subject must be informed of the right to refuse to participate in the study or to withdraw consent to participate at any time without reprisal. Special attention should be given to the specific information needs of individual potential subjects as well as to the methods used to deliver the information. After ensuring that the potential subject has understood the information, the physician or another appropriately qualified individual must then seek the potential subject's freely-given informed consent, preferably in writing. If the consent cannot be expressed in writing, the nonwritten consent must be formally documented and witnessed.

25. For medical research using identifiable human material or data, physicians must normally seek consent for the collection, analysis, storage and/or reuse. There may be situations where consent would be impossible or impractical to obtain for such research or would pose a threat to the validity of the research. In such situations the research may be done only after consideration and approval of a research ethics committee.

26. When seeking informed consent for participation in a research study the physician should be particularly cautious if the potential subject is in a dependent relationship with the physician or may consent under duress. In such situations the informed consent should be sought by an appropriately qualified individual who is completely independent of this relationship.

27. For a potential research subject who is incompetent, the physician must seek informed consent from the legally authorized representative. These individuals must not be included in a research study that has no likelihood of benefit for them unless it is intended to promote the health of the population represented by the potential subject, the research cannot instead be

performed with competent persons, and the research entails only minimal risk and minimal burden.

28. When a potential research subject who is deemed incompetent is able to give assent to decisions about participation in research, the physician must seek that assent in addition to the consent of the legally authorized representative. The potential subjects dissent should be respected.

29. Research involving subjects who are physically or mentally incapable of giving consent, for example, unconscious patients, may be done only if the physical or mental condition that prevents giving informed consent is a necessary characteristic of the research population. In such circumstances the physician should seek informed consent from the legally authorized representative. If no such representative is available and if the research cannot be delayed, the study may proceed without informed consent provided that the specific reasons for involving subjects with a condition that renders them unable to give informed consent have been stated in the research protocol and the study has been approved by a research ethics committee. Consent to remain in the research should be obtained as soon as possible from the subject or a legally authorized representative.

30. *Authors, editors and publishers all have ethical obligations with regard to the publication of the results of research. Authors have a duty to make publicly available the results of their research on human subjects and are accountable for the completeness and accuracy of their reports. They should adhere to accepted guidelines for ethical reporting. Negative and inconclusive as well as positive results should be published or otherwise made publicly available. Sources of funding, institutional affiliations and conflicts of interest should be declared in the publication. Reports of research not in accordance with the principles of this Declaration should not be accepted for publication.*

C. ADDITIONAL PRINCIPLES FOR MEDICAL RESEARCH COMBINED WITH MEDICAL CARE

31. The physician may combine medical research with medical care only to the extent that the research is justified by its potential preventive, diagnostic or therapeutic value and if the physician has good reason to believe that participation in the research study will not adversely affect the health of the patients who serve as research subjects.

32. *The benefits, risks, burdens and effectiveness of a new intervention must be tested against those of the best current proven intervention, except in the following circumstances:*

— The use of placebo, or no treatment, is acceptable in studies where no current proven intervention exists; or

— Where for compelling and scientifically sound methodological reasons the use of placebo is necessary to determine the efficacy or safety of an intervention and the patients who receive placebo or no treatment will not be subject to any risk of serious or irreversible harm. Extreme care must be taken to avoid abuse of this option.

33. *At the conclusion of the study, patients entered into the study are en-*

*titled to be informed about the outcome of the study and to share any benefits
that result from it, for example, access to interventions identified as beneficial
in the study or to other appropriate care or benefits.*

34. The physician must fully inform the patient which aspects of the care
are related to the research. The refusal of a patient to participate in a study
or the patients decision to withdraw from the study must never interfere with
the patient-physician relationship.

35. In the treatment of a patient, where proven interventions do not exist
or have been ineffective, the physician, after seeking expert advice, with informed consent from the patient or a legally authorized representative, may
use an unproven intervention if in the physician's judgment it offers hope
of saving life, re-establishing health or alleviating suffering. Where possible,
this intervention should be made the object of research, designed to evaluate
its safety and efficacy. In all cases, new information should be recorded and,
where appropriate, made publicly available.[7]

Notes

[1]Madison Grant was an American eugenicist, lawyer, and amateur anthropologist. His undergraduate degree was from Yale (Class of 1887); he received
a law degree from Columbia Law School several years later. By 1937, his influential book of scientific racism had sold more than 1.6 million copies in the
United States alone ("Madison Grant," 2011).

[2]To obtain a sense of the 19th century racial philosophy underlying Grant's
formulation, the human species was first divided into three distinct races:
Caucasoid (Europe), Negroid (Africa), and Mongoloid (Asia). The Caucasoids
were further subdivided into Nordics (Northern Europe), Alpines (Central
Europe), Mediterraneans (Southern Europe). The Caucasoid race that needed
to be protected was the Nordic; both the Alpines and the Mediterraneans (as
well as the Negroid and Mongoloid) had to be prevented from United States
immigration ("Madison Grant," 2011).

[3]In a White House ceremony on May 16, 1997 that was attended by five of
the eight remaining study survivors, President Bill Clinton formally apologized
for the Tuskegee study ("Remarks by the President," 1997):

> The eight men who are survivors of the syphilis study at Tuskegee
> are a living link to a time not so very long ago that many Americans
> would prefer not to remember, but we dare not forget. It was a
> time when our nation failed to live up to its ideals, when our nation
> broke the trust with our people that is the very foundation of our
> democracy. It is not only in remembering that shameful past that
> we can make amends and repair our nation, but it is in remembering

that past that we can build a better present and a better future. And without remembering it, we cannot make amends and we cannot go forward.

So today America does remember the hundreds of men used in research without their knowledge and consent. We remember them and their family members. Men who were poor and African American, without resources and with few alternatives, they believed they had found hope when they were offered free medical care by the United States Public Health Service. They were betrayed.

Medical people are supposed to help when we need care, but even once a cure was discovered, they were denied help, and they were lied to by their government. Our government is supposed to protect the rights of its citizens; their rights were trampled upon. Forty years, hundreds of men betrayed, along with their wives and children, along with the community in Macon County, Alabama, the City of Tuskegee, the fine university there, and the larger African American community.

The United States government did something that was wrong— deeply, profoundly, morally wrong. It was an outrage to our commitment to integrity and equality for all our citizens.

To the survivors, to the wives and family members, the children and the grandchildren, I say what you know: No power on Earth can give you back the lives lost, the pain suffered, the years of internal torment and anguish. What was done cannot be undone. But we can end the silence. We can stop turning our heads away. We can look at you in the eye and finally say on behalf of the American people, what the United States government did was shameful, and I am sorry. The American people are sorry—for the loss, for the years of hurt. You did nothing wrong, but you were grievously wronged. I apologize and I am sorry that this apology has been so long in coming.

To Macon County, to Tuskegee, to the doctors who have been wrongly associated with the events there, you have our apology, as well. To our African American citizens, I am sorry that your federal government orchestrated a study so clearly racist. That can never be allowed to happen again. It is against everything our country stands for and what we must stand against is what it was.

So let us resolve to hold forever in our hearts and minds the memory of a time not long ago in Macon County, Alabama, so that we can always see how adrift we can become when the rights of any citizens are neglected, ignored and betrayed. And let us resolve here and now to move forward together.

The legacy of the study at Tuskegee has reached far and deep, in ways that hurt our progress and divide our nation. We cannot be one America when a whole segment of our nation has no trust in America. An apology is the first step, and we take it with a commitment to rebuild that broken trust. We can begin by making sure there is never again another episode like this one. We need to do more to ensure that medical research practices are sound and ethical, and that researchers work more closely with communities.

[4]The Tuskegee study was apparently not the only one of its type that the United States government oversaw. We give a brief excerpt from the Wikipedia entry on the Tuskegee study summarizing the more recently uncovered non-consensual experiments in Guatemala ("Tuskegee syphilis experiment," 2011):

> In October 2010 it was revealed that in Guatemala, the project went even further. It was reported that from 1946 to 1948, American doctors deliberately infected prisoners, soldiers, and patients in a mental hospital with syphilis and, in some cases, gonorrhea, with the cooperation of some Guatemalan health ministries and officials. A total of 696 men and women were exposed to syphilis without the informed consent of the subjects. When the subjects contracted the disease they were given antibiotics though it is unclear if all infected parties were cured.
>
> Wellesley College's historian Susan Reverby discovered records of the experiment while examining archives of John Charles Cutler, a government researcher involved in the now infamous Tuskegee study. In October 2010, the United States formally apologized to Guatemala for conducting these experiments.

[5]The recommendations in this paragraph are in contrast to those of Horrobin's "personal paper" reproduced in Section 7.8.1.

[6]This recommendation was violated by the British government when it kept secret the Bayesian methods pioneered by Alan Turing in breaking the Enigma code in World War II. For the history of this tale, see Sharon Bertsch McGrayne, *The Theory That Would Not Die: How Bayes' Rule Cracked the Enigma Code, Hunted Down Russian Submarines, and Emerged Triumphant From Two Centuries of Controversy* (2011).

[7]It is of interest to note the absence of any recommendations about ending studies early once a treatment group is shown to be superior to a control group.

Chapter 17

The Federal Rules of Evidence

> You have but to hold forth in cap and gown, and any gibberish becomes learning, all nonsense passes for sense.
> — Molière (*The Hypochondriac*, 1673 [*Le Malade imaginaire*])

A common aspect of many modern court proceedings, particularly in litigation involving toxic torts, product liability, contaminating environmental agents, and the like, is the presence of expert witnesses who provide evidence relevant to the matter at hand. Very often such evidence is given through statistical argumentations, possibly through estimated statistical models (that might, for example, assess a probability of causation), meta-analyses, or other methods of data presentation and interpretation developed through graphical or tabular means. To be defensible ethically, such evidence when presented statistically must avoid the various pitfalls documented throughout this book (for example, regression toward the mean, the ecological fallacy, confusing test "impact" with test "bias," and "lurking" third variable confoundings). The issues involved in expert witness admissibility, however, are much broader than just in how data are presented. The scientific reliability and validity of the available evidence are of major interest and are subject to the *Federal Rules of Evidence*. Here, we discuss some of the issues involved in the admissibility of evidence and the proffering of expert witnesses, both as they are currently understood and practiced and how they have evolved historically over the years.[1]

Trial courts have had to evaluate the admissibility of expert testimony ever since a judicial system was established in the United States. Courts in the nineteenth and early twentieth centuries generally asked only whether an expert was "qualified" before expert testimony was considered admissible. Whenever a subject was beyond the ken of an average juror, a qualified expert's opinion was considered crucial to a jury's determination of the facts at issue.

Usually, experts were qualified by dint of success in a relevant profession or occupation. Nothing more than this was required, assuming that what was proffered was relevant to the case at hand; the expertise or body of evidence admitted in trial was generally viewed as inseparable from the expert.

In the early 1920s, a ruling was made, commonly referred to as the *Frye* standard or opinion, that would frame expert witness admissibility for most of the twentieth century. Even now the *Frye* standard may still hold sway, particularly for those states where the current *Federal Rules of Evidence* (FRE) have not been adopted for use in state courts. In *Frye v. United States* (1923), the defendant offered an early form of a polygraph lie detection test (based on systolic blood pressure) to support a plea of innocence to a charge of murder. The relevant part of the ruling, commonly referred to as the "general acceptance" standard, follows:

> Just when a scientific principle or discovery crosses the line between the experimental and demonstrable stages is difficult to define. Somewhere in this twilight zone the evidential force of the principle must be recognized, and while courts will go a long way in admitting expert testimony deduced from a well-recognized scientific principle or discovery, the thing from which the deduction is made must be sufficiently established to have gained general acceptance in the particular field in which it belongs.

The *Frye* opinion did several major new things. First, an explicit separation was made between expertise and the expert, creating the important precedent that a body of asserted knowledge could exist apart from the proferring expert. Second, expert testimony must arise from a knowledge base that has "gained general acceptance in the particular field to which it belongs." Thus, various forms of pseudoscience (or in the modern parlance, "junk science") were inadmissible, irrespective of the person providing such information. A reputation alone as a "hired gun" was not enough; the weapons had to be real and accepted in the specific field of interest. We give the short *Frye* opinion in an appendix to this section (excluding the second-to-last paragraph just quoted). It is a model of brevity given the impact it has had (and is still having) on the judicial system.

The FRE govern how facts are admitted and how parties in federal courts of the United States may prove their cases. These rules were the outcome of a long academic, legislative, and judicial process. They became federal law on January 2, 1975, when President Ford signed the *Act to Establish Rules of Evidence for Certain Courts and Proceedings* ("Federal Rules of Evidence," 2010). Rule 702 governs testimony by experts, and by most accounts, supersedes the *Frye* standard ("Frye standard," 2010). We give Rule 702 below, with a part indicated in brackets that was added (in 2000) in response to a Supreme Court decision that we discuss shortly (*Daubert v. Merrell Dow Pharmaceuticals, Inc.* (1993)):

> Rule 702. Testimony by Experts:
>
> If scientific, technical, or other specialized knowledge will assist the trier of fact to understand the evidence or to determine a fact in issue, a witness qualified as an expert by knowledge, skill, experience, training, or education, may testify thereto in the form of an opinion or otherwise, [if (1) the testimony is based upon sufficient facts or data, (2) the testimony is the product of reliable principles and methods, and (3) the witness has applied the principles and methods reliably to the facts of the case].

In the Supreme Court's decision in the case mentioned above, *Daubert v. Merrell Dow Pharmaceuticals, Inc.*, the Court considered the standard for evaluating the admissibility of scientific expert testimony, and held that under the FRE (superseding *Frye*), trial court judges were the responsible "gatekeepers." These trial judges were to evaluate the validity of the basis for the scientific expertise before the expert would be allowed to testify. These pretrial determinations of admissibility are usually referred to as *Daubert* hearings (or Rule 104(a) hearings), in reference to the Supreme Court opinion (or in reference to the FRE).

The Supreme Court opinion in *Daubert v. Merrell* defined "scientific methodology" as the process of formulating hypotheses and conducting experiments to prove or falsify these hypotheses. In the process, a "flexible" test was provided for establishing "validity" based on four (*Daubert*) factors:

(1) Empirical testing: the theory or technique must be falsifiable, refutable, and testable;

(2) Subject to peer review and publication;

(3) Known or potential error rate;

(4) The degree to which the theory and technique is generally accepted by a relevant scientific community.

The fourth *Daubert* factor looks suspiciously like the *Frye* standard ("Frye standard," 2010). In fact, although *Daubert* may have superseded *Frye*, the *Daubert* standard is not substantially different in practice. Because judges are generally required to consult people in the relevant scientific communities to assess each of the *Daubert* factors, some have referred to the *Daubert* standard merely as "*Frye* in drag." A summary (called a syllabus) of the opinion in *Daubert v. Merrell* is given in an appendix to this section.[2]

Two additional Supreme Court opinions have further articulated the *Daubert* ruling: *General Electric Co. v. Joiner* (1997), and *Kumho Tire Co. v. Carmichael* (1999). The opinion in *General Electric v. Joiner* held that an abuse of discretion standard of review was the correct one for appellate courts to use in the review of a trial court's decision either to admit or not expert testimony. The phrase "abuse of discretion" refers to a trial judge making an error in judgment that is clearly against the evidence or established law ("Abuse of discretion," 2011).

The Supreme Court ruling in *General Electric v. Joiner* held that the trial judge's exclusion of expert witness testimony did not constitute an abuse of discretion. The third opinion in *Kumho Tire v. Carmichael* completes what is commonly named the *Daubert* trilogy. Here, a trial judge's gatekeeping function, identified in *Daubert*, is extended to all expert testimony, including that which is putatively nonscientific. As an aside, the *General Electric v. Joiner* opinion provides one of the better lines in all three opinions from the *Daubert* trilogy:

> Nothing in either Daubert or the Federal Rules of Evidence requires a district court to admit opinion evidence which is connected to existing data only by the *ipse dixit* of the expert.

We provide the syllabi for the *General Electric v. Joiner* and *Kumho Tire v. Carmichael* rulings in an appendix to this section.

Appendix: Daubert v. Merrell Dow Syllabus

SUPREME COURT OF THE UNITED STATES

DAUBERT et ux., individually and as guardians and litem for DAUBERT, et al. v. MERRELL DOW PHARMACEUTICALS, INC.

CERTIORARI TO THE UNITED STATES COURT OF APPEALS FOR THE NINTH CIRCUIT

Decided June 28, 1993

Petitioners, two minor children and their parents, alleged in their suit against respondent that the children's serious birth defects had been caused by the mothers' prenatal ingestion of Bendectin, a prescription drug marketed by respondent. The District Court granted respondent summary judgment based on a well credentialed expert's affidavit concluding, upon reviewing the extensive published scientific literature on the subject, that maternal use of Bendectin has not been shown to be a risk factor for human birth defects. Although petitioners had responded with the testimony of eight other well credentialed experts, who based their conclusion that Bendectin can cause birth defects on animal studies, chemical structure analyses, and the unpublished "reanalysis" of previously published human statistical studies, the court determined that this evidence did not meet the applicable "general acceptance" standard for the admission of expert testimony. The Court of Appeals agreed and affirmed, citing Frye v. United States, for the rule that expert opinion based on a scientific technique is inadmissible unless the technique is "generally accepted" as reliable in the relevant scientific community.

Held: The Federal Rules of Evidence, not Frye, provide the standard for admitting expert scientific testimony in a federal trial.

(a) Frye's "general acceptance" test was superseded by the Rules' adoption. The Rules occupy the field, and although the common law of evidence may serve as an aid to their application, respondent's assertion that they somehow assimilated Frye is unconvincing. Nothing in the Rules as a whole or in the text and drafting history of Rule 702, which specifically governs expert testimony, gives any indication that "general acceptance" is a necessary precondition to the admissibility of scientific evidence. Moreover, such a rigid standard would be at odds with the Rules' liberal thrust and their general approach of relaxing the traditional barriers to "opinion" testimony.

(b) The Rules—especially Rule 702—place appropriate limits on the admissibility of purportedly scientific evidence by assigning to the trial judge the task of ensuring that an expert's testimony both rests on a reliable foundation and is relevant to the task at hand. The reliability standard is established by Rule 702's requirement that an expert's testimony pertain to "scientific ... knowledge," since the adjective "scientific" implies a grounding in science's methods and procedures, while the word "knowledge" connotes a body of known facts or of ideas inferred from such facts or accepted as true on good grounds. The Rule's requirement that the testimony "assist the trier of fact to understand the evidence or to determine a fact in issue" goes primarily to relevance by demanding a valid scientific connection to the pertinent inquiry as a precondition to admissibility.

(c) Faced with a proffer of expert scientific testimony under Rule 702, the trial judge, pursuant to Rule 104(a), must make a preliminary assessment of whether the testimony's underlying reasoning or methodology is scientifically valid and properly can be applied to the facts at issue. Many considerations will bear on the inquiry, including whether the theory or technique in question can be (and has been) tested, whether it has been subjected to peer review and publication, its known or potential error rate, and the existence and maintenance of standards controlling its operation, and whether it has attracted widespread acceptance within a relevant scientific community. The inquiry is a flexible one, and its focus must be solely on principles and methodology, not on the conclusions that they generate. Throughout, the judge should also be mindful of other applicable Rules.

(d) Cross-examination, presentation of contrary evidence, and careful instruction on the burden of proof, rather than wholesale exclusion under an uncompromising "general acceptance" standard, is the appropriate means by which evidence based on valid principles may be challenged. That even limited screening by the trial judge, on occasion, will prevent the jury from hearing of authentic scientific breakthroughs is simply a consequence of the fact that the Rules are not designed to seek cosmic understanding but, rather, to resolve legal disputes.

Appendix: General Electric v. Joiner Syllabus

SUPREME COURT OF THE UNITED STATES
GENERAL ELECTRIC CO. et al. v. JOINER et ux.
CERTIORARI TO THE UNITED STATES COURT OF APPEALS FOR THE ELEVENTH CIRCUIT
Decided December 15, 1997

After he was diagnosed with small-cell lung cancer, respondent Joiner sued in Georgia state court, alleging, *inter alia*, that his disease was "promoted" by his workplace exposure to chemical "PCBs" and derivative "furans" and "dioxins" that were manufactured by, or present in materials manufactured by, petitioners. Petitioners removed the case to federal court and moved for summary judgment. Joiner responded with the depositions of expert witnesses, who testified that PCBs, furans, and dioxins can promote cancer, and opined that Joiner's exposure to those chemicals was likely responsible for his cancer. The District Court ruled that there was a genuine issue of material fact as to whether Joiner had been exposed to PCBs, but granted summary judgment for petitioners because (1) there was no genuine issue as to whether he had been exposed to furans and dioxins, and (2) his experts' testimony had failed to show that there was a link between exposure to PCBs and small-cell lung cancer and was therefore inadmissible because it did not rise above "subjective belief or unsupported speculation." In reversing, the Eleventh Circuit applied "a particularly stringent standard of review" to hold that the District Court had erred in excluding the expert testimony.

Held:

1. Abuse of discretion—the standard ordinarily applicable to review of evidentiary rulings—is the proper standard by which to review a district court's decision to admit or exclude expert scientific evidence. Contrary to the Eleventh Circuit's suggestion, Daubert v. Merrell Dow Pharmaceuticals, Inc., did not somehow alter this general rule in the context of a district court's decision to exclude scientific evidence. Daubert did not address the appellate review standard for evidentiary rulings at all, but did indicate that, while the Federal Rules of Evidence allow district courts to admit a somewhat broader range of scientific testimony than did pre-existing law, they leave in place the trial judge's "gatekeeper" role of screening such evidence to ensure that it is not only relevant, but reliable. A court of appeals applying "abuse of discretion" review to such rulings may not categorically distinguish between rulings allowing expert testimony and rulings which disallow it. This Court rejects Joiner's argument that because the granting of summary judgment in this case was "outcome determinative," it should have been subjected to a more searching standard of review. On a summary judgment motion, disputed issues of fact are resolved against the moving party—here, petitioners. But the question of admissibility of expert testimony is not such an issue of fact, and is reviewable under the abuse of discretion standard. In applying an overly "stringent" standard, the Eleventh Circuit failed to give the trial court the deference that is the hallmark of abuse of discretion review.

2. A proper application of the correct standard of review indicates that the District Court did not err in excluding the expert testimony at issue. The animal studies cited by respondent's experts were so dissimilar to the facts presented here—i.e., the studies involved infant mice that developed alveologenic adenomas after highly concentrated, massive doses of PCBs were injected directly into their peritoneums or stomachs, whereas Joiner was an adult human whose small-cell carcinomas allegedly resulted from exposure on a much smaller scale—that it was not an abuse of discretion for the District Court to have rejected the experts' reliance on those studies. Nor did the court abuse its discretion in concluding that the four epidemiological studies on which Joiner relied were not a sufficient basis for the experts' opinions, since the authors of two of those studies ultimately were unwilling to suggest a link between increases in lung cancer and PCB exposure among the workers they examined, the third study involved exposure to a particular type of mineral oil not necessarily relevant here, and the fourth involved exposure to numerous potential carcinogens in addition to PCBs. Nothing in either Daubert or the Federal Rules of Evidence requires a district court to admit opinion evidence which is connected to existing data only by the *ipse dixit* of the expert.

3. These conclusions, however, do not dispose of the entire case. The Eleventh Circuit reversed the District Court's conclusion that Joiner had not been exposed to furans and dioxins. Because petitioners did not challenge that determination in their *certiorari* petition, the question whether exposure to furans and dioxins contributed to Joiner's cancer is still open.

Appendix: Kumho Tire v. Carmichael Syllabus

SUPREME COURT OF THE UNITED STATES

KUMHO TIRE CO., LTD., et al. v. CARMICHAEL et al.

CERTIORARI TO THE UNITED STATES COURT OF APPEALS FOR THE ELEVENTH CIRCUIT

Decided March 23, 1999

When a tire on the vehicle driven by Patrick Carmichael blew out and the vehicle overturned, one passenger died and the others were injured. The survivors and the decedent's representative, respondents here, brought this diversity suit against the tire's maker and its distributor (collectively Kumho Tire), claiming that the tire that failed was defective. They rested their case in significant part upon the depositions of a tire failure analyst, Dennis Carlson, Jr., who intended to testify that, in his expert opinion, a defect in the tire's manufacture or design caused the blow out. That opinion was based upon a visual and tactile inspection of the tire and upon the theory that in the absence of at least two of four specific, physical symptoms indicating tire abuse, the tire failure of the sort that occurred here was caused by a defect. Kumho Tire moved to exclude Carlson's testimony on the ground that his methodology failed to satisfy Federal Rule of Evidence 702, which says: "If scientific, technical, or other specialized knowledge will assist the trier of fact ... a witness qualified as an expert ... may testify thereto in the form of an opinion." Granting the motion (and entering summary judgment for the defendants), the District Court acknowledged that it should act as a reliability "gatekeeper" under Daubert v. Merrell Dow Pharmaceuticals, Inc., in which this Court held that Rule 702 imposes a special obligation upon a trial judge to ensure that scientific testimony is not only relevant, but reliable. The court noted that Daubert discussed four factors—testing, peer review, error rates, and "acceptability" in the relevant scientific community—which might prove helpful in determining the reliability of a particular scientific theory or technique, and found that those factors argued against the reliability of Carlson's methodology. On the plaintiffs' motion for reconsideration, the court agreed that Daubert should be applied flexibly, that its four factors were simply illustrative, and that other factors could argue in favor of admissibility. However, the court affirmed its earlier order because it found insufficient indications of the reliability of Carlson's methodology. In reversing, the Eleventh Circuit held that the District Court had erred as a matter of law in applying Daubert. Believing that Daubert was limited to the scientific context, the court held that the Daubert factors did not apply to Carlson's testimony, which it characterized as skill- or experience-based.

Held:

1. The Daubert factors may apply to the testimony of engineers and other experts who are not scientists.

(a) The Daubert "gatekeeping" obligation applies not only to "scientific" testimony, but to all expert testimony. Rule 702 does not distinguish between "scientific" knowledge and "technical" or "other specialized" knowledge, but makes clear that any such knowledge might become the subject of expert testimony. It is the Rule's word "knowledge," not the words (like "scientific") that modify that word, that establishes a standard of evidentiary reliability. Daubert referred only to "scientific" knowledge because that was the nature of the expertise there at issue. Neither is the evidentiary rationale underlying Daubert's "gatekeeping" determination limited to "scientific" knowledge. Rules 702 and 703 grant all expert witnesses, not just "scientific" ones, testimonial latitude unavailable to other witnesses on the assumption that the expert's opinion will have a reliable basis in the knowledge and experience of his discipline. Finally, it would prove difficult, if not impossible, for judges to administer evidentiary rules under which a "gatekeeping" obligation depended upon a distinction between "scientific" knowledge and "technical" or "other specialized" knowledge, since there is no clear line dividing the one from the others and no convincing need to make such distinctions.

(b) A trial judge determining the admissibility of an engineering expert's testimony may consider one or more of the specific Daubert factors. The emphasis on the word "may" reflects Daubert's description of the Rule 702 inquiry as "a flexible one. The Daubert factors do not constitute a definitive checklist or test, and the gatekeeping inquiry must be tied to the particular facts. Those factors may or may not be pertinent in assessing reliability, depending on the nature of the issue, the expert's particular expertise, and the subject of his testimony. Some of those factors may be helpful in evaluating the reliability even of experience-based expert testimony, and the Court of Appeals erred insofar as it ruled those factors out in such cases. In determining whether particular expert testimony is reliable, the trial court should consider the specific Daubert factors where they are reasonable measures of reliability.

(c) The court of appeals must apply an abuse-of-discretion standard when it reviews the trial court's decision to admit or exclude expert testimony. That standard applies as much to the trial court's decisions about how to determine reliability as to its ultimate conclusion. Thus, whether Daubert's specific factors are, or are not, reasonable measures of reliability in a particular case is a matter that the law grants the trial judge broad latitude to determine. The Eleventh Circuit erred insofar as it held to the contrary.

2. Application of the foregoing standards demonstrates that the District Court's decision not to admit Carlson's expert testimony was lawful. The District Court did not question Carlson's qualifications, but excluded his testimony because it initially doubted his methodology and then found it unreliable after examining the transcript in some detail and considering respondents' defense of it. The doubts that triggered the court's initial inquiry were reasonable, as was the court's ultimate conclusion that Carlson could not reliably determine the cause of the failure of the tire in question. The question was not the reliability of Carlson's methodology in general, but rather whether he could reliably determine the cause of failure of the particular tire at issue. That tire, Carlson conceded, had traveled far enough so that some of the tread

had been worn bald, it should have been taken out of service, it had been repaired (inadequately) for punctures, and it bore some of the very marks that he said indicated, not a defect, but abuse. Moreover, Carlson's own testimony cast considerable doubt upon the reliability of both his theory about the need for at least two signs of abuse and his proposition about the significance of visual inspection in this case. Respondents stress that other tire failure experts, like Carlson, rely on visual and tactile examinations of tires. But there is no indication in the record that other experts in the industry use Carlson's particular approach or that tire experts normally make the very fine distinctions necessary to support his conclusions, nor are there references to articles or papers that validate his approach. Respondents' argument that the District Court too rigidly applied Daubert might have had some validity with respect to the court's initial opinion, but fails because the court, on reconsideration, recognized that the relevant reliability inquiry should be "flexible," and ultimately based its decision upon Carlson's failure to satisfy either Daubert's factors or any other set of reasonable reliability criteria.

Appendix: Frye v. United States Opinion

FRYE v. UNITED STATES
Court of Appeals of District of Columbia
December 3, 1923, Decided
Appellant, defendant below, was convicted of the crime of murder in the second degree, and from the judgment prosecutes this appeal.

A single assignment of error is presented for our consideration. In the course of the trial, counsel for defendant offered an expert witness to testify to the result of a deception test made upon defendant. The test is described as the systolic blood pressure deception test. It is asserted that blood pressure is influenced by change in the emotions of the witness, and that the systolic blood pressure rises are brought about by nervous impulses sent to the sympathetic branch of the autonomic nervous system. Scientific experiments, it is claimed, have demonstrated that fear, rage, and pain always produce a rise of systolic blood pressure, and that conscious deception or falsehood, concealment of facts, or guilt of crime, accompanied by fear of detection when the person is under examination, raises the systolic blood pressure in a curve, which corresponds exactly to the struggle going on in the subject's mind, between fear and attempted control of that fear, as the examination touches the vital points in respect of which he is attempting to deceive the examiner.

In other words, the theory seems to be that truth is spontaneous, and comes without conscious effort, while the utterance of a falsehood requires a conscious effort, which is reflected in the blood pressure. The rise thus produced is easily detected and distinguished from the rise produced by mere fear of the examination itself. In the former instance, the pressure rises higher than in the latter, and is more pronounced as the examination proceeds, while in the latter case, if the subject is telling the truth, the pressure registers highest at the beginning of the examination, and gradually diminishes as the examination proceeds.

Prior to the trial, defendant was subjected to this deception test, and counsel offered the scientist who conducted the test as an expert to testify to the results obtained. The offer was objected to by counsel for the government, and the court sustained the objection. Counsel for defendant then offered to have the proffered witness conduct a test in the presence of the jury. This also was denied.

Counsel for defendant, in their able presentation of the novel question involved, correctly state in their brief that no cases directly in point have been found. The broad ground, however, upon which they plant their case, is succinctly stated in their brief as follows:

"The rule is that the opinions of experts or skilled witnesses are admissible in evidence in those cases in which the matter of inquiry is such that inexperienced persons are unlikely to prove capable of forming a correct judgment upon it, for the reason that the subject-matter so far partakes of a science, art, or trade as to require a previous habit or experience or study in it, in order to acquire a knowledge of it. When the question involved does not lie within the range of common experience or common knowledge, but requires special experience or special knowledge, then the opinions of witnesses skilled in that particular science, art, or trade to which the question relates are admissible in evidence."

Numerous cases are cited in support of this rule. [deleted paragraph quoted above]

We think the systolic blood pressure deception test has not yet gained such standing and scientific recognition among physiological and psychological authorities as would justify the courts in admitting expert testimony deduced from the discovery, development, and experiments thus far made.

The judgment is affirmed.

17.1 Junk Science

[M]ost men ... can seldom accept even the simplest and most obvious truth if it would oblige them to admit the falsity of conclusions which they have delighted in explaining to colleagues, proudly taught to others, and which they have woven, thread by thread, into the fabric of their lives.

— Leo Tolstoy (*What is Art?*, 1899)

One of the supposed downsides of the *Daubert* standard is that lay judges, now being the "gatekeepers" of scientific evidence, may prevent respected scientists from offering testimony. As a consequence, corporate defendants are increasingly emboldened to accuse their adversaries of merely offering "junk science." This pejorative phrase appears to have been in use at least by the early 1980s. For example, a United States Department of Justice Tort Policy Working Group noted in 1986:

> Another way in which causation often is undermined—also an increasingly serious problem in toxic tort cases—is the reliance by judges and juries on noncredible scientific or medical testimony, studies or opinions. It has become all too common for "experts" or "studies" on the fringes of, or even well beyond the outer parameters of mainstream scientific or medical views, to be presented to juries as valid evidence from which conclusions may be drawn. The use of such invalid scientific evidence (commonly referred to as "junk science") has resulted in findings of causation which simply cannot be justified or understood from the standpoint of the current state of credible scientific and medical knowledge. Most importantly, this development has led to a deep and growing cynicism about the ability of tort law to deal with difficult scientific and medical concepts in a principled and rational way. (p. 35)

The label of "junk science" is easy to apply when there is a need to discount scientific findings that might be an impediment to short-term corporate profit. Thus, Big Tobacco has applied the term to research on the harmful effects of smoking and secondhand smoke; or the Fox News columnist, Steven Milloy, who applies the "junk science" label to research on many topics, including global warming, ozone depletion, DDT, Alar, and mad cow disease. For a snapshot to the pervasiveness of the attacks on behalf of corporate interests, a consult of Milloy's website (junkscience.com) should suffice. The lead comment on this website (as of April 17, 2010) is:

> Now that the most absurd but potentially catastrophic junk science in human history is unraveling and we are preparing to declare victory over gorebull warbling, we can devote more attention to neglected junk.

Several mechanisms might be used to permit sound science to have an appropriate impact on relevant judicial proceedings; this can be done without labeling all such science as "junk" merely because it would have a financial impact on corporate interests. One option for the courts, that might as a matter of course include statisticians, is Rule 706 from the FRE:

Rule 706. Court Appointed Experts
(a) Appointment.
The court may on its own motion or on the motion of any party enter an order to show cause why expert witnesses should not be appointed, and may request the parties to submit nominations. The court may appoint any expert witnesses agreed upon by the parties, and may appoint expert witnesses of its own selection. An expert witness shall not be appointed by the court unless the witness consents to act. A witness so appointed shall be informed of the witness' duties by the court in writing, a copy of which shall be filed with the clerk, or at a conference in which the parties shall have opportunity to participate. A witness so appointed shall advise the parties of the witness' findings, if any; the witness' deposition may be taken by any party; and the witness may be called to testify by the court or any party. The witness shall be subject to cross-examination by each party, including a party calling the witness.

Another strategy, Rule 53 in the *Federal Rules of Civil Procedures*, allows the appointment of special "Masters" to oversee a civil case. We give two excerpts from Rule 53 below:

(a) Appointment
(1) Scope.
Unless a statute provides otherwise, a court may appoint a master only to:
(A) perform duties consented to by the parties;
(B) hold trial proceedings and make or recommend findings of fact on issues to be decided without a jury if appointment is warranted by:
(i) some exceptional condition; or
(ii) the need to perform an accounting or resolve a difficult computation of damages; or
(C) address pretrial and posttrial matters that cannot be effectively and timely addressed by an available district judge or magistrate judge of the district.
. . .
(c) Master's Authority.
(1) In General.
Unless the appointing order directs otherwise, a master may:
(A) regulate all proceedings;

(B) take all appropriate measures to perform the assigned duties fairly and efficiently; and

(C) if conducting an evidentiary hearing, exercise the appointing court's power to compel, take, and record evidence.

A recent set of rulings shows how Masters might work; they were appointed for one particular type of case involving a possible link between vaccines and autism. An article discussing these rulings is from the *New York Times* (March 12, 2010; *3 Rulings Find No Link to Vaccines and Autism*; Donald G. McNeil, Jr.).

A third mechanism for permitting sound science to have an appropriate impact is to allow for the secondary reanalyses of research having regulatory and/or judicial implications but only under guidelines that ensure an ethically motivated process. The International Society for Environmental Epidemiology (ISEE) has issued such a controlling statement entitled *Guidelines for the Ethical Re-analysis and Re-Interpretations of Another's Research* (2009). This is reproduced in an appendix to this section. It is intended to prevent the practice of labeling research "junk" merely by culling through datasets until some anomalous result, possibly due just to chance, is finally identified that would permit a label of "junk" to be applied to the body of research as a whole.

Appendix: ISEE Guidelines for Ethical Re-analysis and Re-interpretation of Another's Research (April 2009)

The following guidelines were adopted by the governing council of the International Society for Environmental Epidemiology in May 2009.

ISEE endorses the following ethics guidelines with regard to the reanalysis and reinterpretation of epidemiological studies. The guidelines address the duties of original investigators, the sponsors of reanalysis and reinterpretation, and the researchers who carry it out. The rights and interests of these parties, research subjects/participants, and the community at large need to be protected. We highlight the issues that each of these parties need to consider.

Reanalysis and reinterpretation occur when a person other than the original investigator obtains an epidemiological dataset and conducts analyses to evaluate the quality, reliability or validity of the dataset, its methods, its results, or the conclusions reported by the original investigator(s). Reanalysis of another's data to explore an unrelated hypothesis which the original author was not interested in pursuing also raises ethical issues, but is not the focus of these guidelines.

Rationale:

Epidemiological studies play a central role in policies and regulations of government and international agencies designed to protect public health. Epidemiological studies are also important in the legal arena, assisting in the determination of the cause of an individual's illness or injury, or quite frequently of injury to a class of individuals, such as a community. Controversy in the public health sciences has more often been resolved through replication than through reanalysis of existing data. However, reanalysis and reinterpretation are becoming increasingly common. Since the results and interpretation of epidemiological studies can have significant implications for stakeholders' interests or ideological commitments, it is not surprising that parties whose policy preferences are challenged by the "facts" revealed by a study could be highly motivated to demand the opportunity to closely examine the data from such a study. They may hope to find results more to their liking or at the minimum to "manufacture doubt" about the original results so as to make them seem less useable for the formation of policy.

Just as one example of regulation associated with reanalyses, under US law, for study data meeting certain conditions around their use in decision making and the nature of the matter under consideration, investigators must provide the raw data from federally-funded studies in response to requests under the Freedom of Information Act. The rationale for this provision, originally passed by the US Congress in 1998, was to enable parties affected by government regulation to examine (and presumably challenge) the data used in developing these regulations. However, in the US there are no equivalent legal requirements for access to raw data of studies that have been paid for with private funding, whether or not they are used in regulatory proceedings, which does open the question of equality of use of available data. For this reason, most reanalyses to date have been performed on publicly-supported studies.

This new environment for reanalyses of data in the US creates new technical and ethical responsibilities for original investigators, those who are tasked with funding and overseeing a reanalysis and reinterpretation, and the reanalyzers and reinterpreters themselves. This situation has prompted the development of the following guidelines. We do recognize that some of these guidelines may not be relevant in countries without such regulation, or with different context and regulations. This may require further development of these guidelines if ethically problematic issues arise from applying them in such countries.

We have organized them according to the stakeholders whose legitimate interests are being protected.

Protecting the Public's Interest in Valid Information:

1) Data from epidemiological studies should be available for properly organized and competent reanalysis and reinterpretation, regardless of whether the study was funded by public monies or by groups with particular interests or ideologies.

2) The original authors have a responsibility to cooperate with, and facil-

itate, properly organized and competent (as indicated below) reanalysis and reinterpretation of their data.

3) The original authors also have a duty to advocate for proper organization and competence in reanalysis/reinterpretation of their data, as specified in these guidelines.

4) A process should be created that leaves the reanalyst free of any conflicting interests or at least declaring them publicly.

5) An independent advisory structure for the reanalysis and reinterpretation should be established that can correct unintended bias in the analysis and in the scientific and public reinterpretations, and assure that the persons chosen to do the reanalysis and reinterpretation are professionally competent to do so. All human beings are prone to have biases that they do not recognize in themselves. The best way to correct for this is to invite advisors with opposing biases based on their public statements or publications to comment on each stage of the process. When funding is available, this advisory process may be quite formal and extensive. However, even when reanalysis is done without special funding, an independent advisory structure, albeit less extensive and perhaps with pro-bono participation, should still be implemented.

6) There should be agreement in advance among the original investigators and reanalysts as to which hypotheses will be explored, and on the extent to which different patterns of evidence would support or not support each of the different hypotheses and why. This may take the form of a formal analytic plan for the reanalysis.

7) Procedures should be established that afford equal opportunity for stakeholders and their scientific advisors on various sides of the issue to comment for the public record on the reanalysis and reinterpretation.

8) The process of reanalysis should assure in advance that the results of the reanalysis and reinterpretation will be widely available regardless of the outcome.

9) The reanalysis and reinterpretation should be published in a way that respectfully clarifies the factual grounds, the scientific claims, the inferential assumptions that warrant those claims and the reasons behind any differences in interpretation.

10) All stakeholders should resist political or other pressures to deviate from these guidelines.

11) Normally, notwithstanding the above guidelines, the reanalyst should recognize that what is being requested essentially constitutes a secondary data analysis. As such, the reanalyst should be required to prepare a formal research proposal including the goals, rationale, hypothesis to be tested and methods of his/her intended reanalysis. This should, like any research proposal, then be submitted to two levels of review:

a) Scientific peer review; and

b) Institutional Review Board (or, a human subjects protection review committee) review through a bona fide institution to ensure the protection of the original subjects. A proposal prepared in this more formal way would also facilitate the work of the independent advisory structure indicated under point #5 above.

Protecting the Rights of Study Participants:

12) All appropriate and necessary measures should be taken to ensure that the confidentiality and privacy of participants in the original study be preserved throughout the reanalysis, including scrupulous adherence to agreements that may have been undertaken between the original investigators and participants or between the original investigators and providers of data.

13) When appropriate and in the case of publicly funded research in the United States, informed consent should mention the possibility of reanalysis by persons other than the original researchers. Protecting the Interests of the Original and Reanalyzing Authors:

14) Rights of data ownership and access to data should be explicitly stated and agreed to in advance by all parties to the reanalysis. In particular, care should be taken to ensure that the original study data not be used for purposes beyond the agreed scope of the reanalysis.

15) Activities should be avoided that obfuscate the facts, harass epidemiologists who have presented unwelcome results, or intimidate future investigators from working in this area.

16) The reanalysts should work to ensure open communication, and fair and respectful dealings with the original investigators. They should create an environment of collegial yet critical truth finding, and provide the original researchers the opportunity to comment on the rationale and methods chosen for reanalysis before it takes place, as well as on the interpretation of the results of any new analysis. If there are differences in interpretation, all parties should respectfully explain the reasons for these differences in written form so that third parties can fairly draw their own conclusions.

17) The reanalysts should acknowledge where appropriate the role of the original investigators in the development of the original methods and instruments and in the collection of the data.

18) The reanalysis process should guarantee the ability of the original authors to publish their work within a reasonable time before any reanalysis and reinterpretation is published.

19) If the reanalysis has dedicated funding, a reasonable budget should be established to facilitate the involvement of the original investigators.

Institutional Changes to Facilitate Proper Reanalysis and Reinterpretation:

The ability to find, access, interpret and reanalyze data according to these guidelines would be facilitated by changes which go beyond the ability of individual researchers to implement. We make three suggestions that ISEE is willing to help realize.

(a) Adequate documentation of data and methods is critical to the conduct of valid scientific research. The sponsors of research should, therefore, provide funds to pay for maintaining adequate documentation of the entire project, including coding manuals and programs used in analysis.

(b) It is in the public interest that certain key research data and their documentation continue to be available for additional scientific research that may be needed, perhaps even far into the future. Government agencies and other funders should work with the research community to develop guidelines for deciding which data to preserve, how and where to preserve them, how the datasets would be indexed, and how and by whom they may be accessed. Approaches to implementing these systems should be broadly consistent with the above guidelines. ISEE stands ready to advise in the development of such systems.

(c) These guidelines could be useful to ISEE members approached by stakeholders who request that a reanalysis of another's work be carried out. In this situation, these guidelines could be incorporated into the contract for the work to be conducted. These guidelines could also help ISEE members whose work is to be reanalyzed in that cooperation beyond what is mandated by the law1 could be contingent on an agreement to adopt all of these guidelines.

17.2 The Consequences of Daubert and the Data Quality Act (of 2001)

> There are in fact, two things, science and opinion; the former begets knowledge, the latter ignorance.
> — Hippocrates of Cos (400BC–370BC)

A spending bill that passed Congress in 2001, named the *Consolidated Appropriations Act*, included a brief two-sentence rider that has since become known (pretentiously) as the *Data Quality Act*. This act directed the Office of Management and Budget (OMB) to develop government-wide guidelines:

> The Director of the Office of Management and Budget shall ... with public and Federal agency involvement, issue guidelines ... that provide policy and procedural guidance to Federal agencies for ensuring and maximizing the quality, objectivity, utility, and integrity of information (including statistical information) disseminated by Federal agencies.

In practice, the Data Quality Act has been used more as a ploy by corporations and their affiliates to suppress and/or delay the release of government reports that would be contrary to their collective economic interests. We quote Chris Mooney from his 2005 book, *The Republican War on Science*:

> As subsequently interpreted by the Bush administration ... the so-called Data Quality Act creates an unprecedented and cumbersome process by which government agencies must field complaints over the data, studies, and reports they release to the public. It is a science abuser's dream come true. (p. 105)

To give a more complete sense of this act's sinister intent, the reader might consult the article written by Chris Mooney for the *Boston Globe* (August 28, 2005). Its title is "Interrogations," and has the teaser line: "Thanks to a little-known piece of legislation, scientists at the EPA and other agencies find their work questioned not only by industry, but by their own government." This article is given as an appendix to this section with the author's permission.[3]

Appendix: *Interrogations*, Chris Mooney

The longstanding fault line in American life between politics and science has become increasingly unstable of late, drawing headlines on divisive issues ranging from stem cell research to evolution. But there's a subterranean aspect of this conflict that rarely makes the news: the fight over how science is used by government to protect us from health and environmental risks—in short, to regulate.

Some time early next year, a federal appeals court in Virginia is expected to decide a pivotal lawsuit concerning the uses of scientific information in the regulatory arena. Brought by the US Chamber of Commerce and the Salt Institute, an industry trade association, the suit challenges a National Institutes of Health study showing that reduced salt intake lowers blood pressure. But there's far more at stake here than government dietary advice or salt industry profits. At a time when science itself has increasingly become the battleground of choice for determining what regulatory actions the government will take, the case turns on whether such fights will ultimately find their resolution in the courtroom, at the hands of nonexpert judges. In the process, the suit will define the scope of the five-year-old Data Quality Act, a below-the-radar legislative device that defenders of industry have increasingly relied upon to attack all range of scientific studies whose results or implications they disagree with, from government global warming reports to cancer research using animal subjects. On its face, the act merely seeks to ensure the "quality, objectivity, utility, and integrity" of government information. In practice, as interpreted by the Bush administration, it creates an unprecedented and cumbersome process that saddles agencies with a new workload while empowering businesses to challenge not just government regulations—something they could do anyway—but scientific information that could potentially lead to regulation somewhere down the road. The Data Quality Act, Chamber of Commerce vice president William Kovacs explained in an interview, allows industry to influence the regulatory process from "the very beginning."

Whether companies can sue agencies that reject their "data quality" complaints, thereby dragging individual studies into the courtroom, is the legal question at the core of the Salt Institute and Chamber of Commerce lawsuit. If the judge in the case writes a precedent-setting opinion, and if higher courts agree, a brand-new body of law could emerge, consisting largely of corporate lawsuits against scientific analyses.

But even if the lawsuit fails to establish judicial review of government regulatory science, Republicans in Congress (who are watching the case) are considering legislation that would ensure it anyway. And lawsuits are just the beginning. The Bush administration has also used the Data Quality Act's language as an excuse for implementing a government-wide, industry-friendly "peer review" system for scientific information—one that scientific societies, including the American Public Health Association and the Association of American Medical Colleges, regarded with serious misgivings when it was first proposed.

In the three decades since the regulatory revolution that created the Environmental Protection Agency and other fixtures of the federal bureaucracy, industry has become increasingly adept both at weighing down the rule-making process with years of preliminaries and at challenging regulations once they are announced. Now, each regulatory decision seems to devolve into a "science" fight—not because we don't have enough qualified scientists in federal agencies, or because they're doing a poor job, but rather because those seeking to avoid regulation constantly seek to raise the burden of proof required for action.

The Data Quality Act provides an ideal means of achieving this goal—all under the guise of an attempt to improve the quality of information. In truth, calling the "Data Quality Act" an "act" smacks of over-glorification. In 2000, a flamboyant industry lobbyist named Jim Tozzi, who had worked for tobacco and chemical companies, and other interests, helped to draft two sentences of legalese that Representative Jo Ann Emerson, Republican of Missouri, then tucked into a massive appropriations bill signed into law late in the Clinton administration.

But to understand where the Data Quality Act comes from, and how it facilitates a veritable war on government science conducted by private industry, you first need some historical context. Over the past several decades, companies subject to government regulation have increasingly sought procedural tools to allow them to challenge and dispute the scientific data forming the basis for such regulation. The overarching strategy—sometimes dubbed "paralysis by analysis"—has been to make the fight turn on the validity of information, rather than moral or policy questions, such as how much of a precautionary approach our government should take in endeavoring to protect us from risks.

One key proponent of challenges to government scientific information was the tobacco industry. In 1998, in a dress rehearsal for the Data Quality Act, lobbyist Tozzi worked on the so-called Shelby Amendment—named after its official "author," Republican senator Richard Shelby of Alabama—for Philip Morris. Also a brief insert to an appropriations bill, and equally loathed by the scientific community, the one-sentence amendment allows for the use of the Freedom of Information Act to obtain "all data produced" by any publicly funded scientific study.

Though subsequently limited in scope by a wary Clinton administration, the Shelby Amendment helps companies conduct their own audits of studies they don't like—as many firms, including tobacco, have often sought to do. In the process, they can reanalyze the data and put a different spin on it. The Data Quality Act has been called the "daughter" of Shelby. Sure enough, Philip Morris documents show that Tozzi circulated language for a "data quality" regulation to the company and received comments in response. While Tozzi agrees there was "definitely" a connection between the tobacco industry and the Data Quality Act, in an interview he insisted it was "not a cause and effect." But he added that if tobacco companies had "continued challenging stuff, data quality would have been a big help to them."

With the Data Quality Act, industry had finally found a means of disputing agency science in the earliest stages of the regulatory process. At the merest public mention that a government agency might be looking at a particular study and wondering if it could compel regulatory action, industry interests could file a complaint challenging the agency's "dissemination" of the information.

"Industry tends to think there's some magic bullet somewhere that's going to protect them from regulation," says Gary Bass, executive director of OMB Watch, a government watchdog group that has monitored the Data Quality Act closely, and a kind of counterpart to Tozzi on the left. "If they can just get there earlier." Now, Bass continues, the strategy is to look "in the science, in the data," for a way of avoiding regulation.

In fact, both the Shelby Amendment and the Data Quality Act represent successful under-the-radar attempts to pass bite-sized pieces of legislation highly reminiscent of a massive, but failed, "regulatory reform" bill pushed by Republicans during the Gingrich Congress in the mid-1990s. "In the end," the Chamber of Commerce's Kovacs says, "what we're going to get is far more than we could have ever gotten by having a comprehensive regulatory law passed."

As soon as the Data Quality Act came into effect in October 2002, corporate interests took it for a test drive. Tozzi's Center for Regulatory Effectiveness teamed up with the Kansas Corn Grower's Association and a group called the Triazine Network to challenge an EPA risk assessment that had discussed evidence suggesting that atrazine, a herbicide in widespread use in US cornfields and on other crops, causes gonadal abnormalities in male frogs (possibly contributing to declining amphibian populations worldwide). Sure enough, citing the lack of an accepted test for measuring endocrine effects, the EPA ultimately concluded that hormone disruption did not constitute a good reason to restrict atrazine's use. Tozzi's petition very likely had an effect on the decision. "You put the sequence together and I think it's common sense to say it had an influence," notes Sean Moulton of OMB Watch.

And the atrazine challenge is just one example. The Data Quality Act has been employed in a wide variety of lesser known industry-friendly campaigns: by a chemical company challenging EPA information on barium, its product; by a law firm that has reportedly represented companies involved in asbestos lawsuits, objecting to an EPA document on asbestos risks from work on vehicle brakes; and by paint manufacturers questioning the basis for regulations on air emissions from their products, to name just a few examples.

All three of these challenges targeted the Environmental Protection Agency. But other agencies have also seen an influx of complaints. Logging interests, for instance, challenged several US Forest Service studies and documents relating to habitat protections for the Northern Goshawk, submitting a massive 281-page petition outlining their objections.

Some liberals and environmental groups have also attempted to employ the Data Quality Act. The libertarian-leaning Tozzi—a jazz aficionado with regulatory "nerd" expertise—has himself teamed up with Americans for Free Access, which used the act to dispute government claims that marijuana has no medical uses. "My musician friends will finally appreciate my work," cracked Tozzi in an e-mail alerting me to the pot petition.

But mid-2004 analyses by the *Washington Post* and OMB Watch showed that the act, in effect for almost two years, had been overwhelmingly used by industry groups. In any all-out war over regulatory science, corporations will ultimately have the financial edge. In effect, then, the Data Quality Act tilts the regulatory playing field still further in industry's favor.

But as successful as Tozzi's and his clients' efforts have been, they might not have gone very far without the help of John Graham. The controversial head of the OMB Office of Information and Regulatory Affairs (and former head of the Harvard Center for Risk Analysis), Graham is responsible for the review of government regulations in the Bush White House.

Not long after coming into office, Graham seized upon the Data Quality Act and instructed federal agencies to draw up their own guidelines for implementing it by October 2002, when the law would go into effect. Then, in September 2003, claiming the Data Quality Act gave his office a newfound role in improving the quality of government science, Graham proposed using the act's thin language to justify an unprecedented government-wide "peer review" system for agency science.

As legal scholar Wendy Wagner of the University of Texas argued in a recent article entitled "The 'Bad Science' Fiction," there's little real evidence to support the notion that government agencies churn out "junk science"—a frequent industry accusation—or that their existing peer review protocols are inadequate. So it's no surprise that scientific heavyweights like the American Public Health Association and American Association for the Advancement of Science announced their concerns over Graham's initial proposal, which would have required review of all "significant regulatory information," and an additionally laborious review process for data with "a possible impact of more than \$100 million in any year."

Most significantly, perhaps, the proposal would have blocked academic scientists from serving as reviewers if they had obtained or were seeking "substantial" research funding from the government agency in question—a condition likely to exclude leading academic experts in a field—yet showed little concern about the participation of industry scientists. Graham's office subsequently softened the proposal, and removed this most objectionable of requirements.

But the administration has still failed to adequately explain why such a "peer review" system was needed in the first place. After all, no government-wide standard for peer review existed in the past—and that may have been a good thing. Different agencies have different needs, just as different scientific disciplines employ different methodologies and standards of proof.

Furthermore, concern about onerous and unnecessary intrusions into the regulatory process remain warranted. The process for vetting "highly influential scientific assessments" under the new peer review plan remains quite burdensome, requiring the preparation of a peer review report that must be made public and a written response from the agency.

Such procedures will only further ossify an already sluggish regulatory process. And as "peer review" critic Sidney A. Shapiro, of the Wake Forest University School of Law, has observed, these procedures are required even for "routine information" that is not "complex, controversial, or novel."

Such objections notwithstanding, in December of last year Graham's office finalized the peer review plan. Its provisions for "highly influential scientific assessments" took effect on June 16. The media hardly noticed. Below the radar, as always, expansion of the Data Quality Act continues apace.

Notes

[1]For a comprehensive resource on evidence and expert witness admissibility, both in a contemporary and historical context, see the *Reference Manual on Scientific Evidence*; a (free) third edition was published in 2011 by the Federal Judicial Center and the National Academies Press).

[2]As can be seen from the meticulous tracings on various websites, the

influence of the *Daubert* trilogy on court proceedings has been enormous. One of the more comprehensive sites is

    www.daubertontheweb.com

This site provides a listing of *Daubert* opinions in federal courts and gives an interesting cumulative classification over thirty-six separate professions for *Daubert* admissibility decisions (Nordberg, 2010).

[3]Based on the Presidential elections of 2008, the cast of characters throughout the various federal agencies has changed. Because of this turnover, the influence of mechanisms such as the *Data Quality Act* and a governmental "peer-review" regulation is likely to change dramatically. These policies were promulgated by John Graham when he was head of the OMB Office of Information and Regulatory Affairs under President Bush (Executive Office of the President, 2004).

Chapter 18

Some Concluding Remarks

> T'aint what a man don't know that hurts him, it's what he know
> that just ain't so.
> – Frank "Kin" Hubbard

To provide a fitting coda to this book, we go back in part to several ideas offered by Kahneman and Tversky about how our reasoning may go astray. In his recent bestseller, *Thinking, Fast and Slow* (2011), Daniel Kahneman relates a cautionary tale about himself and his experiences as a young psychologist in the Israeli Army (pp. 209–211). He was charged with making predictions about future officer candidate performance based on a field "test" of leadership. Although the data did not support the results predicted by Kahneman and his team, they were unshaken in their belief that the field test was valid. This tendency to maintain belief (such as for a test predicting behavior), in the face of evidence to the contrary, has been called the "illusion of validity" by Kahneman and others. There is a corollary that such an illusion usually comes with much (over)confidence in the validity of the prediction regardless of the lack of data to support such confidence.[1]

As we have maintained from the outset, a graduate course in statistics should prepare students in a number of areas that have immediate implications for the practice of ethical reasoning. In this concluding section, we review six broad topics that should be part of any competently taught sequence in the behavioral sciences: (1) formal tools to help think through ethical situations; (2) a basic understanding of the psychology of reasoning and how it may differ from that based on a normative theory of probability; (3) how to be (dis)honest in the presentation of information, and to avoid obfuscation; (4) some ability to ferret out specious argumentation when it has a supposed statistical basis; (5) the deleterious effects of culling in all its various forms (for example, the identification of "false positives"), and the subsequent failures either to replicate or cross-validate; (6) identifying plausible but misguided reasoning from data, or from other information presented graphically.

* * *

One of the trite quantitative sayings that may at times drive individuals "up a wall" is when someone says condescendingly, "just do the math." This saying can become a little less obnoxious when reinterpreted to mean working through a situation formally rather than just giving a quick answer based on first impressions. An example of this may help.[2] In 1990, Craig Whitaker wrote a letter to Marilyn vos Savant's column in *Parade* magazine stating what has been named the Monty Hall problem:[3]

> Suppose you're on a game show, and you're given the choice of three doors. Behind one door is a car; behind the others, goats. You pick a door, say No. 1, and the host, who knows what's behind the doors, opens another door, say No. 3, which has a goat. He then says to you, 'Do you want to pick door No. 2?' Is it to your advantage to switch your choice? (p. 15)

The answer almost universally given to this problem is that switching does not matter, presumably with the reasoning that there is no way for the player to know which of the two unopened doors is the winner, and each of these must then have an equal probability of being the winner. By writing down three doors hiding one car and two goats, and working through the options in a short simulation, it becomes clear quickly that the opening of a goat door changes the information one has about the original situation, and that always changing doors doubles the probability of winning from 1/3 to 2/3.[4]

Any beginning statistics class should always include a number of formal tools to help work through puzzling situations. Several of these have been mentioned in earlier sections: Bayes' theorem and implications for screening using sensitivities, specificities, and prior probabilities; conditional probabilities more generally and how probabilistic reasoning might work for facilitative and inhibitive events; sample sizes and variability in, say, a sample mean, and how a confidence interval might be constructed that could be made as accurate as necessary by just increasing the sample size, and without any need to consider the size of the original population of interest; how statistical independence operates or doesn't; the pervasiveness of natural variability and the use of simple probability models (such as the binomial) to generate stochastic processes; the computations involved in corrections for attenuation; the use of Taylor–Russell charts; the formal distinction between (test) bias and impact, where the latter is not evidence of test "unfairness" per se.

* * *

A second area of interest in developing statistical literacy and learning to reason ethically is the large body of work produced by psychologists. This work compares the normative theory of choice and decisions derivable from probability theory, and how this may not be the best guide to the actual reasoning processes individuals use. The contributions of Tversky and Kahneman (for example, 1971, 1974, 1981) are particularly germane to our understanding of reasoning. People rely on various simplifying heuristic principles to assess probabilities and engage in judgments under uncertainty. We give a classic Tversky and Kahneman (1983) illustration to show how various reasoning heuristics might operate:

> Linda is 31 years old, single, outspoken and very bright. She majored in philosophy. As a student she was deeply concerned with issues of discrimination and social justice, and also participated in anti-nuclear demonstrations. (p. 297)
> Which ... [is] more probable?
> 1. Linda is a bank teller.
> 2. Linda is a bank teller and is active in the feminist movement.
> (p. 299)

Eighty-five percent of one group of subjects chose option 2, even though the conjunction of two events must be less likely than either of the constituent events. Tversky and Kahneman argue that this "conjunction fallacy" occurs because the "representativeness heuristic" is being used to make the judgment; the second option seems more representative of Linda based on the description given for her.

The representativeness heuristic operates where probabilities are evaluated by the degree to which A is representative of B; if highly representative, the probability that A originates from B is assessed to be higher. When representativeness heuristics are in operation, a number of related characteristics of the attendant reasoning processes become apparent: prior probabilities (baserates) are ignored; insensitivity develops to the operation of sample size on variability; an expectation that a sequence of events generated by some random process, even when the sequence is short, will still possess all the essential characteristics of the process itself. This leads to the "gambler's fallacy" (or, "the doctrine of the maturity of chances"), where certain events must be "due" to bring the string more in line with representativeness; as one should know, corrections are not made in a chance process but only diluted as the process unfolds. When a belief is present in the "law of small numbers," even small samples must be highly representative of the parent population; thus, researchers put too much faith in what is seen in small samples and overestimate replicability. Also, people may fail to recognize regression toward the mean because predicted outcomes should be maximally representative of the input and therefore be exactly as extreme.

A second powerful reasoning heuristic is *availability*. We quote Tversky and Kahneman (1974):

> Lifelong experience has taught us that, in general, instances of large classes are recalled better and faster than instances of less frequent classes; that likely occurrences are easier to imagine than unlikely ones; and that the associative connections between events are strengthened when the events frequently co-occur. As a result, man has at his disposal a procedure (the availability heuristic) for estimating the numerosity of a class, the likelihood of an event, or the frequency of co-occurrences, by the ease with which the relevant mental operations of retrieval, construction, or association can be performed. (p. 1128)

Because retrievability can be influenced by differential familiarity and saliences, the probability of an event may not be best estimated by the ease to which occurrences come to mind. A third reasoning heuristic is one of *anchoring and adjustment*, which may also be prone to various biasing effects. Here, estimates are made based on some initial value that is then adjusted (Tversky & Kahneman, 1974).

When required to reason about an individual's motives in some ethical context, it is prudent to remember the operation of the *fundamental attribution error*, where people presume that actions of others are indicative of the true ilk of a person, and not just that the situation compels the behavior. As one example from the courts, even when confessions are extracted that can be demonstrably shown false, there is still a greater likelihood of inferring guilt compared to the situation where a false confession was not heard. The classic experiment on the fundamental attribution error is from Jones and Harris (1967); we quote a summary given in the Wikipedia article on the fundamental attribution error ("Fundamental attribution error," 2010):

> Subjects read pro- and anti-Fidel Castro essays. Subjects were asked to rate the pro-Castro attitudes of the writers. When the subjects believed that the writers freely chose the positions they took (for or against Castro), they naturally rated the people who spoke in favor of Castro as having a more positive attitude toward Castro. However, contradicting Jones and Harris' initial hypothesis, when the subjects were told that the writer's positions were determined by a coin toss, they still rated writers who spoke in favor of Castro as having, on average, a more positive attitude towards Castro than those who spoke against him. In other words, the subjects were unable to see the influence of the situational constraints placed upon the writers; they could not refrain from attributing sincere belief to the writers.

A particularly egregious example of making the fundamental attribution error (and moreover, for nefarious political purposes), is Liz Cheney and her ad on the website "Keep America Safe" regarding those lawyers currently at the Justice Department who worked as advocates for "enemy combatants" at Guantanamo Bay, Cuba. We give an article that lays out the issues by Michael Stone of the *Portland Progressive Examiner* (March 5, 2010; "Toxic Politics: Liz Cheney's Keep America Safe 'Al Qaeda Seven' Ad"):

> Liz Cheney, daughter of former Vice President Dick Cheney and co-founder of the advocacy group "Keep America Safe," is taking heat for a controversial ad questioning the values of Justice Department lawyers who represented Guantanamo Bay detainees.
> Several top political appointees at the Justice Department previously worked as lawyers or advocates for 'enemy combatants' confined at Guantanamo Bay, Cuba. In their ad, Cheney's group derides

the unidentified appointees as the 'Al Qaeda 7.' The ad implies the appointees share terrorist values.

Aside from questioning the values of these Justice Department lawyers, the ad is using fear and insinuations to smear both the Justice Department lawyers and the Obama administration.

Demonizing Department of Justice attorneys as terrorist sympathizers for their past legal work defending Gitmo detainees is wrong. The unfounded attacks are vicious, and reminiscent of McCarthyism.

Indeed, the ad itself puts into question Cheney's values, her patriotism, her loyalty. One thing is certain: her understanding of US history, the founding of our country, and the US Constitution, is left seriously wanting.

John Aloysius Farrell, writing in the Thomas Jefferson Street blog, for *US News and World Report*, explains:

There are reasons why the founding fathers ... in the Bill of Rights, strove to protect the rights of citizens arrested and put on trial by the government in amendments number 4, 5, 6, 7, and 8.

The founders had just fought a long and bloody revolution against King George, and knew well how tyrants like the British sovereign perpetuated power with arbitrary arrests, imprisonments, and executions. And so, along with guarantees like the right to due process, and protection from unreasonable searches and cruel and unusual punishment, the first patriots also included, in the Sixth Amendment, the right of an American to a speedy trial, by an impartial jury, with "the Assistance of Counsel for his defense."

John Adams regarded his defense of the British soldiers responsible for the Boston Massacre as one of the noblest acts of his life for good reason. Our adversarial system of justice depends upon suspects receiving a vigorous defense. That means all suspects must receive adequate legal counsel, including those accused of the most heinous crimes: murder, rape, child abuse and yes, even terrorism.

Defending a terrorist in court does not mean that one is a terrorist or shares terrorist values. Implying otherwise is despicable. Cheney's attacks are a dangerous politicization and polarization of the terrorism issue. Those who would honor our system of law and justice by defending suspected terrorists deserve our respect. Instead Cheney and her group smear these patriots in an attempt to score points against political enemies.

* * *

The presentation of data is an obvious area of concern when developing the basics of statistical literacy. Some aspects may be obvious, such as not making up data or suppressing analyses or information that don't conform to prior expectations. At times, however, it is possible to contextualize (or to "frame") the same information in different ways that might lead to differing interpretations. Chapter 10 on (mis)reporting of data is devoted more extensively to the review of Gigerenzer et al. (2007), where the distinctions are made between survival and mortality rates, absolute versus relative risks, natural frequencies versus probabilities, among others. Generally, the presentation of information should be as honest, clear, and transparent as possible. An example given by Gigerenzer et al. (2007) suggests the use of frequency statements instead of single-event probabilities, thereby removing the ambiguity of the reference class: instead of saying "there is a 30% to 50% probability of developing sexual problems with Prozac," use "out of every 10 patients who take Prozac, 3 to 5 experience a sexual problem." Thus, a male taking Prozac won't expect that 30% to 50% of his personal sexual encounters will result in a "failure."

In presenting data to persuade, and because of the "lead-time bias" medical screening produces (see Chapter 10 or Gigerenzer et al., 2007), it is unethical to promote any kind of screening based on improved five-year survival rates, or to compare such survival rates across countries where screening practices vary. As a somewhat jaded view of our current health situation, we have physicians practicing defensive medicine because there are no legal consequences for overdiagnosis and overtreatment, but only for underdiagnosis. Or, as the editor of the *Lancet* commented (as quoted by R. Horton, *New York Review of Books*, March 11, 2004), "journals have devolved into information laundering operations for the pharmaceutical industry." The ethical issues involved in medical screening and its associated consequences are psychologically important; for example, months after false positives for HIV, mammograms, or prostate cancer, considerable and possibly dysfunctional anxiety may still exist.[5]

When data are presented to make a health-related point, it is common practice to give the argument in terms of a "surrogate endpoint." Instead of providing direct evidence based on a clinically desired outcome (for example, if you engage in this recommended behavior, the chance of dying from, say, a heart attack is reduced by such and such amount), the case is stated in terms of a proxy (for example, if you engage in this recommended behavior, your cholesterol levels will be reduced). In general, a surrogate end point or biomarker is a measure of a certain treatment that may correlate with a real clinical endpoint, but the relationship is not guaranteed. This caution can be rephrased as "a correlate does not a surrogate make."

It is a common misconception that something correlated with the true clinical outcome must automatically then be usable as a valid surrogate end point and can act as a proxy replacement for the clinical outcome of primary interest. As is true for all correlational phenomena, causal extrapolation requires further argument. In this case, it is that the effect of the intervention on the surrogate directly predicts the clinical outcome. Obviously, this is a more demanding requirement.

Outside of the medical arena, proxies play prominently in the current climate-change debate. When actual surface temperatures are unavailable, surrogates for these are typically used (for example, tree-ring growth, coral accumulation, evidence in ice). Whether these are satisfactory stand-ins for the actual surface temperatures is questionable. Before automatically accepting a causal statement (for example, that greenhouse gases are wholly responsible for the apparent recent increase in earth temperature), pointed (statistical) questions should be raised, such as: (a) why don't the tree-ring proxies show the effects of certain climate periods in our history—the Medieval Warm Period (circa 1200) and the Little Ice Age (circa 1600)?; (b) over the last century or so, why has the tree-ring and surface temperature relationship been corrupted so that various graphical "tricks" need to be used to obtain the "hockey stick" graphic demonstrating the apparent catastrophic increase in earth temperature over the last century?; (c) what effect do the various solar cycles that the sun goes through have on our climate; could these be an alternative mechanism for what we are seeing in climate change?; (d) or, is it some random process and we are on the up-turn of something comparable to the Medieval Warm Period, with some later downturn expected into another Little Ice Age?

* * *

A fourth statistical literacy concern is to have enough of the formal skills and context to separate legitimate claims from those that might represent more specious arguments. As examples, one should recognize when a case for cause is made in a situation for which regression toward the mean is as likely an explanation, or when test unfairness is argued for based on differential performance (that is, impact) and not on actual test bias (that is, same ability levels performing differently). A more recent illustration of the questionable promotion of a methodological approach, called "optimal data analysis" (ODA), is given by Yarnold and Soltysik (2004). We quote from their preface:

> [T]o determine whether ODA is the appropriate method of analysis for any particular dataset, it is sufficient to consider the following question: When you make a prediction, would you rather be correct or incorrect? If your answer is "correct," then ODA is the appropriate analytic methodology—by definition. That is because, for any given dataset, ODA explicitly obtains a statistical model that yields the theoretical maximum possible level of predictive accuracy (for example, number of correct predictions) when it is applied to those data. That is the motivation for ODA; that is its purpose. Of course, it is a matter of personal preference whether one desires to make accurate predictions. In contrast, alternative non-ODA statistical models do not explicitly yield theoretical maximum predictive accuracy. Although they sometimes may, it is not guaranteed as it is for ODA models. It is for this reason that we refer to non-ODA models as being *suboptimal*. (p. xi)

Sophistic arguments such as these have no place in the methodological literature (even when a text, such as this one, has been reviewed and published by the American Psychological Association). It is inappropriate to call one's method "optimal" and refer pejoratively to others as therefore "suboptimal." The simplistic approach to classification underlying "optimal data analysis" is known not to cross-validate well (see, for example, Stam, 1997); it is a large area of operations research where the engineering effort is always to squeeze a little more out of an observed sample. What is most relevant in the behavioral sciences is stability and cross-validation (of the type reviewed in Dawes [1979] on proper and improper linear models); and to know which variables discriminate and how, and to thereby "tell the story" more convincingly and honestly.

* * *

The penultimate area of review in this concluding section is a reminder of the ubiquitous effects of searching/selecting/optimization, and the identification of "false positives." We have mentioned some blatant examples in earlier sections—the weird neuroscience correlations; the small probabilities (mis)reported in various legal cases (such as the Dreyfus small probability for the forgery coincidences, or that for the de Berk hospital fatalities pattern, both discussed in Chapter 3); repeated clinical experimentation until positive results are reached in a drug trial—but there are many more situations that would fail to replicate. We need to be ever-vigilant of results obtained by "culling" and then presented as evidence.

A general version of the difficulties encountered when results are culled is labeled the *file-drawer problem* (for example, see "Publication bias," 2010; Rosenthal, 1979). This refers to the practice of researchers putting away studies with negative outcomes (that is, studies not reaching reasonable statistical significance or when something is found contrary to what the researchers want or expect, or those rejected by journals that will consider publishing only articles demonstrating significant positive effects). The file-drawer problem can seriously bias the results of a meta-analysis, particularly if only published sources are used (and not, for example, unpublished dissertations or all the rejected manuscripts lying on a pile in someone's office). We quote from the abstract of a fairly recent review, "The Scientific Status of Projective Techniques" (Lilienfeld, Wood, & Garb, 2000):

> Although some projective instruments were better than chance at detecting child sexual abuse, there were virtually no replicated findings across independent investigative teams. This meta-analysis also provides the first clear evidence of substantial file-drawer effects in the projectives literature, as the effect sizes from published studies markedly exceeded those from unpublished studies. (p. 27)

The general failure to replicate is being continually (re)documented both in the scientific literature and in more public venues. In medicine, there is the work of John Ioannidis: "Contradicted and Initially Stronger Effects in Highly Cited Clinical Research" (*Journal of the American Medical Association*, 2005, *294*, 218–228);

"Why Most Published Research Findings Are False" (*PLoS Medicine*, 2005, *2*, 696–701).

In the popular media, we have the discussion of the "decline effect" by Jonah Lehrer in the *New Yorker* (December 13, 2010), "The Truth Wears Off (Is There Something Wrong With the Scientific Method?)"; or from one of the the nation's national newspapers, "Low-Salt Diet Ineffective, Study Finds. Disagreement Abounds" (*New York Times*, Gina Kolata, May 3, 2011). We give part of the first sentence of Kolata's article: "A new study found that low-salt diets increase the risk of death from heart attacks and strokes and do not prevent high blood pressure."

The subtle effects of culling with subsequent failures to replicate can have serious consequences for the advancement of our understanding of human behavior. A recent important case in point involves a gene–environment interaction studied by a team led by Avshalom Caspi (Caspi et al., 2003). A polymorphism related to the neurotransmitter serotonin was identified that apparently could be triggered to confer susceptibility to life stresses and resulting depression. Needless to say, this behavioral genetic link caused quite a stir in the community devoted to mental health research. Unfortunately, the result could not be replicated in a subsequent meta-analysis (could this possibly be due to the implicit culling over the numerous genes affecting the amount of serotonin in the brain?). Because of the importance of this cautionary tale for behavioral genetics research generally, we reproduce (immediately below and continuing into an endnote), a *News of the Week* item from *Science*, written by Constance Holden (2009), "Back to the Drawing Board for Psychiatric Genetics":

> Geneticists have long been immersed in an arduous and largely fruitless search to identify genes involved in psychiatric disorders. In 2003, a team led by Avshalom Caspi, now at Duke University in Durham, North Carolina, finally landed a huge catch: a gene variant that seemed to play a major role in whether people get depressed in response to life's stresses or sail through. The find, a polymorphism related to the neurotransmitter serotonin, was heralded as a prime example of "gene-environment interaction": whereby an environmental trigger influences the activity of a gene in a way that confers susceptibility.[6]

* * *

Our final statistical literacy issue is the importance of developing abilities to spot and avoid falling prey to the trap of specious reasoning known as an "argument from ignorance," or *argumentum ad ignorantiam*, where a premise is claimed to be true only because it has not been proven false, or that it is false because it has not been proven true ("Argument from ignorance," 2010). Sometimes this is also referred to as "arguing from a vacuum" (paraphrasing from Dawes, 1994, p. 25), where what is purported to be true is supported not by direct evidence but by attacking an alternative possibility. Thus, a clinician might say: "because the research results indicate a great deal of uncertainty about what to do, my expert judgment can do better in prescribing treatment than these results." Or to argue that people "need" drugs just because they haven't solved their problems before taking them.[7]

A related fallacy is *argument from personal incredulity*, where because one personally finds a premise unlikely or unbelievable, the premise can be assumed false, or that another preferred but unproven premise is true instead ("Argument from ignorance," 2010). In both of these instances, a person regards the lack of evidence for one view as constituting proof that another is true. Related fallacies are (a) the *false dilemma* where only two alternatives are considered when there are, in fact, other options ("False dilemma," 2010). The famous Eldridge Cleaver quotation from his 1968 presidential campaign is a case in point: "[Y]ou're either part of the solution or you're part of the problem." Or, (b) the Latin phrase *falsum in uno, falsum in omnibus* (false in one thing, false in everything) implying that someone found to be wrong on one issue, must be wrong on all others as well. In a more homey form, "when a clock strikes thirteen, it raises doubt not only to that chime, but to the twelve that came before." Unfortunately, we may have a current example of this in the ongoing climate-change debate that we have discussed earlier—the one false statistic proffered by a 2007 report from the Intergovernmental Panel on Climate Change (IPCC) on Himalayan glacier melt may serve to derail the whole science-based argument that climate change is real.

There are several fallacies with a strong statistical tinge related to *argumentum ad ignorantiam*. One is the "margin of error folly," (Rogosa, 2005): if it could be, it is. Or, in a hypothesis-testing context, if a difference isn't significant, it is zero. It is important not to confuse a statement of "no evidence of an effect" with one of "evidence of no effect." We now can refer to all of these reasoning anomalies under the umbrella term "truthiness," coined by Stephen Colbert from Comedy Central's, *The Colbert Report* ("Truthiness," 2010; Colbert & Hoskinson, 2005). Here, truth comes from the gut, not books, and refers to the preferring of concepts or facts one wishes to be true, rather than known to be true. For example, in 2009 we had the "birthers," who claimed that Barack Obama was not born in the United States, so constitutionally he cannot be President ("Barack Obama citizenship conspiracy theories," 2011); or that the Health Care Bill included "death squads" ready to "pull the plug on granny," or earlier in the 2000s, there were weapons of mass destruction that justified the Iraq war ("Iraq and weapons of mass destruction," 2011); and on and on.[8]

Notes

[1]The main intent of this book has been to encourage the questioning of any and all situations that may prove to be illusions of validity. One of us was convinced not directly from his doctor's strong recommendation but by a data sheet from the Centers for Disease Control and Prevention that he should have the shingles vaccination. Extending the sobriquet of the state of Missouri, we might adopt the new motto of "show me the data."

[2]An enjoyable diversion on Saturday mornings is the NPR radio show, *Car Talk*, with Click and Clack, The Tappet Brothers (aka Ray and Tom Magliozzi). A regular feature of the show, besides giving advice on cars, is The Puzzler; a recent example on Febuary 12, 2011 gives us another chance to "do the math." It is called the Three Slips of Paper, and it is stated as follows on the Car Talk website (this puzzler is being used by permission of *Car Talk* for fifty bucks: © 2011, Dewey, Cheetham, and Howe):

> Three different numbers are chosen at random, and one is written on each of three slips of paper. The slips are then placed face down on the table. You have to choose the slip with the largest number. How can you improve your odds?
> The answer given on the show:
> Ray: This is from Norm Leyden from Franktown, Colorado. The date on it is 1974—I'm a little behind.

Three different numbers are chosen at random, and one is written on each of three slips of paper. The slips are then placed face down on the table. The objective is to choose the slip upon which is written the largest number.

Here are the rules: You can turn over any slip of paper and look at the amount written on it. If for any reason you think this is the largest, you're done; you keep it. Otherwise you discard it and turn over a second slip. Again, if you think this is the one with the biggest number, you keep that one and the game is over. If you don't, you discard that one too.

Tommy: And you're stuck with the third. I get it.

Ray: The chance of getting the highest number is one in three. Or is it? Is there a strategy by which you can improve the odds?

Ray: Well, it turns out there is a way to improve the odds—and leave it to our pal Vinnie to figure out how to do it. Vinnie's strategy changes the odds to one in two. Here's how he does it: First, he picks one of the three slips of paper at random and looks at the number. No matter what the number is, he throws the slip of paper away. But he remembers that number. If the second slip he chooses has a higher number than the first, he sticks with that one. If the number on the second slip is lower than the first number, he goes on to the third slip.

Here's an example. Let's say for the sake of simplicity that the three slips are numbered 1000, 500, and 10.

Let's say Vinnie picks the slip with the 1000. We know he can't possibly win because, according to his rules, he's going to throw that slip out. No matter what he does he loses, whether he picks 500 next or 10. So, Vinnie loses—twice.

Now, let's look at what happens if Vinnie starts with the slip with the 500 on it. If he picks the 10 next, according to his rules, he throws that slip away and goes to the 1000.

Tommy: Whopee! He wins.

Ray: Right. And if Vinnie picks the 1000 next, he wins again!

Finally, if he picks up the slip with the 10 on it first, he'll do, what?

Tommy: Throw it out. Those are his rules.

Ray: Right. And if he should be unfortunate enough to pick up the one that says 500 next, he's going to keep it and he's going to lose. However, if his second choice is not the 500 one but the 1000 one, he's gonna keep that slip—and he'll win.

If you look at all six scenarios, Tommy will win one in three times, while Vinnie will win three times out of six.

Tommy: That's almost half!

Ray: In some countries.

One particularly rich area in probability theory that extends the type of *Car Talk* example just given is in the applied probability topic known as optimal stopping, or more colloquially, "the secretary problem." We paraphrase the simplest form of this problem from Thomas Ferguson's review paper in *Statistical Science* (1989), "Who Solved the Secretary Problem?": There is one secretarial position to be filled from among n applicants who are interviewed sequentially and in a random order. All applicants can be ranked from best to worse, with the choice of accepting an applicant based only on the relative ranks of those interviewed thus far. Once an applicant has been rejected, that decision is irreversible. Assuming the goal is to maximize the probability of selecting the absolute best applicant, it can be shown that the selection rules can be restricted to a class of strategies defined as follows: for some integer $r \geq 1$, reject the first $r - 1$ applicants and select the next who is best in the relative ranking of the applicants interviewed thus far. The probability of selecting the best applicant is $1/n$ for $r = 1$; for $r > 1$, it is

$$\left(\frac{r-1}{n}\right) \sum_{j=r}^{n} \frac{1}{j-1}.$$

For example, when there are 5 ($= n$) applicants, the probabilities of choosing the best for values of r from 1 to 5 are given in the following table:

r	Probability
1	$1/5 = .20$
2	$5/12 \approx .42$
3	$13/30 \approx .43$
4	$7/20 = .35$
5	$1/5 = .20$

Thus, because an r value of 3 leads to the largest probability of about .43, it is best to interview and reject the first two applicants and then pick the next relatively best one. For large n, it is (approximately) optimal to wait until about 37% ($\approx 1/e$) of the applicants have been interviewed and then select the next relatively best one. This also gives the probability of selecting the best applicant as .37 (again, $\approx 1/e$).

In the *Car Talk* Puzzler discussed above, $n = 3$ and Vinnie uses the rule of rejecting the first "interviewee" but then selects the next that is relatively better. The probability of choosing the best therefore increases from 1/3 to 1/2.

[3]As an interesting historical note, the "Monty Hall" problem has been a fixture of probability theory from at least the 1890s; it was named the problem of the "three caskets" by Henri Poincaré (1896), and is more generally known as (Joseph) Bertrand's Box Paradox (see "Bertrand's box paradox," 2010; "Monty Hall problem," 2010; Aitken & Taroni, 2004, p. 78).

[4]To show the reach of the Monty Hall problem, we give the abstract and

parts of the beginning, discussion, and ending for an article by Herbranson and Schroeder (2010): "Are Birds Smarter Than Mathematicians? Pigeons (*Columba livia*) Perform Optimally on a Version of the Monty Hall Dilemma" (*Journal of Comparative Psychology, 124*, 1–13):

> The "Monty Hall Dilemma" (MHD) is a well known probability puzzle in which a player tries to guess which of three doors conceals a desirable prize. After an initial choice is made, one of the remaining doors is opened, revealing no prize. The player is then given the option of staying with their initial guess or switching to the other unopened door. Most people opt to stay with their initial guess, despite the fact that switching doubles the probability of winning. A series of experiments investigated whether pigeons (*Columba livia*), like most humans, would fail to maximize their expected winnings in a version of the MHD. Birds completed multiple trials of a standard MHD, with the three response keys in an operant chamber serving as the three doors and access to mixed grain as the prize. Across experiments, the probability of gaining reinforcement for switching and staying was manipulated, and birds adjusted their probability of switching and staying to approximate the optimal strategy. Replication of the procedure with human participants showed that humans failed to adopt optimal strategies, even with extensive training. (p. 1)
>
> Animals are frequently presented with choices where the outcomes associated with each option are ambiguous and probabilistically determined. A classic example is the decision of where to forage for food: a location that has the signs of being food-rich might recently have been picked clean, and a location that appears to be food-poor might contain an unseen cache. Thus, an animal cannot be absolutely certain of the outcome of its choice. A low-probability or "bad" choice might yield a favorable outcome, and a high-probability or "good" choice might produce disastrous results. However, even if the chooser cannot be absolutely certain, many of these choices are associated with a particular probability of success. (p. 1)
>
> Taken together, these experiments show that pigeons can learn to respond optimally in a simulation of the MHD. Furthermore, they also suggest that birds learn to respond optimally based on feedback received from completed trials. In particular, a bird's response strategy can be changed by adjusting the relative likelihoods of reinforcement for switching and staying. The surprising implication is that pigeons seem to solve the puzzle, arriving at the optimal solution while most humans do not. (p. 11)
>
> Pigeons, on the other hand, likely use empirical probability to solve the MHD and appear to do so quite successfully. Pigeons might not possess the cognitive framework for a classical probability-based

analysis of a complicated problem like the MHD, but it is certainly not far-fetched to suppose that pigeons can accumulate empirical probabilities by observing the outcomes of numerous trials and adjusting their subsequent behavior accordingly. The MHD then seems to be a problem that is quite challenging from a classical approach, but not necessarily from an empirical approach, and the difference may give pigeons something of an advantage over humans. The aforementioned mathematician Paul Erdös demonstrates this proposition nicely. According to his biography, Erdös refused to accept colleagues' explanations for the appropriate solution to the MHD that were based on classical probability. He was eventually convinced only after seeing a simple Monte Carlo computer simulation that demonstrated beyond any doubt that switching was the superior strategy. Until he was able to approach the problem like a pigeon—via empirical probability—he was unable to embrace the optimal solution. (p. 12)

[5]We list two items in the Suggested Reading related to the drug industry and its propensity to mislead on the way to its primary goal of profit above all else. Both appeared in the *New York Times*: "Menopause, as Brought to You by Big Pharma" (Natasha Singer & Duff Wilson, December 12, 2009, A.3.2); "Sure, It's Treatable. But Is It a Disorder?" (Natasha Singer & Duff Wilson, December 12, 2009, A.3.1).

[6]Continued from the main text:

"Everybody was excited about this," recalls Kathleen Merikangas, a genetic epidemiologist at the National Institute of Mental Health (NIMH) in Bethesda, Maryland. "It was very widely embraced." Because of the well-established link between serotonin and depression, the study offered a plausible biological explanation for why some people are so much more resilient than others in response to life stresses.

But an exhaustive new analysis published last week in *The Journal of the American Medical Association* suggests that the big fish may be a minnow at best.

In a meta-analysis, a multidisciplinary team headed by Merikangas and geneticist Neil Risch of the University of California, San Francisco, reanalyzed data from 14 studies, including Caspi's original, and found that the cumulative data fail to support a connection between the gene, life stress, and depression. It's "disappointing—of all the [candidates for behavior genes] this seemed the most promising," says behavioral geneticist Matthew McGue of the University of Minnesota, Twin Cities.

The Caspi paper concluded from a longitudinal study of 847 New Zealanders that people who have a particular variant of the serotonin transporter gene are more likely to be depressed by stresses, such as divorce and job loss (*Science*, 18 July 2003, pp. 291–293; 386–389). The gene differences had no effect on depression in the absence of adversity. But those with a "short" version of the gene—specifically, an allele of the promoter region of the gene—were more likely to be laid low by unhappy experiences than were those with two copies of the "long" version, presumably because they were getting less serotonin in their brain cells.

Subsequent research on the gene has produced mixed results. To try to settle the issue, Merikangas says, "we really went through the wringer on this paper." The group started with 26 studies but eliminated 12 for various reasons, such as the use of noncomparable methods for measuring depression. In the end, they reanalyzed and combined data from 14 studies, including unpublished data on individual subjects for 10 of them.

Of the 14 studies covering some 12,500 individuals, only three of the smaller ones replicated the Caspi findings. A clear relationship emerged between stressful happenings and depression in all the studies. But no matter which way they sliced the accumulated data, the Risch team found no evidence that the people who got depressed from adverse events were more likely to have the suspect allele than were those who didn't.

Caspi and co-author Terrie Moffitt, also now at Duke, defend their work, saying that the new study "ignores the complete body of scientific evidence." For example, they say the meta-analysis omitted laboratory studies showing that humans with the short allele have exaggerated biological stress responses and are more vulnerable to depression-related disorders such as anxiety and posttraumatic stress disorder. Risch concedes that his team had to omit several supportive studies. That's because, he says, they wanted to focus as much as possible on attempts to replicate the original research, with comparable measures of depression and stress.

Many researchers find the meta-analysis persuasive. "I am not surprised by their conclusions," says psychiatric geneticist Kenneth Kendler of Virginia Commonwealth University in Richmond, an author of one of the supportive studies that was excluded. "Gene discovery in psychiatric illness has been very hard, the hardest kind of science," he says, because scientists are looking for multiple genes with very small effects.

Dorrett Boomsma, a behavior geneticist at Amsterdam's Free University, points out that many people have questioned the Caspi finding. Although the gene was reported to have an effect on depression only in the presence of life stress, she thinks it is "extremely unlikely that it would not have an independent effect" as well. Yet recent whole-genome association studies for depression, for which scientists scan the genomes of thousands of subjects for tens of thousands of markers, she adds, "do not say anything about [the gene]."

Some researchers nonetheless believe it's too soon to close the book on the serotonin transporter. ... geneticist Joel Gelernter of Yale University agrees with Caspi that the rigorous demands of a meta-analysis may have forced the Risch team to carve away too much relevant material. And NIMH psychiatrist Daniel Weinberger says he's not ready to discount brain-imaging studies showing that the variant in question affects emotion-related brain activity.

Merikangas believes the meta-analysis reveals the weakness of the "candidate gene" approach: genotyping a group of subjects for a particular gene variant and calculating the effect of the variant on a particular condition, as was done in the Caspi study. "There are probably 30 to 40 genes that have to do with the amount of serotonin in the brain," she says. So "if we just pull out genes of interest, ... we're prone to false positives." Instead, she says, most geneticists recognize that whole-genome scans are the way to go. McGue agrees that behavioral gene hunters have had to rethink their strategies. Just in the past couple of years, he says, it's become clear that the individual genes affecting behavior are likely to have "much, much smaller effects" than had been thought.

[7]In making policy decisions based on arguments of causality in areas such as medicine and the environment, it is best to remember the *precautionary principle*: whenever a policy or action has the potential of causing harm, say, to the environment or people, and a scientific consensus does not exist that the policy or action is harmful, the burden of proof that it is *not* harmful rests with those wishing to take the action ("Precautionary principle," 2011).

[8]In developing skills to avoid specious reasoning, or to be able to label such reasoning as fallacious, it is helpful to have a number of useful words and phrases at one's disposal. We give a representative few here, with others used in context throughout the book:

ex cathedra: "Spoken with authority; with the authority of the office. ... From Latin *ex cathedra* (from the chair), from *cathedra* (chair). In the Roman Catholic Church, when the Pope speaks *ex cathedra* he is considered infallible. The word cathedral is short for the full-term cathedral church, meaning the principal church of a diocese, one containing a bishop's throne. ... The term is often used ironically or sarcastically to describe self-certain statements, alluding to the Pope's supposed infallibility, as if an office or position conferred immunity from error" (Garg, February 23, 2010).

ad hominem: "Appealing to one's prejudices, emotions, or other personal considerations rather than to intellect or reason. Attacking an opponent personally instead of countering the argument. ... From Latin, literally 'to the person'" (Garg, February 25, 2010).

A Statistical Guide for the Ethically Perplexed

ipse dixit: from the Latin, "he himself said it"; a statement asserted but not proved; to be accepted on faith; a *dictum* ("Ipse dixit," 2010).

obiter dictum: a comment or remark made in passing, particularly by a judge on an issue not directly relevant to the case at hand; from the Latin, "something said in passing" (*Collins English Dictionary*, 2003).

via regia: a reference to an imperial and ancient road; from the Latin, "King's Road" ("Via regia," 2010).

chimerical: created by fanciful imagination; highly improbable (*American Heritage Dictionary of the English Language*, 2000/2006).

equivocal: capable of differing interpretations; ambiguous (*Collins English Dictionary*, 2003).

equivocate: to use equivocal language intentionally, and avoid making an explicit statement (*American Heritage Dictionary of the English Language*, 2000/2006).

Bibliography

Abductive reasoning. (n.d.). In *Wikipedia: The free encyclopedia.* Retrieved on October 29, 2010 from `http://www.en.wikipedia.org/wiki/Abductive_reasoning`

Ablin, R. J. (2010, March 9). The great prostate mistake [Opinion-editorial]. *New York Times.* Retrieved from `http://www.nytimes.com`

Abramson, J. (2011, February 22). The price of colonoscopy [Letter to the editor]. *New York Times.* Retrieved from `http://www.nytimes.com`

Abuse of discretion [Definition]. (n.d.). In *Law.com.* Retrieved on May 18, 2011 from `http://dictionary.law.com/Default.aspx?selected=2291`

An Act Relating to Health; Requiring Informed Consent of a Female Upon Whom an Abortion is Performed; Providing Civil Remedies; Repealing an Obsolete Law; Appropriating Money; Amending Minnesota Statutes 2002, Section 145.4134; Proposing Coding for New Law in Minnesota Statutes, Chapter 145; Repealing Minnesota's Statutes 2002, Section 145.413, Subdivision 1 [Woman's Right to Know Act of 2003], Laws of Minnesota 2003, Chapter 14-S.F.No. 187. Retrieved from Minnesota Office of the Revisor of Statutes [website]: `https://www.revisor.leg.state.mn.us/laws/`

An Act to Amend the Civil Rights Act of 1964 to Strengthen and Improve Federal Civil Rights Laws, to Provide for Damages in Cases of Intentional Employment Discrimination, to Clarify Provisions Regarding Disparate Impact Actions, and for Other Purposes [Civil Rights Act of 1991], Pub. L. No. 102-166 (1991). Retrieved from United States Department of Labor [website]: `http://www.dol.gov/oasam/regs/statutes/Civil-Rights-Act-of-1991.doc`

An Act to Amend the Public Health Service Act to Establish a Program of National Research Service Awards to Assure the Continued Excellence of Biomedical and Behavioral Research and to Provide for the Protection of Human Subjects Involved in Biomedical and Behavioral Research and for Other Purposes [National Research Act of 1974], Pub. L. No. 93-348 (1974). Retrieved from Office of NIH History [website]: `http://history.nih.gov/research/downloads/PL93-348.pdf`

An Act to Close the Achievement Gap with Accountability, Flexibility, and Choice, so That No Child is Left Behind [No Child Left Behind Act of 2001], Pub. L. No. 107-110 (2002). Retrieved from United States Department of Education [website]: `http://www2.ed.gov/policy/elsec/leg/esea02/107-110.pdf`

An Act to Deter and Punish Terrorist Acts in the United States and Around the World, to Enhance Law Enforcement Investigatory Tools, and for Other Purposes [Uniting and Strengthening America by Providing Appropriate Tools Required to Intercept and Obstruct Terrorism (USA PATRIOT Act) Act of

2001], Pub. L. No. 107-56 (2001). Retrieved from United States Government Printing Office [website]: http://www.gpo.gov/fdsys/pkg/PLAW-107publ56/pdf/PLAW-107publ56.pdf

An Act to Establish Rules of Evidence for Certain Courts and Proceedings, Pub. L. No. 93-595 (1975). Retrieved from Federal Evidence Review [website]: http://federalevidence.com/pdf/FRE_Amendments/1975_Orig_Enact/1975-Pub.L._93-595_FRE.pdf

An Act to Limit the Immigration of Aliens Into the United States, and for Other Purposes. [Immigration Act of 1924], Pub. L. No. 68-139 (1924). Retrieved from University of Washington Library [website]: http://library.uwb.edu/guides/USimmigration/43%20stat%2053.pdf

An Act to Preserve Racial Integrity [Racial Integrity Act of Virginia]. (1924). _Acts and joint resolutions (amending the Constitution) of the General Assembly of the State of Virginia_ (S.B. 219, ch. 371, pp. 534–535). Retrieved from the Image Archive on the American Eugenics Movement [website]: http://www.eugenicsarchive.org/eugenics/

An Act to Prohibit Certain Subversive Activities; to Amend Certain Provisions of Law With Respect to the Admission and Deportation of Aliens; to Require the Fingerprinting and Registration of Aliens; and for Other Purposes [Alien Registration Act of 1940; Smith Act of 1940], 18 U.S.C. §2385 (1940; amended 1948, 1956, 1994). Retrieved from United States Government Printing Office [website]: http://www.gpo.gov/fdsys/pkg/USCODE-2009-title18/pdf/USCODE-2009-title18-partI-chap115-sec2385.pdf

An Act to Provide for the Sexual Sterilization of Inmates of State Institutions in Certain Cases [Sterilization Act of Virginia; Virginia Sterilization Act]. (1924). _Acts and joint resolutions (amending the Constitution) of the General Assembly of the State of Virginia_ (S.B. 281, ch. 394, pp. 569–570). Retrieved from the Image Archive on the American Eugenics Movement [website]: http://www.eugenicsarchive.org/eugenics/

Aczel, A. D. (2004). _Chance: A guide to gambling, love, the stock market, and just about everything else._ New York: Thunder's Mouth Press.

Agresti, A., & Presnell, B. (2002). Misvotes, undervotes, and overvotes: The 2000 presidential election in Florida. _Statistical Science, 17,_ 436–440.

Aitken, C. G. G., & Taroni, F. (2004). _Statistics and the evaluation of evidence for forensic scientists_ (2nd ed.). Chichester, England: Wiley.

Alien Registration Act of 1940. See entry for An Act to Prohibit Certain Subversive Activities; to Amend Certain Provisions of Law with Respect to the Admission and Deportation of Aliens; to Require the Fingerprinting and Registration of Aliens; and for Other Purposes.

Allen, M. J., & Yen, W. M. (1979, reissued 2002). _Introduction to measurement theory._ Long Grove, IL: Waveland Press.

Alphonse Bertillon. (n.d.). In _Wikipedia: The free encyclopedia._ Retrieved on February 21, 2012 from http://www.en.wikipedia.org/wiki/Alphonse_Bertillon

American heritage dictionary of the English language (4th ed.). (2000, updated 2006). Boston: Houghton Mifflin.

American heritage science dictionary. (2005). Boston: Houghton Mifflin.

American Psychiatric Association. (1994). _Diagnostic and statistical manual of mental disorders: DSM-IV._ Washington, DC: Author.

American Statistical Association, Committee on Professional Ethics. (1999, August 7). *Ethical guidelines for statistical practice.* Retrieved from http://www.amstat.org/about/ethicalguidelines.cfm

Anderson, J. (2011, November 7). National study finds widespread sexual harassment of students in grades 7 to 12. *New York Times.* Retrieved from http://www.nytimes.com

Andringa, T. (2010, April 22). *Senate subcommittee holds third hearing on Wall Street and the financial crisis: The role of credit rating agencies.* Retrieved from Senate Committee on Homeland Security & Governmental Affairs [website]: http://hsgac.senate.gov/public/index.cfm?FuseAction=Press.MajorityNews&ContentRecord_id=2778a107-5056-8059-7625-aa17151c8b72

Angell, M. (2004). *The truth about the drug companies: How they deceive us and what to do about it.* New York: Random House.

Anrig, G. R. (1987). ETS on "Golden Rule." *Educational Measurement: Issues and Practice, 6*(3), 24–27.

Anthropological criminology. (n.d.). In *Wikipedia: The free encyclopedia.* Retrieved on May 31, 2011 from http://www.en.wikipedia.org/wiki/Anthropological_criminology

Antoine Gombaud. (n.d.). In *Wikipedia: The free encyclopedia.* Retrieved on October 29, 2010 from http://www.en.wikipedia.org/wiki/Antoine_Gombaud

Apophenia. (n.d.). In *Wikipedia: The free encyclopedia.* Retrieved on October 29, 2010 from http://www.en.wikipedia.org/wiki/Apophenia

Arbuthnot, J. (1710). An argument for Divine Providence, taken from the constant regularity observ'd in the births of both sexes. *Philosophical Transactions of the Royal Society of London, 27,* 186–190.

Archibold, R. C. (2010, April 23). Arizona enacts stringent law on immigration. *New York Times.* Retrieved from http://www.nytimes.com

Argument from ignorance. (n.d.). In *Wikipedia: The free encyclopedia.* Retrieved on October 29, 2010 from http://www.en.wikipedia.org/wiki/Argument_from_ignorance

Arizona State University, W. P. Carey School of Business. (2008, February 13). *Ask your doctor if direct-to-consumer healthcare advertising is right for you.* Retrieved from http://knowledge.wpcarey.asu.edu/article.cfm?articleid=1555

Arizona's 'Papers please' law. (2010, June 3; corrected 2010, July 30). Retrieved from FactCheck.org [website]: http://www.factcheck.org/2010/06/arizonas-papers-please-law

Aronowitz, R. (2009, November 19). Addicted to mammograms. *New York Times.* Retrieved from http://www.nytimes.com

Aschwanden, C. (2011, December 19). Studies suggest an acetaminophen–asthma link. *New York Times.* Retrieved from http://www.nytimes.com

Ashby, F. G., Maddox, W. T., & Lee, W. W. (1994). On the dangers of averaging across subjects when using multidimensional scaling or the similarity-choice model. *Psychological Science, 5,* 144–151.

Ashenfelter, O., Ashmore, D., & LaLonde, R. (1995). Bordeaux wine vintage quality and the weather. *Chance, 8*(4), 7–14.

Aspartame. (n.d.). In *Wikipedia: The free encyclopedia.* Retrieved on January 12, 2012 from http://www.en.wikipedia.org/wiki/Apartame

Atrazine. (n.d.). In *Wikipedia: The free encyclopedia.* Retrieved on October 29, 2010 from http://www.en.wikipedia.org/wiki/Atrazine

Austin Bradford Hill. (n.d.). In *Wikipedia: The free encyclopedia.* Retrieved on October 29, 2010 from http://www.en.wikipedia.org/wiki/Austin_Bradford_Hill

Azaria Chamberlain disappearance. (n.d.). In *Wikipedia: The free encyclopedia.* Retrieved on October 29, 2010 from http://www.en.wikipedia.org/wiki/Azaria_Chamberlain_disappearance

BabbleOn5. (2010, January 11). Anecdata [Definition]. In *Urban Dictionary.* Retrieved on April 12, 2011 from http://www.urbandictionary.com/define.php?term=anecdata

Baker, A. (2010, May 12). New York minorities more likely to be frisked. *New York Times.* Retrieved from http://www.nytimes.com

Baker, S. G., & Kramer, B. S. (2001). Good for women, good for men, bad for people: Simpson's paradox and the importance of sex-specific analysis in observational studies. *Journal of Women's Health & Gender-Based Medicine, 10,* 867–872.

Baker, T. B., McFall, R. M., & Shoham, V. (2008). Current status and future prospects of clinical psychology: Toward a scientifically principled approach to mental and behavioral health care. *Psychological Science in the Public Interest, 9,* 67–103.

Bailar, J. C., III. (1997). The promise and problems of meta-analysis [Editorial]. *New England Journal of Medicine, 337,* 559–561. See also LeLorier, J., Grégoire, G., Benhaddad, A., Lapierre, J., & Derderian, F. (1997). Discrepancies between meta-analyses and subsequent large randomized, controlled trials. *New England Journal of Medicine, 337,* 536–542, seven letters to the editor, and the responses of LeLorier & Grégoire and Bailar (1998), *New England Journal of Medicine, 338,* 59–62.

Baldus, D. C., Pulaski, C., & Woodworth, G. (1983). Comparative review of death sentences: An empirical study of the Georgia experience. *Journal of Criminal Law and Criminology, 74,* 661–753.

Baldus, D. C., Woodworth, G. G., & Pulaski, C. A., Jr. (1990). *Equal justice and the death penalty: A legal and empirical analysis.* Boston: Northeastern University Press.

Baldwin, P., & Wainer, H. (2009). A little ignorance: How statistics rescued a damsel in distress. *Chance, 22*(3), 51–55.

Barack Obama citizenship conspiracy theories. (n.d.). In *Wikipedia: The free encyclopedia.* Retrieved on October 24, 2011 from http://www.en.wikipedia.org/wiki/Barack_Obama_citizenship_conspiracy_theories

Barefoot, T. (1984, October 30). *Last statement of Thomas Barefoot.* Texas Department of Criminal Justice, Executed Offenders, Thomas Barefoot, Last Statement, Execution #4. Retrieved from Texas Department of Criminal Justice [website]: http://www.tdcj.state.tx.us/stat/dr_info/barefootthomaslast.html

Barefoot v. Estelle, 463 United States 880 (1983). Retrieved from Justia.com United States Supreme Court Center [website]: http://supreme.justia.com/US/463/880/index.html

Barrett, C. L., Eubank, S. G., & Smith, J. P. (2005). If smallpox strikes Portland . . . . *Scientific American, 292*(3), 54–61.

Becker, R. A., Denby, L., McGill, R., & Wilks, A. R. (1987). Analysis of data from the *Places Rated Almanac. American Statistician, 41*, 169–186.

Begley, S. (2009, October 1). Ignoring the evidence: Why do psychologists reject science? *Newsweek.* Retrieved from http://www.newsweek.com

Belkin, L. (1988, June 10). The law; Expert witness is unfazed by 'Dr. Death' label. *New York Times.* Retrieved from http://www.nytimes.com

The Belmont Report. See entry for National Commission for the Protection of Human Subjects of Biomedical and Behavioral Research.

Benford, F. (1938). The law of anomalous numbers. *Proceedings of the American Philosophical Society, 78*, 551–572.

Benjamini, Y., & Hochberg, Y. (1995). Controlling the false discovery rate: A practical and powerful approach to multiple testing. *Journal of the Royal Statistical Society, Series B, 57*, 289–300.

Bennett, N. G., Bloom, D. E., & Craig, P. H. (1989). The divergence of black and white marriage patterns. *American Journal of Sociology, 95*, 692–722.

Bentham, J. (1827). *Rationale of judicial evidence, specially applied to English practice* (J. S. Mill, Ed.). London: Hunt and Clarke.

Berenson, A., Harris, G., Meier, B., & Pollack, A. (2004, November 14). Despite warnings, drug giant took long path to Vioxx recall. *New York Times.* Retrieved from http://www.nytimes.com

Bernard, C. (1865). *Introduction à l'étude de la médecine expérimentale [Introduction to the study of experimental medicine].* Paris: J. B. Ballière et Fils.

Bernhardt, D., & Heston, S. (2010). Point shaving in college basketball: A cautionary tale for forensic economics. *Economic Inquiry, 48*, 14–25.

Bertin, J. (1983). *Semiology of graphics* (W. J. Berg, Trans.). Madison, WI: University of Wisconsin Press. (Original work published 1973)

Bertrand's box paradox. (n.d.). In *Wikipedia: The free encyclopedia.* Retrieved on October 29, 2010 from http://www.en.wikipedia.org/wiki/Bertrand's_box_paradox

Berwick, D. M. (2005, December 12). Keys to safer hospitals. *Newsweek.* Retrieved from http://www.newsweek.com

Best colleges. (n.d.). *US News & World Report.* Available from http://www.usenews.com/rankings.

Best, J. (2001). *Damned lies and statistics: Untangling numbers from the media, politicians, and activists.* Berkeley and Los Angeles, CA: University of California Press.

Best, J. (2004). *More damned lies and statistics: How numbers confuse public issues.* Berkeley and Los Angeles, CA: University of California Press.

Best, J. (2005). Lies, calculations and constructions: Beyond *How to lie with statistics. Statistical Science, 20*, 210–214.

Best, J. (2008). *Stat-spotting: A field guide to identifying dubious data*. Berkeley and Los Angeles, CA: University of California Press.

Bialik, C. (2009, July 1). Rise and flaw of Internet's election-fraud hunters. *Wall Street Journal*. Retrieved from http://online.wsj.com

Bias (statistics). (n.d.). In *Wikipedia: The free encyclopedia*. Retrieved on October 29, 2010 from http://www.en.wikipedia.org/wiki/Bias_(statistics)

Bickel, P. J., Hammel, E. A., & O'Connell, J. W. (1975). Sex bias in graduate admissions: Data from Berkeley. *Science, 187*, 398–404.

Block, N. (1995). How heritability misleads about race. *Cognition, 56*, 99–128.

Bloodletting. (n.d.). In *Wikipedia: The free encyclopedia*. Retrieved on October 29, 2010 from http://www.en.wikipedia.org/wiki/Bloodletting

Bloom, D. E., & Bennett, N. G. (1985, September). *Marriage patterns in the United States* (Working Paper No. 1701). Retrieved from National Bureau of Economic Research [website]: http://www.nber.org/papers/w1701.pdf

Bloomberg News. (2011, January 28). Pfizer told to pay $142.1 million over marketing of epilepsy drug. *New York Times*. Retrieved from http://www.nytimes.com

Blumenthal, R. (1984, May 6). Vietnam Agent Orange suit by veterans is going to trial. *New York Times*. Retrieved from http://www.nytimes.com

Blyth, C. R. (1972). On Simpson's paradox and the sure-thing principle. *Journal of the American Statistical Association, 67*, 364–366.

Boffey, P. M. (1987, September 1). Lack of military data halts Agent Orange study. *New York Times*. Retrieved from http://www.nytimes.com

Bonastre, J.-F., Bimbot, F., Boë, L.-J., Campbell, J. P., Reynolds, D. A., & Magrin-Chagnolleau, I. (2003, September). *Person authentication by voice: A need for caution*. Paper presented at Eurospeech 2003 - Interspeech 2003, 8th European Conference on Speech Communication and Technology, Geneva, Switzerland, September 1–4, 2003, 33–36. Retrieved from International Speech Communication Association [website]: http://www.isca-speech.org/archive

Boodman, S. G. (2010, January 1). Experts debate the risks and benefits of cancer screenings. *AARP Bulletin*. Retrieved from http://www.aarp.org/health/conditions-treatments/info-12-2009/experts_debate_the_risk_and_benefits_of_cancer_screenings.html

Boston College basketball point shaving scandal of 1978–79. (n.d.). In *Wikipedia: The free encyclopedia*. Retrieved on October 29, 2010 from http://www.en.wikipedia.org/wiki/Boston_College_basketball_point_shaving_scandal_of_1978%E2%80%9379

Bower, B. (2011, February 12). In the zone: Evolution may have trained the mind to see scoring streaks—Even where they don't exist. *ScienceNews*. Retrieved from http://www.sciencenews.org

Box, G. E. P. (1953). Nonnormality and tests on variances. *Biometrika, 40*, 318–335.

Box, G. E. P. (1976). Science and statistics. *Journal of the American Statistical Association*, emph71, 791–799.

Box, G. E. P., & Draper, N. R. (1987). *Empirical model-building and response surfaces*. New York: Wiley.

Bradlow, E. T. (1998). Encouragement designs: An approach to self-selected samples in an experimental design. *Marketing Letters, 9*, 383–391.

Braun, H., & Wainer, H. (2007). Value-added assessment. In C. R. Rao & S. Sinharay (Eds.), *Handbook of statistics: Vol. 26. Psychometrics* (pp. 867–892). Amsterdam: Elsevier.

Breiman, L. (2001). Statistical modeling: The two cultures. *Statistical Science, 16*, 199–215, with comments on 216–226 and a rejoinder on 226–231.

Brief for American Psychiatric Association as Amicus Curiae Supporting Petitioner, Barefoot v. Estelle, 463 United States 880 (1983) (No. 82-6080). Retrieved from http://www.psych.org/lib_archives/archives/amicus_1982_barefoot.pdf

Brief for American Statistical Association as Amicus Curiae, Department of Commerce v. United States House of Representatives, 525 United States 316 (1999). Retrieved from http://www.amstat.org/outreach/pdfs/BriefAmicusCuriae.pdf

Brigham, C. C. (1923). *A study of American intelligence.* Princeton, NJ: Princeton University Press.

Brigham, C. C. (1930). Intelligence tests of immigrant groups. *Psychological Review, 37*, 158–165.

Bright-line rule. (n.d.). *In Wikipedia: The free encyclopedia.* Retrieved on August 1, 2011 from http://www.en.wikipedia.org/wiki/Bright-line_rule

Briscoe v. Virginia, 559 United States _____ (2010). Retrieved from Supreme Court of the United States [website]: http://www.supremecourt.gov/opinions/09pdf/07-11191.pdf

Brown v. Board of Education of Topeka, 347 United States 483 (1954). Retrieved from Justia.com United States Supreme Court Center [website]: http://supreme.justia.com/us/347/483/case.html

Browne, M. W. (1998, August 4). Following Benford's law, or looking out for No. 1. *New York Times.* Retrieved from http://www.nytimes.com

Buchanan, M. (2007, May 16). The prosecutor's fallacy [Web log post]. *New York Times.* Retrieved from New York Times Opinion Pages [website]: http://buchanan.blogs.nytimes.com/2007/05/16/the-prosecutors-fallacy/

Buck v. Bell, 274 United States 200 (1927). Retrieved from Cornell University Law School Legal Information Institute [website]: http://www.law.cornell.edu/supct/html/historics/USSC_CR_0274_0200_ZS.html

Budworth, D. (2009, April 10). Spread-betting fails investors in trouble. *Times (London).* Retrieved from http://www.timesonline.co.uk

Bullcoming v. New Mexico, 564 United States _____ (2011). Retrieved from Supreme Court of the United States [website]: http://www.supremecourt.gov/opinions/10pdf/09-10876.pdf

Bumiller, E. (2010, April 26). We have met the enemy and he is PowerPoint. *New York Times.* Retrieved from http://www.nytimes.com

Bush v. Gore, 531 United States 98 (2000). Retrieved from Cornell University Law School Legal Information Institute [website]: http://www.law.cornell.edu/supct/html/00-949.ZPC.html

Butterfield, F. (2001, April 20). Victims' race affects decisions on killers' sentence, study finds. *New York Times.* Retrieved from http://www.nytimes.com

Butterfly effect. (n.d.). In *Wikipedia: The free encyclopedia.* Retrieved on October 29, 2010 from http://www.en.wikipedia.org/wiki/Butterfly_effect

Buyse, M., George, S. L., Evans, S., Geller, N. L., Ranstam, J., Scherrer, B., ... Verma, B. L., for the ISCB [International Society for Clinical Biostatistics] Subcommittee on Fraud. (1999). The role of biostatistics in the prevention, detection, and treatment of fraud in clinical trials. *Statistics in Medicine, 18,* 3435–3451.

Calame, B. (2007, February 11). Can a 15-year-old be a 'woman without a spouse'? *New York Times.* Retrieved from http://www.nytimes.com

Calculating age-adjusted rates using the direct method: Table 2a. Age-adjusted death rate for diabetes mellitus, State of New Mexico, 2003–2005. (n.d.). Retrieved on October 29, 2010 from New Mexico Indicator-Based Information System (NM-IBIS) [website]: http://ibis.health.state.nm.us/resources/AARate.html

Calculating age-adjusted rates using the direct method: Table 2b. Age-adjusted death rate for diabetes mellitus, Sierra County, New Mexico, 2003–2005. (n.d.). Retrieved on October 29, 2010 from New Mexico Indicator-Based Information System (NM-IBIS) [website]: http://ibis.health.state.nm.us/resources/AARate.html

Callaway, E. (2011). Report finds massive fraud at Dutch universities. *Nature, 479,* 15.

Campbell, D. T. (1975). Assessing the impact of planned social change. In G. M. Lyons (Ed.), *Social research and public policies: The Dartmouth/OECD Conference* (pp. 3–45). Hanover, NH: Dartmouth College, Public Affairs Center.

Campbell, D. T., & Kenny, D. A. (1999). *A primer on regression artifacts.* New York: Guilford Press.

Campbell, D. T., & Stanley, J. C. (1966). *Experimental and quasi-experimental designs for research.* Chicago: Rand McNally.

Campbell, S. K. (1974). *Flaws and fallacies in statistical thinking.* Englewood Cliffs, NJ: Prentice-Hall.

Campbell Systematic Reviews. (n.d.). Retrieved from Campbell Collaboration Library of Systematic Reviews [website]: http://www.campbellcollaboration.org/library.php

Carey, B. (2004, November 9). Long after Kinsey, only the brave study sex. *New York Times.* Retrieved from http://www.nytimes.com

Carey, B. (2010, January 6). Popular drugs may benefit only severe depression, new study says. *New York Times.* Retrieved from http://www.nytimes.com

Carey, B. (2010, June 11). Academic battle delays publication by 3 years. *New York Times.* Retrieved from http://www.nytimes.com

Carey, B. (2011, November 2). Fraud case seen as a red flag for psychology research. *New York Times.* Retrieved from http://www.nytimes.com

Carl Brigham. (n.d.). In *Wikipedia: The free encyclopedia.* Retrieved on October 29, 2010 from http://www.en.wikipedia.org/wiki/Carl_Brigham

Carroll, J. B. (1961). The nature of the data, or how to choose a correlation coefficient. *Psychometrika, 26,* 347–372.

Carroll, J. B. (1982). The measurement of intelligence. In R. J. Sternberg (Ed.), *Handbook of human intelligence* (pp. 29–120). New York: Cambridge University Press.

Carroll, L. (2011, April 1). *1 in 5 US moms have kids with multiple dads, study says: Poverty, lack of education, and divorce perpetuate lack of opportunities.* Retrieved from MSNBC.com [website]: http://www.msnbc.msn.com/id/42364656/ns/health-kids_and_parenting/

Case-control study. (n.d.). In *Wikipedia: The free encyclopedia.* Retrieved on June 1, 2010 from http://www.en.wikipedia.org/wiki/Case-control_study

Caspi, A., Sugden, K., Moffitt, T. E., Taylor, A., Craig, I. W., Harrington H. L., ... Poulton, R. (2003). Influence of life stress on depression: Moderation by a polymorphism in the 5-HTT gene. *Science, 301,* 386–399.

Castle, W. (Producer), & Polanski, R. (Director). (1968). *Rosemary's baby* [Motion picture]. United States: Paramount Pictures. See also entry for Levin, I.

CCNY point shaving scandal. (n.d.). In *Wikipedia: The free encyclopedia.* Retrieved on October 29, 2010 from http://www.en.wikipedia.org/wiki/CCNY_Point_Shaving_Scandal

Chalmers, I. (1990). Underreporting research is scientific misconduct. *Journal of the American Medical Association, 263,* 1405–1408.

Champod, C., Taroni, F., & Margot, P.-A. (1999). The Dreyfus case—An early debate on expert's conclusions. *International Journal of Forensic Document Examiners, 5,* 446–459.

Chapman, L. J., & Chapman, J. P. (1967). Genesis of popular but erroneous psychodiagnostic observations. *Journal of Abnormal Psychology, 72,* 193–204.

Chapman, L. J., & Chapman, J. P. (1969). Illusory correlation as an obstacle to the use of valid psychodiagnostic signs. *Journal of Abnormal Psychology, 74,* 271–280.

Chiu, R. W. K., Akolekar, R., Zheng, Y. W. L., Leung, T. Y., Sun, H., Chan, K. C. A., ... Lo, Y. M. D. (2011). Non-invasive prenatal assessment of trisomy 21 by multiplexed maternal plasma DNA sequencing: Large scale validity study. *British Medical Journal, 342,* c7401. Retrieved from http://www.bmj.com

Christakis, N. A., & Fowler, J. H. (2009). *Connected: The surprising power of our social networks and how they shape our lives.* New York: Little, Brown.

Civil Rights Act of 1991. See entry for An Act to Amend the Civil Rights Act of 1964 to Strengthen and Improve Federal Civil Rights Laws, to Provide for Damages in Cases of Intentional Employment Discrimination, to Clarify Provisions Regarding Disparate Impact Actions, and for Other Purposes.

Claude Bernard. (n.d.). In *Wikipedia: The free encyclopedia.* Retrieved on February 8, 2011 from http://www.en.wikipedia.org/wiki/Claude_Bernard

Clinician's fallacy. See entry for Natural history of disease.

Cohen, P., & Cohen, J. (1984). The clinician's illusion. *Archives of General Psychiatry, 41,* 1178–1182.

Colbert, S. (Writer), & Hoskinson, J. (Director). (2005, October 17). The word—Truthiness [Television series episode]. In J, Stewart, B. Karlin, & S. Colbert (Executive Producers), *The Colbert report.* New York: Comedy Central.

Coleman, J. S., Campbell, E. Q., Hobson, C. J., McPartland, J., Mood, A. M., Weinfield, F. D., & York, R. L. (1966). *Equality of educational opportunity.* Washington, DC: United States Department of Health, Education, and Welfare.

Collins English dictionary: Complete and unabridged (6th ed.) (2003). Glasgow, Scotland: HarperCollins.

Collins, F. L. (1943). *The FBI in peace and war.* New York: G. P. Putnam's Sons. See also entry for Pelletier, L.

Collins, G. (2009, November 18). The breast brouhaha [Opinion-editorial]. *New York Times.* Retrieved from http://www.nytimes.com

Collins, G. (2010, May 7). What every girl should know [Opinion-editorial]. *New York Times.* Retrieved from http://www.nytimes.com

Committee on DNA Forensic Science: An Update, Commission on DNA Forensic Science: An Update, National Research Council. (1996). *The evaluation of forensic DNA evidence.* Washington, DC: National Academy Press.

Committee on DNA Technology in Forensic Science, Board on Biology, Commission on Life Sciences, National Research Council. (1992). *DNA technology in forensic science.* Washington, DC: National Academy Press.

Committee on Identifying the Needs of the Forensic Science Community; Committee on Science, Technology, and Law Policy and Global Affairs; Committee on Applied and Theoretical Statistics; Division on Engineering and Physical Sciences; National Research Council of the National Academies. (2009). *Strengthening forensic science in the United States: A path forward.* Washington, DC: National Academies Press.

Committee on the Framework for Evaluating the Safety of Dietary Supplements, Food and Nutrition Board, Board on Life Sciences, Institute of Medicine and National Research Council of the National Academies, The National Academies. (2005). *Dietary supplements: A framework for evaluating safety.* Washington, DC: National Academies Press.

Committee to Review the Scientific Evidence on the Polygraph; Board on Behavioral, Cognitive, and Sensory Sciences; and Committee on National Statistics; Division of Behavioral and Social Sciences and Education; National Research Council of the National Academies. (2003). *The polygraph and lie detection.* Washington, DC: National Academies Press.

Confirmation bias. (n.d.). In *Wikipedia: The free encyclopedia.* Retrieved on October 29, 2010 from http://www.en.wikipedia.org/wiki/Confirmation_bias

Consolidated Appropriations Act of 2001. See entry for Data Quality Act.

Controlled clinical trial [Definition]. (n.d.). In *National Library of Medicine—Medical subject headings.* Retrieved on October 29, 2010 from http://www.nlm.nih.gov/cgi/mesh/2011/MB_cgi?mode=&term=Controlled+Clinical+Trial&field=entry

Conway, D. A., & Roberts, H. V. (1983). Reverse regression, fairness and employment discrimination. *Journal of Business & Economic Statistics, 1,* 75–85.

Costa, P. T., Jr., & McCrae, R. R. (1992). *NEO PI-R professional manual: Revised NEO Personality Inventory (NEO PI-R) and NEO Five-Factor Inventory (NEO-FFI)* Odessa, FL: Psychological Assessment Resources.

Cox, D. R., & Oakes, D. (1984). *Analysis of survival data.* New York: Chapman & Hall.

COX-2 inhibitor (n.d.). In *Wikipedia: The free encyclopedia.* Retrieved on January 12, 2012 from http://www.en.wikipedia.org/wiki/COX-2_inhibitor

Crews, F. C. (2004, March 11). The trauma trap. *New York Review of Books.* Retrieved from http://www.nybooks.com

Cronbach, L. J. (1982). Prudent aspirations for social inquiry. In W. H. Kruskal (Ed.), *The social sciences: Their nature and uses; Papers presented at the 50th anniversary of the Social Science Research Building, the University of Chicago, December 16–18, 1979* (pp. 61–81). Chicago: University of Chicago Press.

Cronbach, L. J., & Meehl, P. E. (1955). Construct validity in psychological tests. *Psychological Bulletin, 52,* 281–302.

Curzon, F. R. P. [7th Earl Howe]. (2003, February 5). Attention deficit disorders [United Kingdom House of Lords debate]. *Lords Hansard,* Vol. 644, cc. 299–322. Retrieved from UK Parliament [website]: `http://www.publications.parliament.uk/pa/ld200203/ldhansrd/vo030205/text/30205-10.htm`

Custodial Detention Index. See entry for FBI Index.

The dangers of frozen chickens, part two. (1995, November). *Feathers: A Newsletter of the California Poultry Industry Federation,* p. 7.

Data Quality Act. Section 515 of An Act Making Consolidated Appropriations for the Fiscal Year Ending September 30, 2001, and for Other Purposes [Consolidated Appropriations Act, 2001], Pub. L. No. 106-554 (2000/2001). Retrieved from United States Government Printing Office [website]: `http://www.gpo.gov/fdsys/pkg/PLAW-106publ554/content-detail.html`

Daubert v. Merrell Dow Pharmaceuticals, 509 United States 579 (1993). Retrieved from Cornell University Law School Legal Information Institute [website]: `http://www.law.cornell.edu/supct/html/92-102.ZS.html`

Daubert v. Merrell Dow Pharmaceuticals. (n.d.). In *Wikipedia: The free encyclopedia.* Retrieved on November 30, 2010 from `http://www.en.wikipedia.org/wiki/Daubert_v._Merrell_Dow_Pharmaceuticals`

Davenport, C. B. (1916). *Eugenics as a religion* [Typescript of a lecture given at the Golden Jubilee of Battle Creek Sanitarium, Battle Creek, MI]. Charles Benedict Davenport Papers, 1874–1946, Mss. B. D27 (Box 25). Archives of the American Philosophical Society, Philadelphia, PA.

Davey Smith, G., & Ebrahim, S. (2001). Epidemiology—Is it time to call it a day? [Editorial]. *International Journal of Epidemiology, 30,* 1–11.

Davies, S. (2003). Obituary for David Horrobin: Summary of rapid responses. *British Medical Journal, 326,* 1089. Retrieved from `http://www.bmj.com`

Davis, R. H. (2004, March 21). The anatomy of a smear campaign [Opinion-editorial]. *Boston Globe.* Retrieved from `http://bostonglobe.com`

Dawes, R. M. (1975). Graduate admissions variables and future success. *Science, 187,* 721–723.

Dawes, R. M. (1979). The robust beauty of improper linear models in decision making. *American Psychologist, 34,* 571–582.

Dawes, R. M. (1994). *House of cards: Psychology and psychotherapy built on myth.* New York: Free Press.

Dawes, R. M. (2002). The ethics of using or not using statistical prediction rules in psychological practice and related consulting activities. *Philosophy of Science, 69,* S178–S184.

Dawes, R. M. (2005). The ethical implications of Paul Meehl's work on comparing clinical versus actuarial prediction methods. *Journal of Clinical Psychology, 61,* 1245–1255.

De Luca, M., Horovitz, R., Pitt, B. (Producers), & Miller, B. (Director). (2011). *Moneyball* [Motion picture]. United States: Columbia Pictures. See also entry for Lewis, M. M.

Defendant's fallacy. See entry for Prosecutor's fallacy.

Department of Commerce v. United States House of Representatives, 525 United States 316 (1999). Retrieved from Justia.com United States Supreme Court Center [website]: http://supreme.justia.com/us/525/316/case.html

Design of experiments. (n.d.). In *Wikipedia: The free encyclopedia*. Retrieved on July 10, 2010 from http://www.en.wikipedia.org/wiki/Design_of_experiments

Detection bias. See entry for Bias (statistics).

Devlin, K. (2008). *The unfinished game: Pascal, Fermat, and the seventeenth-century letter that made the world modern; A tale of how mathematics is really done*. New York: Basic Books.

Devlin, K., & Lorden, G. (2007). *The numbers behind NUMB3RS: Solving crime with mathematics*. New York: Penguin Group.

Dillon, S. (2011, April 25). High school classes may be advanced in name only. *New York Times*. Retrieved from http://www.nytimes.com

Dirty Harry. See entry for Siegel, D., Daley, R. (Producers), & Siegel, D. (Director).

Dispositive [Definition]. (2010). *Webster's New World college dictionary*. Cleveland, OH: Wiley.

Divination [Definition]. (n.d). Retrieved on October 29, 2010 from Merriam-Webster [website]: http:/www.merriam-webster.com/dictionary/divination

Division of Reproductive and Urologic Products, Office of New Drugs, Center for Drug Evaluation and Research, Food and Drug Administration. (2010, May 20). *Background document for meeting of Advisory Committee for Reproductive Health Drugs (June 18, 2010): NDA 22-236, Flibanserin, Boehringer Ingelheim*. Retrieved from Food and Drug Administration [website]: http://www.fda.gov/downloads/AdvisoryCommittees/CommitteesMeetingMaterials/Drugs/ReproductiveHealthDrugsAdvisoryCommittee/UCM215437.pdf

Doctors' Trial. See entry for United States of America v. Karl Brandt et al.

DoD news briefing—Secretary Rumsfeld and Gen. Myers [News transcript]. (2002, February 12]. Retrieved from U.S. Department of Defense, Office of the Assistant Secretary of Defense (Public Affairs) [website]: http://www.defense.gov/transcripts/transcript.aspx?transcriptid=2636

Dolnick, E. (1991, October). The ghost's vocabulary: How the computer listens for Shakespeare's "voiceprint." *Atlantic*. Retrieved from http://theatlantic.com

Dorfman, D. D. (1978), The Cyril Burt question: New findings. *Science, 201*, 1177–1186.

Draw-A-Person Test. (n.d.). In *Wikipedia: The free encyclopedia*. Retrieved on October 29, 2010 from http://www.en.wikipedia.org/wiki/Draw_a_Person_test

Dred Scott decision. See entry for Scott v. Sandford.

Dror, I. E., Charlton, D., & Péron, A. E. (2006). Contextual information renders experts vulnerable to making erroneous identifications. *Forensic Science International, 156*, 74–78.

Duncan, O. D., & Davis, B. (1953). An alternative to ecological correlation. *American Sociological Review, 18*, 665–666.

Dyson, F. (2004). A meeting with Enrico Fermi. *Nature, 427*, 297.

Edgington, E. S., & Onghena, P. (2007). *Randomization tests* (4th ed.). New York: Chapman & Hall / CRC.

Editorial: Psychology: A reality check [Editorial]. (2009, October 15). *Nature.* Retrieved from http://www.nature.com

Editorial: What about the raters? [Editorial]. (2010, May 1). *New York Times.* Retrieved from http://www.nytimes.com

Efron, B., & Gong, G. (1983). A leisurely look at the bootstrap, the jackknife, and cross-validation. *American Statistician, 37*, 36–48.

Efron, B., & Morris, C. (1977). Stein's paradox in statistics. *Scientific American, 236*(5), 119–127.

Ehrenberg, R. (2010, June 22). The truth hurts: Scientists question voice-based lie detection. *ScienceNews.* Retrieved from http://www.sciencenews.org

Ellard, J. (1989). *Some rules for killing people: Essays on madness, murder, and the mind.* North Ryde, New South Wales, Australia: Angus & Robertson.

Estelle v. Smith, 451 United States 454 (1981). Retrieved from Justia.com United States Supreme Court Center [website]: http://supreme.justia.com/us/451/454

Ethics governing the service of prisoners as subjects in medical experiments: Report of a committee appointed by Governor Dwight H. Green of Illinois [The Green Report]. (1948). *Journal of the American Medical Association, 136*, 457–458.

Eugenics. (n.d.). In *Wikipedia: The free encyclopedia.* Retrieved on February 8, 2011 from http://www.en.wikipedia.org/wiki/Eugenics

Exec. Order No. 9066, F. R. Doc. 42-1563 (1942). Retrieved from Our Documents [website]: http://www.ourdocuments.gov

Executive Office of the President, Office of Management and Budget. (2004, December 16). *Final information quality bulletin for peer review* (Memorandum M-05-03). Retrieved from http://www.whitehouse.gov/omb/memoranda_fy2005_m05-03

Fagan, T. J. (1975). Nomogram for Bayes's Theorem [Letter to the editor]. *New England Journal of Medicine, 293*, 257.

False dilemma. (n.d.). In *Wikipedia: The free encyclopedia.* Retrieved on October 29, 2010 from http://www.en.wikipedia.org/wiki/False_dilemma

The FBI in peace and war. See entries for Collins, F. L., and Pelletier, L.

FBI Index. (n.d.). In *Wikipedia: The free encyclopedia.* Retrieved on April 25, 2011 from http://www.en.wikipedia.org/wiki/FBI_Index

Fed. R. Evid. 403, 702, 706. Retrieved from Cornell University Law School Legal Information Institute [website]: http://www.law.cornell.edu/rules/fre

Federal Judicial Center. (2000). *Reference manual on scientific evidence* (2nd ed.). St. Paul, MN: West Group.

Federal Rules of Civil Procedure, Rule 53. Masters. Retrieved from Cornell University Law School Legal Information Institute [website]: http://www.law.cornell.edu/rules/frcp/Rule53.htm

Federal Rules of Evidence. (2010). Retrieved from Cornell University Law School Legal Information Institute [website]: `http://www.law.cornell.edu/rules/fre`

Federal Rules of Evidence. (n.d.). In *Wikipedia: The free encyclopedia.* Retrieved on October 29, 2010 from `http://www.en.wikipedia.org/wiki/Federal_Rules_of_Evidence`

Feldman, S. R. (2009). *Compartments: How the brightest, best trained, and most caring people can make judgments that are completely and utterly wrong.* Bloomington, IN: Xlibris.

Feller, W. (1968). *An introduction to probability theory and its applications* (3rd ed., Vol. 1). New York: Wiley.

Ferguson, T. S. (1989). Who solved the secretary problem? *Statistical Science, 4,* 282–289.

Feynman, R. P. (1988). *What do you care what other people think? Further adventures of a curious character.* New York: W. W. Norton.

Field, A. P. (2009). Meta-analysis. In R. E. Millsap & A. Maydeu-Olivares (Eds.), *The Sage handbook of quantitative methods in psychology* (pp. 404–422). London: Sage.

Fienberg, S. E. (Ed.). (1988). *The evolving role of statistical assessments as evidence in the courts.* New York: Springer-Verlag.

Fienberg, S. E., & Stern, P. C. (2005). In search of the magic lasso: The truth about the polygraph. *Statistical Science, 20,* 249–260.

Final Rule Declaring Dietary Supplements Containing Ephedrine Alkaloids Adulterated Because They Present an Unreasonable Risk; Final Rule, 69 Fed. Reg. 6788 (2004) (21 C. F. R. pt. 119).

Fisher, R. A. (1925). *Statistical methods for research workers.* Edinburgh, Scotland: Oliver & Boyd.

Fisher, R. A. (1935). *The design of experiments* (1st ed.). New York: Hafner.

Fisher, R. A. (1971). *The design of experiments* (9th ed.). New York: Hafner.

Fournier, J. C., DeRubeis, R. J., Hollon, S. D., Dimidjian, S., Amsterdam, J. D., Shelton, R. C., & Fawcett, J. (2010). Antidepressant drug effects and depression severity: A patient-level meta-analysis. *Journal of the American Medical Association, 303,* 47–53.

Francesco Redi. (n.d.). In *Wikipedia: The free encyclopedia.* Retrieved on May 31, 2011 from `http://www.en.wikipedia.org/wiki/Francesco_Redi`

Francis, T., Jr., Korns, R. F., Voight, R. B., Hemphill, F. M., Boisen, M., Tolchinsky, E., ... Tumbusch, J. J. (1955). An evaluation of the 1954 poliomyelitis vaccine trials: Summary report. *American Journal of Public Health, 45*(5 Pt. 2).

Freedman, D. A. (1983). A note on screening regression equations. *American Statistician, 37,* 152–155.

Freedman, D. A. (1987). As others see us: A case study in path analysis. *Journal of Educational Statistics, 12,* 101–128.

Freedman, D. A. (2001). Ecological inference. In N. J. Smelser & P. B. Baltes (Eds.), *International Encyclopedia of the Social & Behavioral Sciences* (Vol. 6, pp. 4027–4030). Oxford, UK: Elsevier Science.

Friedman, M. (1992). Do old fallacies ever die? *Journal of Economic Literature*, *30*, 2129–2132.

Friedman, R. A. (2010, January 11). Before you quit anti-depressants ... . *New York Times*. Retrieved from http://www.nytimes.com

Friedman, R. A. (2011, May 23). How a telescopic lens muddles psychiatric insights. *New York Times*. Retrieved from http://www.nytimes.com

Froman, T., & Hubert, L. J. (1981). A reply to Moshman's critique of prediction analysis and developmental priority. *Psychological Bulletin, 90*, 188.

Frye standard. (n.d.). In *Wikipedia: The free encyclopedia*. Retrieved on October 29, 2010 from http://www.en.wikipedia.org/wiki/Frye_standard

Frye v. United States, 293 F. 1013 (D.C. Cir. 1923). Retrieved from Daubert on the Web [website]: http://www.daubertontheweb.com/frye_opinion.htm

Fundamental attribution error. (n.d.). In *Wikipedia: The free encyclopedia*. Retrieved on June 2, 2010 from http://www.en.wikipedia.org/wiki/Fundamental_attribution_error

Furman v. Georgia, 408 United States 238 (1972). Retrieved from Cornell University Law School Legal Information Institute [website]: http://www.law.cornell.edu/supct/html/historics/USSC_CR_0408_0238_ZS.html

Galton, F. (1883). *Inquiry into human faculty and its development*. London: Macmillan.

Galton, F. (1886). Regression towards mediocrity in hereditary stature. *Journal of the Anthropological Institute of Great Britain and Ireland, 15*, 246–263.

Galton, F. (1892). *Finger prints*. London and New York: Macmillan.

Galton, F. (1883). *Inquiry into human faculty and its development*. London: Macmillan.

Galton, F. (1908). *Memories of my life*. London: Methuen.

Garg, A. (2010, February 23). Ex cathedra [Definition]. Retrieved from A.Word.A.Day [website]: http://wordsmith.org/words/ex_cathedra.html

Garg, A. (2010, February 25). Ad hominem [Definition]. Retrieved from A.Word.A.Day [website]: http://wordsmith.org/words/ad_hominem.html

Garg, A. (2010, March 1). [Essay preceding definition of "goulash"]. Retrieved from A.Word.A.Day [website]: http://wordsmith.org/words/goulash.html

Gawande, A. (1999, February 8). The cancer-cluster myth. *New Yorker*. Retrieved from http://www.newyorker.com

Gawande, A. (2001, January 8). Under suspicion. *New Yorker*. Retrieved from http://www.newyorker.com

Gelman, A., Park, D., Shor, B., & Cortina, J. (2010). *Red state, blue state, rich state, poor state: Why Americans vote the way they do* (Exp. ed.). Princeton, NJ: Princeton University Press.

Gelman, A., Shor, B., Bafumi, J., & Park, D. (2007). Rich state, poor state, red state, blue state: What's the matter with Connecticut? *Quarterly Journal of Political Science, 2*, 345–367.

Gelman, A., & Stern, H. (2006). The difference between "significant" and "not significant" is not itself statistically significant. *American Statistician, 60*, 328–331.

General Electric Co. v. Joiner, 522 United States 136 (1997). Retrieved from Justia.com United States Supreme Court Center [website]: http://supreme.justia.com/us/522/136/case.html

George Julius. (n.d.). In *Wikipedia: The free encyclopedia.* Retrieved on October 29, 2010 from http://www.en.wikipedia.org/wiki/George_Julius

Gigerenzer, G. (2002). *Calculated risks: How to know when numbers deceive you.* New York: Simon & Schuster.

Gigerenzer, G. (2010). Women's perception of the benefit of breast cancer screening [Editorial]. *Maturitas, 67,* 5–6.

Gigerenzer, G., & Brighton, H. (2009). Homo heuristicus: Why biased minds make better inferences. *Topics in Cognitive Science, 1,* 107–143.

Gigerenzer, G., Gaissmaier, W., Kurz-Milcke, E., Schwartz, L. M., & Woloshin, S. (2007). Helping doctors and patients make sense of health statistics. *Psychological Science in the Public Interest, 8,* 53–96.

Gillis, J. (2010, July 7). British panel clears scientists. *New York Times.* Retrieved from http://www.nytimes.com

Ginzburg, R. (Ed.) (1964). The unconscious of a conservative: Special issue on the mind of Barry Goldwater. *Fact:, 1*(5).

Giuliani, R. (2007, October 29). *Chances* [New Hampshire radio advertisement audio clip]. Retrieved from Washington Post Fact Checker [website]: http://voices.washingtonpost.com/fact-checker/2007/10/rudy_miscalculates_cancer_surv.html

Glaberson, W. (2004, August 8). Agent Orange, the next generation; In Vietnam and America, some see a wrong still not righted. *New York Times.* Retrieved from http://www.nytimes.com

Glaberson, W. (2004, November 19). Agent Orange lawsuits dismissed (G. James, Comp.). *New York Times.* Retrieved from http://www.nytimes.com

Glaberson, W. (2005, March 11). Civil lawsuit on defoliant in Vietnam is dismissed. *New York Times.* Retrieved from http://www.nytimes.com

Gladwell, M. (2010, May 17). The treatment: Why is it so difficult to develop drugs for cancer? *New Yorker.* Retrieved from http://www.gladwell.com

Gladwell, M. (2011, February 14 and 21). The order of things: What college rankings really tell us. *New Yorker.* Retrieved from Colleges That Change Lives [website]: http://www.ctcl.org/files/pdfs/RankingsNewYorkerGladwell-1.pdf

Glass, G. V. (1976). Primary, secondary, and meta-analysis of research. *Educational Researcher, 5*(10), 3–8.

Glass, G. V. (2000, January). Meta-analysis at 25. Retrieved from http://www.gvglass.info/papers/meta25.html

Goddard, H. H. (1912). *The Kallikak family: A study in the heredity of feeblemindedness.* New York: Macmillan.

Golden Rule Insurance Company v. Washburn, No. 419-76 (Ill. Cir. Ct. 7th Jud. Cir. 1984) (Consent Decree).

Goldstein, B. D., & Henifin, M. S. (2000). Reference guide on toxicology. In Federal Judicial Center, *Reference manual on scientific evidence* (2nd ed., pp. 401–437). St. Paul, MN: West Group.

Goode, E. (2004, March 9). Defying psychiatric wisdom, these skeptics say 'prove it.' *New York Times*. Retrieved from http://www.nytimes.com

Goodman, L. A. (1953). Ecological regressions and behavior of individuals. *American Sociological Review, 18*, 663–664.

Goodstein, L. (2011, January 27). Forecast sees Muslim population leveling off. *New York Times*. Retrieved from http://www.nytimes.com

Gottfried, H. (Producer), & Hiller, A. (Director). (1971). *The hospital* [Motion picture]. United States: United Artists.

Gould, S. J. (1996). *The mismeasure of man* (Rev. and exp. ed.). New York: W. W. Norton.

Grant, M. (1916). *The passing of the great race or the racial basis of European history.* New York: Charles Scribner's Sons.

Graphology. (n.d.). In *Wikipedia: The free encyclopedia.* Retrieved on October 29, 2010 from http://www.en.wikipedia.org/wiki/Graphology

Graunt, J. (1662). *Natural and political observations mentioned in a following index, and made upon the bills of mortality.* London: Thomas Roycroft for John Martin, James Allestry, and Thomas Dicas.

Green, M. D., Freedman, D. M., & Gordis, L. (2000). Reference guide on epidemiology. In Federal Judicial Center, *Reference manual on scientific evidence* (2nd ed., pp. 333–400). St. Paul, MN: West Group.

Green report. (n.d.). In *Wikipedia: The free encyclopedia.* Retrieved on February 8, 2011 from http://www.en.wikipedia.org/wiki/Green_report. See also entry for Ethics governing the service of prisoners as subjects in medical experiments: Report of a committee appointed by Governor Dwight H. Green of Illinois.

Greenberg, B. G., Abul-Ela, A.-L. A., Simmons, W. R., & Horvitz, D. G. (1969). The unrelated question randomized response model: Theoretical framework. *Journal of the American Statistical Association, 64*, 520–539.

Gregg v. Georgia, 428 United States 153 (1976). Retrieved from Cornell University Law School Legal Information Institute [website]: http://www.law.cornell.edu/supct//html/historics/USSC_CR_0428_0153_ZS.html

Grier, D. A. (n.d.). *The origins of statistical computing.* Retrieved from American Statistical Association [website]: http://www.amstat.org/about/statisticiansinhistory/index.cfm?fuseaction=paperinfo&PaperID=4

Guidelines for Determining the Probability of Causation and Methods for Radiation Dose Reconstruction Under the [Energy] Employees Occupational Illness Compensation Program Act of 2000; Final Rules, 67 Fed. Reg. 22296 (2000) (42 C. F. R. pts. 81 and 82).

Gulliksen, H. (1950). *Theory of mental tests.* New York: Wiley.

H. R. Rep. No. 109-272, 109th Cong., 1st Sess. (2005) [Conf. Rep.].

Haining, R. (2003). *Spatial data analysis: Theory and practice.* Cambridge, UK: Cambridge University Press.

Hammes, T. X. (2009). Essay: Dumb-dumb bullets—As a decision-making aid, PowerPoint is a poor tool. *Armed Forces Journal, 47*, 12, 13, 28.

Hanson, R. K. (1997). *The development of a brief actuarial risk scale for sexual offense recidivism* (Department of the Solicitor General of Canada Report No. 1997-04). Ottawa, Ontario, Canada: Public Works and Government Services

Canada (Cat. No. JS4-1/1997-4E). Retrieved from Defense for Sexually Violent Predators [website]: http:www.defenseforsvp.com/Resources/Hanson_Static-99/RRASOR.pdf

Hanushek, E. A. (1989). The impact of differential expenditures on school performance. *Educational Researcher, 18*(4), 45–51, 62.

Hardisty, D. J., Johnson, E. J., & Weber, E. U. (2010). A dirty word or a dirty world? Attribute framing, political affiliation, and query theory. *Psychological Science, 21,* 86–92.

Hare, R. D. (2003). *PCL-R* [Psychopathy Checklist—Revised] *technical manual* (2nd ed.). Toronto, Ontario, Canada: Multi-Health Systems.

Harry H. Laughlin. (n.d.). In *Wikipedia: The free encyclopedia.* Retrieved on February 8, 2011 from http://www.en.wikipedia.org/wiki/Harry_H._Laughlin

Hastie, T., Tibshirani, R., & Friedman, J. (2009). *The elements of statistical learning: Data mining, inference, and prediction* (2nd ed.). New York: Springer.

Hays, W. L. (1963). *Statistics for psychologists.* New York: Holt, Rinehart and Winston.

Hays, W. L. (1994). *Statistics* (5th ed.). Belmont, CA: Wadsworth.

Hedges, L. V., Laine, R. D., & Greenwald, R. (1994). An exchange: Part I: Does money matter? A meta-analysis of studies of the effects of differential school inputs on student outcomes. *Educational Researcher, 23*(3), 5–14.

Heingartner, D. (2006, July 18). Maybe we should leave that up to the computer. *New York Times.* Retrieved from http://www.nytimes.com

Henifin, M. S., Kipen, H. M., & Poulter, S. R. (2000). Reference guide on medical testimony. In Federal Judicial Center, *Reference manual on scientific evidence* (2nd ed., pp. 439–484). St. Paul, MN: West Group.

Henry H. Goddard. (n.d.). In *Wikipedia: The free encyclopedia.* Retrieved on February 8, 2011 from http://www.en.wikipedia.org/wiki/Henry_H._Goddard

Herbranson, W. T., & Schroeder, J. (2010). Are birds smarter than mathematicians? Pigeons (*Columba livia*) perform optimally on a version of the Monty Hall Dilemma. *Journal of Comparative Psychology, 124,* 1–13.

Herbst, A. L., Ulfelder, H., & Poskanzer, D. C. (1971). Adenocarcinoma of the vagina—Association of maternal stilbestrol therapy with tumor appearance in young women. *New England Journal of Medicine, 284,* 878–881.

Heterosis. (n.d.). In *Wikipedia: The free encyclopedia.* Retrieved on February 8, 2011 from http://www.en.wikipedia.org/wiki/Heterosis

Higgins, J. P. T., & Altman, D. G. (Eds.). (2008). Assessing risk of bias in included studies. In J. Higgins & S. Green (Eds.), *Cochrane handbook for systematic reviews of interventions* (pp. 187–242). Chichester, England: Wiley.

Higgins, J. P. T., & Green, S. (Eds.). (2008). *Cochrane handbook for systematic reviews of interventions.* Chichester, England: Wiley.

Hill, A. B. (1965). The environment and disease: Association or causation? *Proceedings of the Royal Society of Medicine, 58,* 295–300.

History of phrenology and the psycograph. (n.d.). Retrieved on May 31, 2011 from MuseumOfQuackery.com [website]: http://www.museumofquackery.com/devices/psychist.htm

Holden, C. (2009). Back to the drawing board for psychiatric genetics. *Science, 324,* 1628. Retrieved from http://www.sciencemag.org

Holland, P. W. (1986). Statistics and causal inference. *Journal of the American Statistical Association, 81*, 945–960.

Holland, P. W., & Rubin, D. B. (1983). On Lord's paradox. In H. Wainer & S. Messick (Eds.), *Principals of modern psychological measurement: A festschrift for Frederic M. Lord* (pp. 3–25). Hillsdale, NJ: Erlbaum.

Homer. (1996). *The Odyssey* (R. Fagles, Trans.). New York: Viking Penguin.

Horrobin, D. F. (2003). Are large clinical trials in rapidly lethal diseases usually unethical? *Lancet, 361*, 695–697.

Horton, N. J., & Shapiro, E. C. (2005). Statistical sleuthing during epidemics: Maternal influenza and schizophrenia. *Chance, 18*(1), 11–18.

Horton, R. (2004, March 11). The dawn of McScience [Review of the book *Science in the private interest: Has the lure of profits corrupted biomedical research?*, by S. Krimsky]. *New York Review of Books*. Retrieved from http://www.nybooks.com

The hospital. See entry for Gottfried, H. (Producer), & Hiller, A. (Director).

Houlihan, J., Thayer, K., & Klein, J. (2003, May). *Canaries in the kitchen: Teflon toxicosis—EWG finds that heated Teflon pans can turn toxic faster than DuPont claims.* Retrieved from Environmental Working Group [website]: http://www.ewg.org/reports/toxicteflon

House, MD. See entry for Shore, D. (Creator and Producer).

Howe, Lord. See entry for Curzon, F. R. P. [7th Earl Howe].

Hubert, L. J. (1987). *Assignment methods in combinatorial data analysis.* New York: Marcel Dekker.

Hubert, L., & Wainer, H. (2011). A statistical guide for the ethically perplexed. In A. T. Panter & S. K. Sterba (Eds.), *Handbook of ethics in quantitative methodology* (pp. 61–124). New York: Routledge.

Huettel, S. A., Song, A. W., & McCarthy, G. (2004). *Functional magnetic resonance imaging.* Sunderland, MA: Sinauer Associates.

Huff, D. (1954). *How to lie with statistics.* New York: W. W. Norton.

Hulley, S., Grady, D., Bush, T., Furberg, C., Herrington, D., Riggs, B., & Vittinghoff, E., for the Heart and Estrogen/progestin Replacement Study (HERS) Research Group. (1998). Randomized trial of estrogen plus progestin for secondary prevention of coronary heart disease in postmenopausal women. *Journal of the American Medical Association, 280*, 605–613.

Huygens, C. (1656, 1657). De ratiociniis in ludo aleae [On reasoning in games of chance]. In F. van Schooten, *Exercitationum mathematicarum [Mathematical exercises]* (pp. 517–534). Leiden, The Netherlands: Johan Elsevier.

I led 3 lives. See entries for Philbrick, H. A., and Ziv, F. W. (Producer), Davis, E., Goodwins, L., Herzberg, J., Kesler, H. S., & Strock, H. L. (Directors).

Iatrogenesis. (n.d.). In *Wikipedia: The free encyclopedia.* Retrieved on October 29, 2010 from http://www.en.wikipedia.org/wiki/Iatrogenesis

Idiopathic [Definition]. (n.d.). In *MedicineNet.com: MedTerms Dictionary.* Retrieved on October 29, 2010 from http://www.medterms.com/script/main/art.asp?articlekey=3892

Ignaz Semmelweis. (n.d.). In *Wikipedia: The free encyclopedia.* Retrieved on June 6, 2010 from http://www.en.wikipedia.org/wiki/Ignaz_Semmelweis

I'm not stupid fallacy. See entry for Tavris, C., & Aronson, E.

Immigration Act of 1924. See entry for An Act to Limit the Immigration of Aliens Into the United States, and for Other Purposes.

In re "Agent Orange" Product Liability Litigation, 373 F.Supp.2d 7 (E.D.N.Y. 2005). Retrieved from Fund for Reconciliation and Development [website]: http://www.ffrd.org/AO/10-03-05-agentorange.pdf

In re As.H., 851 A.2d 456 (D.C. 2004). Retrieved from LexisNexis Communities Portal Free Case Law [website]: http://www.lexisone.com/

Information Quality Act. See entry for Data Quality Act.

Intention to treat analysis. (n.d.). In *Wikipedia: The free encyclopedia.* Retrieved on February 21, 2012 from http://www.en.wikipedia.org/wiki/Intention_to_treat_analysis

Intergovernmental Panel on Climate Change Core Writing Team, Pachauri, R. K., & Reisinger, A. (Eds). (2007). *Climate change 2007: Synthesis report. Contribution of Working Groups I, II, and III to the fourth assessment report of the Intergovernmental Panel on Climate Change.* Geneva, Switzerland: Intergovernmental Panel on Climate Change. Retrieved from http://www.ipcc.ch/publications_and_data/ar4/syr/en/contents.html

International Committee of Medical Journal Editors [ICMJE]. (2010, April, updated). *Uniform requirements for manuscripts submitted to biomedical journals: Writing and editing for biomedical publications.* Retrieved on February 8, 2011 from http://www.icmje.org/urm_full.pdf

International Conference on Harmonisation of Technical Requirements for Registration of Pharmaceuticals for Human Use. (1996, June 10). *ICH Harmonised Tripartite Guideline: Guideline for Good Clinical Practice* (E6[R1], Step 4 version). Retrieved from International Conference on Harmonisation [website]: http://ich.org/

International Society for Environmental Epidemiology, Ethics and Philosophy Committee. (April, 2009). *Guidelines for the ethical re-analysis and reinterpretation of another's research.* Retrieved from http://www.iseepi.org/About/Docs/ReAnal_guidelines_Revision2FINAL_April09.pdf

Ioannidis, J. P. A. (2005). Contradicted and initially stronger effects in highly cited clinical research. *Journal of the American Medical Association, 294,* 218–228.

Ioannidis, J. P. A. (2005). Why most published research findings are false. *PLoS Medicine, 2,* 696–701.

Ipse dixit. (n.d.). In *Wikipedia: The free encyclopedia.* Retrieved on October 29, 2010 from http://www.en.wikipedia.org/wiki/Ipse_dixit

Iraq and weapons of mass destruction. (n.d.). In *Wikipedia: The free encyclopedia.* Retrieved on October 24, 2011 from http://www.en.wikipedia.org/wiki/Iraq_and_weapons_of_mass_destruction

Jack B. Weinstein. (n.d.). In *Wikipedia: The free encyclopedia.* Retrieved on October 29, 2010 from http://www.en.wikipedia.org/wiki/Jack_B._Weinstein

Jackson, B., & Jamieson, K. H. (2007). *unSpun: Finding facts in a world of disinformation.* New York: Random House.

Jacobson v. Massachusetts, 197 United States 11 (1905). Retrieved from FindLaw for Legal Professionals [website]: http://caselaw.lp.findlaw.com/cgi-bin/getcase.pl?court=US&vol=197&invol=11

Japanese American internment. (n.d.). In *Wikipedia: The free encyclopedia.* Retrieved on May 27, 2011 from http://www.en.wikipedia.org/wiki/ Japanese-American_internment

Jeon, J. W., Chung, H. Y., & Bae, J. S. (1987). Chances of Simpson's paradox. *Journal of the Korean Statistical Society, 16,* 117–125.

Jeremy Bentham. (n.d.). In *Wikipedia: The free encyclopedia.* Retrieved on October 22, 2010 from http://www.en.wikipedia.org/wiki/Jeremy_Bentham

Johnson, M. (Producer), & Levinson, B. (Director). (1988). *Rain Man* [Motion picture]. United States: United Artists.

Johnson-Reed Immigration Act. See entry for An Act to Limit the Immigration of Aliens Into the United States, and for Other Purposes.

Jones Day®, & Duff & Phelps. (2011, November 7). *Class profile reporting* (Investigative report). Champaign, IL: University of Illinois, College of Law. Retrieved from University of Illinois at Urbana-Champaign, College of Law, Releases and Information [website]: http://www.uillinois.edu/our/news/2011/ Law/index.cfm

Jones, E. E., & Harris, V. A. (1967). The attribution of attitudes. *Journal of Experimental Social Psychology, 3,* 1–24.

Joseph Jagger. (n.d.). In *Wikipedia: The free encyclopedia.* Retrieved on February 19, 2011 from http://www.en.wikipedia.org/wiki/Joseph_Jagger

Joseph Oller. (n.d.). In *Wikipedia: The free encyclopedia.* Retrieved on October 29, 2010 from http://www.en.wikipedia.org/wiki/Joseph_Oller

Kahn, H. A., & Sempos, C. T. (1989). *Statistical methods in epidemiology.* New York: Oxford University Press.

Kahneman, D. (2011). *Thinking, fast and slow.* New York: Farrar, Strauss and Giroux.

Kahneman, D., & Tversky, A. (1979). Prospect theory: An analysis of decision under risk. *Econometrica, 47,* 263–291.

Kansas City preventive patrol experiment. (n.d.). In *Wikipedia: The free encyclopedia.* Retrieved on November 18, 2011 from http://www.en.wikipedia.org/ wiki/Kansas_City_preventive_patrol_experiment

Kaplan, E. L. (1983, June 13 / April 15). This week's citation classic: "Kaplan E L & Meier P. Nonparametric estimation from incomplete observations. *J. Amer. Statist. Assn.,* 53:457–81, 1958." *Current Contents: Life Sciences, 1983*(24), 14.

Kaplan, E. L., & Meier, P. (1958). Nonparametric estimation from incomplete observations. *Journal of the American Statistical Association, 53,* 457–481.

Karl Brandt. (n.d.). In *Wikipedia: The free encyclopedia.* Retrieved on February 8, 2011 from http://www.en.wikipedia.org/wiki/Karl_Brandt

Karl Pearson. (n.d.). In *Wikipedia: The free encyclopedia.* Retrieved on February 8, 2011 from http://www.en.wikipedia.org/wiki/Karl_Pearson

Kaye, D. H., & Freedman, D. A. (2000). Reference guide on statistics. In Federal Judicial Center, *Reference manual on scientific evidence* (2nd ed.) (pp. 83–178). St. Paul, MN: West Group.

Kaye, D. H., & Freedman, D. A. (2011). Reference guide on statistics. In Committee on the Development of the Third Edition of the Reference Manual on

Scientific Evidence; Committee on Science, Technology, and Law, Policy and Global Affairs; Federal Judicial Center; and National Research Council of the National Academies, *Reference manual on scientific evidence* (3rd ed.) (pp. 211–302). Washington, DC: National Academies Press.

Kassin, S. M. (2005). On the psychology of confessions: Does innocence put innocents at risk? *American Psychologist, 60,* 215–228.

Kazdin, A. E. (1982). *Single-case research designs: Methods for clinical and applied settings.* New York: Oxford University Press.

Kelleher, S. (2005, June 27). Rush toward new weight-loss drugs tramples patients' health. *Seattle Times.* Retrieved from http://seattletimes.nwsource.com

Kelleher, S. (2005, June 28). Disease expands through marriage of marketing and machines. *Seattle Times.* Retrieved from http://seattletimes.nwsource.com

Kelleher, S. (2005, June 30). Clash over "little blue pill" for women. *Seattle Times.* Retrieved from http://seattletimes.nwsource.com

Kelley, T. L. (1927). *Interpretation of educational measurements.* Yonkers-on-Hudson, NY: World Book.

Kelley, T. L. (1947). *Fundamentals of statistics.* Cambridge, MA: Harvard University Press.

Kelling, G. L., Pate, T., Dieckman, D., & Brown, C. E. (1974). *The Kansas City preventive patrol experiment: A summary report.* Washington, DC: Police Foundation. Retrieved from http://www.policefoundation.org/pdf/kcppe.pdf

Kenneth and Mamie Clark. (n.d.). In *Wikipedia: The free encyclopedia.* Retrieved on May 31, 2011 from http://www.en.wikipedia.org/wiki/Kenneth_Clark_(psychologist)

Kilborn, P. T. (1991, May 19). 'Race norming' tests becomes a fiery issue. *New York Times.* Retrieved from http://www.nytimes.com

Kimmelman, J., Weijer, C., & Meslin, E. M. (2009). Helsinki discords: FDA, ethics, and international drug trials. *Lancet, 373,* 13–14.

King, G. (1997). *A solution to the ecological inference problem: Reconstructing individual behavior from aggregate data.* Princeton, NJ: Princeton University Press.

Koch's postulates. (n.d.). In *Wikipedia: The free encyclopedia.* Retrieved on May 11, 2010 from http://www.en.wikipedia.org/wiki/Koch's_postulates

Koehler, J. J. (1993–1994). Error and exaggeration in the presentation of DNA evidence at trial. *Jurimetrics Journal, 34,* 21–39.

Kolata, G. (2009, March 19). Studies show prostate test saves few lives. *New York Times.* Retrieved from http://www.nytimes.com

Kolata, G. (2009, October 20). Cancer society, in shift, has concerns on screenings. *New York Times.* Retrieved from http://www.nytimes.com

Kolata, G. (2009, November 16). Panel urges mammograms at 50, not 40. *New York Times.* Retrieved from http://www.nytimes.com

Kolata, G. (2009, November 22). Behind cancer guidelines, quest for data. *New York Times.* Retrieved from http://www.nytimes.com

Kolata, G. (2010, April 19). Cancer fight: Unclear tests for new drug. *New York Times.* Retrieved from http://www.nytimes.com

Kolata, G. (2010, May 11). Doubt is cast on many reports of food allergies. *New York Times*. Retrieved from http://www.nytimes.com

Kolata, G. (2010, May 15). I can't eat that. I'm allergic. *New York Times*. Retrieved from http://www.nytimes.com

Kolata, G. (2010, June 23). Promise seen for detection of Alzheimer's. *New York Times*. Retrieved from http://www.nytimes.com

Kolata, G. (2011, April 11). Screening prostates at any age. *New York Times*. Retrieved from http://www.nytimes.com

Kolata, G. (2011, May 3). Low-salt diet ineffective, study finds. Disagreement abounds. *New York Times*. Retrieved from http://www.nytimes.com

Kolata, G. (2011, August 8). Catching obesity from friends may not be so easy. *New York Times*. Retrieved from http://www.nytimes.com

Kolata, G. (2011, October 29). Considering when it might be best not to know about cancer. *New York Times*. Retrieved from http://www.nytimes.com

Kolker, R. (1999, April 5). High caliber justice. *New York Magazine*. Retrieved from http://www.nymag.com

Korematsu v. United States, 323 United States 214 (1944). Retrieved from Justia.com United States Supreme Court Center [website]: http://supreme.justia.com/us/323/214/case.html

Krämer, W., & Gigerenzer, G. (2005). How to confuse with statistics or: The use and misuse of conditional probabilities. *Statistical Science, 20*, 223–230.

Kumho Tire Co. v. Carmichael, 526 United States 137 (1999). Retrieved from Cornell University Law School Legal Information Institute [website]: http://www.law.cornell.edu/supct/html/97-1709.ZS.html

Lamb, M. (Fall, 2001). Who was Wonder Woman 1? *Bostonia* [Boston University Alumni magazine]. Retrieved from Boston University [website]: http://www.bu.edu/bostonia/fall01/woman/

Larson, A. (2000, March). *Blood alcohol testing in drunk driving cases*. Retrieved from ExpertLaw [website]: http://www.expertlaw.com/library/drunk_driving/Drunk_Blood_Alcohol.html

Laughlin, H. H. (1930). *The legal status of eugenical sterilization: History and analysis of litigation under the Virginia Sterilization Statute, which led to a decision of the Supreme Court of the United States upholding the statute*. Chicago: Fred J. Ringley.

Law, K. R. (2010). *Life Line Screening* [Advertising letter to L. J. Hubert]. Copy in possession of Lawrence J. Hubert.

Law & Order. See entry for Wolf, D. (Creator and Producer), & Stern, J. (Producer).

Layton, J. (2006, May 18). *How police interrogation works*. Retrieved from HowStuffWorks [website]: http://people.howstuffworks.com/police-interrogation.htm.

Lee, J. M. (1993). Screening and informed consent. *New England Journal of Medicine, 328*, 438–440.

Legendre, A. M. (1805). *Nouvelles méthodes pour la détermination des orbites des cometès* [*New methods for the determination of comet orbits*]. Paris: Firmin Didot.

Lehrer, J. (2010, December 13). The truth wears off: Is there something wrong with the scientific method? *New Yorker*. Retrieved from http://www.newyorker.com

LeLorier, J., Grégoire, G., Benhaddad, A., Lapierre, J., & Derderian, F. (1997). Discrepancies between meta-analyses and subsequent large randomized, controlled trials. *New England Journal of Medicine, 337,* 536–542, See also Bailar, J. C., III. (1997). The promise and problems of meta-analysis [Editorial]. *New England Journal of Medicine, 337,* 559–561, seven letters to the editor, and the responses of LeLorier & Grégoire and Bailar (1998), *New England Journal of Medicine, 338,* 59–62.

Levelt Committee. (2011, October 31). *Interim report regarding the breach of scientific integrity committed by Prof. D. A. Stapel.* Tilburg, The Netherlands: Tilburg University. Retrieved from http://www.tilburguniversity.edu/nl/ nieuws-en-agenda/commissie-levelt/interim-report.pdf

Levin, I. (1967). *Rosemary's baby.* New York: Random House. See also entry for Castle, W., (Producer) & Polanski, R. (Director)

Lewis, A. (1987, April 28). Bowing to racism. *New York Times.* Retrieved from http://www.nytimes.com

Lewis, M. M. (2003). *Moneyball: The art of winning an unfair game.* New York: W. W. Norton. See also entry for De Luca, M., Horovitz, R., Pitt, B. (Producers), & Miller, B. (Director).

Lilienfeld, S. O., Wood, J. M., & Garb, H. N. (2000). The scientific status of projective techniques. *Psychological Science in the Public Interest, 1,* 27–66.

Linn, R. L., & Drasgow, F. (1987). Implications of the Golden Rule settlement for test construction. *Educational Measurement: Issues and Practice, 6*(2), 13–17.

Lipson, M. (Producer), & Morris, E. (Director). (1988). *The thin blue line* [Documentary]. United States: Miramax Films.

Liptak, A. (2003, March 16). You think DNA evidence is foolproof? Try again. *New York Times.* Retrieved from http://www.nytimes.com

Liptak, A. (2009, December 19). Justices revisit rule requiring lab testimony. *New York Times.* Retrieved from http://www.nytimes.com

Lisse, J. R., Perlman, M., Johansson, G., Shoemaker, J. R., Schechtman, J., Skalkey, C. S., ... Geba, G. P. (2003). Gastrointestinal tolerability and effectiveness of rofecoxib versus naproxen in the treatment of osteoarthritis: A randomized, controlled trial. *Annals of Internal Medicine, 139,* 539–546.

Loftus, E. F. (2010). Catching liars [Editorial]. *Psychological Science in the Public Interest, 11,* 87–88.

Longley, R. (n.d.) Miranda: Rights of silence—Why the police have to "read him his rights." Retrieved from About.com [website]: http://usgovinfo.about. com/cs/mirandarights/a/miranda.htm

Lord, F. M. (1967). A paradox in the interpretation of group comparisons. *Psychological Bulletin, 68,* 304–305.

Loving v. Virginia, 388 United States 1 (1967). Retrieved from Cornell University Law School Legal Information Institute [website]: http://www.law.cornell. edu/supct/html/historics/USSC_CR_0388_0001_ZO.html

Lucia de Berk. (n.d.). In *Wikipedia: The free encyclopedia.* Retrieved on October 29, 2010 from http://www.en.wikipedia.org/wiki/Lucia_de_Berk

Lucia, S. P. (1929). Invocatio medici: Code of Fushi Ikai No Ryaku, Oath of Hippocrates, and Supplication of Maimonides. *California and Western Medicine, 30,* 117–120.

Mackey, R. (2009, April 30). Lessons from the non-pandemic of 1976 [Web log post]. *New York Times.* Retrieved from New York Times Opinion Pages [website]: http://thelede.blogs.nytimes.com/2009/04/30/lessons-from-the-non-pandemic-of-1976/

Mackey, R. (2010, October 6). Tennessee firefighters watch home burn [Web log post]. *New York Times.* Retrieved from New York Times Opinion Pages [website]: http://thelede.blogs.nytimes.com/2010/10/06/tennessee-firefighters-watch-home-burn/.

Madison Grant. (n.d.). In *Wikipedia: The free encyclopedia.* Retrieved on February 8, 2011 from http://www.en.wikipedia.org/wiki/Madison_Grant

Magliozzi, T., & Magliozzi, R. (February 12, 2011). Puzzler: Three slips of paper. *Car Talk from NPR.* Retrieved from http://www.cartalk.com

Maimonides, M. (1904). *The guide for the perplexed* (M. Friedlander, Trans.) (2nd ed.). London: Routledge & Kegan Paul.

Maimonides, M. (n.d.). Issuing a punitive sentence based on circumstantial evidence: Negative commandment 290. (B. Bell, Trans.). *Sefer hamitzvot [The book of commandments].* Retrieved from Chabad.org [website]: http://www.chabad.org

Mann, C. (1990). Meta-analysis in the breech. *Science, 249,* 476–480.

Marshall, E. (2010). The promise and pitfalls of a cancer breakthrough. *Science, 330,* 900–901.

Marston, J. (2008). Smoking gun [Letter to the editor]. *NewScientist, 197*(2646), 21.

Marston, W. M. (1921). *Systolic blood pressure and reaction-time symptoms of deception and of constituent mental states* (Unpublished doctoral dissertation). Harvard University, Cambridge, MA.

Material [Definition]. (2010). *Webster's New World college dictionary.* Cleveland, OH: Wiley.

Matrixx Initiatives, Inc. v. Siracusano, 563 United States _____ (2011). Retrieved from Cornell University Law School Legal Information Institute [website]: http://www.law.cornell.edu/supct/html/09-1156.ZS.html

Maxwell, M. (1999). A mathematical perspective on gambling. *MIT Undergraduate Journal of Mathematics, 1,* 123–132.

McCleskey v. Kemp, 481 United States 279 (1987). Retrieved from Cornell University Law School Legal Information Institute [website]: http://www.law.cornell.edu/supct/html/historics/USSC_CR_0481_0279_ZO.html

McCullough, B. D. (2008a). Microsoft Excel's 'Not the Wichmann-Hill' random number generators. *Computational Statistics & Data Analysis, 52,* 4587–4593.

McCullough, B. D. (2008b). Special section on Microsoft Excel 2007 [Editorial]. *Computational Statistics & Data Analysis, 52,* 4568–4569.

McCullough, B. D. (2008c). Special section on Microsoft Excel 2007 [Special section]. *Computational Statistics & Data Analysis, 52,* 4568–4606.

McCullough, B. D., & Heiser, D. A. (2008). On the accuracy of statistical procedures in Microsoft Excel 2007. *Computational Statistics & Data Analysis, 52,* 4570–4578.

McDonald, W. A. (1938, December 4). Brigham adds fire to 'War of I.Q.'s.' *New York Times,* p. D10

McGrayne, S. B. (2011). *The theory that would not die: How Bayes' rule cracked the Enigma code, hunted down Russian submarines, and emerged triumphant from two centuries of controversy.* New Haven, CT: Yale University Press.

McKinley, J. (2010, November 22). In California, birth defects show no link. *New York Times.* Retrieved from http://www.nytimes.com

McNally, R. J. (2003). *Remembering trauma.* Cambridge, MA: Belknap Press / Harvard University Press.

McNeil, D. G., Jr. (2010, March 12). 3 rulings find no link to vaccines and autism. *New York Times.* Retrieved from http://www.nytimes.com

Meadow, R. (1977). Munchausen syndrome by proxy: The hinterland of child abuse. *Lancet, 310,* 343–345.

Meadow, R. (1997). Fatal abuse and smothering. In R. Meadow (Ed.), *ABC of child abuse* (3rd ed., pp. 27–29). London: BMJ Books.

Meehl, P. E. (1954). *Clinical versus statistical prediction: A theoretical analysis and a review of the evidence.* Minneapolis: University of Minnesota Press.

Melendez-Diaz v. Massachusetts, 129 S.Ct. 2527 (2009). Retrieved from Cornell University Law School Legal Information Institute [website]: http://www.law.cornell.edu/supct/html/07-591.ZS.html

Messick, D. M., & van de Geer, J. P. (1981). A reversal paradox. *Psychological Bulletin, 90,* 582–593.

Miasma theory. (n.d.). In *Wikipedia: The free encyclopedia.* Retrieved on October 29, 2010 from http://www.en.wikipedia.org/wiki/Miasma_theory

Mill, J. S. [John Stuart]. (1869). [Footnote]. In Mill, J. [James] *Analysis of the phenomena of the human mind* (Vol. II, ch. XIV, p. 5, fn. 2) (J. S. Mill, Ed.). London: Longmans, Reader, Green, and Dyer.

Miller, A. (1996, October 21). Why I wrote "The Crucible." *New Yorker.* Retrieved from http://www.newyorker.com

Miller, G. A., & Chapman, J. P. (2001). Misunderstanding analysis of covariance. *Journal of Abnormal Psychology, 110,* 40–48.

Milloy, S. (n.d.). JunkScience.com [website]: http://www.junkscience.com

Miranda v. Arizona, 384 United States 436 (1966). Retrieved from Justia.com United States Supreme Court Center [website]: http://supreme.justia.com/us/384/436

The Miranda warning. (2010, January 8). Retrieved from U.S. Constitution Online [website]: www.usconstitution.net/miranda.html

Mlodinow, L. (2009). *The drunkard's walk: How randomness rules our lives.* New York: Vintage Books.

Moneyball. See entries for De Luca, M., Horovitz, R., Pitt, B. (Producers), & Miller, B. (Director), and Lewis, M. M.

Monty Hall problem. (n.d.). In *Wikipedia: The free encyclopedia.* Retrieved on October 29, 2010 from http://www.en.wikipedia.org/wiki/Monty_Hall_problem

Mood, A. M., & Graybill, F. A. (1963). *Introduction to the theory of statistics* (2nd ed.). New York: McGraw-Hill.

Mooney, C. (2005, August 28). Interrogations: Thanks to a little-known piece of legislation, scientists at the EPA and other agencies find their work questioned not only by industry, but by their own government. *Boston Globe*. Retrieved from http://www.boston.com/news/globe/ideas/articles/2005/08/28/interrogations/?page=full

Mooney, C. (2005). *The Republican war on science*. New York: Basic Books.

Morrison, P., & Morrison, E. (Eds.) (1961). *Charles Babbage and his calculating machines: Selected writings by Charles Babbage and others*. New York: Dover Publications. [To be reissued by Dover Publications in February 2012 as *On the principles and development of the calculator and other seminal writings*].

Moss, M. (2010, May 29). The hard sell on salt. *New York Times*. Retrieved from http://www.nytimes.com

Moss, P. A. (1994). Can there be validity without reliability? *Educational Researcher, 23*(2), 5–12.

Mosteller, F. & Moynihan, D. P. (Eds.). (1972). *On equality of educational opportunity*. New York: Vintage Books.

Mosteller, F., & Tukey, J. W. (1977). *Data analysis and regression: A second course in statistics*. Reading, MA: Addison-Wesley.

Mosteller, F., & Wallace, D. L. (1964). *Inference and disputed authorship:* The Federalist. Reading, MA: Addison-Wesley.

Mozart effect. (n.d.). In *Wikipedia: The free encyclopedia*. Retrieved on October 29, 2010 from http://www.en.wikipedia.org/wiki/Mozart_effect

Mullis, I. V. S., Dossey, J. A., Owen, E. H., & Phillips, G. W. (1993, April). *NAEP 1992 mathematics report card for the nation and the states: Data from the national and trial state assessments* (Report No. 23-ST02). Princeton, NJ: Educational Testing Service for the National Center for Education Statistics, Office of Educational Research and Improvement, United States Department of Education.

Myers, A. L. (2010, July 10). Seventh lawsuit filed over Ariz. immigration law. *Associated Press*. Retrieved from KOMOnews.com [website]: http://www.komonews.com/news/national/98170869.html

National Commission for the Protection of Human Subjects of Biomedical and Behavioral Research. (1979, April 18). *The Belmont Report: Ethical principles and guidelines for the protection of human subjects of research*. Retrieved from National Institutes of Health, Office of Human Subjects Research [website]: http://ohsr.od.nih.gov/guidelines/belmont.html

National Research Act of 1974. See entry for An Act to Amend the Public Health Service Act to Establish a Program of National Research Service Awards to Assure the Continued Excellence of Biomedical and Behavioral Research and to Provide for the Protection of Human Subjects Involved in Biomedical and Behavioral Research and for Other Purposes.

Natural history of disease. (n.d.). In *Wikipedia: The free encyclopedia*. Retrieved on January 15, 2010 from http://www.en.wikipedia.org/wiki/Natural_history_of_disease

Nazi Doctors' Trial. See entry for United States of America v. Karl Brandt et al.

Newman, J. (2000, October 1). 20 of the greatest blunders in science in the last 20 years: What were they thinking? *Discover Magazine.* Retrieved from http://discovermagazine.com

Nieuwenhuis, S., Forstmann, B. U., & Wagenmakers, E.-J. (2011). Erroneous analyses of interactions in neuroscience: A problem of significance. *Nature Neuroscience, 14,* 1105–1107.

No Child Left Behind Act of 2001. See entry for An Act to Close the Achievement Gap with Accountability, Flexibility, and Choice, so That No Child is Left Behind.

Nordberg, P. (n.d.). *Daubert decisions by field of expertise: Psychologists & psychiatrists.* Retrieved on October 29, 2010 from Daubert on the Web [website]: http://www.daubertontheweb.com/psychologists.htm

Nuremberg Code. (1949). In *Trials of war criminals before the Nuernberg Military Tribunals under Control Council Law No. 10* (Vol. II, pp. 181–182). Washington, DC: United States Government Printing Office.

Nuremberg Doctors' Trial. See entry for United States of America v. Karl Brandt et al.

Nuzzo, R. (2008, June 21). Nabbing suspicious SNPS: Scientists search the whole genome for clues to common diseases. *ScienceNews.* Retrieved from http://www.sciencenews.org

Offit, P. A. (2008). *Autism's false prophets: Bad science, risky medicine, and the search for a cure.* New York: Columbia University Press.

Offit, P. A. (2011). *Deadly choices: How the anti-vaccine movement threatens us all.* New York: Basic Books.

Okie, S. (2010, May 3). Teaching physicians the price of care. *New York Times.* Retrieved from http://www.nytimes.com

P. Lorillard Co. v. Federal Trade Commission, 186 F.2d 52 (1950). Retrieved from Justia.com US Law [website]: http://supreme.justia.com/cases/federal/appelate-courts/F2/186/52/162820

Pakula, A. J., Barish, K., Gerrity, W. C., Starger, M. (Producers), & Pakula, A. J. (Director). (1982). *Sophie's choice* [Motion picture]. United States: ITC Entertainment / Universal Pictures. See also entry for Styron, W.

Pan, Z., Trikalinos, T. A., Kavvoura, F. K., Lau, J., & Ioannidis, J. P. A. (2005). Local literature bias in genetic epidemiology: An empirical evaluation of the Chinese literature. *PLoS Medicine, 2,* 1309–1317.

Paracelsus. (n.d.). In *Wikipedia: The free encyclopedia.* Retrieved on October 29, 2010 from http://www.en.wikipedia.org/wiki/Paracelsus

Pareidolia. (n.d.). In *Wikipedia: The free encyclopedia.* Retrieved on October 29, 2010 from http://www.en.wikipedia.org/wiki/Pareidolia

Parimutuel betting. (n.d.). In *Wikipedia: The free encyclopedia.* Retrieved on October 29, 2010 from http://www.en.wikipedia.org/wiki/Parimutuel_betting

Passell, P. (1990, March 4). Wine equation puts some noses out of joint. *New York Times.* Retrieved from http://www.nytimes.com

Paulos, J. A. (2009, December 10). Mammogram math. *New York Times Magazine.* Retrieved from http://www.nytimes.com

Pear, R. (2010, April 19). President nominates professor [Dr. Donald M. Berwick] to health job. *New York Times.* Retrieved from http://www.nytimes.com

Pear, R. (2010, June 21). Confirmation fight on health chief. *New York Times.* Retrieved from http://www.nytimes.com

Pear, R. (2010, July 6). Obama to bypass Senate to name health official. *New York Times.* Retrieved from http://www.nytimes.com

Pear, R. (2011, November 23). Obama's pick to head Medicare and Medicaid resigns post. *New York Times.* Retrieved from http://www.nytimes.com

Pearson, K., Lee, A., & Bramley-Moore, L. (1899). VI. Mathematical contributions to the theory of evolution.—VI. Genetic (reproductive) selection: Inheritance of fertility in man, and of fecundity in thoroughbred racehorses. *Philosophical Transactions of the Royal Society of London, Series A, 192,* 257–330.

Pearson, K. (with Lee, A., Warren, E., Fry, A., & Fawcett, C. D.). (1901). VIII. Mathematical contributions to the theory of evolution.—IX. On the principle of homotyposis and its relation to heredity, to the variability of the individual, and to that of the race. Part I.—Homotyposis in the vegetable kingdom. *Philosophical Transactions of the Royal Society of London, Series A, 197,* 285–379.

Pelletier, L. (Creator). (1944–1958). *The FBI in peace and war* [Radio series]. New York: Columbia Broadcasting System. See also entry for Collins, F. L.

People v. Collins, 68 Cal.2d 319 (1968). Retrieved from Stanford Law School, Robert Crown Law Library, Supreme Court of California Resources [website]: http://scocal.stanford.edu/opinion/people-v-collins-22583

Per protocol analysis [Definition]. (2005). In *Cochrane Collaboration glossary.* Retrieved on October 29, 2010 from http://www.cochrane.org/glossary/5

Performance bias [Definition]. (2005). In *Cochrane Collaboration glossary.* Retrieved on October 29, 2010 from http://www.cochrane.org/glossary/5

Perron-Frobenius theorem. (n.d.). In *Wikipedia: The free encyclopedia.* Retrieved on October 29, 2010 from http://www.en.wikipedia.org/wiki/Perron%E2%80%93Frobenius_theorem

Petitti, D. (2004). Hormone replacement therapy and coronary heart disease: Four lessons [Commentary]. *International Journal of Epidemiology, 33,* 461–463.

Petitti, D. B., Perlman, J. A., & Sidney, S. (1987). Noncontraceptive estrogens and mortality: Long-term followup of women in the Walnut Creek Study. *Obstetrics & Gynecology, 70,* 289–293.

Phase I, II, III, IV trials [Definition]. (2005). In *Cochrane Collaboration glossary.* Retrieved on October 29, 2010 from http://www.cochrane.org/glossary/5

Philbrick, H. A. (1952). *I led 3 lives: Citizen, 'Communist,' counterspy.* New York: McGraw-Hill. See also entry for Ziv, F. W. (Producer), Davis, E., Goodwins, L., Herzberg, J., Kesler, H. S., & Strock, H. L. (Directors).

Phrenology. (n.d.). In *Wikipedia: The free encyclopedia.* Retrieved on October 29, 2010 from http://www.en.wikipedia.org/wiki/Phrenology

Physiognomy. (n.d.). In *Wikipedia: The free encyclopedia.* Retrieved on October 29, 2010 from http://www.en.wikipedia.org/wiki/Physiognomy

Pileggi, N. (1995). *Casino: Love and honor in Las Vegas.* New York: Pocket Books.

Plead [Definition]. (2010). *Webster's New World college dictionary.* Cleveland, OH: Wiley.

Plessy v. Ferguson, 163 United States 537 (1896). Retrieved from Justia.com United States Supreme Court Center [website]: http://supreme.justia.com/us/163/537/case.html

Poincaré, H. (1896). *Calcul des probabilitées: Leçons professées pendant le deuxième semestre 1893–1894* [*Calculation of probabilities: Lessons given during the second semester 1893–1894*]. Paris: Georges Carré.

Point shaving. (n.d.). In *Wikipedia: The free encyclopedia.* Retrieved on October 29, 2010 from http://www.en.wikipedia.org/wiki/Point_shaving

Police Foundation. (n.d.). *The Kansas City preventive patrol experiment* [Research brief]. Retrieved on November 18, 2011 from Police Foundation [website]: http://www.policefoundation.org/docs/kansas.html

Precautionary principle. (n.d.). In *Wikipedia: The free encyclopedia.* Retrieved on October 24, 2011 from http://www.en.wikipedia.org/wiki/Precautionary_principle

Pressman, E. R. (Producer), & Stone, O. (Director). (1987). *Wall Street* [Motion picture]. United States: 20th Century Fox.

Prisoner malaria: Convicts expose themselves to disease so doctors can study it. (1945, June 4). *Life*, pp. 43, 44, 46.

Problem of points. (n.d.). In *Wikipedia: The free encyclopedia.* Retrieved on October 29, 2010 from http://www.en.wikipedia.org/wiki/Problem_of_points

Prosecutor's fallacy. (n.d.). In *Wikipedia: The free encyclopedia.* Retrieved on April 28, 2010 from http://www.en.wikipedia.org/wiki/Prosecutor's_fallacy

Publication bias. (n.d.). In *Wikipedia: The free encyclopedia.* Retrieved on October 29, 2010 from http://www.en.wikipedia.org/wiki/Publication_bias

Puerperal fever. (n.d.). In *Wikipedia: The free encyclopedia.* Retrieved on June 6, 2010 from http://www.en.wikipedia.org/wiki/Puerperal_fever

Pyromancy. (n.d.). In *Wikipedia: The free encyclopedia.* Retrieved on October 29, 2010 from http://www.en.wikipedia.org/wiki/Pyromancy

Racial Integrity Act of Virginia. See entry for An Act to Preserve Racial Integrity.

Radelet, M. L. (1981). Racial characteristics and the imposition of the death penalty. *American Sociological Review, 46*, 918–927.

Radical mastectomy. (n.d.). In *Wikipedia: The free encyclopedia.* Retrieved on October 29, 2010 from http://www.en.wikipedia.org/wiki/Radical_mastectomy

Rain Man. See entry for Johnson, M. (Producer), & Levinson, B. (Director).

Raloff, J. (2010, January 19). IPCC relied on unvetted Himalaya melt figure. *ScienceNews.* Retrieved from http://www.sciencenews.org

Raloff, J. (2010, June 2). July: When not to go to the hospital. *ScienceNews.* Retrieved from http://www.sciencenews.org

Raloff, J. (2010, July 1). Breast screening tool finds many missed cancers. *ScienceNews.* Retrieved from http://www.sciencenews.org

Randall, T. (2010, May 21). Blood test for early ovarian cancer may be recommended for all. *Bloomberg Businessweek.* Retrieved from http://www.businessweek.com

Randomized controlled trial [Definition]. (n.d.). In *National Library of Medicine—Medical subject headings.* Retrieved on October 29, 2010 from http://www.nlm.nih.gov/cgi/mesh/2011/MB_cgi?mode=&term=Randomized+Controlled+Trial&field=entry

Reasonable/prudent man law & legal definition. (n.d.). In *USLegal Definitions.* Retrieved on May 31, 2011 from http://definitions.uslegal.com/r/reasonable-prudent-man/

Recap of St. Louis Cardinals vs. Pittsburgh Pirates. (2004, April 26). *Associated Press*.

Redi, F. (1909). *Experiments on the generation of insects* (M. Bigelow, Trans.). Chicago: Open Court. (Original work published 1688)

Regulation of Securities Exchanges [Securities Exchange Act of 1934], Pub. L. No. 73-291 (1934), as amended through Pub. L. No. 111-257 (2010). Retrieved from United States Securities and Exchange Commission [website]: http://www.sec.gov/about/laws/sea34.pdf

Remarks by the President [William J. Clinton] *in apology for study done in Tuskegee* [News release]. (1997, May 16). The White House, Office of the Press Secretary. Retrieved from National Archives and Records Administration [website]: http://clinton4.nara.gov/New/Remarks/Fri/19970516-898.html

Research Synthesis Methods. (n.d.). Retrieved from Wiley Online Library [website]: http://onlinelibrary.wiley.com/journal/10.1002/(ISSN)1759-2887

Richmond, C. (2003). David Horrobin [Obituary]. *British Medical Journal, 326*, 885 [correction, *326*, 1091]. Retrieved from http://www.bmj.com. See also entry for Davies, S.

Rind, B., Tromovitch, P., & Bauserman, R. (1998). A meta-analytic examination of assumed properties of child sexual abuse using college samples. *Psychological Bulletin, 124*, 22–53.

Risse, G. (1976). Vocational guidance during the Depression: Phrenology versus applied psychology. *Journal of the History of the Behavioral Sciences, 12*, 130–140.

Roberts, S. (2007, January 16). 51% of women are now living without spouse. *New York Times*. Retrieved from http://www.nytimes.com

Roberts, S., & Pashler, H. (2000). How persuasive is a good fit? A comment on theory testing. *Psychological Review, 107*, 358–367.

Robinson, D. H. (2004). An interview with Gene V. Glass. *Educational Researcher, 33*(3), 26–30.

Robinson, W. S. (1950). Ecological correlations and the behavior of individuals. *American Sociological Review, 15*, 351–357.

Rofecoxib (n.d.). In *Wikipedia: The free encyclopedia*. Retrieved on January 12, 2012 from http://www.en.wikipedia.org/wiki/Rofecoxib

Rogosa, D. (2005). A school accountability case study: California API awards and the *Orange County Register* margin of error folly. In R. P. Phelps (Ed.), *Defending standardized testing* (pp. 205–226). Mahwah, NJ: Erlbaum.

Rorschach test. (n.d.). In *Wikipedia: The free encyclopedia*. Retrieved on October 29, 2010 from http://www.en.wikipedia.org/wiki/Rorschach_test

Rosemary's baby. See entries for Castle, W., (Producer) & Polanski, R., (Director) and Levin, I.

Rosenbaum, P. R. (2002). *Observational studies* (2nd ed.). New York: Springer-Verlag.

Rosenbaum, P. R. (2009). *Design of observational studies*. New York: Springer-Verlag.

Rosenbaum, R. (1995, January 15). The great Ivy League nude posture photo scandal. *New York Times Magazine*. Retrieved from http://www.nytimes.com

Rosenhan, D. L. (1973). On being sane in insane places. *Science, 179,* 250–258.

Rosenthal, R. (1979). The file drawer problem and tolerance for null results. *Psychological Bulletin, 86,* 638–641.

Rossouw, J. E., Anderson, G. L., Prentice, R. L., LaCroix, A. Z., Kooperberg, C., Stefanick, M. L., ... Ockene, J. (Writing Group for the Women's Health Initiative Investigators). (2002). Risks and benefits of estrogen plus progestin in healthy postmenopausal women: Principal results from the Women's Health Initiative randomized controlled trial. *Journal of the American Medical Association, 288,* 321–333.

Royal Statistical Society concerned by issues raised in Sally Clark case [News release]. (2001, October 23). Retrieved from http://www.rss.org.uk/uploadedfiles/documentlibrary/348.doc

Rubin, E. (1971). Quantitative commentary in Thucydides. *American Statistician, 25*(5), 52–54.

Russell, B. (1919). *Introduction to mathematical philosophy.* New York: Macmillan.

S. Rep. No. 109-88, 109th Cong., 1st Sess. (2005).

Sacco and Vanzetti. (n.d.). In *Wikipedia: The free encyclopedia.* Retrieved on February 8, 2011 from http://www.en.wikipedia.org/wiki/Sacco_and_Vanzetti

Sack, K. (2009, November 20). Screening debate reveals culture clash in medicine. *New York Times.* Retrieved from http://www.nytimes.com

Saey, T. H. (2010, June 5). All present-day life arose from a single ancestor. *ScienceNews.* Retrieved from http://www.sciencenews.org

Sager, R. (2009, August 18). Keep off the Astroturf [Opinion-Editorial]. *New York Times.* Retrieved from http://www.nytimes.com

Salholz, E. (1986, June 2). Too late for Prince Charming? *Newsweek,* p. 54.

Sanders, L. (2009, December 19). Trawling the brain: New findings raise questions about reliability of fMRI as gauge of neural activity. *ScienceNews.* Retrieved from http://www.sciencenews.org

Saretzky, G. D. (December, 1982). *Carl Campbell Brigham, the native intelligence hypothesis, and the Scholastic Aptitude Test* [Research memorandum] (Report No. ETS-RM-82-4). Princeton, NJ: Educational Testing Service. Retrieved from Education Resources Information Center [website]: http://www.eric.ed.gov (ED 237516)

Savage, D. G. (2010, June 2). Supreme Court backs off strict enforcement of Miranda rights. *Los Angeles Times.* Retrieved from http://articles.latimes.com

Savage, S. L. (2009). *The flaw of averages: Why we underestimate risk in the face of uncertainty.* Hoboken, NJ: Wiley.

Schwarzenegger, A. (2010, May 10). *Commencement keynote address.* Speech presented at Emory University, Atlanta, GA. Retrieved from http://shared.web.emory.edu/emory/news/releases/2010/05/arnold-schwarzenegger-commencement-speech.html

Scienter law & legal definition. (n.d.). In *USLegal Definitions.* Retrieved on May 31, 2011 from http://definitions.uslegal.com/s/scienter/

Scott v. Sandford, 60 United States 393 (1857). Retrieved from Justia.com United States Supreme Court Center [website]: http://supreme.justia.com/us/60/393/case.html

Sears, Roebuck & Co. v. City of Inglewood (Los Angeles Sup. Ct. 1955). See also entry for Sprowls, R. C.

Securities Exchange Act of 1934. See entry for Regulation of Securities Exchanges.

Securities Exchange Commission Rule 10b-5: Employment of manipulative and deceptive practices, 17 C. F. R. §240.10b-5. Retrieved from http://www.sec.gov/about/laws/secrulesregs.htm

Seife, C. (2010). *Proofiness: The dark arts of mathematical deception.* New York: Viking Penguin.

Selvin, H. C. (1958). Durkheim's *Suicide* and problems of empirical research. *American Journal of Sociology, 63*, 607–619.

Senn, S. (2003). *Dicing with death: Chance, risk, and health.* Cambridge, UK: Cambridge University Press.

Shadish, W. R., Cook, T. D., & Campbell, D. T. (2002). *Experimental and quasi-experimental designs for generalized causal inference.* Boston: Houghton-Mifflin.

Shibboleth. (n.d.). In *Wikipedia: The free encyclopedia.* Retrieved on October 29, 2010 from http://www.en.wikipedia.org/wiki/Shibboleth

Shore, D. (Creator and Producer). (n.d.). *House, MD* [Television series]. Los Angeles, CA: Fox Broadcasting Company.

Siegel, D., Daley, R. (Producers), & Siegel, D. (Director). (1971). *Dirty Harry* [Motion picture]. United States: Malpaso Productions / Warner Bros.

Simmons, J. P., Nelson, L. D., & Simonsohn, S. (2011). False-positive psychology: Undisclosed flexibility in data collection and analysis allows presenting anything as significant. *Psychological Science, 22*, 1359–1366.

Simpson, E. H. (1951). The interpretation of interaction in contingency tables. *Journal of the Royal Statistical Society, Series B, 13*, 238–241.

Singer, N. (2009, July 16). In push for cancer screening, limited benefits. *New York Times.* Retrieved from http://www.nytimes.com

Singer, N. (2009, December 12). Sure, it's treatable. But is it a disorder? *New York Times.* Retrieved from http://www.nytimes.com

Singer, N. (2009, December 18). Cancer center ads use emotion more than fact. *New York Times.* Retrieved from http://www.nytimes.com

Singer, N., & Wilson, D. (2009, December 12). Menopause, as brought to you by Big Pharma. *New York Times.* Retrieved from http://www.nytimes.com

Sizer, N. (1897). *Choice of pursuits; Or, what to do, and why, describing seventy-five trades and professions, and the talents and temperaments required for each; Also, how to educate, on phrenological principles, each man for his proper work. Together with portraits and biographies of more than one hundred successful thinkers and workers* (New ed., rev. and enl.). New York: Fowler & Wells.

Skeem, J. L., & Cooke, D. J. (2010). Is criminal behavior a central component of psychopathy? Conceptual directions for resolving the debate. *Psychological Assessment, 22*, 435–445.

Smith Act of 1940. See entry for An Act to Prohibit Certain Subversive Activities; to Amend Certain Provisions of Law with Respect to the Admission and Deportation of Aliens; to Require the Fingerprinting and Registration of Aliens; and for Other Purposes.

Somatotypes of Sheldon. (n.d.). Retrieved on May 31, 2011 from Human Nature & Human Types [website]: http://somatotypes.human-types.com/

Sommer, J. (2010, March 13). How men's overconfidence hurts them as investors. *New York Times*. Retrieved from http://www.nytimes.com

Sophie's choice. See entries for Pakula, A. J., Barish, K., Gerrity, W. C., Starger, M. (Producers), & Pakula, A. J. (Director), and Styron, W.

Spake, A. (2002, November 10). The menopausal marketplace; Why do we treat change of life as a medical disorder? The answer lies in the interplay of science and marketing. *US News & World Report*. Retrieved from http://health.usnews.com

Specter, M. (2002, May 27). Do fingerprints lie? *New Yorker*. Retrieved from http://www.newyorker.com

Spinney, L. (2008, May 21). Eyewitness identification: Line-ups on trial. *Nature*. Retrieved from http://www.nature.com

Spread betting. (n.d.). In *Wikipedia: The free encyclopedia*. Retrieved on October 29, 2010 from http://www.en.wikipedia.org/wiki/Spread_betting

Sprowls, R. C. (1956–1957). The admissibility of sample data into a court of law: A case history. *UCLA Law Review, 4*, 222–232.

Stacy, M. (2009, December 17). Florida man exonerated after 35 years behind bars. *Associated Press*. Retrieved from USA TODAY [website]: http://www.usatoday.com/news/nation/2009-12-17-exonerated_N.htm

Stam, A. (1997). Nontraditional approaches to statistical classification: Some perspectives on L_p-norm methods. *Annals of Operations Research, 74*, 1–36.

Stampfer, M. J., & Colditz, G. A. (1991). Estrogen replacement therapy and coronary heart disease: A quantitative assessment of the epidemiologic evidence. *Preventive Medicine, 20*, 47–63. (Reprinted 2004 in *International Journal of Epidemiology, 33*, 445–453, with commentaries on 454–467)

Stampfer, M. J., Willett, W. C., Colditz, G. A., Rosner, B., Speizer, F. E., & Hennekens, C. H. (1985). A prospective study of postmenopausal estrogen therapy and coronary heart disease. *New England Journal of Medicine, 313*, 1044–1049.

Stare decisis. (n.d.) In *Wikipedia: The free encyclopedia*. Retrieved on May 27, 2011 from http://www.en.wikipedia.org/wiki/Stare_decisis

Starkie, T. (1833). *A practical treatise of the law of evidence and digest of proofs, in civil and criminal proceedings* (2nd ed.). London: J. and W. T. Clarke.

Stateville Penitentiary Malaria Study. (n.d.). In *Wikipedia: The free encyclopedia*. Retrieved on February 8, 2011 from http://www.en.wikipedia.org/wiki/Stateville_Penitentiary_Malaria_Study

Steinley, D. (2003). Local optima in K-means clustering: What you don't know may hurt you. *Psychological Methods, 8*, 294–304.

Sterilization Act of Virginia. See entry for An Act to Provide for the Sexual Sterilization of Inmates of State Institutions in Certain Cases.

Stevens, J. P. (2010, December 23). On the death sentence [Review of the book *Peculiar institution: America's death penalty in an age of abolition*, by D. Garland]. *New York Review of Books*. Retrieved from http://www.nybooks.com

Stigler, S. M. (1986). *The history of statistics: The measurement of uncertainty before 1900*. Cambridge, MA: Belknap Press / Harvard University Press.

Stigler, S. M. (1990). The 1988 Neyman Memorial Lecture: A Galtonian perspective on shrinkage estimators. *Statistical Science, 5*, 147–155.

Stigler, S. M. (1997). Regression towards the mean, historically considered. *Statistical Methods in Medical Research, 6*, 103–114.

Stone, M. (2010, March 5). Toxic politics: Liz Cheney's Keep America Safe 'Al Qaeda Seven' ad. *Portland Progressive Examiner*. Retrieved from http://www.examiner.com/progressive-in-portland/

Stout, D. (2009, April 2). Obama's census choice unsettles Republicans. *New York Times*. Retrieved from http://www.nytimes.com

Strogatz, S. (2010, April 25). Chances are [Commentary]. *New York Times*. Retrieved from http://www.nytimes.com

Styron, W. (1979). *Sophie's choice*. New York: Random House. See also entry for Pakula, A. J., Barish, K., Gerrity, W. C., Starger, M. (Producers), & Pakula, A. J. (Director).

Su, Y.-S. (2008). It's easy to produce chartjunk using Microsoft® Excel 2007 but hard to make good graphs. *Computational Statistics & Data Analysis, 52*, 4594–4601.

Sulzberger, A. G. (2010, May 21). Defiant judge takes on child pornography law. *New York Times*. Retrieved from http://www.nytimes.com

Sunstein, C. R. (2002). Probability neglect: Emotions, worst cases, and law. *Yale Law Journal, 112*, 61–107.

Sussman, D. (2010, February 27). Opinion polling: A question of what to ask. *New York Times*. Retrieved from http://www.nytimes.com

Swets, J. A, Dawes, R. M., & Monahan, J. (2000a). Better decisions through science. *Scientific American, 283*(4), 82–87.

Swets, J. A, Dawes, R. M., & Monahan, J. (2000b). Psychological science can improve diagnostic decisions. *Psychological Science in the Public Interest, 1*, 1–26.

Talbot, M. (2007, July 2). Duped: Can brain scans uncover lies? *New Yorker*. Retrieved from http://www.newyorker.com

Taubes, G. (2007, September 16). Do we really know what makes us healthy? *New York Times*. Retrieved from http://www.nytimes.com

Tavris, C. (1999, July 19). The politics of sex abuse [Opinion-editorial]. *Los Angeles Times*. Retrieved on October 29, 2010 from Newgon.com [website]: http://newgon.com/prd/lib/Tavris1999.html

Tavris, C., & Aronson, E. (2007). *Mistakes were made (but not by me): Why we justify foolish beliefs, bad decisions, and hurtful acts*. Orlando, FL: Harcourt.

Taylor, H. C., & Russell, J. T. (1939). The relationship of validity coefficients to the practical effectiveness of tests in selection: Discussion and tables. *Journal of Applied Psychology, 23*, 565–578.

Teany, D. R. (2003, January). Pragmatic leadership advice from Donald Rumsfeld. *Ask Magazine*, pp. 30–32.

Temme, L. A. (2003). Ethics in human experimentation: The two military physicians who helped develop the Nuremberg Code. *Aviation, Space, and Environmental Medicine, 74*, 1297–1300.

Terry v. Ohio, 392 United States 1 (1968). Retrieved from Justia.com United States Supreme Court Center [website]: http://supreme.justia.com/us/392/1/case.html

Terry v. Ohio. (n.d.). In *Wikipedia: The free encyclopedia.* Retrieved on August 16, 2011 from http://www.en.wikipedia.org/wiki/Terry_v._Ohio

Texas sharpshooter fallacy. (n.d.). In *Wikipedia: The free encyclopedia.* Retrieved on October 29, 2010 from http://www.en.wikipedia.org/wiki/Texas_sharpshooter_fallacy

Thaler, R. H. (2009, December 19). Gauging the odds (and the costs) in health screening. *New York Times.* Retrieved from http://www.nytimes.com

Thaler, R. H., & Sunstein, C. R. (2008). *Nudge: Improving decisions about health, wealth, and happiness.* New Haven, CT: Yale University Press.

Theil, H. (1971). *Principles of econmeetrics.* New York: Wiley.

The thin blue line. See entry for Lipson, M. (Producer), & Morris, E. (Director).

Thorndike, E. L. (1939). On the fallacy of imputing the correlations found for groups to the individuals or smaller groups composing them. *American Journal of Psychology, 52,* 122–124.

Thorndike, E. L. (1940). *Human nature and the social order.* New York: Macmillan.

Thurstone, L. L. (1933). *The theory of multiple factors.* Ann Arbor, MI: Edwards Brothers.

Thurstone, L. L. (1935). *The vectors of mind: Multiple-factor analysis for the isolation of primary traits.* Chicago: University of Chicago Press.

Tote board. (n.d.). In *Wikipedia: The free encyclopedia.* Retrieved on October 29, 2010 from http://www.en.wikipedia.org/wiki/Tote_board

Trials of war criminals before the Nuernberg Military Tribunals under Control Council Law No. 10. Military Tribunal 1, Case 1, United States v. Karl Brandt et al., October 1946–April 1949 (Vols. I and II). (1949). Washington, DC: United States Government Printing Office.

Trillin, C. (2009, October 13). Wall Street smarts [Opinion-editorial]. *New York Times.* Retrieved from http://www.nytimes.com

Truthiness. (n.d.). In *Wikipedia: The free encyclopedia.* Retrieved on October 29, 2010 from http://www.en.wikipedia.org/wiki/Truthiness

Tufte, E. R. (1983). *The visual display of quantitative information.* Cheshire, CT: Graphics Press.

Tufte, E. R. (1990). *Envisioning information.* Cheshire, CT: Graphics Press.

Tufte, E. R. (1997). *Visual explanations: Images and quantities, evidence and narrative.* Cheshire, CT: Graphics Press.

Tufte, E. R. (2006). *The cognitive style of PowerPoint: Pitching out corrupts within* (2nd ed.). Cheshire, CT: Graphics Press.

Tukey, J. W. (1977). *Exploratory data analysis.* Reading, MA: Addison-Wesley.

Tuna, C. (2009, December 2). When combined data reveal the flaw of averages. *Wall Street Journal.* Retrieved from http://online.wsj.com

Turnock, B. J., & Kelly, C. J. (1989). Mandatory premarital testing for human immunodeficiency virus: The Illinois experience. *Journal of the American Medical Association, 261,* 3415–3418.

Tuskegee syphilis experiment. (n.d.). In *Wikipedia: The free encyclopedia.* Retrieved on February 8, 2011 from http://en.wikipedia.org/wiki/Tuskegee_syphilis_experiment

Tversky, A., & Kahneman, D. (1971). Belief in the law of small numbers. *Psychological Bulletin, 76*, 105–110.

Tversky, A., & Kahneman, D. (1974). Judgment under uncertainty: Heuristics and biases. *Science, 185*, 1124–1131.

Tversky, A., & Kahneman, D. (1981). The framing of decisions and the psychology of choice. *Science, 211*, 453–458.

Tversky, A., & Kahneman, D. (1983). Extensional versus intuitive reasoning: The conjunction fallacy in probability judgment. *Psychological Review, 90*, 293–315.

UK National Screening Committee. (n.d.). *Programme appraisal criteria.* Retrieved on May 27, 2011 from UK Screening Portal [website]: `http://www.screening.nhs.uk/criteria`

United States Department of Justice, Tort Policy Working Group. (1986, February). *Report of the Tort Policy Working Group on the causes, extent and policy implications of the current crisis in insurance availability and affordability* (Report No. 027-000-01251-5). Washington, DC: United States Government Printing Office.

United States Environmental Protection Agency. (2009, October 7). *EPA begins new scientific evaluation of atrazine* [News release]. Retrieved from EPA: United States Environmental Protection Agency [website]: `http://yosemite.epa.gov/opa/admpress.nsf`

United States of America v. Karl Brandt et al. [Nuremberg Doctors' Trial; Doctors' Trial; Nazi Doctors' Trial]. See entry for *Trials of war criminals before the Nuernberg Military Tribunals under Control Council Law No. 10.*

United States v. Fatico, 458 F.Supp. 388 (1978). Retrieved from Leagle [website]: `http://www.leagle.com/XmlResult.aspx?xmldoc=1978846458FSupp388_1773.xml&docbase=CSLWAR1-1950-1985`

Uniting and Strengthening America by Providing Appropriate Tools Required to Intercept and Obstruct Terrorism Act of 2001. See entry for An Act to Deter and Punish Terrorist Acts in the United States and Around the World, to Enhance Law Enforcement Investigatory Tools, and for Other Purposes.

USA PATRIOT Act. See entry for An Act to Deter and Punish Terrorist Acts in the United States and Around the World, to Enhance Law Enforcement Investigatory Tools, and for Other Purposes.

United States Const. amend. IV, VIII, XIV. Retrieved from United States Constitution Online [website]: `http://www.usconstitution.net/const.html`

U.S. Department of Health and Human Services, Agency for Healthcare Research and Quality. (2008). *National healthcare disparities report 2008* (AHRQ Publication No. 09-0002). Retrieved from AHRQ: Agency for Healthcare Research and Quality [website]: `http://www.ahrq.gov/qual/nhdr08/nhdr08.pdf`

U.S. Department of Health and Human Services, Centers for Disease Control and Prevention. (2009, October 6). *Shingles vaccine: What you need to know.* Retrieved from `http://www.cdc.gov/vaccines/pubs/vis/downloads/vis-shingles.pdf`

U.S. Department of Health and Human Services, United States Food and Drug Administration. (2009, August 20). *Adverse Event Reporting System (AERS).* Retrieved from `http://www.fda.gov/Drugs/GuidanceComplianceRegulatoryInformation/Surveillance/AdverseDrugEffects/default.htm`

U.S. Preventive Services Task Force. (2009, November; updated 2009, December). *Screening for breast cancer: Recommendation statement.* Retrieved from United States Preventive Services Task Force [website]: http://www.uspreventiveservicestaskforce.org/uspstf09/breastcancer/brcanrs.htm

Via régia. (n.d.). In *Wikimedia Commons.* Retrieved on October 29, 2010 from http://commons.wikimedia.org/wiki/Category:Via_Regia

Virginia Sterilization Act of 1924. See entry for An Act to Provide for the Sexual Sterilization of Inmates of State Institutions in Certain Cases.

Vogel, G. (2011). Psychologist accused of fraud on 'astonishing scale.' *Science, 334,* 579.

Vrij, A., Granhag, P. A., & Porter, S. (2010). Pitfalls and opportunities in nonverbal and verbal lie detection. *Psychological Science in the Public Interest, 11,* 89–121.

Vul, E., Harris, C., Winkielman, P., & Pashler, H. (2009). Puzzlingly high correlations in fMRI studies of emotion, personality, and social cognition. *Perspectives on Psychological Science, 4,* 274–290.

Wade, N. (2010, May 20). Researchers say they created a 'synthetic cell.' *New York Times.* Retrieved from http://www.nytimes.com

Wagner, E. J. (2006). *The science of Sherlock Holmes: From Baskerville Hall to the Valley of Fear, the real forensics behind the great detective's greatest cases.* Hoboken, NJ: Wiley.

Wagner, S. (1971). *Cigarette country: Tobacco in American history and politics.* New York: Praeger.

Wainer, H. (1976). Estimating coefficients in linear models: It don't make no nevermind. *Psychological Bulletin, 83,* 213–217.

Wainer, H. (1984). How to display data badly [Commentary]. *American Statistician, 38,* 137–147.

Wainer, H. (2000). *Visual revelations: Graphical tales of fate and deception from Napoleon Bonaparte to Ross Perot.* Mahwah, NJ: Erlbaum. (Original work published 1997, New York: Copernicus Books)

Wainer, H. (2005). *Graphic discovery: A trout in the milk and other visual adventures.* Princeton, NJ: Princeton University Press.

Wainer, H. (2009). *Picturing the uncertain world: How to understand, communicate, and control uncertainty through graphical display.* Princeton, NJ: Princeton University Press.

Wainer, H. (2011). How should we screen for breast cancer? Using evidence to make medical decisions. *Significance, 8*(1), 28–30.

Wainer, H. (2011). *Uneducated guesses: Using evidence to uncover misguided education policies.* Princeton, NJ: Princeton University Press.

Wainer, H., & Brown, L. (2004). Two statistical paradoxes in the interpretation of group differences: Illustrated with medical school admission and licensing data. *American Statistician, 58,* 117–123.

Wakley, T. [Editorial]. (1853). *Lancet, 62,* 393–394.

Wall Street. See entry for Pressman, E. R. (Producer), & Stone, O. (Director).

Wallace, H. A. (1923). What is in the corn judge's mind? *Journal of the American Society of Agronomy, 15,* 300–304.

Wallechinsky, D., & Wallace, I. (1981). *People's almanac #3*. New York: William Morrow.

Wallis, W. A., & Roberts, H. V. (1956). *Statistics: A new approach*. Glencoe, IL: Free Press.

Warden, R. (with Radelet, M. L.). (n.d.). *Filmmaker helped free innocent man*. Retrieved from Northwestern University School of Law, Bluhm Legal Clinic, Center on Wrongful Convictions, Meet the Exonerated, Randall Dale Adams [website]: http://www.law.northwestern.edu/wrongfulconvictions/exonerations

Warner, J. (2010, January 8). The wrong story about depression [Opinion-editorial]. *New York Times*. Retrieved from http://www.nytimes.com

Warner, S. L. (1965). Randomized response: A survey technique for eliminating evasive answer bias. *Journal of the American Statistical Association, 60*, 63–69.

Weights for direct age adjustment: Table 1. United States 2000 standard population weights for age standardization. (n.d.). Retrieved on October 29, 2010 from New Mexico Indicator-Based Information System (NM-IBIS) [website]: http://ibis.health.state.nm.us/resources/AARate.html (Original source: Klein, R. J., & Schoenborn, C. A. [2001, January]. Age-adjustment using the 2000 projected United States population. *Healthy People 2010: Statistical Notes* [No. 20]. Hyattsville, MD: Centers of Disease Control, National Center for Health Statistics)

Weinstein, J. B. (1988). Litigation and statistics. *Statistical Science, 3*, 286–297.

Wells, G. L., Memon, A., & Penrod, S. D. (2006). Eyewitness evidence: Improving its probative value. *Psychological Science in the Public Interest, 7*, 45–75.

Whitaker, C. F. (1990, September 9). [Letter to "Ask Marilyn" column]. *Parade Magazine*, p. 15.

Whren v. United States, 517 United States 806 (1996). Retrieved from Justia.com United States Supreme Court Center [website]: http://supreme.justia.com/us/517/806/case.html

Whren v. United States. (n.d.). In *Wikipedia: The free encyclopedia*. Retrieved on August 16, 2011 from http://www.en.wikipedia.org/wiki/Whren_v._United_States

Wilkinson, L., & Friendly, M. (2009). The history of the cluster heat map. *American Statistician, 63*, 179–184.

William Moulton Marston. (n.d.). In *Wikipedia: The free encyclopedia*. Retrieved on October 29, 2010 from http://www.en.wikipedia.org/wiki/William_Moulton_Marston

Wilson, D. (2005, June 26). New blood-pressure guidelines pay off—For drug companies. *Seattle Times*. Retrieved from http://seattletimes.nwsource.com

Wilson, D. (2010, March 16). Many new drugs have strong dose of media hype. *Seattle Times*. Retrieved from http://seattletimes.nwsource.com

Wilson, D. (2010, April 19). Flavored tobacco pellets are denounced as a lure to young users. *New York Times*. Retrieved from http://www.nytimes.com

Wilson, D. (2010, June 18). Drug for sexual desire disorder opposed by panel. *New York Times*. Retrieved from http://www.nytimes.com

Winch, R. F., & Campbell, D. T. (1969). Proof? No. Evidence? Yes. The significance of tests of significance. *American Sociologist, 4*, 140–143.

Wojciechowski, R., Bailey-Wilson, J. E., & Stambolian, D. (2010). Association of matrix metalloproteinase gene polymorphisms with refractive error in Amish and Ashkenazi families. *Investigative Ophthalmology & Visual Science, 51,* 4989–4995.

Wojciechowski, R., Bailey-Wilson, J. E., & Stambolian, D. (n.d.). [Macular degeneration in Mennonites]. Unpublished raw data.

Wolf, D. (Creator and Producer), & Stern, J. (Producer). (n.d.). *Law & Order* [Television series]. New York: NBC/Universal Media.

Wolfers, J. (2006). Point shaving: Corruption in NCAA basketball. *American Economic Review, 96,* 279–283.

Wolfle, D. (1970). Chance, or human judgment? [Editorial]. *Science, 167,* 1201.

Words concerning anti-black discrimination. (2005, December). Retrieved from Wordcraft [website]: `http://wordcraft.infopop.cc/Archives/2005-12-Dec.htm`

World Medical Association. (2008). *Declaration of Helsinki: Ethical principles for medical research involving human subjects* (Rev. 6). Retrieved from `http://www.wma.net/en/30publications/10policies/b3/17c.pdf`

Wright, S. (1921). Correlation and causation. *Journal of Agricultural Research, 20,* 557–585.

Yalta, A. T. (2008). The accuracy of statistical distributions in Microsoft® Excel 2007. *Computational Statistics & Data Analysis, 52,* 4579–4586.

Yarnold, P. R., & Soltysik, R. C. (2004). *Optimal data analysis: A guidebook with software for Windows.* Washington, DC: American Psychological Association.

Yule, G. U. (1903). Notes on the theory of association of attributes in statistics. *Biometrics, 2,* 121–134.

Yule, G. U. (1911). *An introduction to the theory of statistics.* London: Charles Griffin; Philadelphia: Lippincott.

Yule, G. U., & Kendall, M. G. (1950). *An introduction to the theory of statistics* (14th ed.). New York: Hafner.

Zipf's law. (n.d.). In *Wikipedia: The free encyclopedia.* Retrieved on February 21, 2012 from `http://www.en.wikipedia.org/wiki/Zipf's_law`

Ziv, F. W. (Producer), Davis, E., Goodwins, L., Herzberg, J., Kesler, H. S., & Strock, H. L. (Directors). (1953–1956). *I led 3 lives.* Hollywood, CA: Ziv Television Programs. See also entry for Philbrick, H. A.

Zola, É. (1898, January 13). J'accuse … ! Lettre au Président de la République. *L'Aurore,* pp. 1–2.

Author Index

Subject Index

abductive reasoning, 32
absolute risk reduction, 284, 285
abuse of discretion, 452
abuse, childhood sexual, xix, 346
actuarial prediction, xviii, 145
ad hominem, 491
Adams, Randall Dale, 163
additive model, 342
adherer effect, 390
adjusted R^2, 189
adjustment
    direct, 296
    indirect, 296
adverse event, xxi, 274
Adverse Events Reporting System (AERS), 275, 281
affirming the consequent, 30
agent, 414
Agent Orange, 97
aggregation bias, 129
Alexander, Leo, 422
Alien Registration Act (1940), 68
allocation concealment, 405
Alpine race (Central Europe), 242, 445
analysis of covariance, 2, 10, 309, 342
analysis of variance, 10
analyze as randomized, 404
anchoring and adjustment heuristic, 476
anecdata, 302
anecdote, 293
Angell, Marcia, 408
annual mortality rate, 284
anosmia, 274
Anrig, Gregory, 233
anthropological criminality, 280

anthropometry, 80
apophenia, 124
argument from a vacuum, 483
argument from ignorance, 483
argument from personal incredulity, 484
argumentum ad ignorantiam (argument from ignorance), 483
ARIMA model, 203
Arizona Senate Bill 1070, 62
Army Alpha Test, 242
Army Beta Test, 242
as randomized, so analyzed, 404
Ashenfelter, Orley, 159
aspartame, 325
astroturfing, 288, 289
atavistic (throwback), 280
atrazine, 265
attenuation (correlation), 131
attributable proportion of risk, xviii, 94
attributable risk, 94
attrition bias, 405
autism, 291, 312, 313
auto-icon, 116
autocorrelation
    spatial, 202
    temporal, 202
autoregressive-integrated-moving-average (ARIMA) model, 203
availability heuristic, 476
Avendia, 275
average
    mean, 278
    median, 278
    mode, 278
average age (of death), 301

Source Acknowledgments

We gratefully acknowledge the following sources for permission to reprint material.

1 Preamble

2 Introduction

p. 12: Freedman, D. A. (1987). As others see us: A case study in path analysis. *Journal of Educational Statistics, 12*, 101–128. Copyright © 1987 by the American Educational Research Association and the American Statistical Association. Reprinted by permission of SAGE Publications.

p. 12: Breiman, L. (2001). Statistical modeling: The two cultures. *Statistical Science, 16*, 199–215. Reprinted with the permission of the Institute of Mathematical Statistics.

p. 13: From *Statistics: A New Approach* by William A. Wallis and Harry V. Roberts (p. 215). Copyright © 1956, 1962 by The Free Press. Reprinted with the permission of The Free Press, a Division of Simon & Schuster, Inc. All rights reserved.

I Tools from Probability and Statistics

3 Probability Theory: Background and Bayes' Theorem

p. 20; continued on p. 49: *Royal Statistical Society concerned by issues raised in Sally Clark case* [News release]. (2001, October 23). Reprinted with the permission of the Royal Statistical Society. Retrieved from http://www.rss.org.uk/uploadedfiles/documentlibrary/348.doc

p. 27: Koehler, J. J. (1993–1994). Error and exaggeration in the presentation of DNA evidence at trial. *Jurimetrics Journal, 34*, 21–39. Reprinted with the permission of the Author.

p. 28: Committee on DNA Technology in Forensic Science, Board on Biology, Commission on Life Sciences, National Research Council. (1992). *DNA technology in forensic science*. Washington, DC: National Academy Press. Reprinted with the permission of the National Academy of Sciences, courtesy of the National Academies Press, Washington, DC.

p. 29; p. 29: Committee on DNA Forensic Science: An Update, Commission on DNA Forensic Science: An Update, National Research Council. (1996). *The evaluation of forensic DNA evidence*. Washington, DC: National Academy Press. Reprinted with the permission of the National Academy of Sciences, courtesy of the National Academies Press, Washington, DC.

p. 42: Reproduced from Non-invasive prenatal assessment of trisomy 21 by multiplexed maternal plasma DNA sequencing: Large scale validity study, *British Medical Journal*, Chiu, R. W. K., Akolekar, R., Zheng, Y. W. L., Leung, T. Y., Sun, H., Chan, K. C. A., ... Lo, Y. M. D., *342*, c7401, 2011, with permission from BMJ Publishing Group Ltd. Retrieved from http://www.bmj.com

p. 43; continued on p. 78: Marshall, E. (2010). The promise and pitfalls of a cancer breakthrough. *Science, 330*, 900–901. Reprinted with permission from the American Association for the Advancement of Science.

p. 46: UK National Screening Committee. (n.d.). *Programme Appraisal Criteria*. Reprinted with the permission of the UK National Screening Committee. Retrieved on May 27, 2011 from UK Screening Portal [website]: http://www.screening.nhs.uk/criteria

p. 51: Krämer, W., & Gigerenzer, G. (2005). How to confuse with statistics or: The use and misuse of conditional probabilities. *Statistical Science, 20*, 223–230. Reprinted with the permission of the Institute of Mathematical Statistics.

p. 52: Prosecutor's fallacy. (n.d.). In *Wikipedia: The free encyclopedia*. Excerpted from Wikipedia. Retrieved on April 28, 2010 from http://www.en.wikipedia.org/wiki/Prosecutor's_fallacy
Released under the Creative Commons Attribution Share-Alike License: http://creativecommons.org/Licenses/by-sa/3.0/

p. 55: Wells, G. L., Memon, A., & Penrod, S. D. (2006). Eyewitness evidence: Improving its probative value. *Psychological Science in the Public Interest, 7*, 45–75. Copyright © 2006 by the Association for Psychological Science. Reprinted by permission of SAGE Publications.

p. 56: Fienberg, S. E., & Stern, P. C. (2005). In search of the magic lasso: The truth about the polygraph. *Statistical Science, 20*, 249–260. Reprinted with the permission of the Institute of Mathematical Statistics.

p. 58: Larson, A. (2000, March). *Blood alcohol testing in drunk driving cases*. Reprinted with the permission of the Author. Retrieved from ExpertLaw [website]: http://www.expertlaw.com/library/drunk_driving/Drunk_Blood_Alcohol.html

p. 64: Terry v. Ohio. (n.d.). In *Wikipedia: The free encyclopedia*. Excerpted from Wikipedia. Retrieved on August 16, 2011 from http://www.en.wikipedia.org/wiki/Terry_v._Ohio
Released under the Creative Commons Attribution Share-Alike License: http://creativecommons.org/Licenses/by-sa/3.0/

p. 65: Whren v. United States. (n.d.). In *Wikipedia: The free encyclopedia*. Excerpted from Wikipedia. Retrieved on August 16, 2011 from http://www.en.wikipedia.org/wiki/Whren_v._United_States
Released under the Creative Commons Attribution Share-Alike License: http://creativecommons.org/Licenses/by-sa/3.0/

p. 66: FBI Index. (n.d.). In *Wikipedia: The free encyclopedia*. Excerpted from Wikipedia. Retrieved on April 25, 2011 from http://www.en.wikipedia.org/wiki/FBI_Index
Released under the Creative Commons Attribution Share-Alike License: http://creativecommons.org/Licenses/by-sa/3.0/

p. 71: Bonastre, J.-F., Bimbot, F., Boë, L.-J., Campbell, J. P., Reynolds, D. A., & Magrin-Chagnolleau, I. (2003, September). *Person authentication by voice: A need for caution.* Paper presented at Eurospeech 2003 - Interspeech 2003, 8th European Conference on Speech Communication and Technology, Geneva, Switzerland, September 1–4, 2003, 33–36. Copyright © 2003 by the International Speech Communication Association. Reprinted with the permission of the International Speech Communication Association and Jean-François Bonastre representing the Authors. Retrieved from International Speech Communication Association [website]: http://www.isca-speech.org/archive

p. 72; continued on p. 84: Committee on Identifying the Needs of the Forensic Science Community; Committee on Science, Technology, and Law Policy and Global Affairs; Committee on Applied and Theoretical Statistics; Division on Engineering and Physical Sciences; National Research Council of the National Academies. (2009). *Strengthening forensic science in the United States: A path forward.* Washington, DC: National Academies Press. Reprinted with the permission of the National Academy of Sciences, courtesy of the National Academies Press, Washington, DC.

p. 73: Dror, I. E., Charlton, D., & Péron, A. E. (2006). Contextual information renders experts vulnerable to making erroneous identifications. *Forensic Science International, 156,* 74–78. Copyright © 2005 Elsevier Ireland Ltd. Reprinted with permission from Elsevier.

p. 74: Fienberg, S. E. (Ed.). (1988). *The evolving role of statistical assessments as evidence in the courts.* New York: Springer-Verlag. Copyright © 1989 by Springer-Verlag New York Inc. Reprinted with the kind permission of Springer Science+Business Media B. V. and the Author.

p. 79: Curzon, F. R. P. [7th Earl Howe]. (2003, February 5). Attention deficit disorders [United Kingdom House of Lords debate]. *Lords Hansard,* Vol. 644, cc. 299–322. Parliamentary material is reproduced with the permission of the Controller of Her Majesty's Stationery Office on behalf of Parliament. Retrieved from UK Parliament [website]: http://www.publications.parliament.uk/pa/ld200203/ldhansrd/vo030205/text/30205-10.htm

p. 80: Wagner, E. J. (2006). *The science of Sherlock Holmes: From Baskerville Hall to the Valley of Fear, the real forensics behind the great detective's greatest cases.* Hoboken, NJ: Wiley. Reprinted with the permission of the Author.

p. 83: Loftus, E. F. (2010). Catching liars [Editorial]. *Psychological Science in the Public Interest, 11,* 87–88. Copyright © 2010 by the Author. Reprinted by permission of SAGE Publications.

p. 83: Vrij, A., Granhag, P. A., & Porter, S. (2010). Pitfalls and opportunities in nonverbal and verbal lie detection. *Psychological Science in the Public Interest, 11,* 89–121. Copyright © 2010 by the Authors. Reprinted by permission of SAGE Publications.

4 Probability Theory: Application Areas

p. 90: Weinstein, J. B. (1988). Litigation and statistics. *Statistical Science, 3,* 286–297. Reprinted with the permission of the Institute of Mathematical Statistics.

p. 91; p. 98: Maimonides, M. (n.d.). Issuing a punitive sentence based on circumstantial evidence: Negative commandment 290 (B. Bell, Trans.). *Sefer Hamitzvot* [*The Book of Commandments*]. Copyright © by Berel Bell. Reprinted by permission of Berel Bell, Sichos in English, and Chabad.org (http://www.chabad.org). Retrieved from Chabad.org [website]: http://www.chabad.org

p. 93, Table 4.1; p. 96: Fienberg, S. E. (Ed.). (1988). *The evolving role of statistical assessments as evidence in the courts.* New York: Springer-Verlag. Copyright © 1989 by Springer-Verlag New York Inc. Adapted with the kind permission of Springer Science+Business Media B. V. and the Author.

p. 106: Wolfers, J. (2006). Point shaving: Corruption in NCAA basketball. *American Economic Review, 96,* 279–283. Reprinted with the permission of the American Economic Association and the Author.

p. 108: Bernhardt, D., & Heston, S. (2010). Point shaving in college basketball: A cautionary tale for forensic economics. *Economic Inquiry, 48,* 14–25. Copyright © 2009 by the Western Economic Association International. Reprinted with the permission of John Wiley & Sons, Inc.

p. 110: Tversky, A., & Kahneman, D. (1981). The framing of decisions and the psychology of choice. *Science, 211,* 453–458. Reprinted with permission from the American Association for the Advancement of Science and Daniel Kahneman.

p. 111: Hardisty, D. J., Johnson, E. J., & Weber, E. U. (2010). A dirty word or a dirty world? Attribute framing, political affiliation, and query theory. *Psychological Science, 21,* 86–92. Copyright © 2010 by the Authors. Reprinted by permission of SAGE Publications.

p. 112: Reprinted from "Book 12: The Cattle of the Sun" by Homer, from *The Odyssey* by Homer, translated by Robert Fagles, copyright © 1996 by Robert Fagles. Used by permission of Viking Penguin, a division of Penguin Group (USA) Inc.

p. 115: Jeremy Bentham. (n.d.). In *Wikipedia: The free encyclopedia.* Excerpted from Wikipedia. Retrieved on October 22, 2010 from http://www.en.wikipedia.org/wiki/Jeremy_Bentham
Released under the Creative Commons Attribution Share-Alike License: http://creativecommons.org/Licenses/by-sa/3.0/

p. 117: Joseph Jagger. (n.d.). In *Wikipedia: The free encyclopedia.* Excerpted from Wikipedia. Retrieved on February 19, 2011 from http://www.en.wikipedia.org/wiki/Joseph_Jagger
Released under the Creative Commons Attribution Share-Alike License: http://creativecommons.org/Licenses/by-sa/3.0/

5 Correlation

p. 137: Gould, S. J. (1996). *The mismeasure of man* (Rev. and exp. ed.). New York: W. W. Norton. Copyright © 1996, 1981 by Stephen J. Gould. Reprinted with the permission of W. W. Norton & Company, Inc.

6 Prediction

p. 142; p. 154: Friedman, M. (1992). Do old fallacies ever die? *Journal of Economic Literature, 30*, 2129–2132. Reprinted with the permission of the American Economic Association.

p. 143; p. 143; p. 144: Stigler, S. M. (1997). Regression towards the mean, historically considered. *Statistical Methods in Medical Research, 6*, 103–114. Copyright © 1997 by Arnold. Reprinted by permission of SAGE Publications and the Author.

p. 147: Wallace, H. A. (1923). What is in the corn judge's mind? *Journal of the American Society of Agronomy, 15*, 300–304. Reprinted with the permission of the American Society of Agronomy.

p. 149: Grier, D. A. (n.d.). *The origins of statistical computing.* Reprinted with permission from the website of the American Statistical Association (http://www.amstat.org). Copyrighted by the American Statistical Association. All rights reserved. Retrieved from American Statistical Association [website]: http://www.amstat.org/about/statisticiansinhistory/index.cfm?fuseaction=paperinfo&PaperID=4

p. 153: Stigler, S. M. (1990). The 1988 Neyman Memorial Lecture: A Galtonian perspective on shrinkage estimators. *Statistical Science, 5*, 147–155. Adapted with the permission of the Institute of Mathematical Statistics.

p. 155: Reprinted by permission of Waveland Press, Inc., from Allen, M. J., & Yen, W. M., *Introduction to measurement theory.* (Long Grove, IL; Waveland Press, Inc., 1979 [re-issued 2002]). All rights reserved.

p. 164: Warden, R. (with Radelet, M. L.). (n.d.). *Filmmaker helped free innocent man.* Reprinted with the permission of the Authors. Retrieved from Northwestern University School of Law, Bluhm Legal Clinic, Center on Wrongful Convictions. Meet the Exonerated, Randall Dale Adams [website]:
http://www.law.northwestern.edu/wrongfulconvictions/exonerations

p. 168: Daubert v. Merrell Dow Pharmaceuticals. (n.d.). In *Wikipedia: The free encyclopedia.* Excerpted from Wikipedia. Retrieved on November 30, 2010 from http://www.en.wikipedia.org/wiki/Daubert_v._Merrell_Dow_Pharmaceuticals
Released under the Creative Commons Attribution Share-Alike License:
http://creativecommons.org/Licenses/by-sa/3.0/

7 The Basic Sampling Model and Associated Topics

p. 180: Levelt Committee. (2011, October 31). *Interim report regarding the breach of scientific integrity committed by Prof. D. A. Stapel.* Tilburg, The Netherlands: Tilburg University. Reprinted with the permission of the Executive Board of Tilburg University and the consent of Professor Willem Levelt. Retrieved from http://www.tilburguniversity.edu/nl/nieuws-en-agenda/commissie-levelt/interim-report.pdf

p. 196, Figure 7.1; p. 197, Figure 7.2: Adapted with the permission of Taylor & Francis, from Wainer, H. (2000). *Visual revelations: Graphical tales of fate and deception from Napoleon Bonaparte to Ross Perot.* Mahwah, NJ: Erlbaum. (Original work published 1997, New York: Copernicus). © 1997 Howard Wainer. Permission conveyed through Copyright Clearance Center, Inc.

p. 205: Fisher, R. A. (1971). *The design of experiments* (9th ed.). New York: Hafner. Courtesy of the University of Adelaide.

p. 209: McCullough, B. D. (2008a). Microsoft® Excel's 'Not the Wichmann–Hill' random number generators. *Computational Statistics & Data Analysis, 52,* 4587–4593. Copyright © 2008 Elsevier B. V. Reprinted with permission from Elsevier.

p. 210: McCullough, B. D., & Heiser, D. A. (2008). On the accuracy of statistical procedures in Microsoft® Excel 2007. *Computational Statistics & Data Analysis, 52,* 4570–4578. Copyright © 2008 Elsevier B. V. Reprinted with permission from Elsevier.

p. 211: Su, Y.-S. (2008). It's easy to produce chartjunk using Microsoft® Excel 2007 but hard to make good graphs. *Computational Statistics & Data Analysis, 52,* 4594–4601. Copyright © 2008 Elsevier B. V. Reprinted with permission from Elsevier.

p. 212: Yalta, A. T. (2008). The accuracy of statistical distributions in Microsoft® Excel 2007. *Computational Statistics & Data Analysis, 52,* 4579–4586. Copyright © 2008 Elsevier B. V. Reprinted with permission from Elsevier.

p. 216: Horrobin, D. F. (2003). Are large clinical trials in rapidly lethal diseases usually unethical? *Lancet, 361,* 695–697. Copyright © 2003 Elsevier Ltd. Reprinted from the *Lancet* with permission from Elsevier.

p. 224: Garg, A. (2010, March 1). [Essay preceding definition of "goulash"]. Reprinted with the permission of Wordsmith.org. Retrieved from
A.Word.A.Day [website]: `http://wordsmith.org/words/goulash.html`

8 Psychometrics

p. 231; p. 231: Linn, R. L., & Drasgow, F. (1987). Implications of the Golden Rule settlement for test construction. *Educational Measurement: Issues and Practice, 6*(2), 13–17. Copyright © 1987 by the National Council on Measurement in Education. Reprinted with the permission of John Wiley & Sons, Inc.

p. 235; p. 235: Gould, S. J. (1996). *The mismeasure of man* (Rev. and exp. ed.). New York: W. W. Norton. Copyright © 1996, 1981 by Stephen J. Gould. Reprinted with the permission of W. W. Norton & Company, Inc.

p. 238: Karl Pearson. (n.d.). In *Wikipedia: The free encyclopedia.* Excerpted from Wikipedia. Retrieved on February 8, 2011 from `http://www.en.wikipedia.org/wiki/Karl_Pearson`
Released under the Creative Commons Attribution Share-Alike License:
`http://creativecommons.org/Licenses/by-sa/3.0/`

p. 239: Thorndike, E. L. (1940). *Human nature and the social order.* New York: Macmillan. Reprinted with the permission of Robert M. Thorndike.

p. 245: Saretzky, G. D. (December, 1982). *Carl Campbell Brigham, the native intelligence hypothesis, and the Scholastic Aptitude Test* [Research memorandum] (Report No. ETS-RM-82-4). Princeton, NJ: Educational Testing Service.

Reprinted with the permission of the Educational Testing Service. Retrieved from Education Resources Information Center [website]: http://www.eric.ed.gov (ED 237516)

p. 246: Heterosis. (n.d.). In *Wikipedia: The free encyclopedia*. Excerpted from Wikipedia. Retrieved on February 8, 2011 from http://www.en.wikipedia.org/wiki/Heterosis

Released under the Creative Commons Attribution Share-Alike License: http://creativecommons.org/Licenses/by-sa/3.0/

p. 250: Jones Day®, & Duff & Phelps. (2011, November 7). *Class profile reporting* (Investigative report). Champaign, IL: University of Illinois, College of Law. Reprinted with the permission of Jones Day®, Duff & Phelps, the University Ethics Officer, and the Campus Legal Counsel. Retrieved from University of Illinois at Urbana-Champaign, College of Law, Releases and Information [website]: http://www.uillinois.edu/our/news/2011/Law/index.cfm

p. 255: Sacco and Vanzetti. (n.d.). In *Wikipedia: The free encyclopedia*. Excerpted from Wikipedia. Retrieved on February 8, 2011 from http://www.en.wikipedia.org/wiki/Sacco_and_Vanzetti

Released under the Creative Commons Attribution Share-Alike License: http://creativecommons.org/Licenses/by-sa/3.0/

II Data Presentation and Interpretation

9 Background: Data Presentation and Interpretation

p. 266: United States Environmental Protection Agency. (2009, October 7). *EPA begins new scientific evaluation of atrazine* [News release]. Reprinted with the permission of the United States Environmental Protection Agency. Retrieved from EPA: United States Environmental Protection Agency [website]: http://yosemite.epa.gov/opa/admpress.nsf

p. 268: Committee on the Framework for Evaluating the Safety of Dietary Supplements, Food and Nutrition Board, Board on Life Sciences, Institute of Medicine and National Research Council of the National Academies, The National Academies. (2005). *Dietary supplements: A framework for evaluating safety*. Washington, DC: National Academies Press. Reprinted with the permission of the National Academy of Sciences, courtesy of the National Academies Press, Washington, DC.

p. 271: Mozart effect. (n.d.). In *Wikipedia: The free encyclopedia*. Excerpted from Wikipedia. Retrieved on October 29, 2010 from http://www.en.wikipedia.org/wiki/Mozart_effect

Released under the Creative Commons Attribution Share-Alike License: http://creativecommons.org/Licenses/by-sa/3.0/

p. 274: Kenneth and Mamie Clark. (n.d.). In *Wikipedia: The free encyclopedia*. Excerpted from Wikipedia. Retrieved on May 31, 2011 from http://www.en.wikipedia.org/wiki/Kenneth_Clark_(psychologist)

Released under the Creative Commons Attribution Share-Alike License: http://creativecommons.org/Licenses/by-sa/3.0/

p. 280: Risse, G. (1976). Vocational guidance during the Depression: Phrenology versus applied psychology. *Journal of the History of the Behavioral Sciences*, *12*, 130–140. Copyright © 1976 by Wiley Periodicals, Inc. Reprinted with the permission of John Wiley & Sons, Inc.

10 (Mis)reporting of Data

p. 285: Best, J. (2005). Lies, calculations and constructions: Beyond *How to lie with statistics*. *Statistical Science*, *20*, 210–214. Reprinted with the permission of the Institute of Mathematical Statistics.

p. 290; p. 294; p. 295; p. 295: Seife, C. (2010). *Proofiness: The dark arts of mathematical deception*. New York: Viking. Reprinted with the permission of the Author.

p. 300: *Cigarette country: Tobacco in American history and politics*, Susan Wagner. © 1971 by Praeger Publishers, Inc. Reproduced with permission of ABC-CLIO, Santa Barbara, CA.

p. 303: Houlihan, J., Thayer, K., & Klein, J. (2003, May). *Canaries in the kitchen: Teflon toxicosis—EWG finds that heated Teflon pans can turn toxic faster than DuPont claims*. Reprinted with the permission of the Environmental Working Group. Retrieved from Environmental Working Group [website]: http://www.ewg.org/reports/toxicteflon

p. 304: Morrison, P., & Morrison, E. (Eds.) (1961). *Charles Babbage and his calculating machines: Selected writings by Charles Babbage and others*. New York: Dover Publications. [To be reissued by Dover Publications in February 2012 as *On the principles and development of the calculator and other seminal writings*]. Reprinted with the permission of Dover Publications.

11 Inferring Causality

p. 307: Fisher, R. A. (1971). *The design of experiments* (9th ed.). New York: Hafner. Courtesy of the University of Adelaide.

p. 308: Winch, R. F., & Campbell, D. T. (1969). Proof? No. Evidence? Yes. The significance of tests of significance. *American Sociologist*, *4*, 140–143. Copyright © 1969, American Sociological Association. Reprinted with the kind permission of Springer Science+Business Media B. V.

p. 316: Miasma theory. (n.d.). In *Wikipedia: The free encyclopedia*. Excerpted from Wikipedia. Retrieved on October 29, 2010 from
http://www.en.wikipedia.org/wiki/Miasma_theory
Released under the Creative Commons Attribution Share-Alike License:
http://creativecommons.org/Licenses/by-sa/3.0/

p. 319: Ignaz Semmelweis. (n.d.). In *Wikipedia: The free encyclopedia*. Excerpted from Wikipedia. Retrieved on June 6, 2010 from http://www.en.wikipedia.org/wiki/Ignaz_Semmelweis
Released under the Creative Commons Attribution Share-Alike License:
http://creativecommons.org/Licenses/by-sa/3.0/

p. 324; continued on p. 329: Berwick, D. M. (2005, December 12). Keys to safer hospitals. *Newsweek*. When this article was written, Dr. Donald M. Berwick was the President and CEO of the Institute for Healthcare Improvement. Reprinted with the permission of the Author. Retrieved from `http://www.newsweek.com`

p. 327: Puerperal fever. (n.d.). In *Wikipedia: The free encyclopedia*. Excerpted from Wikipedia. Retrieved on June 6, 2010 from `http://www.en.wikipedia.org/wiki/Puerperal_fever`

Released under the Creative Commons Attribution Share-Alike License: `http://creativecommons.org/Licenses/by-sa/3.0/`

p. 328: Koch's postulates. (n.d.). In *Wikipedia: The free encyclopedia*. Excerpted from Wikipedia. Retrieved on May 11, 2010 from `http://www.en.wikipedia.org/wiki/Koch's_postulates`

Released under the Creative Commons Attribution Share-Alike License: `http://creativecommons.org/Licenses/by-sa/3.0/`

12 Simpson's Paradox

p. 340, Figure 12.2; p. 341, Table 12.1: WAINER, HOWARD; *GRAPHIC DISCOVERY: A TROUT IN THE MILK AND OTHER VISUAL ADVENTURES.* © 2005 by Princeton University Press. Adapted by permission of Princeton University Press.

13 Meta-Analysis

p. 347; continued on p. 355: Tavris, C. (1999, July 19). The politics of sex abuse [Opinion-Editorial]. *Los Angeles Times*. Reprinted with the permission of the Author. Retrieved on October 29, 2010 from Newgon.com [website]: `http://newgon.com/prd/lib/Tavris1999.html`

p. 348: Warner, J. (2010, January 8). The wrong story about depression [Opinion-Editorial]. *New York Times*. Reprinted with the permission of the Author. Retrieved from `http://www.nytimes.com`

p. 353: Cronbach, L. J. (1982). Prudent aspirations for social inquiry. In W. H. Kruskal (Ed.), *The social sciences: Their nature and uses: Papers presented at the 50th anniversary of the Social Science Research Building, the University of Chicago, December 16–18, 1979* (pp. 61–81). Chicago: University of Chicago Press. Copyright © 1982 by the University of Chicago. All rights reserved. Published 1982. Adapted with the permission of the University of Chicago Press.

14 Statistical Sleuthing and Explanation

p. 363; p. 364, Table 14.1: Feller, W. (1968). *An introduction to probability theory and its applications* (3rd ed., Vol. 1). New York: Wiley. Copyright, 1950 by William Feller. Copyright © 1957, 1968 by John Wiley & Sons, Inc. Adapted with the permission of John Wiley & Sons, Inc.

p. 365: Mlodinow, L. (2009). *The drunkard's walk: How randomness rules our lives*. New York: Vintage Books. Reprinted with the permission of the Author.

Source Acknowledgments

III The Production of Data and Experimental Design

15 Background: Experimental Design and the Collection of Data

16 Ethical Considerations in Data Collection

p. 426: International Committee of Medical Journal Editors (ICMJE) (2010, April, update). II. Ethical considerations in the conduct and reporting of research. E. Protection of human subjects and animals in research. *Uniform requirements for manuscripts submitted to biomedical journals: Writing and editing for biomedical publication*. Reprinted with permission of the International Committee of Medical Journal Editors. Retrieved on February 8, 2011 from `http://www.icjme.org/ethical_6protection.html`

p. 426: Kimmelman, J., Weijer, C., & Meslin, E. M. (2009). Helsinki discords: FDA, ethics, and international drug trials. *Lancet, 373*, 13–14. Copyright © 2009 Elsevier Ltd. Reprinted from the *Lancet* with permission from Elsevier.

p. 440: World Medical Association. (2008). *Declaration of Helsinki: Ethical principles for medical research involving human subjects* (Rev. 6). Copyright, World Medical Association. All rights reserved. Reprinted with the permission of the World Medical Association. Retrieved from `http://www.wma.net/en/30publications/10policies/b3/17c.pdf`

17 The Federal Rules of Evidence

p. 462: International Society for Environmental Epidemiology, Ethics and Philosophy Committee. (April, 2009). *Guidelines for the ethical re-analysis and re-interpretation of another's research*. Reprinted with the permission of Dr. Raymond Neutra and the International Society for Environmental Epidemiology. Retrieved from `http://www.iseepi.org/about/ReAnal_guidelines_Revision2FINAL_April09.pdf`

p. 467: Mooney, C. (2005). *The Republican war on science*. New York: Basic Books. Reprinted with the permission of the Perseus Books Group.

p. 467: Mooney, C. (2005, August 28). Interrogations: Thanks to a little-known piece of legislation, scientists at the EPA and other agencies find their work questioned not only by industry, but by their own government. *Boston Globe*. Reprinted with the permission of the Author. Retrieved from `http://www.boston.com`

18 Some Concluding Remarks

p. 474: Whitaker, C. F. (1990, September 9). [Letter to "Ask Marilyn" column]. *Parade Magazine*, p. 15. Copyright © Marilyn vos Savant. Initially published in *Parade Magazine*. All rights reserved. Reprinted with permission.

p. 476: Tversky, A., & Kahneman, D. (1974). Judgment under uncertainty: Heuristics and biases. *Science, 185*, 1124–1131. Reprinted with permission from the American Association for the Advancement of Science and Daniel Kahneman.

p. 477: Fundamental attribution error. (n.d.). In *Wikipedia: The free encyclopedia*. Excerpted from Wikipedia. Retrieved on June 2, 2010 from `http://www.en.wikipedia.org/wiki/Fundamental_attribution_error` Released under the Creative Commons Attribution Share-Alike License: `http://creativecommons.org/Licenses/by-sa/3.0/`

p. 477: Stone, M. (2010, March 5). Toxic politics: Liz Cheney's Keep America Safe 'Al Qaeda Seven' ad. *Portland Progressive Examiner*. Reprinted with the permission of the Author. Retrieved from `http://www.examiner.com/progressive-in-portland/`

Printed in the United States
by Baker & Taylor Publisher Services